# Agriculture and Fisheries Management

# Agriculture and Fisheries Management

Edited by Geoffrey Gilbert

SYRAWOOD
PUBLISHING HOUSE
New York

Published by Syrawood Publishing House,
750 Third Avenue, 9th Floor,
New York, NY 10017, USA
www.syrawoodpublishinghouse.com

**Agriculture and Fisheries Management**
Edited by Geoffrey Gilbert



International Standard Book Number: 978-1-68286-474-6 (Hardback)



**Cataloging-in-Publication Data**

Agriculture and fisheries management / edited by Geoffrey Gilbert.
p. cm.
Includes bibliographical references and index.
ISBN 978-1-68286-474-6
1. Agriculture. 2. Agricultural systems. 3. Farm management. 4. Agricultural engineering.
5. Fishery management. I. Gilbert, Geoffrey.
S439 .A37 2017
630--dc23

Printed in China.

# TABLE OF CONTENTS


Preface ........ IX

Chapter 1 **Substitution of wheat bran by cassava flour in the diet: Effect on the growth of chickens** ........ 1
Beugré Grah Avit Maxwell, Gbogouri Grodji Albarin, Kimsé Moussa, Gnakri Dago

Chapter 2 **Effect of integrated use of lime, manure and mineral P fertilizer on bread wheat (Triticum aestivum) yield, uptake and status of residual soil P on acidic soils of Gozamin District, north-western Ethiopia** ........ 8
Mekonnen Asrat, Heluf Gebrekidan, Markku Yli-Halla, Bobe Bedadi, Wakene Negassa

Chapter 3 **Principal-agent relationship in policy implementation of the use of forest area for mining activity, Indonesia** ........ 18
Manifas Zubayr, Dudung Darusman, Bramasto Nugroho, Dodik Ridho Nurrochmat

Chapter 4 **Green economy: Challenges and prospects for improved aquatic agricultural system (AAS) in Niger delta communities** ........ 26
Nenibarini Zabbey, Ijeoma Favour Vincent-Akpu, Ibisime Etela

Chapter 5 **Development of farmer support system on dairy and meat industry of goat utilizing ICT** ........ 34
Teppei Hirata, Takeshi Miyagi, Yasunori Nagata, Shiro Tamaki, Tsutomu Omatsu, Tetsuya Mizutani

Chapter 6 **Food and feeding habits of the red-belly tilapia (*Tilapia zillii* Gervais, 1848) (Pisces: Cichlidae) in Lake Ziway, Ethiopia** ........ 41
Elias Dadebo, Negesse Kebtineh, Solomon Sorsa, Kassaye Balkew

Chapter 7 **Effect of alternate wetting and drying (AWD) irrigation for *Boro* rice cultivation in Bangladesh** ........ 48
Md. Redwanur Rahman, Sheikh Helena Bulbul

Chapter 8 **Evapotranspiration and water stress estimation from TIR and SWIR bands** ........ 55
Girolimetto Daniela, Venturini Virginia

Chapter 9 **Local soil knowledge and its use in crop allocation: Implications for landscape-scale agricultural production and conservation efforts in Talamanca, Costa Rica** ........ 65
Leigh A. Winowiecki, Matthew P. Whelan, Paul A. McDaniel, Marilyn Villalobos, Eduardo Somarriba

Chapter 10 **Effects of mineral nitrogen and phosphorus fertilizers on yield and nutrient utilization of bread wheat (*Tritcum aestivum*) on the sandy soils of Hawzen District, Northern Ethiopia** ........ 74
Bereket Haileselassie, Dawit Habte, Mehretab Haileselassie, Gebremedhin Gebremeskel

Chapter 11 **Farmers' perception on agricultural technologies a case of some improved crop varieties in Ghana** ........ 84
Emmanuel Asiedu - Darko

Chapter 12 **Mercury stress resistances in *Nauclea orientalis* seedlings inoculated with arbuscular mycorrhizal fungi** ........ 88
Hanna Artuti Ekamawanti, Yadi Setiadi, Didy Sopandie, Dwi Andreas Santosa

Chapter 13 **Predicting growth and yield models for *Eucalyptus* species in Aek Nauli, North Sumatera, Indonesia** ........ 96
Siti Latifah, Teodoro Reyes Villanueva, Myrna Gregorio Carandang, Nathaniel Cena Bantayan, Leonardo M. Florece

Chapter 14 **Study of phosphorus adsorption and its relationship with soil properties, analyzed with Langmuir and Freundlich models** ........ 104
Fisseha Hadgu, Heluf Gebrekidan, Kibebew Kibret, Birru Yitaferu

Chapter 15 **Decomposition rate of pigs' manures and nutrient release pattern in wetland condition** ........ 116
Bokossa Hervé Kouessivi Janvier, Saïdou Aliou, Fiogbé Emile Didier, Kossou Dansou

Chapter 16 **Asymmetry effect of intercropping non host crops between cabbage and climatic factor on the population of the diamondback moth (*Plutella xylostella* L.) and yield** ........ 124
Mohammad Feizal Daud, Fauziah I., Mohd Rasdi Z., Fairuz K., Abu Zarim U., Syed Abdul Rahman S. A. R., Ismail R., Mohd Hanysyam M. N., Norazliza R.

Chapter 17 **The most important factors affecting yield of the *Eucalyptus* species** ........ 131
Siti Latifah, Teodoro Reyes Villanueva, Myrna Gregorio Carandang, Nathaniel Cena Bantayan, Leonardo M. Florece

Chapter 18 **An assessment of the socio-economic and ecological impacts of environmental changes on rural livelihood: A study across Addado, Buhodle and Northern Galkaayo of central and northern Somalia** ........ 138
Badal Ahmed Hassan, Edinam K. Glover, Olavi Luukkanen, Ramni Jamnadass, Ben Chikamai

Chapter 19 **Strategies of eco-friendly stimulant use policy in Perum Perhutani, Indonesia** ........ 151
Sukadaryati

Chapter 20 **Connectivity and ecological indicators analysis of tropical forest landscape in Batang Toru watershed - Indonesia** ........ 160
Samsuri, I. Nengah Surati Jaya, Cecep Kusmana, Kukuh Murtilaksono

Chapter 21 **Spatiotemporal distribution of peatland fires in Kapuas District, Central Kalimantan Province, Indonesia** ........ 168
Achmad Siddik Thoha, Bambang Hero Saharjo, Rizaldi Boer, Muhammad Ardiansyah

Chapter 22 **Nutritional evaluation of varying levels of cooked flamboyant seed meal (*Delonix regia*) on the growth performance and body composition of Nile tilapia (*Oreochromis niloticus*) fingerlings** ........ 176
Gabriel Gana Bake, Elizabeth Itunu Martins, Suleiman Omeiza Eku Sadiku

Chapter 23 **Analysis of rules in use the formation of committee boundary (PTB) state forest area in Indonesia** ..... **183**
Pernando Sinabutar, Bramasto Nugroho, Hariadi Kartodihardjo, Dudung Darusman

Chapter 24 **Methane emission on intensive rice farming with water frequency and fertilizer management in North Sumatera** ..... **191**
Khadijah EL Ramija, Zulkifly Nasution, M. Zarlis, Retno Widiastuti

Chapter 25 **Wetland management in Bangladesh: A study on Beel Bakar** ..... **199**
Shoumo Khondoker, Md. Lokman Hossain, Khondoker Azizul Hoque Moni

Chapter 26 **Influence of small cereal intercropping and additive series of seed proportion on the yield and yield component of lupine (*Lupinus Spp.*) in north western Ethiopia** ..... **208**
Yayeh Bitew

Chapter 27 **Forest offence types, trends and lapses: Case study of the Nkawie Forest District of Ghana from 2005-2010** ..... **217**
Mercy A. A. Derkyi, Ton Dietz

Chapter 28 **Sustainable harvesting of *Alstonia boonei* De Wild. (Apocynaceae) in the community forest of Mbeth II, East Region of Cameroon** ..... **226**
Jean Lagarde Betti, Joseph Ambara

**Permissions**

**List of Contributors**

**Index**

# PREFACE

This book on agriculture and fisheries management deals with the scientific practice of maintaining a fishery with respect to sustainable and equitable development. Other natural resources that must be monitored are soil quality and fresh water. Agribusinesses utilize fisheries to uphold sustainable yield, land quality and food security practices. The contents in this text seek to provide methods that take into account practical concerns such as ecosystem development and conservation as well as technical and technological approaches such as population dynamics and monitoring. Through this book, we attempt to further enlighten the readers about the new concepts in this field. Coherent flow of topics, student-friendly language and extensive use of examples make this book an invaluable source of knowledge.

The researches compiled throughout the book are authentic and of high quality, combining several disciplines and from very diverse regions from around the world. Drawing on the contributions of many researchers from diverse countries, the book's objective is to provide the readers with the latest achievements in the area of research. This book will surely be a source of knowledge to all interested and researching the field.

In the end, I would like to express my deep sense of gratitude to all the authors for meeting the set deadlines in completing and submitting their research chapters. I would also like to thank the publisher for the support offered to us throughout the course of the book. Finally, I extend my sincere thanks to my family for being a constant source of inspiration and encouragement.

**Editor**

# Substitution of wheat bran by cassava flour in the diet: effect on the growth of chickens

**Beugré Grah Avit Maxwell[1], Gbogouri Grodji Albarin[2], Kimsé Moussa[3], Gnakri Dago[1]**

[1]Laboratoire de Nutrition et de Sécurité Alimentaire, U.F.R. des Sciences et Technologie des Aliments, Université Nangui Abrogoua, 02 BP 801 Abidjan 02, Côte d'Ivoire
[2]Laboratoire de Biochimie et de Microbiologie, Université Jean Lorougnon Guedé, BP 150 Daloa, Côte d'Ivoire
[3]Laboratoire de Biologie et de Cytologie Animale, UFR des Sciences de la Nature, Université Nangui Abrogoua, 02 BP 801 Abidjan 02, Côte d'Ivoire

**Email address:**
maxwellfrb@yahoo.fr (B. G. A. Maxwell)

---

**Abstract:** Chicken diet cost represents up to 70 % of the production charges. Inputs, which are mostly imported products such as lysine, methionine, wheat bran are the cause of the expensive cost of foods. Wheat bran is not available in Côte d'Ivoire and could be efficiently substituted by cassava. Cassava is affordable and available all year round. The objective of this work was to determine the effects of the substitution of wheat bran by cassava flour on the growth of chickens. After administering *ad libitum* during 8 weeks the different food categories (starting and growing food), results showed that protein, fat, ash and calcium contents improved more with the addition of cassava for both starting and growing periods. The energy value of cassava-based feed was higher than that of foods containing wheat bran: 12 155 kJ / kg against 11 428 kJ / kg for the starting food and 12 586 kJ / kg against 11 577 kJ / kg for growth, respectively. Throughout the rearing period, the growth of chicken fed with the cassava-based diet was much higher than that of the chicken fed with the control diet. In the growing food, palmitic acid was the dominant saturated fatty acids with the cassava-based diet exhibiting higher values of saturated fatty acids. The content of unsaturated fatty acids from cassava (48.6 %) was slightly higher than the control diet (47.6 %). Except magnesium, macro-element concentrations were higher in the cassava-based diet than those of food control. Micronutrients such as manganese and zinc had a high value in cassava-based diet than the control diet. Hydrogen cyanide which is in the fresh pulp at 100 mg / kg, decreased at 1.8 mg / kg of dry matter in the start food and at 1.9 mg / kg of dry matter in the growth food. The substitution of wheat by cassava improves feed nutritional profile, also reduces the production cost, food industry will benefit from this up to 13, 36 % in start-up period and up to 13, 25 % in growing period.

**Keywords:** Cassava, Growth, Hydrogen Cyanide, Substitution

---

## 1. Introduction

One of the major challenges of the Ivory Coast is to obtain and ensure food production for a growing population. This food production will be achieved through the enhancement of agricultural products, including livestock. That concern led the government to initiate in 1972 a vigorous policy of food self-sufficiency. That policy focused on short-term livestock-based [1].

The broiler of intensive and egg production have very quickly had the support of farm operators to the point that in 2000, Ivory Coast produced 22 841 tons of meat and 41,19 tons of eggs covering 80 % and 100 % of the national consumption of meat and eggs, respectively [2]. However, the cost of chickens feed alone accounts for 70 % of the costs of production [3, 4].

Inputs, which are mostly imported products such as lysine, methionine, wheat bran are the cause of the expensive cost of foods [5]. If we substitute these inputs for local products, we will bring down production costs. Wheat bran, a by-product of imported wheat is not available throughout the country. It may well be replaced by cassava, which has apparently equivalent energy, is cheaper and easily accessible anywhere any season [6, 7].

The objective of this work was to determine the effect of the substitution of wheat bran by cassava flour.

## 2. Material and Methods

### 2.1. Animals

For the study, we had 350 chicks, *Hubbard* strain, aged 1 day and weighing 42 ± 1 g, purchased from FOANI company (Abidjan), 350 traditional breeding chicks from the Sahel countries and from the forest of Ivory Coast, conventionally called African chicks and baptised *saiguè-Sissè*.

### 2.2. Poultry Diets

The composition of start-up and growth food used in this study is recorded in Tables 1 and 2. Chicks received food and drinking water *ad libitum*, in feeders and drinking trough every day according to the method of [8, 9]. The food of start-up (Table 1) was fed to chicks during the first 3 weeks of rearing while; the food for growth (Table 2) was administered to chickens after 3 weeks to, significantly, increase the energy.

### 2.3. Measurements

#### 2.3.1. Weighing Poultry

The chicks were weighed using an electronic balance before feeding. They were then weighed weekly, at the same time (9 h), throughout the period of breeding.

### 2.4. Feeding

During the 8 weeks of rearing, rations were weighed and remnant feed are evaluated.

### 2.5. Study of Nutritional and Energy Values of Formulated Food

Amino acids content was analysed using the method of liquid chromatograph type BECKMAN 420 coupled with an integrator CR3A SHIMADZU [10]. Starch content was determined using the modified technique of Mc Cready [11], Moisture was determined gravimetrically after drying sample overnight at 105 °C. Lipid content was determined using the method of soxhlet [12, 13], fatty acids content was quantified using gas chromatography method [14, 15]. Ash was quantified after incinerating the sample in the muffle furnace at 550 °C overnight and minerals (potassium, phosphorus, iron, manganese sodium, calcium, magnesium, zinc, copper and silica) were analysed by atomic absorption spectrophotometry (Perkin Elmer, Model 1100, Paris, France). The following method used by Young G [16]. Soluble sugars content was determined through the method described by Ashwell M and A Diplock [17]. Hydrogen cyanide content was determined using the method of Food and Agriculture Organisation, Metabolizable energy was calculated using the technique indicated by BIPEA [18].

### 2.6. Production Cost of the Food

The prices of raw materials were collected in several markets of Ivory Coast: the region of "Bas-Sassandra", the region of Lagoons and the region of "Agneby". Prices were collected from 10 markets in each region, a total of 30 markets.

### 2.7. Statistical Analysis

Statistical techniques were used to analyse the results [19]. Results are expressed as the mean ± standard deviation of several values of samples with Kyplot (version 2.0 beta 15, ©1997- 2001, Koichi Yoshioka) statistical software. The data were statistically analyzed by one way analysis of variance (ANOVA). Means were compared by Turkey's test. Differences were considered statistically significant at $*P < 0.05$. All measures were done in triplicate.

## 3. Results

### 3.1. Nutritional and Energy Value of Diets

#### 3.1.1. Chemical Composition and Energy Value of Diets

The chemical composition and energy of the control diet (wheat-based) and the cassava-based feed are shown in Tables 3 and 4.

Rates of protein, fat and ash of the cassava-based foods improved more compared to those of control foods both for the start-up and for the growing period:

In the start-up food, the difference in the nutrients content reached to 1% for protein and cellulose, this difference was about more than 2% concerning fatty matter, for the others nutrients, values were more or less close.

In the growth food, the differences were noted only for protein, cellulose and fatty matter (value close to 1%).

Three nutrients drawn differences between food in this two periods of life. Energy value was higher in the cassava feeds [12 155 ± 8 kJ / kg and 12 586 ± 18 kJ / kg] compared to, the control diet [11 428 ±9 kJ/kg and 11 577 ±129 kJ / kg].

### 3.2. Amino Acid Content of Diets

The amino acid composition of the control diet and substituted diet based cassava are shown in Tables 5 and 6.

Fourteen amino acids were identified including 7 essential amino acids (Tables 5 and 6) which are noted with an asterisk (*). Alanine, aspartic acid and methionine contents were higher in the wheat bran based starter diet than in the new cassava-based starter diet.

At the same time in the growth phase, the levels of these amino acids were high in the control diet (Table 6). However, the substituted diet based cassava appeared rich in essential amino acids.

### 3.3. Fatty Acid Content of Diets

The fatty acid content of diet fat are shown in tables 7

and 8 into two broad classes:

- four saturated fatty acids : lauric acid (C12: 0), myristic acid (C14: 0), palmitic acid (C16: 0) and stearic acid (C18: 0) ;
- two unsaturated fatty acids, including oleic acid (C18: 1n9) and one polyunsaturated fatty acid: linoleic acid (C18: 2n6).

Saturated fatty acids in the start-up food were strongly dominated by palmitic acid (between 30 and 45 %) in both control and cassava based diet. Unsaturated fatty acids in the start-up food of cassava had a higher amount (48.9 %) than the control diet based wheat (38.5%) in the same phase. Meanwhile, in the growing food, palmitic acid was the dominant saturated fatty acid. The content of unsaturated fatty acids from cassava (48.6 %) was slightly higher than the control diet (47.6 %).

### *3.4. Mineral Content of Diets*

There are major elements and micronutrients in the diets (see Tables 9 and 10). Macro-elements, with the exception of magnesium, were higher in the cassava based diet than those in the food control. Cobalt and silica, with a weak value in the food, had identical values (10 ± 1 g / kg). Micronutrients such as manganese and zinc had a high value in cassava based diet around 300 and 400 mg/kg as a difference between starting food and growing food respectively.

### *3.5. Hydrogen Cyanide in the Diets*

Hydrogen cyanide rate in cassava based diet is shown in Table 11. That presence was respectively of 1.8 ± 0.2 mg / kg in the start-up food and 1.9 ± 0.1 mg / kg in food for growth.

### *3.6. Growth Performance of Food*

Figures 1 and 2 shows the body growth curves of the broiler and african chicken fed with the two types of foods. Throughout the rearing period, the growth of chicken fed with the cassava-based diet was much higher. In eight weeks, broilers fed with the cassava-based diet weight 1882 ± 12 g, while the broiler fed with the control diet, weight 1730 ± 15 g.

African chicken were heavier on cassava based diet (890 ± 11 g) than on wheat bran diet (796± 9 g)

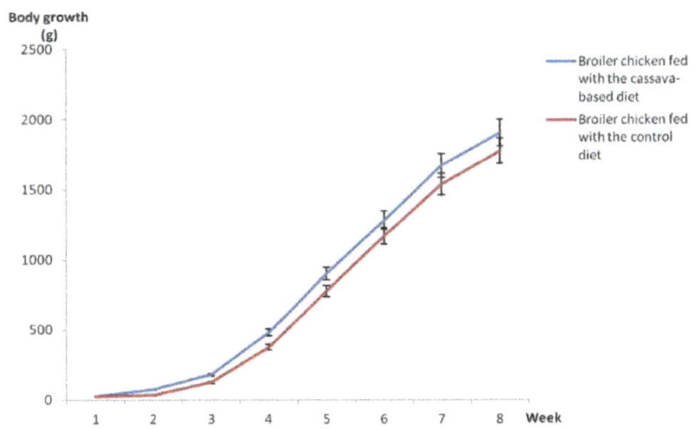


***Figure 1.*** *Growth in weight of broiler chicken submitted to the control diet and cassava-based diet.*

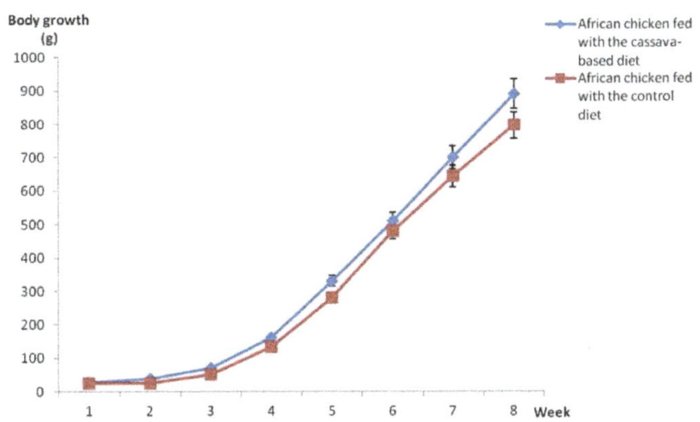


***Figure 2.*** *Growth in weight of African chicken submitted to the control diet and cassava-based diet.*

### *3.7. Production Cost of the Diet*

Table 12 present the production costs of starting foods: the control diet based on the wheat (industrial type) and the cassava-based diet.

The control diet (industrial type) cost is up to 152 350 F CFA per tonne, while the cost of the food start-up food of cassava amounted to 132 050 F CFA per tonne, a difference of 20 300 F CFA, representing 13.36 % discount.

The growing food for its part is up to 145 250 F CFA per

tonne for the blank feed against

126 900 F CFA for food made of cassava, a difference of 18 350 F CFA representing 13.25 % cost reduction.

### *3.8. Health Status of Chickens*

The health status of chickens subjected to the experiment was generally acceptable. The cumulative mortality rate for both tests is reported in Table 13. The mortality rate for birds on the control diet (industrial) and the cassava food vary between 3% and 6%:

- with the control diet, the mortality rate of African chicken is 4%, while the broiler is 6%;
- for food made of cassava, the mortality rate for African chicken is 3%, with 6% for the broiler.

## 4. Discussion

Mortality rates of birds during the study period are very low. These rates are well below the 10% limit recommended by [20]. These results could be explained by the good nutritional characteristics offered by the food given to animals. Indeed the food must provide some energy, the plastic elements and factors operating to maintain life [21].

Also, the elimination of hydrogen cyanide cassava by means of various treatments makes the cassava food safe for consumption [22, 23].

Thus, the weight gain of chickens fed with different diets based on cassava and chickens subject to the control diet show a significant difference both in the two phases: the start-up and the growth.

Diets of starting and growing at 20 % incorporation of cassava have advantages compared to the control. Indeed, both the broiler chicken and the African chicken type, the curve of chicken fed with the diet containing cassava is above that of chicken fed with the control diet.

This rapid growth of chicken fed with the cassava food is due to the energy brought by cassava and element supplied by the palm oil and soybean meal. Energy needs taking precedence over all other material requirements, any deficiency leads necessarily to a limitation of plastics and other needs regulators [24].

Soybean meal is important because of its richness in essential amino acids including lysine [25]. Also, it should be noted that palm oil is characterized by its high content of palmitic acid and energy [26]. It is well known that carbohydrate and fat provide energy. But in this growth process, the unsaturated fatty acids, including linoleic acid, play an essential role in body weight gain by acting preferentially in the fortifications of the body. So his absence leads a general weakness [27, 28].

It should be noted that the cassava based diet provides better benefits to the animals. Compared to the control diet (based on the wheat), the cassava based diet is cheaper. This difference in the production cost is greater with the start up food. The low cost of the cassava based diet compared to the control diet was due to the low price of cassava tuberous roots. The incorporation of cassava, however, prompted the provision of certain raw materials, namely palm oil, soybean meal and copra. These raw materials have led to the adequacy of nutritional needs and cost.

With the contributions of soybean meal, the protein is relatively higher in the control diet. It is the same energy that would result from supplementation of the meal, coconut, oil palm and even cassava flour.

## 5. Conclusion

Concerning nutritional and energy value of diets, protein, cellulose and fatty matter drew differences between food types in the starting up as in the growth period. The energy value is higher for food based on cassava. Also the substituted diet based cassava was rich in essential amino acids, oligo-elements. Fed with the cassava-based diet induces a gain of weight for the two types of chicks. It is easy to conclude that the food based on cassava is a good option for contents and energy. In more, food based on cassava reduces the production cost by 15% in start-up and 18% in growing feeding period of chickens. The effects of food based on cassava were positives about contents, energy and costs. The substitution of wheat by cassava is a productive option. This option could be improve and become one of the best solutions to breed chickens in guaranties.

***Table 1.*** *Composition of diets for poultry. Source : [9]*

| Ingredient | Starting food (g/100g) | Growing food(g/100g) |
|---|---|---|
| Maize | 59 | 59 |
| Wheat bran | 14 | 16 |
| Cottonseed meal | 13 | 13 |
| Fishmeal | 12.45 | 10.45 |
| Shell | 1 | 1 |
| Salt | 0.3 | 0.3 |
| Vitamin complex* | 0.15 | 0.15 |
| Lysine | 0.05 | 0.05 |
| Methionine | 0.05 | 0.05 |

***Table 2.*** *Composition of diets substituted cassava*

| Ingredients | Starting food (g/100g) | Growing food (g/100g) |
|---|---|---|
| Maize | 39 | 39 |
| Cassava | 20 | 20 |
| Copra meal | 1.5 | 1.45 |
| Cotton seed meal | 13 | 14 |
| Soy bean meal | 13.45 | 14 |
| Crude palm oil | 1.5 | 2 |
| Fishmeal | 10 | 8 |
| Shell | 1 | 1 |
| Salt | 0.3 | 0.3 |
| Vitamin complex | 0.15 | 0.15 |
| Lysine | 0.05 | 0.05 |
| Methionine | 0.05 | 0.05 |

***Table 3.*** *Chemical composition and energy value of starting foods*

| | Food control | Cassava-based food |
|---|---|---|
| Protein (%) | 20.01 ± 0.7 | 21 ± 0.6 |
| Starch (%) | 42.10 ± 2.2 | 42.70 ± 2.1 |
| Sugar (%) | 2.70 ± 0.1 | 2.30 ± 0.1 |
| Cellulose (%) | 8.10 ± 0.5 | 7.20 ± 0.4 |
| Fatty matter (%) | 4.28 ± 0.2 | 6.01 ± 0.3 |
| Ash (%) | 8.90 ± 0.7 | 9 ± 0.8 |
| Metabolizable energy (kilojoules /kg) | 11428 ± 9 | 12155 ± 8 |

***Table 4.*** *Chemical composition and energy value of growth food*

| | Wheat-based food (control)* | Cassava-based food* |
|---|---|---|
| Protein (%) | 19.04 ± 0.8 | 20.03 ± 0.9 |
| Starch (%) | 43.5 ± 3.1 | 43.6 ± 2.9 |
| Sugar (%) | 2.8 ± 0.1 | 2.5 ± 0.1 |
| Cellulose (%) | 8.3 ± 0.6 | 7.7 ± 0.5 |
| Fatty matter (%) | 4.3 ± 0.2 | 5.2 ± 0.4 |
| Ash (%) | 8.5 ± 0.5 | 8.8 ± 0.4 |
| Metabolizable energy (KJ /kg) | 11577 ± 12 | 12586 ± 18 |

*The results are the average of 10 tests

***Table 5.*** *Amino acids content (g / 100 g) of the starting food*

| Amino acid | Wheat-based food (control) | Cassava-based food |
|---|---|---|
| Alanine | 10.2 ± 0.4 | 5.4 ± 0.3 |
| Arginine | 5.4 ± 0.2 | 6.1 ± 0.3 |
| Aspartic acid | 6.4 ± 0.2 | 2.5 ± 0.01 |
| Glutamic acid | 1.52 ± 0.06 | 2.4 ± 0.2 |
| Glycine | 1.51 ± 0.04 | 3.1 ± 0.1 |
| Histidine | 3.7 ± 0.1 | 6.3 ± 0.1 |
| Isoleucine * | 8.1 ± 0.1 | 8.8 ± 0.5 |
| Leucine * | 10.2 ± 0.7 | 12.8 ± 0.8 |
| Lysine * | 4.4 ± 0.5 | 10.1 ± 0.3 |
| Methionine * | 3.20 ± 0.1 | 1.92 ± 0.02 |
| Phenylalanine+tyrosine* | 10.2 ± 1 | 13.8 ± 1.3 |
| Serine | 1.52 ± 0.08 | 2.2 ± 0.5 |
| Threonine * | 1.64 ± 0.02 | 5.03 ± 0.44 |
| Valine * | 6.5 ± 0,3 | 8.6 ± 0.4 |

Amino acids marked (*) are essential amino acids
The results are the average of 10 tests

***Table 6.*** *Amino acids content (g/100 g) of the growing food*

| Amino acid | Wheat-based food (control) | Cassava-based food |
|---|---|---|
| Alanine | 9.4 ± 0.7 | 3.8 ± 0.3 |
| Arginine | 5.7 ± 0.2 | 7.5 ± 0,4 |
| Aspartic acid | 6.8 ± 0.3 | 2.07 ± 0.01 |
| Glutamic acid | 1.62 ± 0.08 | 2.71 ± 0.02 |
| Glycine | 2.23 ± 0.05 | 4.52 ± 0.07 |
| Histidine | 4.14 ± 0.11 | 5.52 ± 0.12 |
| Isoleucine * | 7.8 ± 0.1 | 9.2 ± 0.7 |
| Leucine * | 9.40 ± 0.2 | 13.4 ± 0.2 |
| Lysine * | 4.7 ± 0.5 | 9.4 ± 0.7 |
| Methionine * | 4.2 ± 0.2 | 2.02 ± 0.04 |
| Phenylalanine+tyrosine* | 10.17±0.9 | 13.57±1.2 |
| Serine | 2.14 ± 0.07 | 3.41 ± 0.04 |
| Threonine * | 1.11 ± 0.03 | 2.08 ± 0.02 |
| Valine * | 5.02 ± 0.5 | 8.4 ± 0.2 |

Amino acids marked (*) are essential amino acids
The results are the average of 10 tests

***Table 7.*** *Fatty acids content of the starting food*

| Fatty acid | Wheat-based food (control) | Cassava-based food |
|---|---|---|
| Lauric acid | 4.91 ± 0.13 | 1.04 ± 0.02 |
| Myristic Acid | 3.84 ± 0.15 | 2.36 ± 0.04 |
| Palmitic Acid | 43.7 ± 3.8 | 35.9 ± 3.2 |
| Stearic Acid | 8.5 ± 0.8 | 9,7 ± 0,3 |
| Oleic Acid | 33.1 ± 2.4 | 33.4 ± 3.8 |
| Linoleic acid | 5.4 ± 0.8 | 15.5 ± 2.1 |
| Saturated fatty acids | 60.95 ±1.5 | 49 ± 2.05 |
| Unsaturated fatty acid | 38.5± 1 | 48.9 ± 1.85 |

The results are the average of 10 tests

***Table 8.*** *Fatty acids content of the growing food*

| Fatty acid | Wheat-based food (control) | Cassava-based food |
|---|---|---|
| Lauric acid | 1.88 ± 0.01 | 1.48 ± 0.07 |
| Myristic Acid | 3.74 ± 0.05 | 2.1 ± 0.4 |
| Palmitic Acid | 31.7 ± 2.6 | 36.7 ±1.8 |
| Stearic Acid | 14.92 ±1.07 | 10.9 ± 0.8 |
| Oleic Acid | 31.6 ± 2.3 | 39.2 ± 2.7 |
| Linoleic acid | 16.04 ±1.05 | 9.4 ± 0.4 |
| Saturated fatty acids | 52.24 ± 1.2 | 51.18 ± 1 |
| Unsaturated fatty acid | 47.6 ± 2 | 48.6 ± 1.15 |

The results are the average of 10 tests

***Table 9.*** *Minerals content of the starting food*

| | Wheat-based food (control) | Cassava-based Food |
|---|---|---|
| | ***Macro-elements (g / kg)*** | |
| Calcium | 10.6 ± 0.7 | 11.9 ± 0.5 |
| Phosphorus | 3.3 ± 0.5 | 4.7 ± 0.1 |
| Potassium | 51.3 ± 1.1 | 88.2 ± 1.5 |
| Sodium | 1.7 ± 0.1 | 2.3 ± 0.1 |
| Magnesium | 30.3 ± 0.9 | 27.5 ± 1 |
| Silica | 0.01 ± 0.0 | 0,01 ± 0,001 |
| | ***Micronutrients (mg / kg)*** | |
| Iron | 8100 ± 100 | 7800 ± 100 |
| Curare | 150 ± 10 | 110 ± 10 |
| Zinc | 1400 ± 100 | 1700 ± 100 |
| Manganese | 1100 ± 100 | 1500 ± 100 |
| Cobalt | 10 ± 1 | 10 ± 1 |

***Table 10.*** *Minerals content of the growing food*

| | Wheat-based food (control) | Cassava-based Food |
|---|---|---|
| | ***Macro-elements (g / kg)*** | |
| Calcium | 10.7 ± 0.9 | 12.4 ± 0.7 |
| Phosphorus | 3.1 ± 0.5 | 4.8 ± 0.3 |
| Potassium | 53.1 ± 1.1 | 84.2 ± 4.9 |
| Sodium | 1.5 ± 0.1 | 2.4 ± 0.1 |
| Magnesium | 31.5 ± 1.3 | 23.5 ± 1.1 |
| Silica | 0.01 ± 0.0 | 0.01 ± 0.0 |
| | **Oligo-elements (mg / kg)** | |
| Iron | 8100 ± 100 | 7800 ± 100 |
| Curare | 150 ± 10 | 110 ± 10 |
| Zinc | 1400 ± 100 | 1700 ± 100 |
| Manganese | 1100 ± 100 | 1500 ± 100 |
| Cobalt | 10 ± 1 | 10 ± 1 |

***Table 11.*** *Residual cyanide in cassava-based food*

| | **Cassava- based food *(starting food)**** | **Cassava- based food *(growing food)**** |
|---|---|---|
| Residual cyanide (mg / kg) | 1,8 ± 0,2 | 1,9 ± 0,1 |

***Table 12.*** *Production cost of one tone of food*

| | ***Cost (franc CFA / Tone)*** | |
|---|---|---|
| | **Start up** | **Growth** |
| Food control (based on wheat) | 152 350 | 145 250 |
| Cassava- based food | 132 050 | 126 900 |
| Gap | 20 300 | 18 350 |
| (franc CFA/Tonne) | (13.36 %) | (13.25 %) |

***Table 13.*** *Health status of chickens*

| | | **Wheat-based food (control)** | | **Cassava -based food** | |
|---|---|---|---|---|---|
| | **Number of breeding Chicken** | **Chicken died** | **Mortality rates (%)** | **Chicken died** | **Mortality Rates (%)** |
| African Chiken | 350 | 14 | 4 | 11 | 3 |
| Broiler | 350 | 20 | 6 | 21 | 6 |

## Acknowledgements

We were grateful to the University of Nangui Abrogoua, Abidjan (Côte d'Ivoire) and his Department of Nutrition for providing the experimental sites and other facilities to undertake this study. Thanks to anonymous referees who had tremendously improved the earlier version of the manuscript.

## References

[1] Akassey PR (1992) Le guide de l'agriculture en Côte d'Ivoire. Volume IV. Edition, PRAT/EUROPA, Paris 274-286.

[2] Anonyme 4 (2000) Bilan des ressources animales. Editeur, DGRA. Abidjan. pp 46.

[3] Ferrando R (1969) Alimentation du poulet et de la poule pondeuse. Editeur, Vigot Frère, Paris, pp 200.

[4] Garoud R (1992) Nutrition et alimentation des animaux d'élevage. Editeur: Foucher, Paris; pp 286.

[5] Smith AJ (1992) L'élevage de la volaille. Volume. 1 Editeur : Maisonneuve et Larose, ACCT, Editeur, Paris; pp 183.

[6] Jalloh A, Dahniya MT, Gbani A B et HC Ezumah (1995) L'effet de la date d'introduction du Riz (Oryza Saliva) dans la culture du manioc (Manihot esculenta, CRANTZ) sur leur production en Sierra Leone. Institute of Agricultural Research, P.M.B.450, Freetown, Sierra Leone; p 215-221.

[7] Kraisid T (2002) Agriculture, Alimentation, Nutrition en Afrique, FAO. pp 412.

[8] Ocho AL (1998) Valorisation de la graine d'hévéa (hevea brasiliensis) : production, composition chimique de la graine et efficacité alimentaire du tourteau chez le poulet de chair et la pondeuse. Mémoire pour l'obtention du grade de Docteur 3ème cycle. L'UFR de Biosciences de l'Université de Cocody, Côte d'Ivoire , M (811) pp 94.

[9] FAO (1977) L'Alimentation des volailles dans les pays tropicaux et subtropicaux. (*Coll. F.A.O. Progrès et mise en valeur. Agriculture n° 82*). Rome, 103 p.

[10] Liaset B, Julshamm K and M Epse (2003) Chemical composition and theoretical nutrition evaluation of produced fraction from enzymatic hydrolysis of salmon frames with Protamex TM. *Process Biochem.*, 38 (12): 1747-1759.

[11] Mc Cready (1999) Sugars of citrus juices, Food technology, p 19-20.

[12] Rauscherk K, Engste R and U Freimuth (1986) Untersuchung Von Lebensmitteln Veb, *Fauchbuch verlag. Leipzig*, 939 p.

[13] Rosignol CA (1998) Un préliminaire important: l'extration analytique des lipides. *Oléagineux, Corps Gras, Lipides*, 5 (4) : 301-307.

[14] Morrisson WC (1978) Cereal lipids in «Advance in cereal science and technology », *Edition Pomeranz Y. AACC*. Vol II, 221-348.

[15] Uceda et Hermoso (1996) Méthodes d'analyse des matières grasses et dérivés; ETIG, Editeur, Paris - France sixième édition; p 238.

[16] Young G (1985) Dosage de la silice par spectrophotomètre d'absorption atomique. *Analytical Chemistry.* NORME AFNOR N F t 90005, p.1951.

[17] Ashwell M and A Diplock (1999) Scientific concepts of functional foods in Europe. Consensus Document.. *British Journal of Nutrition,* 81(4) : S1-S27.

[18] B.I.P.E.A. (1976) Bureau Interprofessionnel d'Etudes Analytiques. Recueil des Méthodes d'Analyses des Communautés Européennes,

[19] Danielie P (2001) Théorie et méthodes statistiques. Application agronomique. Les presses agronomiques de Gembloux. Editeur, Gembloux, Belgique; pp 463.

[20] Anonyme 1 (1983) Manuel d'Aviculture en zone tropicale, Ministère des Relations, Extérieures, Coopération et Développement. 2éme Edition, Abidjan; 160-185.

[21] Pagot J, Bres P et QL Lecler (1983) Manuel d'Aviculture en zone tropicale, 2ème de la série Manuel et Précis d'Elevage, réalisée par l'Institut d'Elevage et de Médecine vétérinaire des Pays tropicaux, publiée par le Ministère de la Coopération ISNB : 2 :11.084499-X.; pp 185.

[22] Bruijin GH (1971) A study on the cyanogenetic character of cassava. Meded. Lanbouwhogesch., Wageningen; (73): 1-140

[23] Nartey F (1978) Studies on cassava, Manihot Utilissima pohl. Cyanogenesis: The biosynthesis, linamarin and lotaustralin in etiolated seedlings. Phytochemistry, (7): 1307-1312.

[24] Barbato GF, Cherry JA, Siegel PB and HP Van Krey (1980) Quantitative analysis of the feeding behaviour of four populations of chickens. Physiology and Behaviour, (25): 885-891.

[25] Anonyme 3 (1992) *Le guide de l'agriculture en Côte d'Ivoire*. Vol 4, éditeur, PRAT / EUROPA , Paris. 286 p.

[26] Anonyme 2 (1991) Mémento de l'agronome 4ème édition, Edition CIRAD, Paris. p 1635.

[27] Hargis P S and M Van Elswyk (1993) Fatty acid modification of poultry products. World's Poultry. Science; (49): 251-264.

[28] Leskanich C O and RC Noble (1950) Manipulation of the n-3 polyunsaturated fatty acid composition of avian egg and meat. World's Poultry Science, 1997; (53): 155-184

# 2

# Effect of integrated use of lime, manure and mineral P fertilizer on bread wheat (Triticum aestivum) yield, uptake and status of residual soil P on acidic soils of Gozamin District, north-western Ethiopia

**Mekonnen Asrat[1, *], Heluf Gebrekidan[1], Markku Yli-Halla[2], Bobe Bedadi[1], Wakene Negassa[3]**

[1]School of Natural Resources Management and Environmental Sciences, Haramaya University, Haramaya, Ethiopia
[2]Environmental Soil Science, University of Helsinki, Helsinki, Finland
[3]Department of Plant Soil and Microbial Sciences, Michigan State University, East Lansing, USA

**Email address:**
mek.asrat@yahoo.com (M. Asrat), helufgebrekidan@yahoo.com (H. Gebrekidan), mylihall@mappi.helsinki.fi (M. Yli-Halla), bobedadi2009@gmail.com (B. Bedadi), wakenechewaka@yahoo.co.uk (W. Negassa)

**Abstract:** Soil acidity problem is one of the bottlenecks to improve crop production in high rainfall regions of Ethiopia in general and in Gozamin District of Amhara region in particular. The aim of this study was to examine the effects of integrated use of lime, manure and mineral P fertilizer on acid soils for wheat production and status of residual soil P. The treatments were factorial combinations of lime, manure and P fertilizer which were laid down in a randomized complete block design with three replications. The field study was conducted on *Dystric Nitisols* in the 2011 and 2012 main cropping seasons at Enerata Kebele, Gozamin District. Lime application hastened early germination while plant height was enhanced by interaction of lime and P. Most parameters were significantly ($p < 0.01$) affected by two-way interactions while three-way interactions effect increased grain and straw yield at non-significant ($p > 0.05$) level. In these interactions lime rates were related in quadratic trend while manure and mineral P related linearly. The combined application of 5 t manure and 2.2 t $ha^{-1}$ lime increased grain and straw yield by 279% and 187%, respectively over the control treatment through economical analysis result 515 Eth. birr ($28) $ha^{-1}$ profit obtained due to additional yield obtained. Although all treatments residual soil P were categorized under very low status, it has strong positive correlation ($r^2 = 0.79$) with lime application. The present study showed the combined application of 5 t manure and 2.2 t $ha^{-1}$ lime was found to be economical feasible to improve wheat yield and yield components and residual soil P of acidic soils of the study area.

**Keywords:** Bread Wheat Yield, Dystric Nitisols, Integrated, Lime, Manure

## 1. Introduction

Soil fertility degradation has been described as the most important constraint to food security in sub-Saharan Africa [1]. High population growth of sub-Saharan Africa led to removal of vegetation cover and continuous cultivation of lands with low nutrient replenishment results in depletion of soil nutrients. Hence, crop productivity of sub-Saharan African countries can be increased with supplemental nutrients [2]. However, the cost of fertilizers has been increasing and has reached unaffordable level to small-scale farmers [3].

The Ethiopian highlands have a great potential for crop production, contributing 95% of the cultivated land and 90% of the economic activities of the country [4]. However, the potential in most of these areas is heavily affected by soils degradation. On the other hand, the problem has been aggravated in magnitude and degree from year to year in the last three decades. Most cultivated lands of the Ethiopian highlands are prone to soil acidity due to removal of ample amount of nutrients by leaching, crop mining and runoff as compared with grazing and forest lands. A recent study also showed that soil acidity increased on cultivated land as compared with forest and grazing lands in western part of Ethiopia [5]. Moreover, continuous application of acid forming chemical fertilizers on highly weathered tropical soils increase soil acidity problem [6]. Rampant acidic soils covered 40.9% of Ethiopia [7] which extend from southwestern to northwestern with east-west distribution but are concentrated in the western part of the country [8]. On the other hand, highly weathered tropical soils have strong P sorption capacities which intensify limitation of land suitability. Therefore, these cultivated lands require integrated soil fertility management when they are put under cultivation of economic important crops.

About 24% of the Amhara National Regional State is affected by acidity while 23% of Amhara region has been severely eroded [9]. The cultivated lands of Chemoga watershed (the study area) lost about 79 t soil $ha^{-1}$ per year [10] which is higher than the country's 42 t soil $ha^{-1}$ in a year as in [11]. The previous studies unequivocally, revealed severity of soil related problems that can affect agricultural sustainability of the Gozamin District. Soil acidity problem has forced most farmers to grow acid tolerant crops at the expense of economically important crops or to allocated their cultivated lands to eucalyptus plantation for the last three decades. The changes in land use and cover that occurred in Chemoga watershed between 1957 and 1998 have deteriorated soil nature which makes succession of eucalyptus plantation leading to its counteracting soil acidity and nutrient deficiency problems [12].

Wheat is one of the most staple food and economically important crop in Amhara region. However, wheat productivity has been declining from year to year in northwestern Ethiopian highlands, in general and in the study area in particular mainly due to intensive soil degradation and depletion of plant nutrients. Similarly, wheat productivity in northwestern Ethiopian highlands is 1.38 t $ha^{-1}$ (very low) as compared with 1.58 t $ha^{-1}$ of Ethiopian highlands and 2.5 t $ha^{-1}$ of the global productivity [13, 14].

Crop productivity of marginally fertile acidic soils can be improved through integrated use of soil ameliorants and fertilizers. In line with this, soil fertility management in Ethiopia has begun to shift toward integrated approach [15]. Liming of acid soils supplemented by fertilizers and small-scale irrigation under properly planned efficient water management can help to increase land use efficiency [8]. Liming increases soil pH and availability of P, Mo and N nutrient while it reduces exchangeable acidity [16, 6]. Similarly, application of manure on acidic soils reduces $Al^{3+}$ toxicity and increases nutrient content of the soils [17, 18]. However, application of manure should not be considered as an alternative or total substitute for lime [8]. Therefore, to improve crop yield on such soils external P is vital because soil P deficiency is a major production constraint in soils of tropical highlands [19].

Crop yield is significantly increased by integrated use of lime and fertilizer [15]. However, meager information is available about the management effects on crop yield which limits adoption of integrated fertility management in Amhara region in general and Gozamin District in particular. Therefore, generating information on the management effect on wheat yield may help to design strategies for future research and development interventions. The aim of this study was to explore effect of integrated use of lime, manure and mineral P fertilizer on wheat and status of residual soil P.

## 2. Materials and Methods

### *2.1. Description of the Study Site*

The experiment was conducted at Enerata Kebele, Gozamin District, northwestern Ethiopian highlands. The site is located at 10°23'43-45" N and 37°44'29-34" E and at altitude of 2501 masl. According to National Metrological Service Agency, the mean annual rainfall of three decades before study was 1,344 mm while mean, minimum and maximum air temperatures were 16.4, 10.3 and 22.5 °C, respectively (Fig. 1).

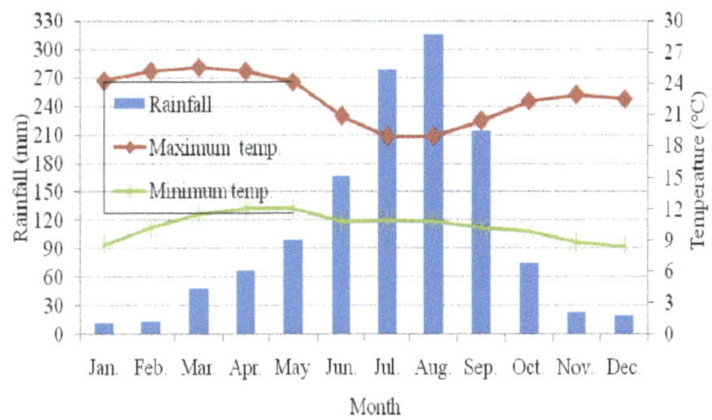


***Figure 1.*** *Mean monthly rainfall, minimum and maximum temperature of the study area based on three decades records at the Debre Maroks Meteorological Station.*

The area is characterized by uni-modal rainfall pattern and its main rainfall extends from mid of June to October with peak rainfall in mid of July to end of August. The study site is classified as cool thermal zone and moist moisture regime or moist cool (M4) [20]. The study was conducted under rain-fed condition. The rainfall distribution of 2011 cropping season was optimum. However, the subsequent season's rainfall was variable with higher rainfall from July to August and lower rainfall from mid-September to mid-November. Effective rainfall of 2012 cropping season was decreased by 82 mm as compared to that of 2011 cropping season (Table 1).

***Table 1.*** *Metrological conditions of the cropping seasons.*

| Month | Effective rainfall (mm) | | Mean temp. (°C) | |
|---|---|---|---|---|
| | 2011 | 2012 | 2011 | 2012 |
| July | 146 | 160 | 15.7 | 15.5 |
| August | 154 | 158 | 15.5 | 15.5 |
| September | 153 | 137 | 15.8 | 15.5 |
| October | 36 | 33 | 19.3 | 16.8 |
| November | 82 | 1 | 16.3 | 16.2 |
| Total/mean | 571 | 489 | 16.5 | 15.9 |

Source: National Metrological Service Agency, Bahir Dar Branch Office

### 2.2. Sampling, Analysis and Characterization of Experimental Soil, Grain, Straw, Lime and Manure

Two composite soil samples were collected to determine selected physiochemical properties before treatment application and one composite sample pre treatment after harvest of the second season crop to evaluate effect of the treatments on status of residual soil P. From each treatment one composite wheat grain and straw samples were collected after harvesting the first season using standard procedure to evaluate the treatment effect on P uptake.

The collected samples were air-dried, ground and sieved with 2 mm sieve size. A morphological characteristic of experimental soil was described by FAO procedure [21]. Soil texture was analyzed by the hydrometer method [22]. Soil color was determined using Munsell soil color chart while bulk density was estimated from undisturbed soil samples. Porosity was determined using the following equation.

$$f = 1 - Pb/Ps \tag{1}$$

Where, f = total porosity, Pb bulk density and Ps = particle density (2.65 g $cm^{-3}$).

Soil pH in water and 1M KCl a solution was measured at 1:2.5 soils to water ratio and the value was read using a combined glass electrode pH meter. Exchangeable acidity was determined saturating the soil samples with 1M KCl solution and titrating with 0.01 M NaOH described as in [23]. Exchangeable Al was determined from aqueous solutions extracted by 1M KCl and NaF and titrated with 0.01M HCl

Exchangeable bases (Ca, Mg, Na and K) were extracted with 1M $NH_4OAc$ at pH 7 and then Ca and Mg were determined by atomic absorption spectrophotometer, while K and Na were determined using flame photometry. Cation exchange capacity was determined by extracting the soil by 1M NH4OAc at 7 pH; while the adsorbed ammonium ions in soil were displaced with NaCl. Then the ammonium liberated from the distillation was titrated using 0.1M NaOH for CEC determination. Effective cation exchange capacity was calculated as the sum of exchangeable basic cations and Al [24]. Soil total N and organic carbon were determined at using dry combustion method in Vario MAX elemental apparatus. Available P was extracted by Mehlich solution after shaking for 5 min on a reciprocating shaker according to Mehlich No. 3 method [25] and then quantified colorimetrically with spectrometer. Plant tissue's P content was determined as in [22]. Wheat grain and straw P uptake were determined by multiplying P concentrations of each plant part by their respective dry matter weights.

The study was conducted on *Dystri-HaplicNitisols*. The experimental field has not received lime or/and P fertilizer before this study. The soil was clay in texture and developed from intensively weathered basalt parent material. It has low bulk density. The soil was very strongly acidic [26]. The net negative charge of the soil as identified by Δ pH = pH ($H_2O$) - pH (KCl) equation was 0.82. Soil colloids have great potential in weather-able acidic cations [27]. The soil is medium in cation exchange capacity and the soil mineralogical family classification, based on a ratio of CEC - 7 to clay was 0.29 and grouped under kaolinite clay or mixed mineralogical family [28]. It has low Ca and Na, very low available P and medium total N [29]. Selected soil physicochemical properties of the study site determined pre-planting are presented in Table 2.

***Table 2.*** *Selected soil physicochemical properties before planting.*

| Soil properties | Unit | Value |
|---|---|---|
| Sand | % | 41.0 |
| Silt | % | 15.0 |
| Clay | % | 44.0 |
| Textural class | - | Clay |
| Bulk density | g $cm^{-3}$ | 1.17 |
| Porosity (f) | % | 56.0 |
| pH 1:2.5 ($H_2O$) | | 4.62 |
| pH 1:2.5 (KCl solution) | | 3.80 |
| Organic carbon | % | 1.63 |
| Total N | % | 0.17 |
| Available P | mg $kg^{-1}$ | 7.79 |
| Exchangeable acid | $cmol_{(+)} kg^{-1}$ | 1.75 |
| Exchangeable Al | $cmol_{(+)} kg^{-1}$ | 1.20 |
| Exchangeable Ca | $cmol_{(+)} kg^{-1}$ | 5.13 |
| Exchangeable Mg | $cmol_{(+)} kg^{-1}$ | 2.88 |
| Exchangeable K | $cmol_{(+)} kg^{-1}$ | 0.8 |
| Exchangeable Na | $cmol_{(+)} kg^{-1}$ | 0.2 |
| Total exchangeable bases | $cmol_{(+)} kg^{-1}$ | 9.01 |
| Cation exchange capacity (CEC) | $cmol_{(+)} kg^{-1}$ | 19.95 |
| Effective Cation Exchange Capacity (ECEC) | $cmol_{(+)} kg^{-1}$ | 10.21 |
| Base Saturation | % | 45 |

One composite sample of the liming material was prepared in each season. The lime purity was estimated by thoroughly mixing at adjusted moisture content of 2g of Dejen lime in 50 ml of 1 M HCl and 70 ml of deionized water solution for an hour until it stopped bubbling which was then mildly heated on hot plate for 45 minutes. The mixed solution was transferred into 500 ml volume flask and filled with deionized water. Then 25 ml of the mixed solution was taken and 2-3 drops of Tashiro's indicator added to it which was then titrated by 0.1 M NaOH until the red colour of the indicator disappeared. Fineness factor of Dejen lime as estimated from percentage of lime that passed through 20, 60, 80 and 100 mesh sieves were 98.92, 48.44, 24.97 and 21.71%, respectively. Its relative neutralizing value in the first year application was determined by multiplying the lime fineness factor (LFF) with its purity. Accordingly, the lime material was calcitic material with moisture content of 1.0562%. Dejen lime has purity of 0.91 and fineness factor of 0.52. Therefore, in the first year application its relative neutralizing value (NV) (effective calcium carbonate equivalent) was 0.47.

The field lime requirement was estimated based as in [30] through exchangeable acidity, permissible acid saturation percentage for wheat cultivation (PASP), effective cation exchange capacity (ECEC) and lime requirement factor values, accordingly the predetermined rates during the experimental set up were 0, 2.2 and 3.3.t $ha^{-1}$ lime.

In each cropping season, two composite manure samples were collected from decomposed manure for the determination of its moisture, total N and available P content. Moisture content of the manure was estimated through oven-drying at 105 $^0$C for 24 hours. Total N and available P content of the manure were determined through wet digestion and Olsen method [31], respectively. The manure had mean moisture content of 31%, total N 1.2% and available P of 0.46 % (4.6g $kg^{-1}$).

### *2.3. Treatment and Experimental Design*

The experiment treatments consisted of three levels of manures (0, 5 and 10 t $ha^{-1}$), mineral P fertilizer (0, 20 and 30 kg P $ha^{-1}$ applied as triple superphosphate) and lime (0, 2.2 and 3.3 t $ha^{-1}$). The treatments were arranged in factorial combination and replicated three times which were laid out in randomized complete block design (RCBD).

### *2.4. Field Management*

The field was ploughed three times with local conventional oxen driven practices followed by manual seed-bed preparation and laid out according to experimental scheme. Dejen lime was broadcasted and thoroughly mixed into soil 50 days before wheat planting. Manure was decomposed under shade for two months and incorporated into soil 20 days before planting. The whole doses of mineral P fertilizer were applied in band at planting. Bread wheat (*Triticum aestivum* L) variety ETBW-4621 (*Kulkulo*) was used as test crop which was hand drilled at 150 kg $ha^{-1}$ rate on July 21 and 24 of 2011 and 2012 cropping seasons, respectively. Bread wheat was planted on gross plot of 2 m x 1.25 m (2.5 $m^2$) with 20 cm inter row spacing and the net plot size was 0.6 $m^2$. All recommended cultural practices of wheat production were adopted for the management of the experiment.

### *2.5. Data Collection*

All plant data collected from crop emergence to harvest were treated as growth parameters. Growth parameters were days to emergence, heading and maturity, number of tillers, plant height, spike length and kernels per spike. However, thousand seeds and hectoliter weights, harvest index and economical yields were considered as yield parameters. Days to emergence and heading were recorded when 50% of the crop emerged and headed, respectively while days to physiological maturity was measured when 90% of the crop was matured.

Number of tillers was counted at late tillering stage (40-45 days after planting) from randomly sampled plants at 0.5 m row length of the net plot. Plant height, spike length and kernels per spike were measured from ten randomly sampled plants at 90% physiological maturity. Plant height was measured from the ground level to the tip of spike excluding awns. Similarly, spike length was measured from the base to the tip of spike by excluding the awns. Grain yield was estimated by adjusting moisture content to 12.5%. The economic analysis was carried out by considering estimated prices of lime, manure and mineral P fertilizer during the application season while value of bread wheat grain at harvesting season. Accordingly, the prices of manure, lime and wheat were 500 ($ 28), 1,400 ($ 78), 7,500 ($ 417) Eth. birr per ton, respectively while mineral P fertilizer price was 102.33 Eth. birr per kg P ($ 5.7).

### *2.6. Statistical Analysis*

The data were subjected to analysis of variance by following the standard procedure [32] then analyzed using Statistical Analysis System SAS software version 9.1 [33]. Significant mean of the treatments were delineated by Duncan's Multiple Range test (DMRT) comparison at $p < 0.01$ or $p < 0.05$. Correlation analysis was carried between the parameters to determine magnitude and degree of their relation.

## 3. Results and Discussion

### *3.1. The Crop Parameter Response*

The response of most growth and yield parameters of the crop were affected by cropping season due to the variability of rainfall in 2012 season (Table 3). In the subsequent season, delayed germination, early maturity, short height and low number of kernels per spike of wheat were recorded as compared to that of the 2011 cropping season. Moreover, in 2012 cropping season grain and straw yield decreased by 29 and 36%, respectively.

***Table 3.*** *Analysis of variance (ANOVA) for bread wheat growth and yield parameters.*

| Parameter | Sources of variations mean squares with significance level | | | | | | | | CV (%) |
|---|---|---|---|---|---|---|---|---|---|
| | L (2) | M (2) | P (2) | L*M (4) | L*P (4) | M*P (4) | L*M*P (8) | Error (96) | |
| Days to emerge | $2.117^{*}$ | $0.673^{ns}$ | 0.117ns | $0.302^{ns}$ | $0.247^{ns}$ | $0.886^{ns}$ | $0.28^{ns}$ | 0.439 | 11.53 |
| No. tillers | $69.57^{ns}$ | $479.6^{**}$ | $93.17^{ns}$ | $77.04^{ns}$ | $65.41^{ns}$ | $65.24^{ns}$ | $56.7^{ns}$ | 46.52 | 16.77 |
| Days to heading | $14.19^{*}$ | $19.34^{**}$ | $86.67^{**}$ | $2.47^{ns}$ | $2.77^{ns}$ | $1.87^{ns}$ | $1.3^{ns}$ | 3.02 | 2.41 |
| Days to mature | $0.35^{ns}$ | $5.57^{ns}$ | $4.24^{ns}$ | $0.31^{ns}$ | $10.7^{*}$ | $1.51^{ns}$ | $2.46^{ns}$ | 3.96 | 1.59 |
| Plant height (cm) | $371.7^{**}$ | $331.2^{**}$ | $1558.2^{**}$ | $19.35^{ns}$ | $66.77^{*}$ | $22.62^{ns}$ | $17.0^{ns}$ | 19.29 | 6.69 |
| Spike length (cm) | $4.52^{**}$ | $2.86^{*}$ | $19.38^{**}$ | $1.72^{*}$ | $2.28^{**}$ | $0.32^{ns}$ | $0.29^{ns}$ | 0.60 | 12.3 |
| Kernels per spike | $5.44^{ns}$ | $22.52^{ns}$ | $369.6^{**}$ | $46.12^{ns}$ | $24.21^{ns}$ | $22.95^{ns}$ | $10.4^{ns}$ | 21.72 | 16.69 |

| Parameter | Sources of variations mean squares with significance level | | | | | | | | CV (%) |
|---|---|---|---|---|---|---|---|---|---|
| | L (2) | M (2) | P (2) | L*M (4) | L*P (4) | M*P (4) | L*M*P (8) | Error (96) | |
| Thousand seeds weight | $6.17^{ns}$ | $181.9^{**}$ | $7.74^{ns-}$ | $2.71^{ns}$ | $4.99^{ns}$ | $5.8^{ns}$ | $0.774^{ns}$ | 2.546 | 4.8 |
| Hectoliter weight | $158.5^{ns}$ | $54.34^{ns}$ | $7.3^{ns}$ | $224.78^{*}$ | $181.14^{ns}$ | $194.3^{ns}$ | $68.5^{ns}$ | 82.62 | 11.13 |
| Grain yield (t $ha^{-1}$) | $9.76^{**}$ | $10.29^{**}$ | $10.84^{**}$ | $0.96^{**}$ | $0.02^{ns}$ | $0.538^{**}$ | $0.24^{ns}$ | 0.126 | 17.36 |
| Straw yield ( t $ha^{-1}$) | $14.56^{**}$ | $13.74^{**}$ | $18.21^{**}$ | $1.137^{**}$ | $0.21^{ns}$ | $0.93^{*}$ | $0.28^{ns}$ | 0.27 | 17.52 |
| Biomass yield (t $a^{-1}$) | $48.17^{**}$ | $47.81^{**}$ | $57.03^{**}$ | $4.136^{**}$ | $0.33^{ns}$ | $2.85^{**}$ | $1.02^{ns}$ | 0.7 | 16.68 |
| Harvest index | $0.008^{**}$ | $0.01^{**}$ | $0.004^{*}$ | $0.005^{**}$ | $0.001^{ns}$ | $0.00004^{ns}$ | $0.0006^{ns}$ | 0.001 | 7.46 |

### *3.1.1. Wheat Growth*

Most growth parameters were significantly affected by lime, manure and P application. However, days to maturity, plant height and spike length were significantly influenced by interaction of lime and P (Table 3). Wheat emerged between 5 and 7 days of planting and days to emergence were significantly affected by 2.2 t $ha^{-1}$ lime. Lime hastened early emergence of wheat. Liming reduce Al concentration thus increase the availability of certain nutrient thereby improving plant growth. Total root length of plant restricted at lower pH, and thus the unamended soil showed reduced root length as compared with lime-amended soil [34].

The number of tillers per $m^2$ was in ranges between 132 and 354, being significantly ($p < 0.01$) increased by application of manure. Manure improves physicochemical properties of soils thereby enhancing water infiltration and root penetration [35, 36] and thus promoting plant tillering. On the other hand, manure amendment in acidic soils increases soil pH, mineral N, available P, K and Ca [18, 37]. The number of tillers was affected by N rate [14, 38]. High N rate increased the number of tillers by 11-15% over the control. The crop reached to heading within ranges of 69 to 78 days after planting. Lime, manure and P applications were affected the number of days to heading (Table 3).

Two-way interaction of lime and P showed significant ($p < 0.05$) response to days to maturity (Table 3). On these treatments, early maturity of the crop was promoted at zero level of one factor and with increase in the rate of the other factor. Therefore, early maturities of non-limed treatments were hastened with P rates while early maturities of non-P applied treatments were hastened through liming (Table 4). Plant growth affected by P availability and P deficient crop delayed in maturity. These results are line with the reports of other researchers. For instance [39] reported that acidic soils adsorbed 70–90% of an applied P into various low soluble P compounds without giving any immediate contribution to crop growth, thus acidic sensitive and low acid tolerant crops usually delayed in maturity. Similarly, phosphorus deficient rice was delayed by 10-12 days to reach maturity [40]. However, liming reduces P sorption capacity of acidic soil thereby increasing P availabilities of native soil and fertilizer, consequently it enhances early maturity [41, 42]. In contrast, on plots that received both lime and mineral P fertilizer were delayed in days of maturity can probably be associated with the P adsorbed by high lime rate (Table 4).

***Table 4**. Lime and P interaction effect of on maturity days.*

| Lime (t $ha^{-1}$) | kg P $ha^{-1}$ | | |
|---|---|---|---|
| | 0 | 20 | 30 |
| 0 | $126.06^{b}$ | $125.39^{ab}$ | $124.83^{ab}$ |
| 2.2 | $124.88^{ab}$ | $125.56^{ab}$ | $125.78^{ab}$ |
| 3.3 | $124.5^{a}$ | $126.11^{b}$ | $126.06^{b}$ |

Interaction means within rows and columns with the same letter are not significantly different at $p < 0.01$

Interaction of lime and mineral P fertilizer showed significant ($p < 0.05$) response to plant height (Table 3 and 5). In line with this, heights of okra plant *(Abelmoschus esculentus* L.) were increased by application of 500 kg lime and 10 kg P $ha^{-1}$ [43] besides groundnut height significantly increased by integrated application of 3.7 t lime and 49 kg P $ha^{-1}$ [44]. Longer wheat height was obtained with combined application of P and 2.2 t $ha^{-1}$ lime as compared with 3.3 t $ha^{-1}$ combinations. Phosphorus application increased crop height; however, crop height per unit P was decreased with P rate (Table 5). Low rate fertilization induces sharp plant growth but without increasing nutrient concentration of the tissue; however, marginal applications probably induce nutrient concentration increment of tissue rather than physical growth.

***Table 5.** Main and interaction of lime and P effects on plant height (cm).*

| Lime(t $ha^{-1}$) | kg P $ha^{-1}$ | | | *Lime |
|---|---|---|---|---|
| | 0 | 20 | 30 | |
| 0 | $54.56^{e}$ | $66.21^{bc}$ | $68.02^{abc}$ | $62.93^{C}$ |
| 2.2 | $64.22^{c}$ | $69.43^{ab}$ | $70.86^{a}$ | $68.17^{A}$ |
| 3.3 | $59.68^{d}$ | $68.38^{ab}$ | $69.37^{ab}$ | $65.81^{B}$ |
| *P | $59.49^{B}$ | $68.0^{A}$ | $69.42^{A}$ | |

Interaction mean within rows and columns with the same letter are not significantly different at $p < 0.01$, lower case superscripted = significance of two-way interaction, upper case superscripted = significance of main effect, *= main effect mean

Spike lengths ranged between 8.4 and 3.0 cm with a mean of 6.3 cm. It was significantly affected by two-way interactions of lime (lime and manure; lime and P) (Table 3). Interaction effects of lime and manure, lime and P were significant at $p < 0.01$ and $p < 0.05$, respectively to spike length. Spike lengths of non-limed combinations increased with rates of manure or P, in contrary, spike length of zero manure or P combinations increased through lime rate (Table 6). Each spike on average consisted of 28 kernels. Application of 20 kg P $ha^{-1}$ increased number of kernels by

15% against the control treatment. However, P had no effects on number of kernels [44].

### 3.1.2. Wheat Yield

Most parameters were significantly affected by two-way interactions of manure (lime and manure; P and manure) (Table 3). Manure and P combinations were linearly related to the response of most parameters while lime combinations were related in quadratic trend. Mean weight of thousand wheat seeds was 33.27g. Seeds weight significantly increased by application of 10 t $ha^{-1}$ manure, it showed 11% more weight as compared to the control treatment's seeds weight. In line with this, application of P and lime did not significantly affect seeds weight [14, 45-47]. An application of manure improved N content of soils through its decomposition and creating favorable environmental conditions for the occurrence of organisms. The crop mean grain hectoliter weight was 81.7 kg $hl^{-1}$. Well-filled and plump wheat grains were obtained on lime and manure applied plots as compared to the other treatments.

***Table 6.*** *Main and two-way interactions effects of lime, manure and P on spike length (cm).*

| Lime t $ha^{-1}$ | Manure t $ha^{-1}$ | | | P kg $ha^{-1}$ | | | *Lime |
|---|---|---|---|---|---|---|---|
| | 0 | 5 | 10 | 0 | 20 | 30 | |
| 0 | $5.69^{d}$ | $5.94^{cd}$ | $6.33^{abcd}$ | $4.84^{C}$ | $6.49^{AB}$ | $6.63^{A}$ | $5.99^{B}$ |
| 2.2 | $6.03^{cd}$ | $6.88^{a}$ | $6.77^{ab}$ | $6.11^{AB}$ | $6.79^{A}$ | $6.78^{A}$ | $6.56^{A}$ |
| 3.3 | $6.47^{abc}$ | $6.14^{bcd}$ | $6.47^{abc}$ | $5.88^{B}$ | $6.73^{A}$ | $6.47^{AB}$ | $6.36^{AB}$ |
| * Manure and P | $6.06^{B}$ | $6.32^{AB}$ | $6.52^{A}$ | $5.61^{B}$ | $6.67^{A}$ | $6.63^{A}$ | |

Interaction mean within rows or columns with the same letter are not significantly different at $p < 0.01$, lower case superscripted = significance of interaction of manure and lime, upper case superscripted = significance of interaction of lime and P and individual main effects, *= main effect mean

***Table 7.*** *Main and two-way interactions of manure effects on grain yield (t $ha^{-1}$)*

| Manure t $ha^{-1}$ | Lime t $ha^{-1}$ | | | P kg $ha^{-1}$ | | | *Manure |
|---|---|---|---|---|---|---|---|
| | 0 | 2.2 | 3.3 | 0 | 20 | 30 | |
| 0 | $0.898^{d}$ | $1.98^{bc}$ | $1.787^{c}$ | $1.191^{D}$ | $1.654^{BC}$ | $1.820^{B}$ | $1.56^{C}$ |
| 5 | $1.741^{c}$ | $2.689^{a}$ | $2.17^{b}$ | $1.466^{CD}$ | $2.501^{A}$ | $2.619^{A}$ | $2.2^{B}$ |
| 10 | $2.215^{b}$ | $2.736^{a}$ | $2.212^{b}$ | $1.958^{B}$ | $2.515^{A}$ | $2.704^{A}$ | $2.39^{A}$ |
| * Lime and P | $1.618^{C}$ | $2.468^{A}$ | $2.056^{B}$ | $1.538^{B}$ | $2.223^{A}$ | $2.381^{A}$ | |

All denoted figures are the same as Table 6

The mean grain yield was 2.05 t $ha^{-1}$ while mean maximum and minimum grain yields were 3.13 and 0.71 t $ha^{-1}$, respectively. Both main effects and two-way interactions of manure significantly ($p < 0.01$) affected the grain yield. In line with this, it has been reported that in highly weathered tropical soils, wheat grain yield improved by application of lime and fertilizers [48]. Sole application of 3.7 t $ha^{-1}$ lime has also been reported to have increased wheat grain yield by twice of the control treatment [49].

Integrated liming and fertilization through manure and mineral P fertilizer management brought more grain yield than their respective main effects. Significant mean of the interaction was obtained on combined 5 t $ha^{-1}$ manure and 2.2 t $ha^{-1}$ lime treatment which brought 0.26 to 0.85 t $ha^{-1}$ more grain yield than respective individual effect.

Lime rates were quadratically increased grain yield while the yield increased linearly by manure rates. Application of 2.2 t $ha^{-1}$ lime in combination with fertilizer's treatments gave more grain yields than 3.3 t $ha^{-1}$ combinations (Table 7). Even though the three-way interactions were not-significant to wheat yield, combined application of 2.2 t lime, 5 t $ha^{-1}$ manure and 30 kg P $ha^{-1}$ showed the highest grain and straw yield. The statistical analysis revealed that integrated application of 5 t manure and 2.2 t $ha^{-1}$ lime gave 2.7 t $ha^{-1}$ grain yield with additional yield of 2 t $ha^{-1}$ (279%) over the control treatment. Meanwhile the combined application of 5 t manure and 20 kg P $ha^{-1}$ gave 2.5 t $ha^{-1}$ grain yield with additional yield of 1.8 t $ha^{-1}$ (252%) over the control treatment. According to the crop grain and straw yield response, the combined application of 5 t manure and 2.2 t $ha^{-1}$ lime applications in *Dystric Nitisols* of northwestern Ethiopian improved crop yield.

### 3.1.3. Economic Analysis for Grain Yield

Marginal rate of return of the combined 2.2 t lime and 5 t $ha^{-1}$ manure treatment was 51.50 which gave net benefit of 515 Eth. birr ($28) $ha^{-1}$ as compared with combined 5 t manure and 20 kg P $ha^{-1}$ due to 70 kg $ha^{-1}$ grain yield increment (Table 8).

***Table 8***. *Economical analysis of two-way interactions of manure effects for bread wheat grain yield.*

| Treatment (combination) | Cost of the input (Eth, birr) | | | | Grain yield (t $ha^{-1}$) | Total value of grain (Eth. birr) | Net return (Eth. birr) | Marginal rate of return (%) |
|---|---|---|---|---|---|---|---|---|
| | Manure (M) | Lime (L) | mineral P (P) | Total | | | | |
| $M_{10\,t\,ha^{-1}} \times P_{30\,kg\,ha^{-1}}$ | 5,000 | - | 3,070 | 8,070 | 2.704 | 20,280.00 | 12,210.00 | |
| $M_{10\,t\,ha^{-1}} \times L_{2.2\,t\,ha^{-1}}$ | 5,000 | 3,080 | - | 8,080 | 2.736 | 20,520.00 | 12,444.00 | 2340 |
| $M_{5\,t\,ha^{-1}} \times P_{20\,kg\,ha^{-1}}$ | 2,500 | - | 3,070 | 5,570 | 2.619 | 19,642.50 | 14,072.50 | |
| $M_{5\,t\,ha^{-1}} \times L_{2.2\,t\,ha^{-1}}$ | 2,500 | 3,080 | - | 5,580 | 2.689 | 20,167.50 | 14,587.50 | 5150 |

***Table 9.** Main and two way interactions of manure effects on straw yield (t ha$^{-1}$).*

| Manure (t ha$^{-1}$) | Lime (t ha$^{-1}$) | | | P (kg ha$^{-1}$) | | | *Manure |
|---|---|---|---|---|---|---|---|
| | 0 | 2.2 | 3.3 | 0 | 20 | 30 | |
| 0 | 1.666$^{e}$ | 2.897$^{bcd}$ | 2.628$^{cd}$ | 1.909$^{D}$ | 2.501$^{CD}$ | 2.781$^{BC}$ | 2.397$^{B}$ |
| 5 | 2.509$^{d}$ | 3.777$^{a}$ | 3.076$^{bc}$ | 2.173$^{CD}$ | 3.44$^{AB}$ | 3.75$^{A}$ | 3.121$^{A}$ |
| 10 | 3.137$^{b}$ | 3.754$^{a}$ | 3.212$^{b}$ | 2.864$^{BC}$ | 3.452$^{AB}$ | 3.787$^{A}$ | 3.368$^{A}$ |
| * Lime and P | 2.438$^{C}$ | 3.476$^{A}$ | 2.972$^{B}$ | 2.316$^{C}$ | 3.131$^{B}$ | 3.439$^{A}$ | |

All denoted figures are the same as Table 6

Like the grain yield, the straw yield was significantly affected by two-way interactions of manure (Table 3 and 9). This in agreement with another report where combined application of lime and mineral P fertilizer significantly increasing straw yield of rice [50]. The combined application of 5 t manure and 2.2 t lime ha$^{-1}$ gave 3.8 t straw ha$^{-1}$ which increased the yield by 2.5 t ha$^{-1}$ (187%) over the control. Meanwhile the combined application of 5 t manure and 20 kg P ha$^{-1}$ application gave 3.4 t straw ha$^{-1}$ which increased the yield by 2.1 t ha$^{-1}$ (162%) over the control treatment.

***Table 10.** Wheat P uptake (kg ha$^{-1}$) of the treatments.*

| Lime t ha$^{-1}$ | Manure t ha$^{-1}$ | kg P ha$^{-1}$and uptake | | |
|---|---|---|---|---|
| | | 0 | 20 | 30 |
| 0 | 0 | 1.44 | 1.91 | 3.28 |
| 0 | 5 | 2.53 | 4.60 | 7.58 |
| 0 | 10 | 5.6 | 8.61 | 8.96 |
| 2.2 | 0 | 7.03 | 6.37 | 11.86 |
| 2.2 | 5 | 8.46 | 10.00 | 12.4 |
| 2.2 | 10 | 14.67 | 16.09 | 17.74 |
| 3.3 | 0 | 7.20 | 10.45 | 10.45 |
| 3.3 | 5 | 7.31 | 14.96 | 14.46 |
| 3.3 | 10 | 11.73 | 12.73 | 12.95 |

The highest biomass was 7.59 t ha$^{-1}$ which was obtained with combined application of 2.2 t lime, 5 t manure and 30 kg P ha$^{-1}$ whereas the lowest biomass yield was 2.02 t ha$^{-1}$ in control treatment. Similarly, biomass yield was significantly ($p < 0.01$) affected by two-way interactions of manure (Table 3). The combined application of 5 t manure and 2.2 t ha$^{-1}$ lime treatment increased the biomass by 152% over the control treatment while the combined application of 5 t manure and 30 kg P ha$^{-1}$ increased the yield by 105% over the control yield.

The highest wheat harvest index was obtained on combined 2.2 t lime and 10 t ha$^{-1}$ manure treatment, the harvest index increased by 19% against the control treatment. The harvest index of wheat observed in the present study was in the range of 0.35 to 0.44 that agreed with similar findings elsewhere [51]. The interaction of lime and manure significantly affected wheat harvest index.

Most parameters were correlated positively ($r^2 = 0.24$ to 0.95) with each other except days to emergence and heading. However, number of tillers, plant height, spike length, kernels per spike, thousand seeds weight, grain and straw yield correlations followed similar trends.

### *3.1.4. Nutrient P Uptake*

Wheat P uptake was influenced by the application of lime and manure. Crop grown in acidic soils showed more P uptake in response to liming than manure application. Phosphorus uptake values were recorded in the ranges between 1.44 and 17.74 kg ha-1. Wheat absorbed high P on combined 2.2 t lime, 10 t ha-1 manure and 20-30 kg P ha-1 treatments (Table 10). Liming increased soil pH as a consequent fixed P released to plant absorption. Hence, slight decrease in P uptake has been observed at high rate of liming (3.3 t ha-1). In line with this, liming increased P uptake of bread wheat by 91-93% over the control treatment [52]. On the other hand, manure increased P availability through decomposition and by having liming effect. However, P applications non-significantly increased wheat P uptake in acidic soils.

### *3.2. Status of Residual Soil P*

Before planting, soil P was 7.79 mg kg$^{-1}$(13.67 kg P ha$^{-1}$ in 15 cm soil depth), however, at the post-harvest of 2012 season the treatments residual P records were between 2.58 and 29.69 kg ha$^{-1}$ or 1.47 and 16.92 mg kg$^{-1}$. The lowest residual soil P was recovered from combined zero lime and manure treatment, whereas the highest on high manure and lime (Table 11).

***Table 11.** Soil P in 15 cm soil depth at the post-harvest of 2012 season.*

| Lime t ha$^{-1}$ | Manure t ha$^{-1}$ | kg P ha$^{-1}$ and residual soil P (kg ha$^{-1}$) | | |
|---|---|---|---|---|
| | | 0 | 20 | 30 |
| 0 | 0 | 9.55 | 2.58 | 4.33 |
| 0 | 5 | 6.46 | 12.81 | 14.39 |
| 0 | 10 | 9.55 | 18.25 | 16.16 |
| 2.2 | 0 | 16.74 | 14.99 | 10.34 |
| 2.2 | 5 | 15.11 | 22.10 | 19.16 |
| 2.2 | 10 | 10.42 | 22.04 | 14.78 |
| 3.3 | 0 | 26.75 | 26.75 | 21.39 |
| 3.3 | 5 | 19.13 | 27.85 | 25.47 |
| 3.3 | 10 | 29.69 | 27.01 | 27.69 |

Residual soil P of lime received treatments were increased by 55 to 60% while manure received treatments were increased by 8 to 22% as compared with non-treated plots (Table 11). Residual soil P was positively correlated ($r^2 = 0.79^{**}$) to lime. In line with this, as in [41, 42] reported liming of acidic soils reduce the P sorption thus increasing its availabilities. However, over all residual soil P of the treatments were categorized under very low status.

## 4. Conclusions

Most wheat growth and yield parameters were

significantly ($p < 0.01$) affected with the interaction of lime and manure, lime and P and interaction of manure and P. However, the crop economical yields were significantly affected by interaction of manure and lime; manure and P. The combined application of 5 t manure and 2.2 t lime $ha^{-1}$ increased wheat grain and straw yield by 2 and 2.5 t $ha^{-1}$ over the control treatment, respectively. Meanwhile, the combined application of 5 t manure and 20 kg P $ha^{-1}$ increased grain and straw yields by 1.8 and 2.1 t $ha^{-1}$ over the control treatment, respectively. On the other hand, the combined application of 5 t manure and 2.2 t $ha^{-1}$ lime gave more net benefit of 515.50 Eth. birr ($27.6) as compared with the combined application of 5 t manure and 30 kg P $ha^{-1}$ since the small-scale farmers in the study area have more access to lime than mineral P fertilizer. Wheat absorbed 10.29 kg $ha^{-1}$ P on combined 5 t manure and 2.2 t $ha^{-1}$ lime treatment while slightly lower (9.85 kg $ha^{-1}$) in combined 5 t manure and 30 kg P $ha^{-1}$ application. Although soil P was very low in the study areas the combined application of lime and manure increased available P that attributed to high P desorbing capacity of lime. Overall, the combined application of 5 t manure and 2.2 t lime $ha^{-1}$ has improved wheat yield on marginally fertile acidic soils of northwestern Ethiopian highlands.

## Acknowledgments

We acknowledge the Ministry of Education for Federal Democratic Republic of Ethiopia to the financial grant. Our especial thanks extended to the University of Helsinki School of Natural Resources Management and Environmental Sciences for lime analysis and part of the soil analyses.

## References

[1] World Bank. Natural resource degradation in Sub Saharan Africa. Restoration of soil fertility. A concept paper and action plan. World Bank, Washington, USA. 1996.

[2] C. Marc, S. Abebe and H. Mitiku. Farmers' knowledge of soil fertility and local management strategies in Tigray, Ethiopia. Managing Africa's Soils, 2000, 10: pp 1-30.

[3] A. Maatman, M. Wopereis, K. Debrah and J. Groot. From Thousands to Millions: Accelerating Agricultural Intensification and Economic Growth in Sub-Saharan Africa. *In:* Bationo (Ed.), Springer. Advances in Integrated Soil Fertility Management in Sub-Saharan Africa: Challenges and Opportunities, 2007, pp 75–84.

[4] Food and Agriculture Organization. Ethiopian Highland Reclamation Study, Ethiopia. Final Report (Volume I and II). Food and Agricultural Organization of the United Nations, 1986, Rome, Italy.

[5] C. Achalu, G. Heluf, K. Kibebew and T. Abi. Response of barley to liming of acid soils collected from different land use systems of western Oromia, Ethiopia. Journal of Biodiversity and Environmental Sciences, 2012, 2(5): pp 37-49.

[6] A. Nekesa. Effect of Minjingu phosphate rock and agricultural lime in relation to maize, groundnut and soybean yields on acid soils of western Kenya. MSc Thesis, Moi University, Eldoret, Kenya. 2007.

[7] H. Schlede. Distribution of acid soils and liming materials in Ethiopia. Ethiopian Institute of Geological Surveys, Ministry of Mines and Energy. Addis Ababa, Ethiopia. 1989.

[8] A. Mesfin. Nature and management of acid soils in Ethiopia, 2007. www.eiar.gov.et/ Soil/ soils acid.pdf.

[9] Amhara National Regional State's Investment office. Potential survey, identification of opportunities and preparations of projects profiles and feasibility studies, part one potential assessment. Unpublished soil survey study report by Development Studies Association (DSA) and Shawel Consult International (SCI). 2006.

[10] B. Woldeamlak and S. Geert. Assessment of soil erosion in cultivated field using a survey methodology for rill in the Chemoga watershed, Ethiopia. Unpublished doctoral dissertation, Wageningen University, The Netherlands. 2003.

[11] H. Hurni. Land degradation, famines and resource scenarios in Ethiopia. In: Pimentel, D. (Ed). World Soil Erosion and Conservation. Cambridge University Press, pp 27–62. 1993.

[12] B. Woldeamlak and L. Stroonsijder. Effects of agro-ecological land use succession on soil properties in Chemoga watershed, Blue Nile basin, Ethiopia. Geoderma, 2003, 111: pp 85-98.

[13] G. Ashenafi. Triticale production in Ethiopia- its impact on food security and poverty alleviation in the Amhara Region, Doctoral thesis, Kassel University, Germany. 2008. urn:nbn:de 0002-4115no. 2007:03.

[14] A. Worku Effects of nitrogen and seed rates on yield and yield components of bread wheat (*Triticum aestivum* L.) in Yelmana Densa District, northwestern Ethiopia. MSc. Thesis, Haramaya University, Haramaya, Ethiopia. 2008.

[15] International Food Policy Research Institute. Fertilizer and soil fertility potential in Ethiopia, Constraints and opportunities for enhancing the system, working paper, July, 2010. www.ifpri.org. 2010.

[16] E. Caires, L. Alleoni, M. Cambri and G. Barth. Surface application of lime for crop grain production under a no-till system. American Society of Agronomy J., 2005, 97: pp791–798.

[17] K. Hati, A. Swarup, D. Singh, A. Misra and P. Ghos. Long-term continuous cropping, fertilization and manure effects on physical properties and organic carbon content of a sandy loam soil. Australian Journal of Soil Research, 2006, 44: pp 487–495.

[18] J. Whalen, C. Chang, G. Clayton and J. Carefoot. Cattle manure amendments can increase pH of acid soils. Soil Science Society of America Journal, 2000, 64: pp 962–966.

[19] R. Onwonga, J. Lelei, B. Freyer, J. Friedel, S. Mwonga and P. Wandahwa. Low cost techniques for enhancing N and P availability and maize (*Zea mays L.*) performance on acid soils. World Journal of Agricultural Sciences, 2008, 4: pp 862-873.

[20] Ministry of Agriculture and Rural Development. Agro-Ecological Zones of Ethiopia. Addis Ababa, Ethiopia. 2005.

[21] Food and Agriculture Organization (FAO). Guidelines for Soil Description. Food and Agriculture Organization of the United Nations, 2006, Rome, Italy.

[22] S. Sahlemedhin and B. Taye. Procedure for soil and plant analysis. National Soil Research Centre, Ethiopian Agricultural Research Organization, Addis Ababa, Ethiopia. 2000.

[23] D. Rowell. Soil Science: Method and Applications. Addison, Wesley, England: Longman Scientific and Technical, Longman Group UK Limited. 1994.

[24] United States Department of Agriculture Natural Resources Conservation Services (USDANRCS). Soil Survey Laboratory Information Manual. National soil Survey Laboratory, Soil Survey Investigation Report No. 45 Version 1, pp 1-305. 1995.

[25] M. Mehlich. Mehlich 3 soil test extractant: a modification of the Mehlich 2 extractant. Communications in Soil Science and Plant Analysis, 1984, 15: pp 1409–1416.

[26] R. Bruce and G. Rayment. Analytical methods and interpretations used by the Agricultural Chemistry Branch for Soil and Land Use Surveys. Queens land Department of Primary Industries. 1982.

[27] T. Barauah and H. Barthakur. A Text Book of Soil Analyses. Vikas Publishing House Pvt. Ltd. New Delhi, India. 1997, pp 334.

[28] National Soil Survey Laboratory Staff. Principles and procedures for using soil survey laboratory data. USDASCS, Lincoln, NE. 1983.

[29] P. Hazelton and B. Murphy. Interpreting Soil Test Results: What do all the numbers mean? 2nd (Ed). NSW (New South Wales Department) Department of Natural Resources, Collingwood, Australia: CSIRO Publishing, 2007. http://www.publish. csiro.au.

[30] B. Taye, M. Zebene and E. Abayneh. Laboratory procedure for the determination of lime requirement and calcium carbonate content in lime. National Soil Research Center Ethiopian Institute of Agricultural Research, unpublished laboratory manual. 2007.

[31] S. Olsen, C. Cole, F. Watanab and L. Dean. Estimation of available phosphorous in soils by extraction with sodium bicarbonate. USA, Circular 939: pp 1-19. 1954.

[32] Gomez and A. Gomez. Statistical Procedure for Agricultural Research 2nd (Ed.). Wiley, A., Inter-Science Publication, New York, USA. 680. 1984.

[33] SAS Institute Inc.SAS User's Guide, Statistics version 9.1 (Ed). SAS Inst. Cary, NC, USA. 2002.

[34] R. Haling, R. Simpson, R. Culvenor, H. Lambers and A. Richardson, Effect of soil acidity, soil strength and macropores on root growth and morphology of perennial grass species differing in acid-soil resistance. Blackwell Publishing Ltd, Plant Cell and Environment, 2011, 34, pp 444–456. doi: 10.1111/j.1365-3040.2010.02254.x.

[35] T. Butler and J. Muir. Dairy manure compost improves soil and increases tall wheatgrass. Agronomy J., 2006, 98: pp 1090–1096.

[36] G. McAndrews, M. Liebman, C. Cambardella and T. Richard. Residual effects of composted and fresh solid swine (*Sus scrofa* L.) manure on soybean (*Glycine max* L.) growth and yield. Agronomy J., 2006, 98: pp 873–882.

[37] F. Naramabuye, R. Haynes and A. Modi, Cattle manure and grass residues as liming materials, in a semi-subsistence farming system. Agriculture, Ecosystems and Environment, 2008, 124: pp 136-141.doi:10.1016/j.agee.2007.08.005.

[38] H. Melesse. Response of bread wheat (*Triticum aestivum* L.) varieties to N and P fertilizer rates in Ofla District, southern Tigray, Ethiopia. MSc Thesis, Haramaya University, Haramaya, Ethiopia. 2007.

[39] I. Holford. Soil phosphorus measurements and its uptake by plants. Aust. J. Soil Res., 1977, 35: pp 227–239.

[40] N. Fageria, N. Slaton and V. Baligar. Nutrient management for improving lowland rice productivity and sustainability. Adv. Agronomy, 2003, 80: pp 63–152.

[41] M. Osundwa, J. Okalebo, W. Ngetich, J. Ochuodho, C. Othieno, B. Langat and V. Omenyo. Influence of agricultural lime on soil properties and wheat (*Triticum aestivum* L.) yield on acidic soils of Uasin Gishu County, Kenya. American Journal of Experimental Agriculture, 2013, 3(4): pp 806-823.

[42] J. White, M. Bell and N. Marzies. Effect of subsoil acidity treatments on the chemical properties of Ferrosol. *Proceeding agronomy Conference,* 10–15 September 2006, Australia. 2006.

[43] F. Oluwatoyinbo, M. Akande and J. Adediran. Response of okra (*Abelmoschus esculentus*) to lime and phosphorus fertilization in an acid soil. World Journal of Agricultural Sciences, 2005, 1(2): pp 178-183.

[44] R. Ranjit. Response of groundnut genotypes to lime and phosphorus levels in coastal alluvial soil of north Karnataka. MSc Thesis, Dharwad University, India. 2005.

[45] K. Endalkachew. Effects of rates and methods of phosphorus placement on residual soil P yield and P uptake of wheat in *Nitosols* of Kulumsa area, Arsi zone. MSc Thesis, Alemaya University, Haramaya, Ethiopia. 2006.

[46] N. Fageria and A. Santos. Yield physiology of dry bean. *J. Plant Nutr*ition, 31: 983- 1004. 2008.

[47] J. Nakagawa, I. Imaizumi and C. Rosseto. Effects of some phosphorus sources and of liming on groundnut seed quality. Pesquisa Agropecuaria Brasileria, 1990, 25(4): pp 505-512.

[48] N. Fageria and V. Baligar. Fertility management of tropical acid soils for sustainable crop production. *In*: Rengel, Z. (Ed.) Handbook of Soil Acidity, Marcel Dekker, New York, USA. 2003. pp 359–385.

[49] D. Guangdi, P. Rajinder, P. John and R. Keith. A financial analysis of lime application in a long-term agronomic experiment on the south-western slopes of New South Wales. Crop and Pasture Science, 2009, 61: pp 12–23.

[50] A. Mongia, N. Singh and A. Guha. Response of rice to liming and P application in acid soils of South Andaman. Journal of the Indian Society of Soil Science, 46(4): pp 697-700. 1998.

[51] D. Hussins and W. Pan. Nitrogen efficiency components analysis: an evaluation of cropping system difference in productivity. Agronomy Journal, 85: pp 898-905. 1993.

[52] K. Abreha, G. Heluf, M. Tekalign, and T. Kindie. Wheat crop response to liming materials and N and P fertilizers in acidic soils of Tsegede highlands, northern Ethiopia. Agriculture, Forestry and Fisheries. 2, 3: pp 126-135. 2013doi: 10.11648/aff.20130203.12.

# 3

# Principal-agent relationship in policy implementation of the use of forest area for mining activity, Indonesia

**Manifas Zubayr[1], Dudung Darusman[2], Bramasto Nugroho[2], Dodik Ridho Nurrochmat[2]**

[1]Forestry Ministry of Indonesia, Jakarta, Indonesia
[2]Department of Forest Management, Bogor Agricultural University, Bogor, Indonesia

**Email address:**
manifas_zaf@yahoo.co.id (M. Zubayr)

**Abstract:** The use of forest area (UFA) is defined as utilization of a portion of forest land for any development purposes outside forestry without changing its function and designation. This study is aimed to explore the relationship of principal and agent in the UFA implementation. Based on the agency theory and using descriptive-qualitative methods, this study identified specific relationship characteristics, i.e. the lack of authority in agent selection, the domination of principal on information, the moral hazard behavior of both principal and agent, the higher risk assumed by the principal, and the poor implementation of incentive structure. The relationship patterns did not occur only between principal and agent, but also among different principals and between one agent and another especially in cooperation and conflict relationship. This study confirmed that better implementations of UFA are strongly influenced by the efficiency of contractual agreement, the principal control on the process of UFA implementation, the agent commitment on the agreed contract and the amount of transaction costs.

**Keywords:** The Use of Forest Area, Agency Theory, Principal, Agent

## 1. Introduction

Forest management and mining in forest resource management perspective and mine resources, has undergone a paradigm struggle due to strong demands of all parties to fulfill the needs of both the raw material resources of the community both local, national and international. Those fulfillment is related to environmental and economic issues which then leads to a difference in opinions when states allow some licenses/concessions of mining operations in forest areas [19].

The most obvious problem is deforestation and degradation due to activities of forest concessions, encroachment and conversion of land, while mining generally revolve around the issue of land and environmental degradation. Reference [13] state that deforestation in various simulation models which conducted by that the dominant stimultant of deforestation are logging activities and mining (large scale), and also potentially illegal activities.

To accommodate conflict interests of the management of natural resources, in particular between the interests of management and utilization of forest resources from mining, the government has devised UFA policies. UFA is defined as the use of a portion of forest land for development purposes outside forestry without changing its function and designation (Regulation of the Minister of Forestry No. P.16/Menhut-II/2014).

UFA policies are set forth in the legislation, in the form of laws, regulations, President's instructions until ministerial regulations as implementation of technical regulations. In essence, those laws and regulations set operation of activities outside the forestry sector in the forest area in the mechanism of leasehold of the forest area license (LFAL). Up to now, the Ministry has issued a license to use forest land for mining activities as many as 1,025 units, with an area of 3,392,898.87 hectares[1].

Basically, LFAL is a principal-agent (PA) relationship as a form of authority (contract) forest management from the government as the principal (P) to the mining company as

[1] General Directorate of Planology, Ministry of Forestry. 2014. Progress of leasehold of the forest area license for mining activity until January 2014. Accessed at http://www.dephut.go.id/index.php/news/details/9468 on March 15th 2014

an agent (A). With those LFAL allowed to use and utilize+
e energy and mineral resources in the forest area for both public inquiry, survey, exploration and exploitation (production operations). Contractual relationship between P and A is not always run smoothly, many problems occurred when implemented in the field as a result of policy imperfections. Some researchers argue that PA approach can be done in the realm of policy (Table 1).

***Table 1.*** *Some researchers argumentations about the use of PA relationship in policy*

| No | References | Principal-agent in policy |
|---|---|---|
| 1. | [1] | Principal-agent model is an analytical framework that is very useful in explaining incentives problem in the public institutions with two possible conditions, i.e,(1) there are several principals with each goals and interests are not coherent and (2) the principal can not act in accordance with public interest, but give priority to interests that are more narrow. |
| 2. | [2] | PA framework is an approach to analyze commitment of public policy for the manufacture and implementation involves contractual issues relating to assymetric information, moral hazard, bounded rationality, and adverse selection. |
| 3. | [10] | Agency theory is able to explain the potential conflict of interest between the various interested parties. |
| 4. | [14] | Principal-agent model can be used to explain the central problems in the interaction between the principal-agent in formulation and implementation of public policy related to the performance and service. Then the reciprocal relationship between the principal-agent on the policy cycle, from formulation to implementation of the policy and vice versa: 1) government as a principal to agent in the public service; 2) the society as a principal to political agent in various forms of government. |

Agency theory is intended to solve two problems that can occur in agency relationships. First, the agency problem arises when: a) the desires or goals of the P and A opposite; and b) is a costly affair for P to verify what was actually done by A. Secondly, the problem of risk sharing arises when the P and A have different attitudes towards risk [22]. Agency theory focuses on the problem of asymmetric information, where the agent has more information about the actual performance, motivation, and goals, which could potentially create moral hazard and adverse selection. P itself must pay (costs) to monitor the performance of A and determine the structure of incentives and monitoring efficiency [20]

The purpose of this study was to analyze the implementation of UFA policy approach in relationship between the government (the principal) and the company of LFAL holders (agents) as well as identify problems and obstacles.

# 2. Method

This study used descriptive qualitative approach. Qualitative approach look at reality as a result of reconstruction of researchers who are directly involved in social situations. One of fundamental characteristic of this research method is involvement and interaction of the researchers with the reality observed [24].

Data will be collected by the method of document review and regulation analysis, in-depth interviews with some experts and informants, and field observations at several mining companies having LFAL. Preferred observation form is unstructured observation that is carried out without using a guide. In this observation, the researcher must be able to develop the power of observation in observing an object and the object must master the science of research/observation [3]. The sampling was choosed using purposive and snowball sampling technique. However, approach methode conducted by [17] was used to know the transaction costs issued by the agent in UFA implementation.

Primary data were obtained from experts and informants of parties involved directly in the process of UFA implementation namely Ministry of Forestry, Provincial and District Forestry, District Department of Mines and Energy, Technical Implementation Unit in the center region, companies holding of LFAL and companies holding of the license of forest production utilization (LFPU). Secondary data was obtained from the documents and reports belong to companies as well as government agencies, books, magazines, newspapers or research reports and other relevant and reliable published sources.

Data collection was conducted in Bogor, Jakarta and several observation locations in Kutai Kartanegara regency of East Kalimantan, Tanah Bumbu regency of South Kalimantan, North Konawe regency and Kolaka of Southeast Sulawesi. Data and information collecting was conducted in 2006. Furthermore, field observation was carried out within Juli - December 2013.

# 3. Results and Discussion

Various issues in agency theory are described by [15] namely: moral hazard, adverse selection, asymmetric information, risk preferences, transaction costs, monitoring, incentives and contract design. While the issues or characteristics of PA relationship in UFA implementation are as follows:

### *3.1. Adverse Selection*

Selection in PA relationship is an important early stage for the implementation and sustainability of the contract which will be agreed together. Selection of A determine the success of P in achieving goals or importances. However, in practice, P often made a mistake in choosing A (adverse selection). References [20 , 15] defined miscast as P mistake in identifying A expertise and inability of P to verify completely A capabilities before making decision to give authority to A.

In UFA implementation, A is a company that has

received mining exploitation permit from the central government (Minister of Energy and Mineral Resources) or from the local government (governor or regent). So the selection of A is in other sectors. P does not have the option of selecting A given authority to manage the forest. P accepts a mining company (as A) that asked LFAL if A has all the requirements.

Adverse selection has a huge impact on the contract implementation. Especially mismatch in contract implementation, P has responsibility and must struggle to achieve goals. Furthermore, P also should be able to eliminate the moral hazard that is done by A and alleviate costs in monitoring the implementation of the contract (monitoring costs).

Generally, the PA relationship is developed on the basic of needs of P that have the importances or the goals of the organization then gives trust to the other party (A) to conduct the work for achieving the goals or importances of P. While the PA relationship in UFA implementation is more on a request from A who have an importance or purpose to exploit mineral with P. Situations of PA relationship bring two (2) consequences, i.e. P does not have opportunity to choose and select A and A does not have opportunity or 'right' to reject the contract. The lack of choice for P has large impact on the performance of institutions which formed during UFA implementation, especially on the fulfillment obligation of A during implementation of the contract.

### 3.2. Contractual Relationship

Agency problems triggered by difference of interest between the principal and agent. These problems can be eliminated by arranging an efficient contract to juxtapose P and A importances. The contract should be arranged by considering all aspects that can boost enhancement performance and reinforce A commitment to fulfill the goals of P. Contracts can also minimize the possibility of A deviating behavior in its performance. P always hoped A can fulfill its importances. While the failure in arranging contract would cause opportunistic attitude on A. Opportunistic attitude of A arises in form of neglect obligations and responsibilities or perform other illegal activities.

To resolve the agency problem, agency relationship theory identify and determine the choice between the two types of formal contract to control PA relationship, first is behavioral-based contract, and second is outcome-based contract.

***Table 2.*** *Comparison of characteristics of PA relationship in UFA implementation with contract based on the behavior and the final result.*

| No | Indicators that determining Principal-Agent Relationship | Behavior Based Contract* | Outcome Based Contract* | UFA's Contract** |
|---|---|---|---|---|
| 1 | *Asymmetric information* | Low | High | High |
| 2 | *Outcome Uncertainty* | High | Low | High |
| 3 | *Outcome Measurement* | Relatively Difficult | Relatively Easy | Relatively Easy |
| 4 | *Risk aversion* | High | Low | High |
| 5 | *Conflict of importance* | Low | High | High |
| 6 | *Incentive approach* | Principal and Agent have the view that the agent entitled to a reward | Principal give reward to Agent based productivity capabilities | Principal does not give any reward for the ability of agent productivity |
| 7 | *Transaction cost* ** | - | - | Relatively High |
| 8 | *Moral hazard*** | - | - | Relatively High |
| 9 | *Contract implementing* ** (*contract implementation*) | - | - | Difficult |

Sources:
* = Adapted from Rungtusanatham et al (2007)
** = Assessment is the result of researcher interpretation during the UFA policy implementation research and observation

When analyzing and comparing LFAL contract with indicators which proposed by [22], LFAL contract tend to avoid model of the second contract. The data show that P does not have capability to oversee A behavior, as well as the desired outcome is not as desired by P. An Assessment of A responses in implementation of the contract shows that almost all A obligation is not properly implemented according to LFAL contract. This condition is not as proposed by [23] where when P has the sufficiency to acquire and analyze information that is used to monitor agents behavior with efficient cost, then behavioral-based contract become an option.

According to [13], P can arrange contract where they can do a tactic or strategy to modify A behavior. To design an efficient P-A relationship, the contract should be self-enforcing. That is, the contract can force or press selfishness in A to obey with what is expected of P. P must have at least four (4) ways to design an efficient contract and have self-enforcing to limit the possible fraud that will be done by A. First, P can be carefully filter the potential of A when they want to do agreement. Second, P can monitor A activities. Third, P can arrange the contract in which include provisions that may enhance P credibility and A commitment (rewards and punishment approach). Fourth, P can compare the performance between two or more A. If the design is appropriate, to compare how they can collect information about A behavior and also can prevent A behavior that would ruin his performance.

### *3.3. Asymmetric Information*

Asymmetric information that occurs in the process of UFA implementation can be identified in two (2) cells of four (4) cells according to [26] were related with differences in goals between P and A. In the UFA implementation only identified B and C cells with different material information related to goal differences between P and A. If material information was based on P goal, then the mismatch relationship information such as cell C, where P control the information completely. Whereas if the material information is based on A goal, then the situation will be opposite to be B, where information is widely controlled by A. While A and D cells were not identified in this study. In various studies and literature, mismatch information B cell type were much found and discussed. Mismatch type such those information by [26] called as a 'classic case' (Fig. 1).

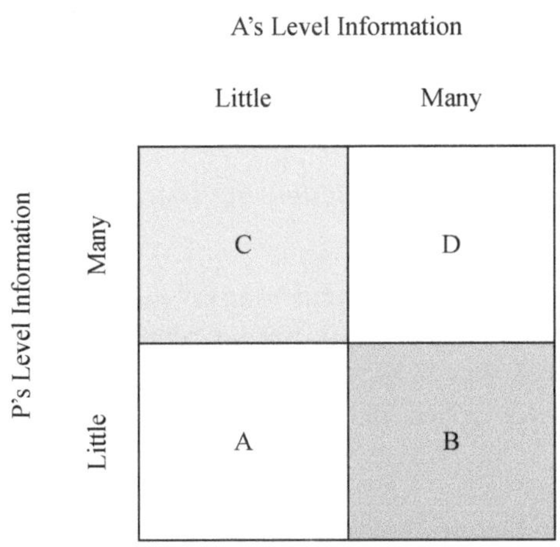


***Figure 1.*** *Quadrant of mismatch information related to the goal differences between P and A (adapted from Waterman and Meier 1998)*

Differences goals between P and A which caused by differences goals generate the difference in material information that is had by both parties. That differences creates the existence of two types of mismatch information in the study of the UFA implementation. Types were referred as ' classic case ' in agency theory above refers to the goal to be achieved by A. P did not have any information, knowledge or experience about A goal, so P could not oversee activities related to mine production was done by A. so, P did not know A performance and very difficult to know whether A did deviation or not from the contract (moral hazard). In contrast, in the UFA implementation has a C cell type. Almost all information about forest management was the domain of the Ministry of Forestry as P. While LFAL holder companies hardly have such information. The absence of forest management information had by A cause a wide gap between P goal and A performance.

### *3.4. Lack of Incentives*

Reference [18] stated that the amount of compensation received by A depends on the amount of profit generated in accordance with the P contract agreement. Reference [4] argued that in countries with high levels of corruption, incentive mechanism greatly help to improve the motivation and worker performance. Base on both opinions, P should give incentives to A for successfully managing forests comparable with fruition of P. In the practice, the incentives were not applied.

Specifically, in the context of the UFA implementation incentives is an award or ease given by the government to LFAL holder for their performance in carrying out their obligations in accordance with their LFAL. The purpose of these incentives include: respect agent performance achievement, increase motivation and agents commitment, ensure fairness, maintain performance, and reduce risk of failure.

According to some officials in the Ministry of Forestry, non-tax government revenues (NTGV) through success reduction in the reclamation and revegetation was the incentive form for A. As for A, reduction in the NTGV was not attractive, especially with the discretion of the Director General of Planology about NTGV actually be a disincentive for A. A is more interested in legal certainty and continuity of their business in the forest area. On the other hand, the received of forest area which has reclaimed and revegetated did not guarantee A to get replacement area, because there has no regulations about it.

Incentives in the form of law certainty, area certainty and guarantee the continuity of the business as well as in other forms was expected by A, Therefore, P need to make restrictions or rules about incentives instrument in LFAL to increase motivation and commitment of A in doing the contract properly and accordance with P desire.

### *3.5. Risk Preferences*

None of the people who want the risk to be assumed in relationship because it connotes to loss that should be accepted as result of taken decisions. In P-A relationship, the risk can certainly be felt by both parties. However, the agency theory both P and A will struggle to avoid the risk of implementation of the agreed contract. UFA implementation is essentially a risk management policy caused by other operational sectors in forest area. The most rational option, is giving other sectors chance to use forest as well in synergy with other sectors to restore forest condition.

Risk received (either by P or A) is the result of incomplete and weak supervision enforcement contract by P. The risk will be bigger assumed by P, especially in law consequency. The implementation of contract process was quite long, should provide an opportunity for P doing intensive supervision and continuous. Effectiveness factors and continuity of supervision become greatest challenge during implementation of the contract. If we follow the implementation process of the contract, the risks assumed by P was accumulation of A performance and result of the effectiveness and continuity of P supervision. The biggest risk will be assumed by P was the destruction of forest, abandoned mine pits without reclamation and revegetation, so the purpose of forest recovery have failed.

### 3.6. Moral Hazard

In agency relationship, the two parties (P-A) will struggle to maximize their utility with the mutual benefit principle. However, because one party controls information better, will lead to opportunistic behavior of one party did not act for other party's importance. This situation caused the appearance of temptation for one or more persons to deviate in order to maximize [5, 8].

Opportunistic behavior performed by A to achieve the goals, reduce risk and transaction costs in implementation the contract. Lack of information that had about policies and regulations in forest management, it triggered A to do moral hazard. This behavior was supported by inability of P in controlling function and enforcement of A behavior. On the other hand, A precisely controlled the information, knowledge and experience techniques to achieve the same goal which was not owned by P.

Many cases that occurred in East Kalimantan and South Kalimantan is let the former mine that was not longer used. The holes of mine left open without reclamation and revegetation. Behavior without reclamation and revegetation or 'mine-and-run' would be the greatest risk for P if neglect LFAL contract allowed to happen.

Greater moral hazard was illegal mining caused by several factors, namely financial inability of A to fulfill the transaction costs for LFAL request, long permit bureaucracy, weak supervision and enforcement, market demand and mining commodity prices in the market as well as the strength backing of companies.

In the UFA implementation, moral hazard was not only done by A, but also by P. The collusion practice and gratification indicated in the UFA implementation process. Lack information of A was used by the person who becomes a free rider for profit. In LFAL submission process, A spent transaction costs between 7-15 billion. According [9] self-importanced behavior was not a moral issue. In the organization context, it was normal behavior because an opportunity to do so was available. However, it should be assessed whether it supports the implementation of the UFA policy actually towards forest destruction

### 3.7. Control

Control in the contract implementation of LFAL to be one thorny issue in the UFA implementation. To overcome the agency relationship problems between P and A, the required management control systems which is means to align goals between agent and principal [6]. Important aspects in control management systems there are existence a system and information exchange process, internal controls and audits, performance measurement and evaluation, compensation and incentive.

Although the contract in LFAL is based on combination of behavior and outcomes, but control is done by P toward A performance tends to be based on behavior. However, these controls can not be implemented properly by P. Inability of P in control function becomes the cause of failure achieve that has been established. Budget and personnel resources constraints to be the reason for P performing control functions. Decentralization also has not implemented properly. Decentralization did not automatically result better forest resource management methods [18, 19]. Meanwhile, P could not simply rely on A commitment to carry out the contract properly, moreover A has not knowledge of forest management.

Low effort of A is the moral hazard problem in agency theory. Moral hazard problems may occur at the ex ante that is when LFAL submission and ex post at the time of contract execution, fulfillment the obligations, monitoring and control of post-mining results. In the stages of agency theory, the second problem was an error in A selection. A may not have the ability and P did not observe on A ability. Unknowing of P toward A ability requires P to control behavior of A tightly. In such situations the behavior-based control become the first option, although it occured higher transaction as a logical consequence that must be accepted.

### 3.8. Principal-Agent Relationship Issues in the UFA Implementation

Agency theory gives attention to the settlement of two problems that can occur in agency relationship [15] A, was always positioned as party who tend to prioritize their importances rather than P importances. This tendency at last raises its own problems (agency problems). Agency theory that has developed is generally directed to solve any problems that may occur in agency relationship.

#### 3.8.1. Principal-Agent Problems

PA relationship in the UFA implementation has different characteristics and problems with the PA in theory of its agency relationship in general. As an illustration to make it easier to compare different problems that occur in two types of PA relationship are summarized in the following table:

*Table 3. Comparison of PA relationship problems in general with P-A relationship in the UFA implementation*

| Problem Identification | Principal-Agent Relationship | |
|---|---|---|
| | Generally | UFA implementation |
| Initiative | Principal | Agent |
| Agent Selection | Applied | Not Applied |
| Contract Formulating | Together (*bargaining*) | Principal *(absolute)* |
| Mastery of information | Agent | Principal |
| Moral Hazard | Agent | Principal andAgent |
| Incentives application | Applied | Not Applied |
| Control | Strong | Weak |
| Risk Distribution | Principal | Principal |

### 3.8.2. Principal-Principal Problems

Recent studies show that P-P problem can occur in an institution if there is a mismatch goals between P in that institutions [25]. In particular, it will happen because of difference importances between big P (superior) and small P (inferior). Previous studies examined the problem of P-P have solid foundation of economic problems which the majority of P doing conspiracy with the company's owner to take over the resources / assets of the company [12].

In the context of the UFA implementation, P-P relationship was due to the two delegation of authority to the LFAL holder that were the Ministry of Forestry (P-for) and the Ministry of Energy and Mineral Resources or Regent (P-mine). Reference [11] defined the P-P problem as a conflict between two individuals or groups P, big P as the party that controls and small P as controlled. [25] developed the definition by stating that the problem of P-P refers to the adjustment of the value of small P becomes P big, it often affects decisions such as asset sales and purchases. While in this study, the P-P problem is defined as a relationship between two or more P who gave delegates to the A to achieve the goals of each level of P. P can be parallel or there are different levels, both positions in government, politics, economics, social and culture.

The existence of more than one P by [16] referred to as multiple principals. Each P delegate accordance with their authority. P should solve internal problems between them then collectively (collective principal) motivate A to improve their performance to fulfill P importance. The establishment of collective principal will make it easier for A to implement contract that has been formulated. However, the collective principal in the UFA implementation is difficult to formed due to lack of coordination both P.

The fundamental problem that arises between P-for and P-mine is the difference importance in natural resource management. Until now there is unavailable policy or rule that can bridge two importances. These conditions make the forestry sector's bargaining position is very weak in determining policy the existence of mining in forest area. The forestry sector is almost always defeated by other sectors with the reason of national importance and economic growth.

### 3.8.3. Agent-Agent Problems

AA problem also identified in the UFA implementation. P-for having 2A, that were mining company as LFAL holder (A-mine) and forestry company as LFPU holder (A-for). Problems appear as a result of the P-for giving LFAL area to A-mine overlapping with LFPU concessions given to A-for. Conflict between A-mine and A-for still occur both in the domain of administration, coordination and in the field. Some problems between A-mine and A-for resolved with business to business (B to B). Most other cases are forced to be reported to the P-for.

In general, the conflict appears because A-for does not agree with A-mine activities considered disruptive the preservation of A-for forests status. The other reason, there is no agreement on stands compensation and infrastructure that has been built. While A-mine feel has had LFAL lawful to perform activities in forest areas that have been set. A-mine also complains to irrational demands of stands compensation and infrastructure of A-for. In situations of unresolved conflict, identified that A-for trying to do rent seeking/opportunistic behavior in conflict. As a result, transaction costs of A-mine are rising. Some advanced problems due to overlapping licenses in the same area is unclear boundaries of forest management authority, the opening access to the LFPU territory belonging to the A-for uncontrolled.

***Table 4.*** *Some cases A-for and A-mine conflicts in forest area*

| No | A-for | A-mine | Conflict Matter | Resolution | Progress |
|---|---|---|---|---|---|
| 1. | Intracawood Manufacturing PT | Jelai Cahaya Mineral, PT | Forest area management problem | Mediation by Ministry of Forestry | Unfinished |
| 2. | Adindo Hutani Lestari, PT | Pipit Mutiara Jaya, PT | Forest area management problem and compensation fee | Mediation by Ministry of Forestry | Unfinished |
| 3. | ITCI Hutani Manunggа, PT | Gunungbayan Pratamacoal, PT | Forest area management problem and compensation | Mediation by Ministry of Forestry | Unfinished |
| 4. | Kirana Chatulistiwa, PT | Anugerah Daya Gemilang, PT | Compensation fee | Business to Business | Finished/Clear |
| 5 | Karda Traders, PT Amprah Mitra Jaya, PT. and Sari Bumi Kusuma, PT | Kapuas Prima Coal, PT | Compensation fee | Business to Business | Finished/Clear |

As a brief overview of PA relationship problem in the study of UFA policy implementation were identified as quadrant in Fig. 2.

| Mining \ Forestry | Principal (P-for) | Agent (A-for) |
|---|---|---|
| Principal (P-mine) | - Differences motivation and natural resource management objectives (self-importance conflict)<br>- Lack of coordination between sectors, and between central and regional<br>- The exchange of information is limited<br>- *Bargaining position (Politic, Power)* | Unidentified |
| Agent (A-mine) | *Adverse Selection*<br>*Moral hazard*<br>P's Lack of Control<br>A's low commitment<br>*Transaction Cost* A<br>The absence of incentives | - Interrupting forest governance performance of A-for<br>- Unclear boundary management authority<br>*Compensation cost*<br>- The lack of coordination and cooperation<br>*Moral hazard* (*rent seeking, opportunistic behavior*) A-for<br>*High transaction cost* to A-mine<br>*Forestland access* |

***Figure 2.*** *Quadrant of P-A relationship variations and problem identification.*

### *3.8.4. Transaction Cost*

In the UFA implementation process, A can not be separated from cost consequences, start from information retrieval, LFAL submission, implement their rights and obligations (contract). Transaction costs in the UFA implementation context is a cost to be assumed by P and A as implementation result of the agreed contract. Transaction costs include the obtaining cost up to the LFAL publication (contract), implementation of contract along with other costs as a result of the contract implementation and contract settlement costs. As an illustration of the estimated transaction costs issued by A is as follows:

***Table 5.*** *Transaction cost of agent in the UFA implementation*

| No | Type | Transaction Cost (Rp) |
|---|---|---|
| 1. | Submission of LFAL application | 7 billion |
| 2. | Conflict resolution | ≥ 8 billion |
| 3. | Non Taxes Government Revenue | 1,76 billion/year |
| 4. | Agreement with the LFPU holder | 66,5 billion/year |

Some characteristics of agency theory in the UFA implementation that studied conical on four important factors that affect A performance in implement contract as detailed in LFAL. Five supporting factors are effective contract, high commitment from both parties, strict control of P to A performance, low transaction costs and appropriate incentive for A.

## 4. Conclusion

PA relationship in the UFA implementation had distinctive characteristics. The initiative relationship comes from A, there was no Agent's election, P mastering more information, moral hazard was done by both parties, and there has been no incentive. Agency relationship identified not only PA relationship happened, but also the relationship of PP (P-for and P-mine) and the relationship of AA (A-for and A-mine) in cooperation and conflict form. Conflict relationship between P-for and P-mine caused by differences goals and motivation in resource management and lack of coordination. While the conflict relationship between A-for and A-mine caused by overlapping licenses in the same forest area.

PA relationship in the UFA implementation could walk better if there were at least five supporting factors, namely; effective contract, high commitment from both sides, strict controls, low transaction costs and appropriate incentive

In general, agency theory was a simple analytical tool and relatively applied easier to unravel the problems of principal-agent relationship in public policy. The use of agency theory in the study of public policy make it easier to analyze the relationship, roles, responsibilities, behaviors, risks and prediction results (outcomes) to be obtained. Thus, in the public policy of implementation process, the principal had the opportunity to do remediation efforts on policies to improve performance of the institution.

## References

[1] Andvig, J.C., O.H. Fjeldstad, Amundsen, I., T. Sissener, T. Søreide, "Corruption A Review of Contemporary Research", Chr. Michelsen Institute Development Studies and Human Rights Report R, 7, 2001.

[2] Bergman, M. and J.E. Lane, "Public Policy in a Principal-Agent Framework", Journal of Theoretical Politics, 2(3), pp, 339-352, 1990.

[3] Bungin, B. "Penelitian Kualitatif: Komunikasi, Ekonomi, Kebijakan Publik, dan Ilmu Sosial Lainnya", Jakarta, Kencana Prenada Media Grup, 2009.

[4] Calhoun MA. "An agency assessment if host country employee goal misalignment and information asymmetry", International Business Reseacrh Teaching and Practice, 1(1), pp, 1-15, 2007.

[5] Eisenhardt, K. "Agency theory: an assesment and review", Academy of Management Review, 14, pp, 57-74, 1989.

[6] Ekanayake, S. "Agency theory, national culture and management control system", Journal of American Academy of Business", 4, pp, 49-54, 2004.

[7] Fox, G. "The Real Coase Theorems", Cato Journal, 27(3), pp, 373-396, 2007.

[8] Gibbons, R, "Incentives between firms (and within)", Management Science, 51(1), pp, 2–17, 2005.

[9] Hasanbasri, M. "Maksimasi, free rider dan kegagalan implementasi kebijakan", Jurnal Kebijakan Kesehatan Indonesia, 1(3), pp, 121-124, 2012.

[10] Jensen, M. and W. Meckling, "Theory of the firm: managerial behaviour, agency cost, and ownership structure. Jurnal of Financial Economics", 3(4), pp, 305-360, 1976.

[11] Jiang, Y. and M.W. Peng, "Principal-principal conflicts during crisis", Asia Pacific Journal of Management, pp, 1-13, 2010.

[12] Kahle, M.I.M. "What is influencing financially driven shareholder activism in the US and the UK-P-A or principal-principal problem?", Thesis, Faculty of Old Dominion University, University of Massachusetts. Amherst, 2010.

[13] Khan, A., A.B. Gunawan, and A. Smajgl, "Impacts of non-forestry policy on deforestation and poverty in East Kalimantan: An agent-based analysis", Tropical Forest Management Journal, 16(1), pp, 41-52, 2010.

[14] Lane, J.E. "The principal-agent approach to politics: policy implementation and public policy-making", Open Journal of Political Science, 3(2), pp, 85-89, 2013.

[15] Murphy, D.N.P, "Product reliability and warranty: an overview and future research", Produção, 17(3), pp, 426-434, 2007

[16] Nielson, DL, and M.J. Tierney, "Delegation to International Organizations: Agency Theory and World Bank Environmental Reform", International Organization, 57, pp, 241-276, 2003.

[17] Nurrochmat, D.R., M.F Hasan, D. Suharjito, A. Hadianto, M. Ekayani, Sudarmalik, H. Purwawangsa, Mustaghfirin, and ED Ryandi, in Nurrochmat & Hasan (Ed.), "Ekonomi Politik Kehutanan. Mengurai Mitos dan Fakta Pengelolaan Hutan", Ed$^{nd}$ , Jakarta, INDEF, 2012.

[18] Nurrochmat, D.R. and H. Purwandari, " Desentralisasi Pemerintahan Desa", Bogor, Kemitraan UNDP-PSP3 IPB, 2006.

[19] Nurrohmat, D.R., M. Krott, and R. Birner, "Decentralization policy and the struggle for authority over forest resources in Tebo Regency, Jambi". Tropical Forest Management Journal, 12(2), pp, 27-35, 2006.

[20] Petrie, M. A. "Framework for public sector performance contracting", OECD Journal on Budgeting, 2, pp, 117-153, 2002.

[21] Rajan, M.V. and R.E Saouma, "Optimal information asymmetry", The Accounting Review, 81(3), pp, 677–712, 2006.

[22] Rungtusanatham, M., Rabinovich E., Ashenbaum B, Wallin C, "Vendor-Owned Inventory Management Arrangements in Retail: An Agency Theory Perspective", Journal of Business Logistics, 28(1), pp, 111-135, 2007.

[23] Sarwoko, H. Analisis Hubungan antara Prinsipal dan Agen pada Perguruan Muhammadiyah di Jakarta [disertasi]. Jakarta: Fakultasi Ilmu Sosial dan Ilmu Politik. Program Pascasarjana. Universitas Indonesia, 2010.

[24] Somantri, G.R. "Memahami Metode Kualitatif", Makara, Sosial Humaniora, 9(2), pp, 57-65, 2005.

[25] Su, Y., D. Xu, P.H. Phan, "Principal-principal conflict in the governance of the Chineses public corporation", Management and organization review 4(1), pp, 17-38, 2008.

[26] Waterman, R.W. and K.J. Meier, "P-A Models: An Expansion? Journal of Public Administration Research and Theory", 8(2), pp: 173-202, 1998.

# Green economy: Challenges and prospects for improved aquatic agricultural system (AAS) in Niger delta communities

**Nenibarini Zabbey[1, *], Ijeoma Favour Vincent-Akpu[2], Ibisime Etela[3]**

[1]Department of Fisheries, Faculty of Agriculture, University of Port Harcourt, PMB 5323, East-West Road, Choba, Rivers State, Nigeria
[2]Department of Animal and Environmental Biology, Faculty of Biological Sciences, College of Natural and Applied Sciences, University of Port Harcourt, PMB 5323, East-West Road, Choba, Rivers State, Nigeria
[3]Department of Animal Science, Faculty of Agriculture, University of Port Harcourt, PMB 5323, East-West Road, Choba, Rivers State, Nigeria

**Emails address:**
zabbeyn@yahoo.com (N. Zabbey), nenibarini.zabbey@uniport.edu.ng (N. Zabbey),
ijeoma.vincent-akpu@uniport.edu.ng (I. F. Vincent-Akpu), ibisime.etela@uniport.edu.ng (I. Etela)

**Abstract:** There is dire need for deliberate green action plan to solve the endemic environmental crises occasioned by unsustainable utilization of natural resources and environmental degradation in the Niger Delta; a region with its characteristic terrain and eco-fragility that is highly susceptible to adverse environmental and climate change-related disasters. But, to successfully implement a green economy plan in the Niger Delta region, there has to be a well-thought out strategy with due appraisal of baseline environmental issues and concerns. The people of the delta depend, primarily, on the exploitation of the region's rich bio-resources with fishing and farming forming the major sources of livelihoods. Observations by independent and intergovernmental global environmental protection players indicate that, the creeks, lands, mangroves, forests, etc provide inestimable goods and ecological services to the region's inhabitants. Yet, they have been consistently degraded over the past four decades, mainly by industrial activities. Some identified challenges in greening the region and viable alternatives for sustainable agricultural development are, also, discussed. The authors have, therefore, proposed strategies such as consideration of complex biophysical and socio-economic factors, multi-stakeholder engagement, and a paradigm shift from the contemporary fossil energy-based agriculture to the more sustainable practices of biofuels and organic agriculture. A modelled paradigm for transforming current improved aquatic agricultural systems (AAS) to become more productive while, maintaining ecosystem services is proposed. Therefore, this paper examines current environmental issues and concerns as well as efforts variously made to address them in the Niger Delta against the backdrop of existing economic activities.

**Keywords:** Agro-Ecology, Ecological Services, Wetlands, Niger Delta, Livelihood

## 1. Introduction

The Niger Delta, which plays host to majority of the crude oil and gas facilities in Nigeria, borders the Atlantic coast of southern Nigeria where River Niger system divides into numerous tributary outlets that discharge the latter's waters into the Atlantic Ocean. The geographic Niger Delta spans over 20,000 km$^2$ and has a coastline of about 450 km. It has been described as the largest wetland in Africa and among the three largest in the world [1]. About 2,370 km$^2$ of the Niger Delta area consists of rivers, creeks and estuaries while, stagnant swamp covers about 8,600 km$^2$. Also, the mangrove swamps of the delta span over 10,000 km$^2$ and are considered the largest mangrove swamps in Africa [2, 3]. The region falls within the tropical rainforest zone in the tropics lying between the Tropic of Cancer at the northern boundaries (latitude 23.5° N) and the Tropic of Capricorn at the southern boundaries (latitude 23.5° S); specifically, described as moist deciduous and semi-evergreen seasonal forests. The ecosystems of the area are highly diverse with nutrient-rich alluvial soil that support numerous species of terrestrial and aquatic flora and fauna and human life [4].

### *1.1. Characteristics of the Natural Resource-Base of the Niger Delta*

The Niger River Delta region is highly endowed with renewable and non-renewable resources. Oil and gas from the region have since become the main sources of revenue for the Nigerian state; accounting for about 97 percent of the country's total export earnings. The wetland environments of the region are the foundation pillars that support livelihood of the indigenous people who depend on their natural resources and ecology for a living [5]. As a result of man's desire to meet these social needs, the Niger Delta has remained a highly susceptible region to adverse environmental changes enhanced by its location in the coastal region and various incidental human activities. Therefore, the present paper x-rays the various environmental, social and food needs met by the region and their varying impacts.

### *1.2. Socioeconomic and Political Hindrances to Green Economy in the Niger Delta*

The importance of agriculture in the economy of most countries, including Nigeria, cannot be over-emphasized. It constitutes the backbone for economic growth and industrial development in such countries and regions such as the Niger Delta region. Until recently, over 65 percent of the people of the delta were sustained by agriculture-related activities [6], mostly at the subsistence level. However, Government's direct agricultural production policy has resulted in the establishment of agricultural parastatals and companies such as Risonpalm Limited, Delta Rubber Company, Pabod Food Company, School-to-Land Authority (vocational institution) and the Agricultural Development Programme (ADP), and so on [6]. However, these hardly had systemic sustainability plans, and were easily overwhelmed by the desire for quick gains from the oil and gas sector resulting to manpower drain to the latter sector. The result is a continuous decline in the number of persons participating in agriculture and hence dwindling productivity.

Other predisposing factors include the interplay of plummeting fish stocks, extensive removal of mangroves (which are the key fish breeding sites) and poverty (living below the World Bank benchmark of USD $1.25 per day) have made some unscrupulous artisanal fisher folks to adopt destructive fishing methods while, in the industrial fisheries subsector, over-capitalisation is the primary cause of over-exploitation. Inappropriate provision of subsidies contributes to widespread over-fishing and distortion of trade in fisheries products. Industrial fishing in the tropics is, largely, characterised by huge harvests of immature fish and non-target species, which are illegally sold to bycatch traders on the high sea or simply thrown overboard as discards and thereby risking further aquatic pollution resulting from reduced dissolved oxygen in the water [7].

However, in terms of priority, most inhabitants of the Niger Delta region believe that, the major socioeconomic challenge in the region is poverty, since majority (over 60 percent) of the working class population are either unemployed or under-employed. There are reports in the literature to suggest that, there is a direct and critical link between environmental degradation and rural poverty levels [8]. For any transition to green economy to work, it must be preceded by deliberate and workable sustainable development packages that would, ultimately, result in poverty reduction by reversing the ongoing destruction of natural resources hotspots and by creating employment. As analysed by the UNDP [9], ecosystems and natural resources are vital for the survival and economic development of the inhabitants of the Niger Delta.

### *1.3. Natural and Man-Made Environmental Problems in the Niger Delta*

#### *1.3.1. Oil Exploration and Exploitation*

Oil was first discovered in commercial quantities in Nigeria's Niger Delta region in 1956 and since the early 1970s, oil has dominated the country's economy. While the Niger Delta is, globally, heralded as the 13$^{th}$ largest producer of petroleum in the world on the one hand, it has and continues to suffer serious environmental damages within the same period due to the heavy exploitation of these resources causing it to be ranked as one of the five most polluted and degraded locations on planet earth [10]. Severe environmental problems resulting from oil and gas exploration and development activities have rendered majority of the lands unproductive and the water too polluted for fish (and other aquatic lives) or human use.

#### *1.3.2. Oil Spills*

Available records indicate that, the Niger Delta region of Nigeria experiences, on the average, 273 oil spills resulting to about 115,000 barrels of crude oil worth US$5.64 million spilled annually from 1976-2001(valid at the average prevailing rate of US$49 per barrel within the period under review), making the region most vulnerable to oil spill than anywhere else in the world [11]. Moreover, oil spills impact significantly on farming and fisheries livelihoods and general water quality of the Niger Delta [9,12,13].

#### *1.3.3. Gas Flaring*

The flaring of gas has been ongoing in the Niger Delta region for over four decades yet, an immediate end to this daily nightmare seems not at sight. Today, there are about 123 flaring sites in the region [2]. With an estimated average of 17.2 billion $m^3$ $y^{-1}$ of associated gas emitted in the Niger Delta [14], amounting to 16 percent of global emissions. This huge volume of gas emission makes Nigeria one of the highest emitters of associated gas in the world [15]. Consequent upon this, acid rains have remained one of the effects of gas flaring, which occurs when polluting chemicals are absorbed by water droplets in the clouds. The acidity in rain water appears to be higher in the Niger Delta region and decreases further away from the region. These acidic droplets can increase the acidity of the soil and affect the chemical balance of aquatic ecosystems. Acid rains lead to loss of biodiversity, corroded roofs, destruction of forests and economic crops as well as lung diseases in mammals [16,14].

#### *1.3.4. Pipeline Crossing*

Forests and other vegetal formations are fragmented or totally removed when oil and gas pipelines are being laid. Similarly, rivers and swamps are blocked or sand filled in the process of laying the pipelines. With oil canals and network of pipelines traversing the delta land- and water-scapes, the inhabitants find it very challenging to, freely, undertake their conventional economic activities. This is because, embarking on these activities such as fishing and farming is almost impossible without standing the risk of, inadvertently, causing accidents when the pipelines are unwittingly damaged or even get damaged naturally due to loss of integrity resulting from age-related corrosion.

#### *1.3.5. Dredging and Sandfilling*

Intensive dredging activities are going on in the Niger Delta basin as a prerequisite for one project or the other. The main goal of such dredging operations has been canalization for transportation purposes, reclamation of land for construction of harbour and collection of sand for building purposes [10]. Dredging effects include destruction of spawning, breeding and feeding grounds of aquatic organisms; reduced number of benthic organisms; increased turbidity of water bodies; mobilization and bioavailability of sediment trace metals [17].

#### *1.3.6. Waste Discharges*

The water systems of the Niger Delta are used, extensively, for the disposal of various types of wastes, without conventional treatments [9]. Discharge or dumping of wastes such as sewage sludge, industrial wastes and dredged materials, municipal solid wastes and effluents and non-point pollution from agricultural and urban runoffs are common place in the region [18,19].

#### *1.3.7. Deforestation*

One recent important observed feature in the region is the almost complete absence of primary forests, majorly attributable to unsustainable human activities. Such practices include uncontrolled logging, slash-and-burn farming approach, oil exploration and exploitation, urbanization and mining activities, and so on. Deforestation has since been identified as a serious problem in Nigeria, which currently has one of the highest rates (3.3 percent) of forest erosion in the world [20], having temporal increment. Oladipo [21] estimated deforestation rate of tropical rainforest (not wetlands forest) in Nigeria as approximately 3.5 percent annually, translating to a loss of 350,000-400,000 ha of forest land per year. As forest cover declines, desertification becomes more intense, wildlife populations plummet due to poaching, habitat loss, and deepened soil erosion. There has also been a drop in the productivity of coastal and inland fisheries due to soaring mangrove losses [22]. The regeneration rate of biomass may also decline significantly affecting the amount of fuel wood available for local people. In addition, extensive mangrove loss in the Niger Delta has implications for global climate change. Though mangroves, salt marshes and seagrasses make up only less than 0.5 percent of the coastal bed, they sequester as much as 71 percent of the carbon stored by ocean sediment [23]

#### *1.3.8. Erosion and Floods*

The occurrence of coastal erosion has been reported in the Niger Delta by Okon and Egbon [24]. Most coastal regions of the world are already experiencing flooding due to rise in sea levels and extreme precipitation. The rises in sea levels have been linked to the impacts of global warming by the Inter-governmental Panel on Climate Change (IPCC). According to the report of IPCC [25], working with records over the last 100 years, a strong correlation exists between global temperature and sea level rise.

Also, Ogba and Utang [26] reported that, the predicted rises in sea levels and ocean surges will worsen the problems of coastal erosion that is already a menace in some parts of the globe, especially, the Niger Delta. The associated inundations will increase problems of sea-water intrusion into fresh water sources and ecosystems, destroying natural habitats stabilizing systems such as mangrove forests, and would affect agriculture, fisheries and general livelihoods. The salinization of groundwater will lead to shortage of freshwater, and thus have profound implications in regions occupied by wetlands and swamps such as the Niger Delta region of Nigeria where the inhabitants mostly depend on underground water through dugout wells as their main source of water supply for drinking as well as for other domestic uses.

## 2. Action to be Taken

For any green economic proposal to succeed, the various facets of environmental degradation in the Delta region must be addressed. This requires strategies that would integrate the complexes of biophysical and socioeconomic factors, and multi-stakeholder involvement. With the present economic situation, agricultural sector needs to be revived in order to make room for green alternatives in the area of food production, resource management, contamination and degradation of the environment and green employment.

### *2.1. Food Production*

Reliable food supply is an essential prelude to rural economic stability. The government at local levels can expand local small-scale agriculture and support transition to organic agriculture rather than subsidizing costly agro-chemicals. This way, these subsistence farmers would have their capacities built to the extent that they would naturally graduate to become big farmers who engage in industrial food production through individual efforts or as farmers' cooperatives or organisations capable of positively utilizing available genetically modified crops. Family owned and operated farms of small to medium sizes constitute the most reliable, high quality and economical food production systems [27]. Thus, agricultural policies must be designed to make existing and emerging family farms, economically, viable. The government should encourage agricultural practices that are, economically, viable without compromising the integrity and stability of the environment [1]. Industrial agriculture is characterized by large-scale, monoculture production and high levels of external inputs such as pesticides and fertilizers and thus, reduces the positive impacts of biodiversity and ecosystems integrity. For such large-scale farming ventures, strategic environmental impact assessment might be helpful in forestalling possible negative environmental impacts of such projects.

Ecosystem restoration and conservation, sustainable agriculture and non-agricultural production practices need to be developed. There is a large amount of evidence to show that, a farmer-led approach, utilizing known and proven agricultural techniques and practices, can transform the livelihoods of farmers, increase food security and reduce malnutrition while, preserving the environment [28]. Organic agriculture utilizes the application of traditional and scientific information to decrease the usage of sprays and chemical substances in crop production. Organic agriculture, which is based on ecosystem management, is recommended for small-scale farmers in the Niger Delta, as they require minimal external inputs and use naturally available materials. The practice involves integrating crops and animals in a single farm production system, inclusion of cover and cash crops, selective breeding of locally adapted crops and livestock, production of on-farm organic fertilizers, use of inexpensive botanicals, and so on. It is, also, about the application of indigenous knowledge of the farmers themselves as a vital tool for building resilience and adaptation to climate change [29].

### *2.2. Environmental Degradation*

Proper Environmental Impact Assessment (EIA) should be carried out on future developmental projects in the region while, an Environmental Evaluation Report (EER) should be conducted on impacted environment with adequate mitigation strategies put in place. Monitoring should be participatory between the relevant regulatory authority in collaboration with other stakeholders and external development institutions working with local communities. Improvement in natural resource management and conservation should be encouraged and rewarded, especially, when efficient resource exploitation and utilization methods are practised.

### *2.3. Employment*

In the Niger Delta region of Nigeria, agricultural employment has not kept pace with increasing population. There is a dearth of occupational farming in the Niger Delta since, most of the farmers do so on part-time basis or at best in conjunction with other perceived priority non-farm enterprises. Organic and small-scale farm based agriculture, which tends to utilise human labour, farmers expertise and community experience, could create a lot of job opportunities for the local people. Considerable hectares of land in the Niger Delta, including areas already established for oil palm, rubber and rice plantations, with improved techniques and better working conditions will reduce unemployment rate in the region. Climate change adaptations and mitigation measures, such as soil conservation, rehabilitation of degraded farm lands, afforestation and reforestation projects promise to create green employment and sustain rural livelihoods. Thus, deliberate efforts toward transition to a green economy will create demand for workers, which will require training and expertise in maintaining or restoring environmental quality and avoiding future damage to the ecosystem [30]. To this end, multinational companies, NGOs and governments should build the capacity of the local populace in green environment-related technologies and activities for a more sustainable impact of such intervention.

### *2.4. Social Protection*

Social understanding should exist between the companies and their host communities, with the latter having a meaningful say in social and environmental matters that affect them as well as access to environmental benefit sharing and rights.

## 3. Maintaining Ecosystem Services for Sustainable Rural Livelihoods in Aquatic Agriculture Systems in the Niger Delta

In most riparian societies, aquatic resources play critical role in rural livelihoods and poverty reduction. The cardinal objective of agro-ecology or sustainable agriculture is to reduce production input and reliance on fossil fuel, while scaling up output and maintaining ecosystem balance and service viability for multiple livelihoods dependents [31]. It entails improved farm gate prices for rural food products, output value addition, adequate flow of market information and extension education.

The mangrove belt of the Niger Delta in Nigeria is the

largest in Africa and the fourth largest in the world [32]. According to Spalding *et al.* [3], the mangrove area of Nigeria covers 10,515 km$^2$. The broad deltaic ecological zone is home to over 21 million persons that mainly rely on mangrove ecosystem goods and services. Ronnback *et al.* [33] listed nine of such services, grouped under three broad functions of mangroves, namely: regulatory, reproductive and cultural functions.

Aquatic agricultural systems (AAS) are diverse natural resource exploitation and farming systems where dependent households cultivate a range of crops, raise livestock, farm or catch fish, gather fruits and other tree crops and harvest resources such as timber, reeds and wildlife [34]. They are systems in which the annual production dynamics of freshwater and/or coastal ecosystems contribute significantly to total household income [35]. For AAS outputs to be sustainable, it must be a system in which biodiversity and environmental conservation is an explicit objective of agriculture and rural development, and the latter are explicitly considered in shaping conservation strategies.

Hitherto, there had been segregation of agricultural areas from conservation areas. But, with expanding agricultural footprints, environmental conservation practitioners have since the beginning of the 21$^{st}$ century identified the need for agriculture to be integrated with biodiversity conservation at landscapes and seascape [36]. Managing and maintaining AAS for different stakeholder production outputs and increased livelihood returns by rural poor households require continuous supply of priceless environmental goods and services. This can be guaranteed by investment in sustainable production methods, consensus building amongst the diverse stakeholders, reduction in postharvest losses and by dismantling market barriers that undermine farm gate prices of rural smallholder producers. It also entails concerted efforts aimed at protecting vulnerable and marginalised groups.

## 4. Connectedness of Sustainable Aquatic Production and Ecosystem Integrity: A Case Model for the Bonny River Estuary

Though there are broad elements that could be shared by different sustainable AAS, there exists no unilateral framework or approach that can solve the myriad of AAS improvement challenges globally. This applies to the Niger Delta region with different ecozones (floodplains, freshwater basins and inshore coastal water belts) and variation of socio-economic and cultural contexts [4]. Here, the Bonny River estuary is used as a representative case study of how ecosystem services can be managed and sustained for improved rural livelihoods of AAS in the Niger Delta. The Bonny estuary occupies between latitudes 4$^0$25' and 4$^0$50N and longitudes 7$^0$00' and 7$^0$15"E [37]. It has an estimated 66,000 hectares made up of elevated beach ridges, mangrove swamps, rivers and creeks. Figure 1 shows the present production system in the Bonny River estuary.

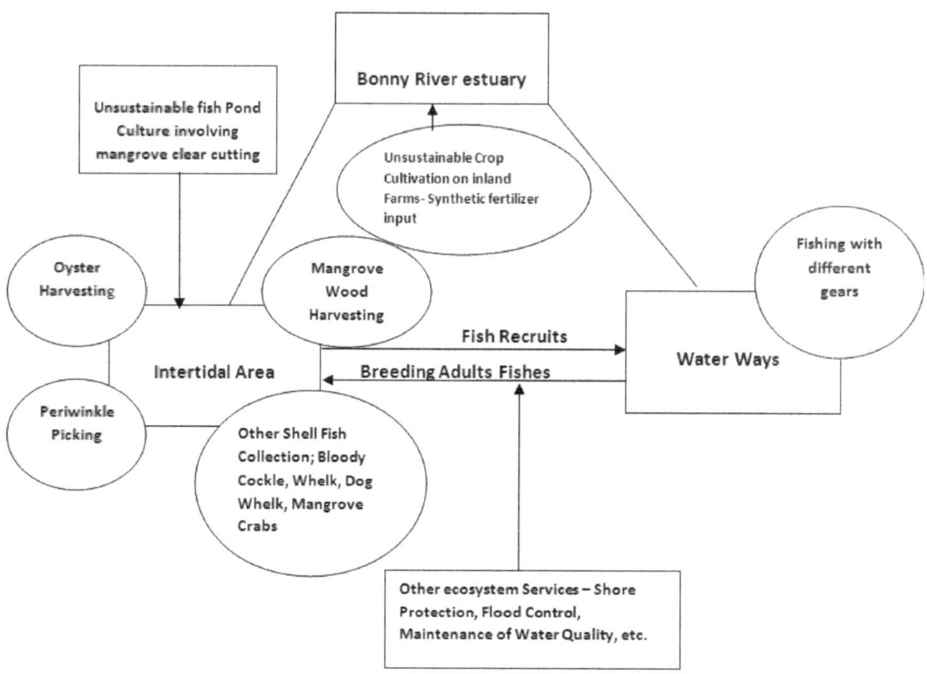


***Figure 1.*** *Flow chart showing AAS at Bonny River estuary.*

The figure suggests that, the main factors undermining sustainability are clear cutting of mangrove for establishment of fish ponds and uncontrollable use of inorganic fertilizers and pesticides to boost crop production on island farms that dot the estuary. There are also flashpoints of domestic and industrial waste dumpsite and outfall along the community's coastal corridors, in addition to threats of oil spillage from a network of oil pipelines traversing the river system to storage reservoirs at Bonny oil terminal.

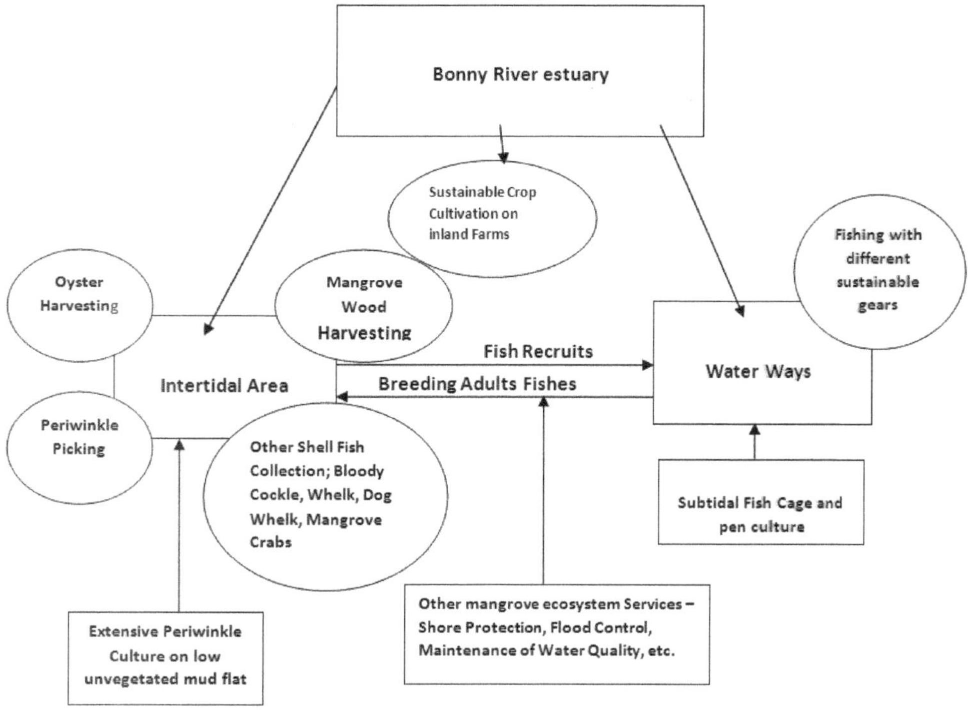


***Figure 2.*** *Model of sustainable AAS for the Bonny River estuary.*

Given the above identified potential and real threats, the model below (Figure 2) is proposed as a sustainable paradigm for managing and maintaining ecosystem services for the Bonny River estuary AAS. Fish pen and cage culture in the subtidal waterways is suggested instead of earthen fishpond culture in the intertidal area; the latter requires removal of mangrove cover and excavation of mangrove peaty sediment to create ponds and for building embankments. The former will protect the integrity of mangroves to continue to deploy their inherent services for the overall maintenance and productivity of the seascape. Improving oil pipeline integrity and maintenance regime, followed by the entrenchment of a robust and proactive contingency response strategy will reduce the incidence of oil spills and environmental impact on ecosystem quality, biodiversity and livelihoods.

To improve the livelihood of women, an extensive system of periwinkle farming on low unvegetated intertidal mudflats is recommended. Here, the farmers can use low level (in terms of tidal heights) submersible, water-permeable lateral separators to demarcate periwinkle culture beds from the mangrove margin of the mid-intertidal to the low water tidal mark of the subtidal creek. This is more or less a semi natural flow-through system as both the outer and inner edges of the rearing compartments would be opened for normal tidal communication at high tide between the permanently submerged waterways and the upper intertidal area. Overtime, exploitation of the cultured stocks will reduce harvest pressure on the wild populations. The periwinkles will grow to attain reproductive age, thereby acting as recruitment brood stock, and the recruits would, in turn, act as buffer stocks for the entire system. In addition, the enhanced size quality of farmed periwinkles will assure high farm gate pricing, thereby increasing the income of the rural operators.

While there are functional fishermen associations in the Bonny River estuary catchment, women periwinkle pickers are not organised into unions. This lack of networking and representative platform, together with customary barriers, make women participation in AAS governance and involvement in decision making next to nothing. Meanwhile, good governance regime is critical for direct resource user groups and other stakeholders to build capacities to manage resilience [34].

### *4.1. How to Integrate AAS and Ecosystem Services for Rural Livelihood Improvement: A General Summary*

1. Strengthening co-benefits of ecosystem livelihood dependents,
2. Detailed consideration of specific socio-cultural and economic contexts through multi-stakeholder engagement and analysis,
3. Mainstreaming gender equality and transforming traditional and social norms that create gender inequality,
4. Strengthening governance across various user groups,
5. By encouraging targeted investment to improve productivity for smallholder farmers – such as the diversification of income earning sources for traditional periwinkle scavengers,

6. Research and education on improved approaches and protection of vulnerable and marginalised groups, particularly women,
7. Integrating conservation into the fabrics of AAS productive and management mechanisms, and
8. Investment in methods that will minimize postharvest losses.

# 5. Conclusions

It would appear that the challenges of transition to green economy in the Niger Delta are more formidable than the opportunities are promising. However, some of the environmental challenges can be transformed to opportunities and livelihood gains. For any green economy project to work, environmental protection, restoration of degraded environment and careful management of natural resources are required while, investing a higher proportion of the national budget (minimum of 10% as recommended by the Comprehensive Africa Agriculture Development Programme; CAADP) in the agricultural sector. Much of this will depend on governments making the right policies and setting up requisite institutional frameworks from national to the local levels, which will link agricultural production to the requirements for conservation of the environment and biodiversity. The productive outputs of AAS dependent households can be enhanced and secured through deliberate efforts at maintaining ecosystem services by conserving environmental integrity, and by closing gender and poverty gaps between men and women in riparian rural communities. Understanding local environmental dynamics, stakeholder spatio-temporal resource use, gender mainstreaming, research, networking and community-level management framework are the basics for achieving sustainable agro-ecological resource exploitation and farming in wetlands, freshwater and marine coastal rural poor areas of the Niger Delta – for increased production outputs and poverty reduction.

# References


[1] Environmental Justice Foundation 2004, Farming the Sea, Costing the Earth: Why We Must Green the Blue Revolution, Environment Justice Foundation (EJF), UK. 77 pp.

[2] Awosika, L. F., 1995, 'Impact of Global Climate Change and Sea Level Rise on Coastal Resource and Energy Development in Nigeria, In: Umolu, J. C. (Ed.). Global Climate Change: Impact on Energy Development'. DAMTECH Nigeria Limited, Nigeria.

[3] Spalding, M., Kainuma, M., Collins, L., 2010, World Atlas of Mangroves. London, UK: Washington, DC: Earthscan.

[4] Aston-Jones, N. 1998, 'The Human Ecosystems of the Niger Delta – An ERA Handbook', Kraft Books Limited, Ibadan.

[5] Adams, W. M., 1993, Indigenous use of Wetlands and Sustainable Development in West Africa, The Geographical Journal, 159(2), 209-218.

[6] Allison-Oguru, E. A., Berepubo, N. A., Kalio, A. E., 2002, 'Agriculture, In: Alagoa, E. J. and Derefaka, A. A. (Eds), The land and people of Rivers State: Easthen Niger Delta', 361-386.

[7] Earys, S., 2005, A Guide to Bycatch Reduction in Tropical Shrimp-trawl Fisheries. Food and Agriculture Organisation (FAO) of the United Nations, Rome, Italy, 110 pp.

[8] Amnesty International (AI) 2009, Nigeria: Petroleum, pollution and poverty in the Niger Delta, AI International Secretariat, London, UK.

[9] UNDP, 2006, Niger Delta Human Development Report. UNDP. <http://hdr.undp.org/en/reports/nationalreports/africa/nigeria/name,3368,en.html> (accessed 22.02.13).

[10] Aaron, K. K., 2005, Perspective: Big Oil, Rural Poverty, and Environmental Degradation in the Niger Delta Region of Nigeria, Journal of Agricultural Safety and Health, 11(2), 127-134.

[11] Steiner, R., 2008, Double Standards? International Standards to Prevent and Control Pipeline Oil Spills, Compared with Shell Practices in Nigeria. A Report Submitted to Friends of the Earth, The Netherlands. 61pp.

[12] Pegg, S., Zabbey, N., 2013, Oil and water: the Bodo spills a nd the destruction of traditional livelihood structures in the Niger Delta, Community Development Journal 48, 391 – 405.

[13] Zabbey, N., Uyi. H., 2014, Community responses of intertidal soft-bottom macrozoobenthos to oil pollution in a tropical mangrove ecosystem, Niger Delta, Nigeria, Marine Pollution Bulletin 82, 167 -174.

[14] Dung, E. J., Bombom, L. S., Agusomu, T. D., 2008, The effects of gas flaring on crops in the Niger Delta, Nigeria, Geojournal, 73, 293-305.

[15] Environmental Rights Action and Climate Justice Programme 2005, 'Gas Flaring in Nigeria: A human rights, environmental and Economic Monstrosity', 33pp

[16] Nduka, J. K. C., Orisakwe, O. E., Ezenweke, L. O., Ezenwa, T. E., Chendo, M. N., Ezeabasili, N. G. (2008). Acid rain phenomenon in Niger Delta region of Nigeria: Economic, biodiversity, and public health concern, The Scientific World Journal, 8, 811-818.

[17] Ajugwo, A. O., 2013, Negative effects of gas flaring: the Nigerian experience, Journal of Environment Pollution and Human Health, 1(1), 6-8.

[18] Hodgson, G., 1994, The environmental impact of marine dredging in Hong Kong. Coastal Management in Tropical Asia 2, 1 - 8.

[19] Arimoro, O. A., Ikomi, R. B., 2008, Response of macroinvertebrate communities to abattoir wastes and other anthropogenic activities in a municipal stream in the Niger Delta, Nigeria, Environmentalist 28, 85–98.

[20] Zabbey, N., Etela, I., 2011, Impact of abattoir waste on Woji Creek, Port Harcourt, Nigeria, using physicochemistry and macrobenthic diversity indices, African Journal of Aquatic Science, 36, 279-287.

[21] Butler, R. A., 2006, Diversities of Image – Rainforest Biodiversity, Retrieved 9 January 2006, from Mongabay.com / A place out of time: Tropical Rainforests and the Perils they face, http://www.rainforest.mongabay.com/0305.htm.

[22] Oladipo, D., 2010, The state of Nigeria's forests, International Institute of Tropical 4, 24-25.

[23] Zabbey, N., Hart, A. I., Erondu, E.S., 2010, Functional roles of mangroves of the Niger Delta to the coastal communities and national economy, In: Ansa, E. J., Fashina, B., Ndimele, P. E. (Eds) Proceedings of the 25th Annual Conference and Fair of the Fisheries Society of Nigeria (FISON), held 25-29 October, 2010 at the Administrative Staff College of Nigeria (ASCON), Topo, Badagry, Lagos, 115-118.

[24] Nellemann, C., Corcoran, E., Duarte, C. M., Valdes, L., De Young, C., Fonseca, L., Grimsditch. E. (Eds.). (2009). Blue carbon. A rapid response assessment. United Nations Environment Programme. GRID-Arendal, www.grida.no.

[25] Okon, R. N. and Egbon, P. C., 1999, Fiscal Federalism and revenue allocation, The poverty of the Niger Delta, In: Aigbokhan, B. E. (Ed.) Fiscal Federalism and Nigeria's Economic Development, Annual Conference of the Nigerian Economics Society (NES), Ibadan.

[26] IPCC 1990, Working Group 111, 1990, Strategies for Adaptation to Sea Level Rise. Report of the coastal zone management subgroup, Intergovernmental Panel on Climatic Change Working Group, Rikkswaterstatt, Netherlands, 122 pp.

[27] Ogba, C. O., Utang, P. B., 2007, Vulnerability and adaptations of Nigeria's Niger Delta coast settlements to sea level rise. TS 7B Coastal Zone Management and Environmental Issues. Strategic Integration of Surveying Service, FIG Working Week, Hong Kong SAR, China, 15.

[28] Willler, H., Kilcher, L., (Eds) 2009, The World of Organic Agriculture, Statistic and Emerging Trend IFOAM, DE-Bonn and FiBL, CH-frick. 339 pp.

[29] Niggli, U., Flieβbach, A., Hepperly, P., Scalabba, N., 2009, Low greenhouse gas agriculture: mitigation and adaptation potential of sustainable farming system FAO, April 2009, Rev. 2 – 2009, Rome. Italy.

[30] Binns, P., 2012, Opportunities and Challenges Facing Farmers in Transitioning to Green Economy Agriculture Practice. Perspectives (Issue № 5), United Nations Environment Programme (UNEP), 24 pp.

[31] UNEP 2008, Green Jobs: towards decent work in a sustainable, low carbon world, UNEP/ILO/IOE/ITUC, Available: www.unep.org/civil-society/publications/index.asp.

[32] Nandy, S., Mitra, S. 2004, Features of Indian Sunderbans mangrove swamps, Environment and Ecology, 22, 339-344.

[33] Ronnback, P., Crona, B., Ingwall, L., 2007, The return of ecosystem goods and services in replanted mangrove forests: perspective from local communities in Kenya. Environmental Conservation, 34(4), 313-324

[34] Lebel, L., Anderies, J. M., Campbell, B., Folke, C., Hatfield-Dodds, S., Hughes, T. P., Wilson, J., 2006, Governance and the capacity to manage resilience and regional social-ecological systems. Ecology and Society, 11(1), 19.

[35] Weeratunge, N., Chiuta, T. M., Choudhury, A., Ferrer, A., Husken, S. M. C., Kura, Y., Kusakabe, K., Madzudze, E., Maetala, R., Naved, T. T., Scgwarz, A., Kantor, P., 2012, Transforming aquatic agriculture systems towards gender equality: a five country review. CGIAR Working paper AAS-2012-21.

[36] Scherr, S. J., McNeely, J. A., 2008, Biodiversity conservation and agricultural sustainability: towards a new paradigm of 'ecoagriculture' landscapes. Philosophical Transactions of The Royal Society 163, 477-494.

[37] Hart, S.A., Hart, A.I., 1998, Aspects of the food and feeding biology of Mugil cephalus (Linnaeus, 1758) (Perciformes: Mugilidae) in Bonny Estuary, Nigeria. African Journal of Applied Zoology 1, 39 – 54.

# 5

# Development of farmer support system on dairy and meat industry of goat utilizing ICT

**Teppei Hirata[1,*], Takeshi Miyagi[2], Yasunori Nagata[2], Shiro Tamaki[1], Tsutomu Omatsu[3], Tetsuya Mizutani[3]**

[1]Department of Information Engineering, Faculty of Engineering, University of the Ryukyus, Nishihara town, Japan
[2]Department of Electrical and Electronic Engineering, Faculty of Engineering, University of the Ryukyus, Nishihara town, Japan
[3]Research and Education Center for Prevention of Global Infectious Disease of Animals, Faculty of Agriculture, Tokyo University of Agriculture and Technology, Fuchu-shi, Tokyo, Japan

**Email address:**

teppei@neo.ie.u-ryukyu.ac.jp (T. Hirata), tmiyagi1130@gmail.com (T. Miyagi), ngt@eee.u-ryukyu.ac.jp (Y. Nagata), shiro@neo.ie.u-ryukyu.ac.jp (S. Tamaki), tomatsu@cc.tuat.ac.jp (T. Omatsu), tmizutan@cc.tuat.ac.jp (T. Mizutani)

**Abstract:** Recently, due to the rapid growth of population and food demand in the Asian region, animal feeding industry is more prosperous, so that animal food production is increasing remarkably in the region. That is, the livestock movement is getting more activated and expanding rapidly, accompanied by the enlarged globalization of economy and distribution in the Asian region. This movement of the livestock industry in the region will be expectedly accelerated by "Trans-Pacific Strategic Economic Partnership Agreement (TPP)". On the other hand, however, this rapidly accelerated activation and enlargement of livestock movement and transportation, contains an internal risk that the infectious diseases such as foot-and-mouth disease may break out and prevail widely in the region. In order to reduce such a risk of spreading of livestock epidemics, it is necessary to establish a global livestock information system with up-to-date technologies to achieve the food safety of developed and relieved livestock industry in the region. However, due to the budget problem for many animal farmers and the present high-priced information and communication devices, utilization movement of the modern gadget for the small-to-medium scale animal feeding management is substantially stagnant in this region. In this paper, we propose a brand new designed system with the method of ICT applications, devices and techniques to seek for the solution to this problem, concerned with the small-to-medium scale livestock feeders. Furthermore, we examine the relationships between Machine to Machine (M2M) system, database system, and Web system; those are fundamental keystones of this newly proposed system.

**Keywords:** Animal Industry, Infectious Diseases of Livestock, Goat, M2M, Web System, Mid-to-Small Scale Farmers

## 1. Introduction

Catastrophically widespread infectious disease of bovine spongiform encephalopathy (BSE), foot-and-mouth disease, and avian influenza gave rise to a big wave of social critical controversy for food safety of agricultural animal food products.Especially, BSE broke out in Britain in early 2000's by which livestock were widely contaminated to the other areas outside of the country. It was pointed out that recording management of livestock information, especially about, to which organization or where it belongs, was absentminded those days. It was attributable to poor traceability information management. That sensational traceability related controversy caused a movement to construct traceability management system around the world. By such disgraceful events for animal feeding industry, cautiousness about animal food safety among consumers has been provoked, and has been widely spread. On the side of the agricultural animal farmers, it has been more natural and accelerated to disclose the important traceability results to alleviate wariness of the consumers, since then.

Therefore, it is required to establish a new information management system for animal industry to solve these problems. Our objective is to develop a new livestock information system for food production and animal management. This is a new information management system to recognize the feeding condition or situation by visual data for farmers, especially for mid-to-small scale animal food producers, by utilizing Information and Communication Technology. This system is designed to provide support for livestock feeders and animal food producers, besides, to analyze from diversified perspectives via big data computing, and to make advanced use on their own.

This paper illustrates one new project called "PINZA" cloud project which has been kicked off already. This is oriented to develop a new cloud system to support for agricultural animal food producers including traceability system by utilizing ICT in form of consortium mainly led by University of the Ryukyus. This project has been supported by Strategic Information and Communications R & D Promotion Programme (SCOPE) by Ministry of Internal Affairs and Communications of Japan. The project by this consortium has now been in progress of which members are the Faculty of Engineering at University of Ryukyus, Okinawa National College of Technology, and other institutes, starting from 2011. The goal of this project is information sharing by various related parties from animal husbandry field to make scientific analysis and to disclose information.

In this project, goat is selected as model of farm animal which is one of the most common livestock in Okinawa, and is considered to be important domestic animal in other areas in Asia as well.

The project acronym, PINZA, stands for our goal, "Perfect INformation management system for Zoological Analysis (Figure.1)." PINZA also means goat in Miyakojima Island dialect of Okinawa Prefecture, where goat feeding is an important sector of agriculture, and where eating goat meat is recognized as the local custom as well.

The most important goal for this project is to build an information sharing platform for multilateral or advanced use of animal feeding information obtained through daily feeding routine of each farm animal. Many mid-to-small scale agricultural animal feeders commonly recognized in Asia, where demand for food has shown a rapid growth these years, have not obtained any proper traceability checking system. Therefore, once spread and popularized over Asia, this information access platform is expected to turn into a great potential engine for efficient feeding and faster distribution of animal food.

The basic design for this system and its applied business model were already illustrated on the previous report (Hirata et. al. [1, 2]). The importance of that business model is to take a strategy of cost reasonableness such as inexpensive device applied to data collecting of large amount from individual animal. Besides, user friendliness of access to needed information will be the strong point that any related parties including farmers, restaurants, and distributors in the line of value chain, and also, including local government officers, veterinarians, and livestock administrations, can access to such information in need under the information sharing objective.

This paper debates the structure of the research for the management efficiency of the animal feeding; purpose and developing of the database and the Web system as traceability information sharing platform on goat feeding, utilizing ICT to be oriented for our information system. This research's fruit will be, expectedly, 'prospective economic growth engine' to support the mid-to-small scale livestock farmers' management in Asia.

**P** Perfect
**IN** INformation management system for
**Z** Zoological
**A** Analysis

***Figure 1.*** *Pinza Cloud project.*

# 2. M2M Interface and the Basic Design of the System

This chapter illustrates the functions and designs of machine to machine (M2M) interface as the keystone of the system.

## *2.1. Input of Information of Animal Management through M2M (Machine to Machine) Interface*

We built the system to collect various information of the animal management in the farm through M2M (machine to machine) interface semi-automatically. M2M interface is a microcomputer for various signals transformation to transmit from the measuring devices to the database for the Internet and mobile phone terminals.

The entities of the structure of the M2M interface are illustrated as shown in Figure. 2. We made designs of the information system in which one signal from the measuring devices widely used in livestock industry (such as flow meter and load cell, or other sensors) can be transformed into another signal, for the use of the Internet of mobile phone terminals. The transformed signal was transmitted through the smartphone, then, to the database by 3G or Wi-Fi network.

In the tagged microchips on the goat ears, the digitally recorded ID number of individual goat is recognized by RFID reader. In this system, RFID reader and smartphone terminal are connected by Bluetooth. Then the ID number recognized by the ear tag will be matched up to the values such as milk yield and body weight from the measuring devices. These data processing will be conducted by an exclusively designed application, installed in the smartphone terminal. The Basic design specification of processing of M2M interface is shown as following (Table. 1). The signals that can be transformed by M2M interface is analog (voltage) or pulse signals. And these signals will be transformed into the adaptable signal to the smartphone with Android.

### *2.2. Basic Specification of the System*

Goat is our selected pilot livestock as a model of mid-to-small scale farm. Goat is commonly disseminated as a domestic animal in agricultural areas in Asia. Its meat, milk, leather, and hair are all to become products for commercial use. Also, its excrement is utilized for compost. That is, goat is recognized as very functional domestic animal from commercial use to compost.

Furthermore, goat is a small-sized domestic animal and its price is reasonable, so it is still vital source of income in not-rich or pinched agricultural areas still. Especially, our research focused on milk and meat production of goats as commonly seen in the Ryukyu Islands and the other Asian areas.

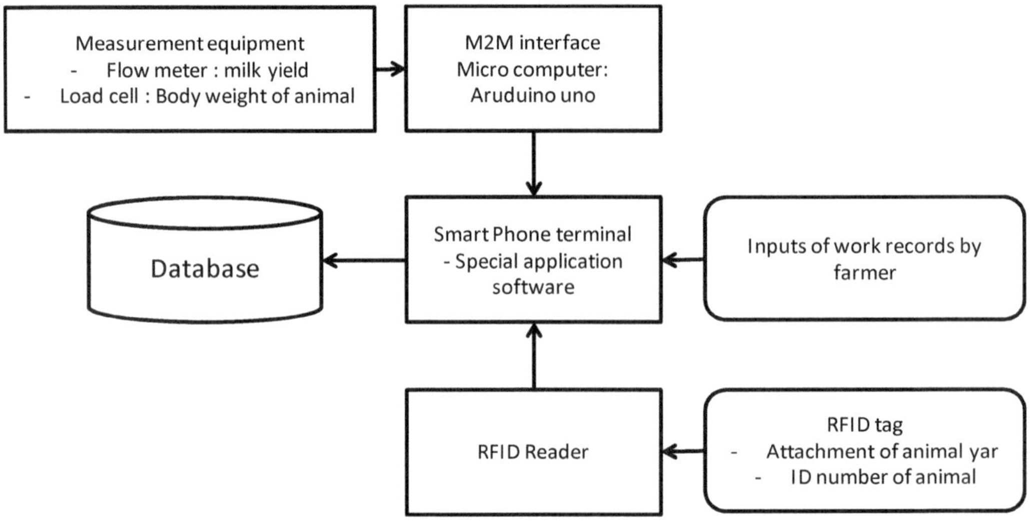


***Figure 1.*** *System construction utilizing M2M interface structure.*

***Table 1.*** *The basic specifications of M2M interface.*

| Item | Specification |
|---|---|
| Microcomputer | Arduino Uno |
| Embedded software | Application development language:<br>Application Development Environment |
| Input signal | Number of channels: 8Chs<br>Input signal : Analog and Pulse |
| Output signal | Adapted for Android |

#### *2.2.1. System of Efficient Support for Meat Goat Farmers*

There are some important points for goat feeder to realize efficient management. Goat feeder should increase fine goats by selective breeding, with lower cost, and to make them grow faster. Also it is necessary to predict the change of the goat body weight properly, in order to forward at the best timing. Another important point for the improvement of goat farming management is, proper management of reproduction; efficient increase of the female goat. For meat goat feeding management, three basic functions were activated such as measuring and predicting of body weight, and reproduction management. The detail of the system was illustrated by previous work (Oshiro et. al. [3]).

#### *2.2.2. System of Efficient Support for Dairy Goat Farmer*

For dairy goat farming, it is required to properly manage the calving and reproduction to periodically keep the track of milk yield. Flow meter is used as measuring method of milk, put in the pipeline from the teat cup to the bucket. The unsteady flow current of milk in the pipeline is caused by the negative pressure of the milk tank. To overcome this problem, the centrifugal machine should be placed before the flow meter to remove the air from milk current. However, such apparatus leads to budget problem. It would not be possible to obtain high-priced apparatus for the mid-to-small scale individual farmer. To seek for the solution to this budget problem, we placed a load cell under the milk tank that could measure the weight of milk to make an approximate milk yield estimate of each goat instead of using flow meter. This low cost device is a great progress we made in measuring method of the milk yield, even though it may be a little inferior in terms of precision.

The user scans the microchip in the goat ear at the time of starting and finishing of milking since this scanning works as trigger of on and off of measuring in this system. Time spent of the milking and ID numbers of the milking workers are also recorded. The measured data were matched up to the ID of the goat by the application installed in the smartphone. Length of milking time is an important indicator for dairy goat management concerned with labor cost. Milking is one of the most time consuming task of daily routines, which is conducted in the early morning and the evening. Therefore, the labor cost is occupied to a considerable degree compared to the desirable working hours of the goat farming management. Therefore, proper selection of the goats, depending on the milking time, leads to the efficient management. Such related important information is also contained in the database of this system.

Furthermore, the function of the milk yield prediction was added to the system to support the efficient management. The graph of lactation in milk yield shows growths to a higher peak of the milk yield but reduce persistency as a gamma curve (4). The examples of lactation curve is shown as Figure. 8.

As the function of milk yield prediction, the Wood model was used as algebraic model [4].

By this function of the milk yield prediction, the dairy goat farmers can obtain the essential estimate or indicators of the milk yield changes, or the characteristics of the dairy goat for the selective breeding later on. The basic functions of the system are shown at Table 3. The details about this system were reported by Ijichi et. al. [5].

### *2.2.3. Structure and Workflow of the System*

We, here, discuss the structure and workflow of the whole system. The system was built in the server of the Faculty of Engineering, University of the Ryukyus. The server contains the database system, the program for analysis on animal farming industry, and Web system for the users who can view various data.

Data from measuring devices are input via M2M interface to the database automatically. The details of the database will be discussed in the next Chapter. The system contains the database and data analysis program so that the needed information is picked out for analysis. Workflow of the whole system is illustrated at Figure. 4.

***Table 2.*** *Basic functions of the control system for meat goat breeding.*

| Function | Specification |
|---|---|
| Measurement of body weight | - Measuring the weight of the individual livestock.<br>- During the measurement of body weight, individual number and measurement data are recorded to the database. |
| Body weight prediction | - Make predictions of future changes in the individual's body weight. |
| Breeding management | - Recording mating ( Records of parents) .<br>- Alarm function of the expected date of birth. |

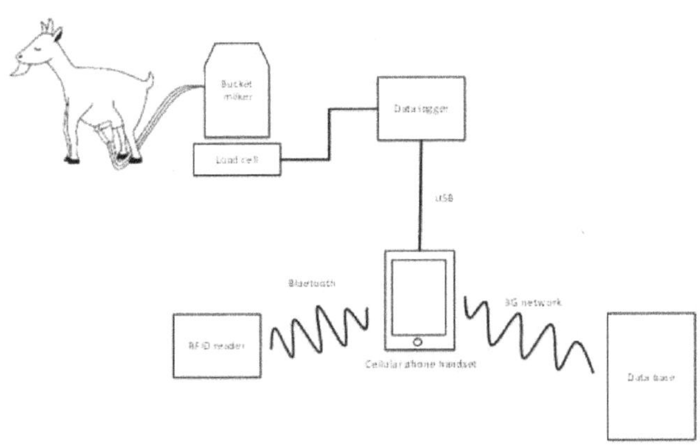


***Figure 3.*** *System configuration diagram of milk yield measuring.*

***Table 3.*** *Basic functions of the daily goat management.*

| Function | Specification |
|---|---|
| Milk yield measurement | - Recording the milk yield and time required for the milking<br>- Measured by a load cell placed at the bottom of the bucket milker |
| Milk yield prediction | - Make predictions of future changes in the individual's milk yield |
| Breeding management | - Record of mating<br>- Calculate the number of days after delivery |

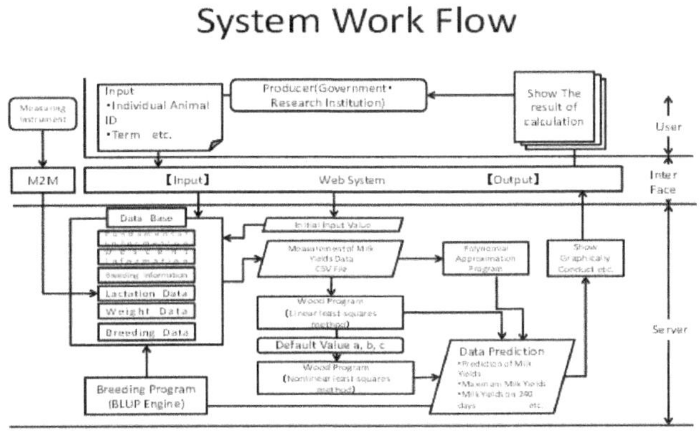


***Figure 4.*** *Overall configuration of the system.*

# 3. Building of the Database

This system is aimed at information sharing of various livestock by related parties such as farmers, veterinarians and administrative organizations. The system covers various livestock information of basic health, mating, milking, and any other related data. And these data are centrally controlled for management. The database for this system was designed as a relational database. The main entities of designed table are as follows.

- Basic livestock information: identification number of the animal, birth date, sex, ID number of the parents, nickname, the name of owner, etc.
- Organization information: name of the organization, address and geographical information of organization, name of representative, access authorization.
- User's information: name of the user, birth date, name of affiliation, qualifications (veterinarian, artificial inseminator)
- Body weight information: measuring date, measuring value, ID number of device
- Milking information: milking date, name of milking person, quantity of milk yield, length of milking time

As an example of building the relational database, ER (Entity Relationship) diagram of farm animal management is illustrated at Figure. 5. MySQL was chosen as the Database Management System (DBMS) for development.

# 4. Development and Implementation of Web System

## 4.1. Basic Design of Web System

Web system was chosen for user interface to operate various information managements, instead of exclusive application software. Types of the users' OS or terminals are miscellaneous, depending on each person or company. Because Web system is effective under such changeable environment, it is more attractive than exclusive use applications.

By this Web system, the user logged in from top page once, the main page will appear. The main page shows lists of goats kept in the same organizations. On this main page, each goat's name shows up by order of numbers of references reproductive management such as mating, delivery, and lactation from top to bottom. Since the delivery date is one of the most important items for livestock management, this page was also designed that the listed order can be adjustable to delivery date order. Furthermore, the goat ID number is clicked, each animal's information will appear in detail (the date of birth, pedigree information for three generations, reproductive information, body weight information, etc.) on the screen.

On the main page, also, tabs for click of reproduction, lactation, body weight management are prepared. On each individual goat page, the same tabs of the main page are placed as users can view. Example of the main page is shown at Figure. 6, and the system chart of structural page views is illustrated at Figure7.

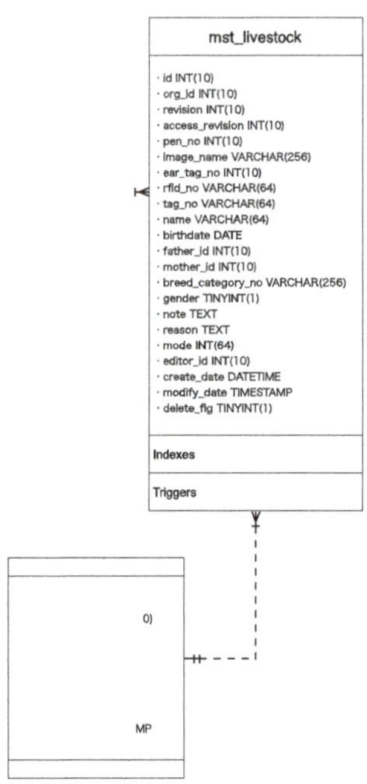


***Figure 5.*** *Entity relationship diagram of database on body weight management system.*

## 4.2. Details of Each Function and Page Design of the System

### 4.2.1. Page Design and Function of Reproduction Management

Subpages of management for mating and delivery were also established. On the page of mating, there show up the goat IDs list by the order of probabilities dates of mating, which are confirmed in estrus, and are likely to copulate within sooner days. The goat ID button should be clicked, when the user wants to input mating data.

Also, natural mating or artificial insemination can be chosen then. The record of mating was once saved, and alarm was set, then a reminder would be showing up after 150 days of average pregnancy period of goat. The delivery management page shows the list by order of the expected delivery dates. Clicked the goat ID button, the user could keep the record of delivery information such as the ID number of the kid, each sex of kids, and the number of the neck tug placed for tentative use.

***Figure 6.*** *Example of user interface of this system.*

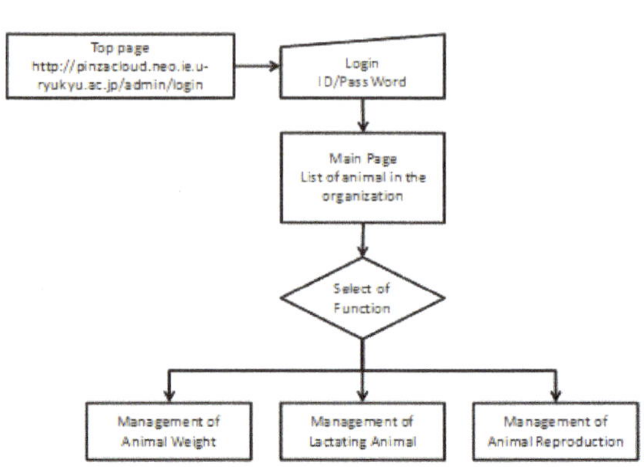


***Figure 7.*** *Schematic diagram showing the transition page of Web System.*

### *4.2.2. Page Design and Function of Lactation Management*

In this page, the goat IDs are arranged in the list by order of the lapse days after its delivery. Male goat data are not showing up here; only female ones appear. Clicking of female goat ID number will move to the page of detailed lactation information. On the detailed lactation information page, the prediction of milk yield would be demonstrated as graph. Also, the lapse days, details of the goat's historical data of lactation would appear. An example of the detailed page is shown at Figure.8.

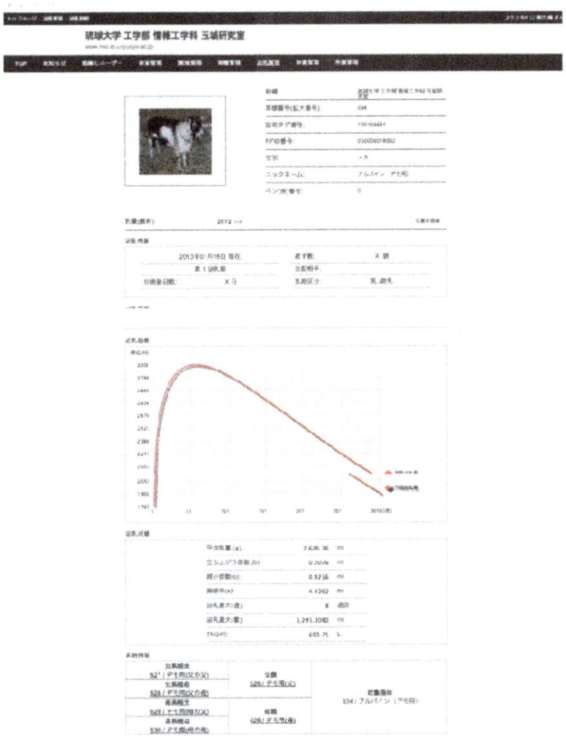

***Figure 8.*** *Design of lactation management screen.*

### *4.2.3. Environment of System Development*

This section explains the environment of system development for Web system. The details are described at Table 4.

***Table 4.*** *Development environment of the web system.*

| Server | Development environment |
|---|---|
| Database server | OS : CentOS 6.3<br>DBMS※ : MySQL 5 |
| Web application server<br>Web API server | OS : CentOS 6.3<br>Development language:<br>PHP 5.2.6, Python 2.6, C/C++<br>Web service : apache 2.2 |

※DBMS : database management system

### *4.3. Development of the Database*

Our database was built as relational database by utilizing of free soft MySql as a database management system (DBMS). Usage of free software leads to easy modification of the system to answer to various requests by users.

The business model which we aimed at is that, not limited to, goat feeders but any other various related parties such as veterinarians and governmental officers can make use of the database[1]. Provided that all of the goats in Japan were enrolled in this system, then, miscellaneous kinds of researches and analysis could be possible from various points of perspectives or concepts. It is due to livestock goat regulations under different local circumstances. For example, goats are kept as livestock for dairy use in Hokkaido, and milk are used as raw material for cheese. In this case, if the goat farming is needed to be put under the local government management, then the regulations of milk or lactations are concerned to obey. On the other hand, in Okinawa, meat goat feeding is more popular. Then, the goat's birth dates are very important. The strict management of the birthdates is imposed to the meat goat feeders. It is because Okinawa prefectural regulation requires every each goat of 12 month years old, or older, to take inspections of prions of BSE; that is, goats under 12 month old are free from cerebral inspections.

As such above mentioned, the way of using the database will differ from one to another prefecture or area. So, it would be difficult to meet with the each different need by building each different database. We chose to arrange the open source for the database management to satisfy different needs. Adopting other new technologies, like Open ID as solution for rearrangement of access authorization, would be the key to further progress in system building and advanced use.

### *4.4. Development of Web System*

We adopted Web system by cloud system to provide the users with data service for comfortable and friendly use, which any related users can operate under any numerous communication environments. For cloud system operation, as a matter of course, environment of the Internet connection is needed around the users. However, from the average mid-to-small scale farmers' view of the world, seemingly, there are very limited numbers of farmers who have been provided such environment. As we can see, the mobile phone disseminated rapidly nowadays. The Internet environment are widely spreading with accelerated speed than used to be, even in the agricultural areas also, by worldwide distribution of the smartphone which has an important role. Under this research, we constitute and offer a brand new proposal, faced by such a rapid dissemination of up-to-date communication style at present and even more advanced style coming in the future. However, yet for now, the simulated tests of the actual operation for Web system were conducted only in the browser installed in the PC. Therefore, the system should be improved, to be applicable to more miscellaneous kinds of the Internet environment including smartphones. Also, operational abilities of the smartphone seem to be limited with the built-in browser for now, so that a new exclusive-use application for the system should be taken steps for development for further progressed utilization.

## 5. Future Development

This final chapter debates our foreseeable future development from this project's results. Efficient management of livestock farmers' daily work was our focus for this project. Progress in communication of farm animal information will bear great fruits, we assured, not limited to daily animal management efficiency, but prevention of infectious disease, further progress in public health, and other various advantageous merits. For example, It is known that the measurement of milk yield plays an important roles in management of mastitis and milk fever; disease specific to dairy animals. Also, It would be considered that the management of body weight also becomes easier the discovery of diseases, including zoonotic diseases. For this reason, we need to proceed in this research further, as veterinarians can easily connect to the information system, or the other important related parties can do so for further improvement of daily management by our future study.

This database also may be useful for animal quarantine procedure. Hirata et al. [6] subjected that mathematical model of infectious disease and its application (simulation of transmission of pathogens) is useful for animal husbandry industry. However, many of case, basically data are lucking to conduct simulation. Our database would be complement the luck of data.

## 6. Conclusions

In this study, the design of information management system to support the dairy and meat goat farmers have been performed. The functions of this system are weight management of animals and milk yield management. In addition, the system also has the ability to integrate information of these.

In this study, the design of system was performed, but the verification test was not performed. It is required that the verification of stability of the system, and statistical research of processing. To be complete this study, it is necessary that these problems can be solved.

## Acknowledgement

This study was conducted with the assistance of the Ministry of Internal Affairs and Communications of Japan by SCOPE scheme. We were able to achieve full cooperation of Hagoromo Bokujo, Co. Ltd. RFID tag and readers were provided by Surge MIYAWAKI Co. Ltd. And this study was supported by Okinawa Cross Head Co., Ltd. and NTTDATA CUSTOMER SERVICE Corporation. We would like to presence the appreciation.

## References

[1] T. Hirata, K. Kuwae, A. Shimabukuro and S. Tamaki, "Development of quality control and breading management system of goats based on information and communication technology", ASTL Vol. 16 ASAP, ISSN: 2287-1233, pp85-89, 2012

[2] T. Hirata, K. Kuwae, A. Shimabukuro and S. Tamaki, "Development of quality control and breading management system of goats based on information and communication technology", Studies in Informatics and Control, pp. 81-90, 2013.

[3] M. Oshiro, Y. Shiroma, A. Ijichi, T. Hirata, C. Gakiya, K. Kuwae, M. Shinjo and S. Tamaki, "Development of information management system for livestock production by using Information and Communication Technology" Record of 2012 Conference of IEEE in Okinawa, pp80-82,2012

[4] Wood PDP: ``Algebraic of the lactation curve in cattle", Nature 216, 164-165,1967

[5] A. Ijichi, Y. Shiroma, M. Oshiro, T. Hirata, C. Gakiya, K. Kuwae S. Tamaki and M. Shinjo, "Development of automatic measurement systems of goat milk yield and milk yield prediction system using RFID tags and mobile phone terminal", Record of 2012 Conference of IEEE in Okinawa, pp83-86,2012

[6] Teppei Hirata, Yoshihito Yonahara, Faramarz Asharif, Tsutomu Omatsu, Takeshi Miyagi, Yasunori Nagata, Tetsuya Mizutani, and Shiro Tamaki, "A Study on Mathematical Model of Caprine Arthritis Encephalitis (CAE) and Development of Animal Quarantine Information System Adapted for Small Island" IEEK Transactions on Smart Processing and Computing, vol. 2, no. 4, August 2013

# Food and feeding habits of the red-belly tilapia (*Tilapia zillii* Gervais, 1848) (Pisces: Cichlidae) in Lake Ziway, Ethiopia

**Elias Dadebo[1, *], Negesse Kebtineh[2], Solomon Sorsa[1], Kassaye Balkew[1]**

[1]Department of Biology, Hawassa University, P. O. Box 5, Hawassa, Ethiopia
[2]Department of Wild Life, Wetland and Fishery Management, Wondo Genet College of Forestry and Natural Resources, Hawassa University, Shashemene, Ethiopia

**Email address:**
edadebo@yahoo.com(E. Dadebo), tamenege@gmail.com (N. Kebtineh), sorsasota@yahoo.com (S. Sorsa), kassayebalkew@gmail.com (K. Balkew)

**Abstract:** Food and feeding habits of *T. zillii* were studied from 572 fish samples collected from April to May 2011 (dry season) and July to August 2011 (wet season) from Lake Ziway. Stomach content analysis was conducted using frequency of occurrence and volumetric methods of analyses. Macrophytes, detritus and phytoplankton were the dominant food categories occurring in 94.9%, 94.2% and 82.5% of the total stomachs examined and constituting 45.2%, 29.4% and 16.8% of the total volume, respectively. The contributions of insets, nematodes, zooplankton and ostracods were relatively low. Macrophytes (50.4%), phytoplankton (21.8%) and detritus (18.5%) constituted the bulk of the food volume during the dry season. In the wet season detritus (40.8%), macrophytes (37.8%), phytoplankton (12.5%) and insects (6.5%) contributed the bulk of the food categories consumed. Phytoplankton, detritus and insects were important food categories of juveniles (5.0-9.9 cm TL) whereas macrophytes, detritus and phytoplankton were important food categories of adults. The importance of phytoplankton, detritus and insects declined with size of fish whereas the importance of macrophytes and nematodes increased with fish size. Based on the results of the stomach contents it was concluded that the species is an herbivorous feeding mainly on macrophytes, detritus and phytoplankton. The contribution of animal origin food was low.

**Keywords:** Diet composition, feeding, Lake Ziway, *T. zillii*

## 1. Introduction

The red-belly tilapia *Tilapia zillii* (Gervais, 1848) has wide spread distribution in tropical and sub-tropical regions of Africa and Eurasia (35°N-10°S, equating to temperature range of 7-43°C). In Africa, its distribution extends from Morocco and Egypt in the North, Côte d'Ivoire and Nigeria in the West to Democratic Republic of Congo in Central Africa [1]. It has been introduced in many countries including Ethiopia, Eritrea, Madagascar, Tanzania, Hawaii, Japan, Philippines, Singapore and United States of America [2]. It reaches a maximum length of 26.0 cm, 289 g weights and can live for about seven years [3]. It usually lives in water depth of up to 1m [3].

*T. zillii* is economically and ecologically important as food fish, for aquaculture, commercial aquarium trade, weed control and recreational fishery in its native range and in many countries it has been introduced [4]. The species was introduced in Lake Ziway in 1980s with the intentions of increasing productivity of the lake [2].

Various authors have studied the food and feeding habits of *T. zillii* in the middle east and Africa, and reported that macrophytes, phytoplankton, insects (Chironomidae pupa and Chironomidae larvae), zooplankton and detritus constituted the major food categories of the species [2, 5-7]. According to Adeyemi et al. [8], the major food items of juveniles comprised of phytoplankton (green algae, blue green algae and diatoms), zooplankton, fish scales and bottom deposits. The same authors reported the importance of phytoplankton, zooplankton, benthic insects and fish scales for adult *T. zillii* in Gbedikere Lake, Nigeria. Negassa and Padanillay [7] and Mohamoud et al. [9] found aquatic macrophytes and vegetation matters of terrestrial origin as the most important food items of *T. zillii* in Lake Ziway, Ethiopia and in Lake Timsah, Egypt, respectively.

Few studies are available on the food and feeding habits of *T. zillii* in Lake Ziway. Negassa and Getahun [2] and Negassa and Padanillay [7] studied the feeding habits of the species and emphasized the importance of macrophytes, detritus and phytoplankton. Markos et al. [10] studied the feeding habits of *T. zillii* in Lake Ziway in connection with the mercury concentration and reported macrophytes, detritus and phytoplankton as important food categories.

According to Markos et al. [10], *T. zillii* is found at the lowest trophic level in Lake Ziway. Negassa and Padanillay [7] conducted comparative study on reproductive biology and feeding habits of *T. zillii* and *O. niloticus* in Lake Ziway. According to these authors, the dominant food categories of *T. zillii* are macrophytes, phytoplankton (green algae, blue green algae and diatoms), zooplankton, fish scales and bottom deposits. Since the information available on the food and feeding habits of *T. zillii* in Lake Ziway is scanty, detailed study is required to guide the management of the stock in the lake. The aim of this work was therefore, to study the food and feeding habits (diet composition, seasonal variation in diet and ontogenetic diet shifts) of the species in Lake Ziway.

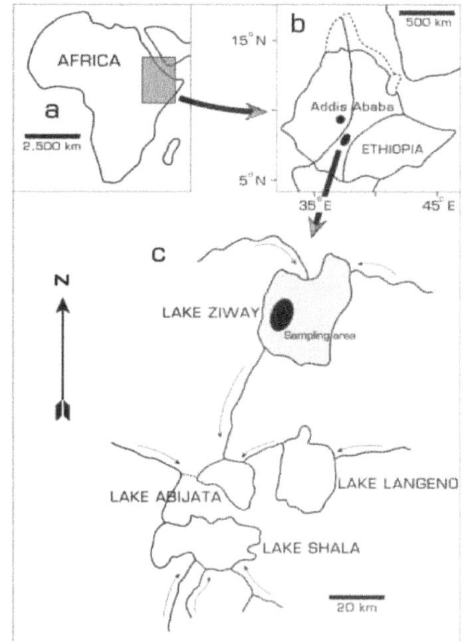


***Figure 1.*** *Map of Africa with horn of Africa region highlighted (a), map of Ethiopia with the southern rift valley lakes area indicated (b) and map of the Ziway-Shala basin lakes (with the sampling area in Lake Ziway indicated (c)*

## 2. Materials and Methods

### *2.1. Description of the Study Area*

Lake Ziway (Latitude: 7° 52'-8° 8'N and Longitude: 38° 40'-38° 56'E) is the most northerly lake of the four lakes in the Ziway- Shalla basin of the central part of the Ethiopian Rift Valley. Some evidences indicate that the four lakes, namely Lakes Langeno, Abijata, Shalla and Ziway formed one big lake when the water level was much higher than the present about 6,000 years ago [11]. The similarities of fossilized diatoms on the upper banks of the lakes and the present interconnections they have also suggest the possibility of being one lake in the past [11].

Lake Ziway is located about 145 km south of the capital city, Addis Ababa at an altitude of 1,636 m, with a surface area of 434 km$^2$ [11]. The maximum depth of the Lake is 7 m, mean depth is 2.5 m and the shoreline length is 137 km [12]. Two major inflows namely Rivers Meki and Katar enter the lake from the north western and southeastern plateaus, respectively. The lake has an outflow, Bulbula River, which is the major source of freshwater to otherwise highly saline Lake Abijata (Figure 1). The lake water is suitable for irrigation purpose and as a result, much of the land around it is under continuous cultivation for vegetables, fruits production and commercial floriculture. The climate of Lake Ziway region is semi-arid to sub-humid with mean temperature and precipitation of 25°C and 650 mm per year, respectively [13].

Because of the shallowness of the lake the littoral zone extends several hundreds of meters into the lake and fringed by thick stands of emergent, submergent and floating vegetation (*Cyperus, Papyrus, Phragmites mauritianus, Typha domingensis, Eicchorina crassipes, Hydrilla verticulata and Alisma plantogo* [14]. The phytoplankton community is dominated by blue green algae (*Microcystis, Lyngbya, Coelospharium, Merismopedia, Chroococcus, Anabaena*), diatoms (*Navicull a, Cymbella, Surirella, Gyrosigma, Nitzschia, Synedra, Pinnularia*) and green algae (*Scenedesmus, Spirogyra, Pediastrum, Ankistrodesmus, Cosmarium and Botryococcus*) [7].

The zooplankton community of the lake is composed of cyclopoid copepods (*Thermocyclops, Mesocyclops*), cladocerans (*Moina, Diaphanosoma, Cerriodaphnia, and Bosmina*) and rotifers (*Brachionus and Keratella*). The bottom fauna comprises gastropods (*Anisus natalensis, Biomphalaria sudanica, Bullinus forskahlii, Lymnea natalensis* and *Mellanoides tuberculata*), different kinds of insects, spiders, ostracods, and nematodes [7].

The indigenous fish fauna of the lake include: The Nile tilapia (*Oreochromis niloticus)*, the African big barb (*Labeobarbus intermedius)*, the straightfin barb (*Barbus paludinosus)*, the stone lapping (*Garra quadrimaculata*) and the black lampeye (*Aplocheilichthys antinorii*). The barbus species could be more diverse than mentioned above. The crucian carp (*Carassius carassius*) and *T. zillii* are exotic species introduced into the lake several decades ago with the aim of boosting the productivity of the lake [15]. The African catfish (*Clarias gariepinus*) was introduced accidentally as live fish were being transported from Lake Langeno to the cold storage site at the vicinity of Lake Ziway [15].

### *2.2. Sampling and Measurements*

Fish samples were collected using experimental monofilament gillnets (40 mm, 50 mm, 60 mm and 80 mm stretched mesh size). In addition, fish were purchased from the commercial landings of the fishermen. In order to obtain juvenile fish a 6 mm mesh beach seine was used in the shallower part of the lake. Samples were collected in April and May (dry season) and July and August (wet season) in the year of 2011. Total length (TL) was measured to nearest millimeter and the total weight (TW) of each fish was weighed to the nearest gram using a digital balance (Scaltec Model 23565, USA).

### *2.3. Diet Composition*

The stomach content of each fish was kept in separate sampling bottles containing 5% formalin solution. The stomach contents were then examined using a compound microscope (Leica, DME, magnification 100x). The relative importance of the different food items found in the stomach contents were determined using frequency of occurrence and volumetric methods [16]

In frequency of occurrence method, the number of stomach samples in which a given food item was found was expressed as a percentage of all non-empty stomachs examined. This method gives an estimate of the proportion of the population that feeds on a particular food item. This method is advantageous to establish relative abundances and requires less time and apparatus. It is however inadequate when a significant component of the diet does not occur in discreet units of uniform size. It provides little information on the food values of different items. In volumetric analysis, food items were sorted into different taxonomic categories, and the water displaced by a group of items in each category was measured in a partially filled graduated cylinder [16]. The volume of water displaced by each category of food items was expressed as a percentage of the total volume of the stomach contents [16].

In quantifying the volume of microscopic food items, the stomach content from each sample was diluted with tap water to a known volume. After thoroughly mixing the samples, one drop was taken on a microscope slide. The numbers of different food items were counted from 3 field of visions at different parts of the cover slip, and the total number of each food item per stomach sample was calculated by multiplying the mean number of each food items in a field of vision by the total number of field of vision under a cover slip area, and by total number of drops in the diluted stomach sample. The relative volume of each food item in a stomach was computed by multiplying the proportion of each food item in a drop by the total volume of the stomach content. Mean volume percentage of food items was calculated using the method of Wallace [17]

### *2.4. Statistical Analysis*

A chi-square ($\chi^2$) test was employed to compare the variations of the frequency of occurrence of the different food categories during the dry and wet seasons [19]. Similarly, the volumes of the different food categories were compared using Mann-Whitney's U test during the dry and wet seasons [18]. This non-parametric test was used because the data did not satisfy the assumption of equal variance to employ parametric test.

Dietary overlap between different length-classes was calculated using Schoener Diet Overlap Index (SDOI) [19], using the following formula:

$$\alpha = 1\text{-}0.5\ (\textstyle\sum_{i=1}^{n}|pxi - pyi|)$$

where α is percentage overlap, SDOI, between length classes *x* and *y*, *pxi* and *pyi* are proportions of food category (type) *i* used by length classes *x* and *y*, and *n* is the total number of food categories. Overlap in the index is generally considered to be biologically significant when the value α exceeds 0.60 [20].

## 3. Results

### *3.1. Diet Composition*

Out of the total number of 612 fish samples collected, 572(93.5%) had some food in their stomachs whereas the remaining 40 (6.5%) were empty. The size range of fish used for stomach content analysis was 5.4-24.0 cm TL and 3.2-266.6 g TW. *T. zillii* in Lake Ziway consumed different food categories including macrophytes, detritus, phytoplankton, insects, nematodes, ostracods and zooplankton. Food of plant origin, namely, macrophytes, detritus and phytoplankton constituted the bulk of the food of *T. zillii* in Lake Ziway occurring in 94.9%, 94.2% and 82.5% of the stomachs and comprising 45.2%, 29.4% and 16.8% of the total volume, respectively (Table 1).

Food of animal origin comprised insects (Diptera, Ephemeroptera, Plecoptera and Hemiptera), nematodes, zooplankton (Copepoda and Cladocera) and ostracods. The importance of these food categories was relatively low in the diet. Among food of animal origin, insects were relatively important occurring in 38.3% of the stomachs and comprising 6.0% of the total volume (Table 1). Other animals were of little importance.

### *3.2. Seasonal Variation in the Diet*

The frequency of occurrence of macrophytes and phytoplankton significantly varied during the dry (n= 272) and wet (n= 300) seasons ($\chi^2$ test, P<0.01). Similarly the volumetric contributions of mcarophytes, phytoplankton and detritus significantly differed during the two seasons of the year (U test, P<0.01). Macrophytes were the most important food items during the dry season, occurring in 51.4% of the stomachs and constituting 50.4% of the total volume of the food categories (Table 2). During the wet season, their contribution declined slightly occurring in 43.5% and comprising 37.8% of the total volume (Table 2).

The contribution of phytoplankton was higher during the dry season than the wet season (Table 2). They occurred in

85.7% and 74.9% of the stomachs during the dry and wet seasons and their volumetric contribution was 21.8% and 12.5%, respectively (Table 2). Diatoms and blue green algae were of comparable importance during both dry and wet seasons (Table 2). The frequency of occurrence of detritus was comparable during the dry (47.7%) and wet (46.5%) seasons but its volumetric contribution was higher during the wet season (40.8%) than the dry season (18.5%) (Table 2).

Insects and nematodes occurred in relatively fewer number of stomachs and their volumetric contribution was low (Table 2). The contribution of insects was more or less comparable during both seasons. During the dry season, insects occurred in 22.4% of the stomachs and constituted 5.7% of the total volume of food items (Table 2). During the wet season, insects occurred in 15.9% of the stomachs and accounted for 6.5% of the total volume of food items (Table 2).

Nematodes occurred in 7.3% and 9.7% of the stomachs during the dry and wet months, respectively. Their volumetric contribution was 2.2% and 1.8% of the total volume of food items during the dry and wet seasons, respectively (Table 2). Ostracods and zooplankton were of little importance during both dry and wet periods, because they occurred in only few stomachs and their volumetric contribution was insignificant (Table 2).

***Table 1:*** *Frequency of occurrence (%) and volumetric contribution (%) of various food items in the diet of T. zillii sampled from Lake Ziway (n=572). Note that the sum of the food categories in bold adds up to 100% in volumetric analysis.*

| Food item | Frequency of occurrence | | Volumetric contribution | |
|---|---|---|---|---|
| | Frequency | Percent | Volume (ml) | Percent |
| **Macrophytes** | **543** | **94.9** | **94.3** | **45.4** |
| **Detritus** | **539** | **94.2** | **61.4** | **29.6** |
| **Phytoplankton** | **472** | **82.5** | **35.3** | **17.0** |
| Diatoms | 449 | 78.5 | 23.1 | 11.1 |
| Blue green algae | 367 | 64.2 | 8.2 | 3.9 |
| Green algae | 213 | 37.2 | 3.3 | 1.6 |
| Euglenoids | 49 | 8.6 | 0.6 | 0.3 |
| **Insects** | **219** | **38.3** | **12.6** | **6.1** |
| Diptera | 200 | 35 | 9.8 | 4.7 |
| Ephemeroptera | 39 | 6.8 | 1.8 | 0.9 |
| Plecoptera | 18 | 3.1 | 0.5 | 0.24 |
| Hemiptera | 9 | 1.6 | 0.5 | 0.24 |
| **Nematodes** | **98** | **17.1** | **3.2** | **1.5** |
| **Ostracods** | **44** | **7.7** | **0.6** | **0.29** |
| **Zooplankton** | **12** | **2.1** | **0.3** | **0.14** |
| Copepods | 12 | 2.1 | 0.3 | 0.14 |
| Cladocerans | 2 | 0.35 | 0.004 | 0.002 |

***Table 2:*** *Relative contribution (%) of different food items in the diet of T. zillii during dry (n= 272) and wet (n= 300) season from Lake Ziway. Note that the sum of the major categories of food items in bold adds up to 100% in volumetric analysis.*

| Food item | Frequency of occurrence (%) | | Volumetric contribution (%) | |
|---|---|---|---|---|
| | Dry season | Wet season | Dry season | Wet season |
| **Macrophytes** | **51.4** | **43.5** | **50.4** | **37.8** |
| **Detritus** | **47.7** | **46.5** | **18.5** | **40.8** |
| **Phytoplankton** | **85.7** | **74.9** | **21.8** | **12.5** |
| Diatoms | 43.2 | 35.3 | 15.2 | 7.7 |
| Blue green algae | 37.1 | 27.1 | 4.7 | 2.9 |
| Green algae | 21.7 | 15.6 | 1.7 | 1.5 |
| Euglenoids | 2.3 | 6.3 | 0.2 | 0.4 |
| **Insects** | **32.4** | **15.9** | **5.7** | **6.5** |
| Diptera | 30.6 | 15.8 | 4.1 | 5.3 |
| Ephemeroptera | 2.9 | 3.9 | 0.9 | 0.9 |
| Plecoptera | 2.1 | 1.0 | 0.5 | 0.2 |
| Hemiptera | 0.0 | 1.5 | 0.2 | 0.1 |
| **Nematodes** | **7.3** | **9.7** | **2.2** | **1.8** |
| **Ostracods** | **2.6** | **6.5** | **0.3** | **0.5** |
| **Zooplankton** | **1.4** | **0.7** | **0.2** | **0.1** |
| Copepods | 1.4 | 0.7 | 0.2 | 0.1 |
| Cladocerans | 0.3 | 0.0 | 0.003 | 0.0 |

### *3.3. Ontogenetic Diet Shifts*

There was no significant variation in the diet of individuals in the size classes I and II (α= 0.88), I and III (α= 0.64), I and IV (α= 0.60), II and III (α= 0.74), II and IV (α= 0.68) and III and IV (α= 0.95). The variation in diet with size of *T. zillii* was slight in the present study (Figure 2). In general, the importance of macrophytes increased with size whereas the importance of detritus, phytoplankton and insects declined with fish size (Figure 2). In size class 5.0-9.9 cm TL *T. zillii* fed on macropytes (37.8%), detritus (36.6%), phytoplankton (18.8%) and insects (9.6%) of the total volume of food items. The contribution of nematodes (1.2%) was insignificant. When *T. zillii* attained 10.0-14.9 TL size class the contribution of macrophytes increased to 47.4% while the importance of detritus, phytoplankton and insects decreased to 29.5%, 14.4% and 4.6%, of the mean volume in the size class, respectively (Figure 2).

In the size class 15.0-19.9 cm TL the importance of macrophytes further increased to 71.8%. The importance of the remaining food items, namely, detritus, phytoplankton and insects declined to 12.5%, 8.6 and 4.0%, respectively (Figure 2). When *T. zillii* attained the size class 20.0-24.9 cm TL, the volumetric contribution of macrophytes increased to 76.1%. Detritus, phytoplankton and insects declined to 10.9%, 7.2% and 2.6% of the total volume in the size class, respectively (Figure 2).

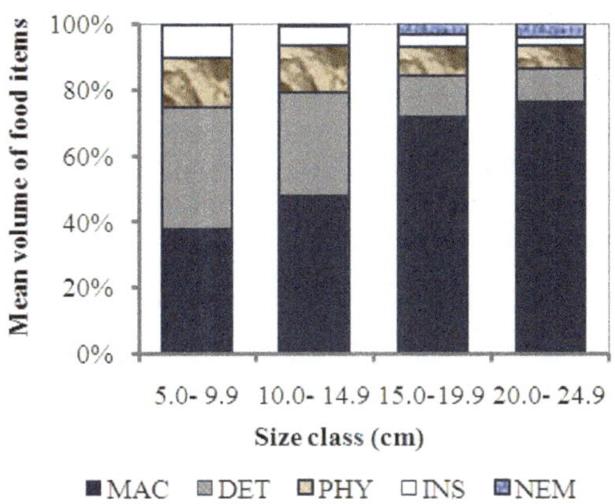


***Figure 2.*** *Percent mean volume of food items consumed by different size classes of T. zillii sampled from Lake Ziway (MAC- Macrophytes, PHY-Phytoplankton, DET- Detritus, INS- Insects, NEM- Nematodes).*

Generally, the importance of macrophytes increased with fish size whereas the importance of detritus, phytoplankton and insects declined with size of fish. The contribution of nematodes was insignificant in all size classes (Figure 2).

## 4. Discussion

Analysis of stomach contents during this study indicated that *T. zillii* fed on a variety of food categories including macrophytes, detritus, phytoplankton, insects, zooplankton, nematodes and ostracods. Various authors studied the feeding habits of *T. zillii* in different parts of African water bodies and reported its herbivorous feeding habits [3, 5-6, 21].

During the present work *T. zillii* was found to be confined to the littoral habitat. This preference of littoral areas could be due to its herbivorous feeding habits. This is in agreement with the findings of Siddiqui [21] in Lake Naivasha (Kenya) and Markos et al. [10] in Lake Ziway (Ethiopia). Spataru [5] studying *T. zillii*, in Lake Kinneret (Israel) and Negassa and Padanillay [7] working with the same fish in Lake Ziway (Ethiopia), both reported the importance of macrophytes in the diet of the species. In addition to macrophytes *T. zillii* also consumed other food categories, including phytoplankton, detritus, insects and zooplankton [2-3, 6].

Negassa and Padanillay [7] reported that the importance of food of animal origin such as insects (Diptera and Ephemeroptera), zooplankton mainly Copepoda and Cladocera and fish scales in the diet of *T. zillii*. Spataru [5] and Negassa and Getahun [2] indicated the importance of phytoplankton consisting of diatoms, blue green algae, green algae and euglenoids in Lake Kinneret (Israel) and Lake Ziway (Ethiopia). Negassa and Padanillay [7] also reported the importance of detritus, nematodes and ostracods in the diet of *T. zillii* in Lake Ziway. Since the works of Negassa and Getahun [2] and Negassa and Padanillay [7] used only frequency of occurrence of different food items, it was not possible to compare the volumetric importance of the different food categories to the species.

Monthly fluctuation has great effect on the availability and contribution of different food categories. In this study, the most important food categories that constituted the bulk of the food during the dry season were macrophytes, detritus, phytoplankton and insects. Macrophytes were the most important food items both during the dry season (50.4% by volume) and the wet season (51.4% by volume). The present study is in agreement with the findings of Adama et al. [22] in Sahelo-Sudanian Rivers (Burkina Faso) and Negassa and Padanillay [7] in Lake Ziway (Ethiopia). The contributions of detritus and insects during the dry season were relatively lower than the wet season. Various authors have reported the lower importance of detritus and insects to the diet of tilapia species such as *T. zillii* and *O. niloticus* during dry season [23-25].

According to Njiru et al. [26] the importance of phytoplankton was relatively low during the wet season than the dry season in Lake Victoria (Kenya) in the diet of *O. niloticus*. In the present study, among the phytoplankton groups, diatoms contributed the highest volume during the dry season than the wet season. The reason for the abundance of this food items during the dry season could be autotrophic nature of the phytoplankton using light as their energy input for their growth [5]. During the dry season, the water of Lake Ziway may not be as turbid as in the wet season because during the wet season large quantities of silt is brought into the lake through runoff and tributary rivers such as River Meki and River Katar. The silt brought during the wet season increases the turbidity of the lake and reduces algal growth. The importance of minor food categories such as nematodes, ostracods and zooplankton was also relatively high during the dry season in the diet of *T. zillii* in the present study. Legner [27] reported that the importance of zooplankton during dry season in California. Several other investigators also reported the relative importance of nematodes and ostracods during dry season than the wet season in Lake Naivasha (Kenya) [22], Lake Kinneret (Israel) [5], Lake Ziway [2] and in two West African reservoirs, Selingue and Manantai (Mali) [28].

During the wet season, detritus was the most important food item of *T. zillii* in Lake Ziway (40.8% by volume). The source of this food item could be the floods that introduce different plant materials into the Lake and the leaves of plants that fall into the lake from the emergent macrophytes. Negassa and Getahun [2] reported high contribution of plant materials and detritus in the diet of *T. zillii* during the wet season. The same authors reported that these food items are highly dispersed along the surface of the water column at wet period of the year due to floods. In the present study, macrophytes constituted 37.8% of the total volume in Lake Ziway during the wet season. Tadesse [29] also reported that macrophytes and detritus were the dominant food items during the wet season than the dry season in the diet of *O. niloticus* in Lake Ziway. The present work revealed that the contribution of insects was relatively high in the wet season than the dry season. This is in agreement with the results of Akinuwmi [6] where the availability of insects was more important in the wet season than the dry season.

From the results of the present study, it was evident that *T. zillii* undergoes slight ontogenetic diet shift during its life cycle. Based on the volumetric contributions of different food items the relative importance of detritus, phytoplankton and insects was high in the stomachs of juveniles (5.0-9.9 cm TL and 10.0-14.9 cm TL), whereas, in the stomachs of adults (15.0-19.9 cm TL and 20.9-24.9 cm TL), macrophytes were the dominant food category. Other investigators have also reported the importance of detritus, phytoplankton, insects and nematodes in the diet of juvenile *T. zillii* in Lake Naivasha (Kenya) [28], Lake Kennert (Israel) [5] and Lake Ziway (Ethiopia) [2]. The most probable reason why juvenile *T. zillii* feed on these food items could be that these food items are abundant in the littoral part of the lake where weed-bed fauna and decomposing plant materials are found in higher quantities. Juvenile *T. zillii* normally live in this environment in order to protect themselves from the risk of predators [4].

The results of this study also clearly showed that as fish increased in size, the importance of macrophytes increased while the importance of phytoplankton, detritus and insects declined sharply. This is in agreement with the findings of Wanink and Joordens [30] (Lake Victoria), Negassa and Getahun [2] and Negassa and Padanillay [7] in Lake Ziway. Mosaad [31] reported that adult *T. zillii* ingest benthic invertebrates and insects in Lake Quarn (Egypt). The results of the present study are also in agreement with these findings, but the volumetric contribution of these macro-invertebrates was comparatively low in the present study. Meyer [32] pointed out the change in composition of the diet as the fish grows with an increase in the minimum size of organisms eaten. Akinuwmi [6] reported that adult *T. zillii* mainly feed on macrophytes in Ondo State University Fish Farm, Nigeria. Various workers [5, 23, 31] in Lake Kinneret (Israel), Lake Victoria (Kenya) and Sahelo-Sudanian Reservoirs (Burkina Faso), respectively pointed out the significant contributions of nematodes in the diet of adult *T. zillii*. In contrast to the above findings, nematodes played insignificant role in the diet of *T. zillii* in Lake Ziway.

# 5. Conclusion

The results of the present study have clearly indicated that the most important food categories of *T. zillii* in Lake Ziway were macrophytes, phytoplankton, detritus and insects. Food of minor importance were nematodes, ostracods and zooplankton. During the dry season, macrophytes and phytoplankton were the most important food categories whereas their contributions declined sharply during the wet season. On the other hand, detritus was the most important food category during the wet season but its contribution declined during the dry season. Ontogenetic diet shift was slight in the present study. The relative importance of detritus, phytoplankton and insects was high in the diet of juveniles. As the size of fish increased the importance of macrophytes increased whereas the importance of phytoplankton, detritus and insects declined sharply.

# Acknowledgements

We thank fishermen Fikre Dolie and Birhanu Bante for their assistance during sample collection. The Biology Department of Hawassa University deserves special thanks for providing laboratory materials and other facilities during the study period. We are grateful to Mrs. Yirgashewa Bekele who assisted us during stomach content analysis. Ministry of Education, Addis Ababa, Ethiopia, provided financial support for the second author.

# References

[1] El- Shazly A. 1993. Biological studies on four cichlid fishes (*Tilapia nilotica, Tilapia galilae, Tilapia zillii, Tilapia aurea*). MSc Thesis, Zagazig University, Egypt.

[2] Negassa A. and Getahun A. 2004. Breeding season, length-weight relationship and condition factor of introduced fish, *Tilapia zillii* (Gervais, 1848) (Pisces: Cichlidae) in Lake Ziway, Ethiopia. SINET: Ethiop. J. Sci. 26(2): 115-122.

[3] Kariman A.S.H. and Nadhan G. 2009. Some observations on fisheries biology of *Tilapia zillii* (Gervais, 1848) and *Solea vulgaris* (Quensel, 1806) in Lake Qarun, Egypt. World Fish Mari. Sci. 1(1): 20-28.

[4] Mehanna S.F. 2004. Population dynamics of two cichlids, *Oreochromis aureus* and *Tilapia zillii*, from Wadi El-Raiyan Lakes. Egypt. Agri. Marine Sci. 9(1):9-16.

[5] Spataru P. 1978. Food and feeding habits of *Tilapia zillii* (Gervais, 1848) (Cichlidae) in Lake Kinneret (Israel). Aquaculture 14(4): 327-338.

[6] Akinuwmi F.O. 2003. Food and feeding habits of *Tilapia zillii* (Pisces: Chichlidae) in Ondo State University Fish Farm. Proceeding of the 16th Annual Conference of the Fisheries Society of Nigeria (FISON), Department of Environmental Biology and Fisheries (Nigeria).

[7] Negassa A. and Padanillay C. 2008. Abundance, food habits, and breeding season of exotic *Tilapia zillii* and native *Oreochromis niloticus* fish species in Lake Ziway, Ethiopia, Mj. Int. J. Sci. Tech. 2: 345-359.

[8] Adeyemi S.O., Bankole N.O., Adikwu I.A. and Akombu P.M. 2009. Food and feeding habits of some commercially important fish species in Gbedikere Lake, Bossa, Kogi State, Nigeria. Int. J. Lakes and Rivers 2(1): 31-36.

[9] Mohamoud F.W., Amin M.A., Elboray F.K., Ramadan M.A. and Halfawy O.K. 2011. Reproductive biology and some observation on the age, growth, and management of *Tilapia zillii* (Gervais: 1848) from Lake Timsah, Egypt. Int. J. Fish. Aquac. 3(2): 16-26.

[10] Markos T., Reidar B. and Bjørn O.R. 2011. Mercury concentrations are low in commercial fish species of Lake Ziway, Ethiopia, but stable isotope data indicated biomagnifications. Ecotox. Environ. Safety 74: 953-959.

[11] Makin M.J., Kingham T.J., Waddans A.E., Birchall C.H. and Tamene T. 1975. Development prospects in the southern rift valley, Ethiopia. Land Resources Division, Ministry of Overseas Development 142: 324-567.

[12] Ayenew T. 1998. The hydro biological system of the Lake District basin, central main Ethiopian rift valley, International Institute of Aerospace Survey and Earth Sciences. PhD Thesis, Earth Sciences, the Netherlands.

[13] Lemma A. 2005. Site action plan for conservation and sustainable use of the Lake Ziway biodiversity, Rift Valley Lakes Project, Institute of Biodiversity, Addis Ababa.

[14] Admassu D. and Ahlgren I. 2000. Growth of juvenile tilapia, *Oreochromis niloticus* L. from Lakes Ziway, Langeno and Chamo (Ethiopian rift valley) based on otolith micro-increment analysis. Ecol. Freshwat. Fish 9(3): 127-137.

[15] Lake Fisheries Development Project (LFDP) (1998) Fisheries development in Ethiopia: which way now? Project. News. *Bulletin.*11 no. 1. Lake Fisheries Development Project, Addis Ababa, Ethiopia.

[16] Hyslop E.J. 1980. Stomach contents analysis-a review of methods and their application. J. Fish Biol. 17: 411-429.

[17] Wallace R.K. 1981. An assessment of Diet-Overlap Indexes. Trans. Am. Fish. Soc. 110: 72-76.

[18] Sokal R.R. and Rohlf F.J. 1995. Biometry: The principles and practices of statistics in biological research 3rd ed. New York, Freeman and Company, pp. 859.

[19] Schoener T.W. 1970. Non-synchronous spatial overlap of lizards in patchy habitats. Ecology 51: 408-418.

[20] Mathur D. 1977. Food habits and competitive relationships of Bandfin shiner in Halawakee Creek, Alabama. Am. Midland Natura. 97: 89-100.

[21] Siddiqui A.Q. 1977. Reproductive biology, length-weight and relative condition of *Tilapia leucostica* (Trewaeva) in Lake Naivasha, Kenya. J. Fish Biol. 10: 351-260.

[22] Adama O., Wendengouing G., Allassane O., Germain G. and Brnerd H. 2008. Seasonal diet shift of the most important fish species in a Sahelo-Sudanian Reservoirs (Burkina Faso). J. Fish. Aquat. Sci. 3(4): 240-251.

[23] Brown J. A. 1986. The development of feeding behavior in the Cichlidae. J. Fish Biol. 29: 17-178.

[24] Layman C.A. and Silliman B.R. 2002. Preliminary survey and diet analysis of juvenile fishes of an estuarine creek on Andros Island, Bahamas. Bull Mar. Sci. 70(1): 199-210.

[25] Layman C.A., Quattrochi J.P., Peyer C.M. and Allgeier J.E. 2007. Niche width collapse in a resilient top predator following ecosystem fragmentation. Ecol. Lett. 10(10): 937-944.

[26] Njiru M., Okeyo-Owuor J.B., Muchiri M. and Cowx I.G. 2004. Shifts in the food of Nile tilapia, *Oreochromis niloticus* (L.) in Lake Victoria, Kenya. Afr. J. Ecol. 42: 163-170.

[27] Legner E.F. 1978. Mass culture of *Tilapia zillii* (Cichlidae) in pond ecosystems. Entomopora 23: 51-55.

[28] Perga M.E., Arfi R. and Gerdeaux F. 2005. Seasonal variations in fish d13C and d15N in two West African reservoirs, Selingue and Manantali (Mali): modification of trophic links in relation to water levels. Isotoes. Environ. Health Stud. 41(2): 1-15.

[29] Tadesse Z. 1988. Studies on some aspects of the biology of *Oreochromis niloticus* L. (Pisces: Cichlidae) in Lake Ziway, Ethiopia. MSc Thesis, Addis Ababa University, Addis Ababa.

[30] Wanink H. and Joordens C.A. 2003. Dietary shifts in *Brycinus sadleri* (Pisces: Characidae) from Southern Lake Victoria. Inst. Evolut. Ecol. Sci. 12: 230-350.

[31] Mosaad M.N.M. 1990. Biological studies on five fish species from Lake Qarun, Egypt: length-weight relationship and condition factor. Proc. Zool. Soc. Egypt 21: 331-344.

[32] Meyer D.E. 2002. Technology for successful small-scale tilapia culture (CRSP Research Report 02- 179). CRSP (Aquaculture Collaborative Research Support Program). Tilapia Sessions 2: 97-106.

# Effect of alternate wetting and drying (AWD) irrigation for *Boro* rice cultivation in Bangladesh

**Md. Redwanur Rahman, Sheikh Helena Bulbul**

Institute of Environmental Science, University of Rajshahi, Rajshahi, Bangladesh

**Email address:**
redwan_rahman@lycos.com (Md. R. Rahman), hbsplendid07@gmail.com (S. H. Bulbul)

**Abstract:** A field experiment was conducted at the Bangladesh Rice Research Institute to find out possible effects of alternate wetting and drying irrigation (AWDI) on the yield, water use and water use efficiency (WUE) of *Boro* rice. The experimental plots were laid out with 2 factors RCBD combining two modern varieties of rice *viz.*, BRRI *dhan29* and BRRI hybrid2, which received four irrigation treatments randomly and was replicated thrice. The treatments ranged from continuous submergence ($T_1$) of the field to a number of delayed irrigations ($T_2$, $T_3$ and $T_4$) denoting application of 5cm irrigation water when water level in the perforated PVC pipe fell 15, 20 and 25cm below ground level (G.L.), respectively. The study revealed that treatment $T_1$ attributed by the highest total water use (122.2cm) and the lowest WUE (84.34kg/ha/cm) produced the lowest grain yield (4.71t/ha). Treatment $T_2$, on the contrary, gave the highest yield (5.69 t/ha) and consequently the second highest WUE (85.55 kg/ha/cm) indicating quite a large water saving (15cm) compared to treatment $T_1$. The yields in treatments $T_3$ (5.45 t/ha) and $T_4$ (5.27 t/ha) were significantly lower at 1% level of significance compared to that of treatment $T_1$. Significant effect was found either for the treatment or for the varieties on the number of effective and total tillers hill -1, plant height, number of effective tillers hill-1, grain yield, straw yield, biological yield and harvest index.

**Keywords:** Alternate Wetting and Drying Irrigation, *Boro* Rice, Yield, Water Use Efficiency

## 1. Introduction

In agriculture sector rice is the only major grain crop that is grown almost exclusively as food needs mush water to produce. Rice grown under traditional practices in the Asian tropics and subtropics requires between 700-1500mm of water per cropping season depending on soil texture [1]. However, this conventional water management method leads to a high amount of surface runoff, seepage, and percolation that can account for between 50–80% of the total water input [2].

The interactions between water use for rice cultivation, surface water and groundwater resources are often very close-such that active cross-sector dialogue and integrated vision are also needed to promote sustainable water use.

Sustainability is often referred to as a process which leads to better relationships between humans and the natural environment and between themselves. [3] emphasizes this relationship and reinforces that Sustainability is the doctrine that economic growth and development must take place, and be maintained over time, within the limits set by ecology in the broadest sense by the interrelations of human beings and their works, the biosphere and the physical and chemical laws that govern it. It follows that environmental protection and economic development are complementary rather than antagonistic processes, (*Scientific American*: September, 1989). The outcome of this interrelated process is balanced development. This is supported by the Florida Centre for Community Design and Research (2010) which states that Sustainability is the optimal balance of natural, economic and social systems overtimes.

The sustainability concept argues for a holistic and balanced approach to life where economic prosperity, nature conservation and social justice are given equal weight in any long term strategies of [4]. New definitions of sustainability are constantly emerging, however they all share common aspects. The Research Group on the Global Futures provides an array of definitions for sustainability but concludes that most definitions have three aspects in common. These are living within limits; understanding the interconnections between the economy, society and the environment and equitable distribution of resources and opportunities (Research Group on Global Futures 2005). [5] see sustainability as a matter of making adjustments to present human activities, to sustain twentieth century natural resources largely unchanged and unchallenged into the twenty-first century. Hence, the main feature of sustainability is the direct practical changes that it requires. Sustainability is very important for this research as it deals with the way the sustainability concepts in terms of water use for rice cultivation are being transferred and reinforced in young people, researcher, farmers, and policy makers through the educational system.

Many researchers indicate that rice is the major consumer of irrigation water in Bangladesh. It is grown under two distinct water regions, continuous standing and alternate wetting. The conventional method of rice planting requires continuous pond water on the field, which is possible where irrigation water is abundant and cheap. In this method irrigation water is used for evapotranspiration (ET) and seepage-percolation (S&P). But in reality, only ET is the true water requirement for crop growth and S&P are the unavoidable losses. However, rice can be grown under alternate wetting and drying conditions with necessarily sacrificing yields and adoption of such practices may allow savings of costly water. Alternate Wetting and Drying (AWD) involves technology that tackles water scarcity in irrigated rice cultivation and has the potential to contribute to a more sustainable and effective water and energy use. This AWD tool is a single device designed to observe water level in rice field for deciding the time of irrigation. It involves installation of a perforated pipe (preferably PVC) in rice field to allow observation of water level. In one part, such pipe of 10cm diameter and 30cm long is installed having 10cm above and 20cm below the ground surface.

By applying AWD, farmers or pump-owners are able to save 15 to 30% of their irrigation water. Water productivity, *i.e.* the volume of irrigation water required to produce a certain amount of rice, increases compared to conventional cultivation [6], [7]. To identify the sustainable water management for *boro* rice cultivation by practicing alternate wetting and drying (AWD) was the core objective of this study. More specifically, the objective could be outlined as to find out, from a number of AWDI irrigation treatments, the best one with the highest water use efficiency that would result in an insignificant yield loss and ensure the best use of the available water resources.

## 2. Materials and Methods

The Bangladesh Rice Research Institute (BRRI) Farm of Gazipur was selected as the experimental site. Topography of the land being plain was suitable for check basin irrigation. Individual plots were located inside a close growing rice field so that actual growing condition (reception of the direct and diffused fluxes) prevails in the site. Soil texture of the experimental site was found to be silty loam. The upper root zone of the experimental field was tilled with high puddling intensity. The experimental plots (4×2.5m) were laid out with 2 factors RCBD combining two modern varieties of rice (BRRI*dhan* 29 and BRRI hybrid2) and four irrigation treatments that were replicated thrice. This resulted in a total of 24 plots in the field with 8 plots in a row. Each of the plots was separated by 1m of transition zone while each of the replications was demarcated by a buffer zone of 1.5m in between. To prevent seepage, polythene sheets were pushed into the edges of the levees along the inner perimeter of all plots. PVC pipes of 4cm in diameter and 40cm in length were installed in the field keeping 7cm above the soil and the remaining 33cm which was perforated underneath to measure the depletion of soil water in the field. Irrigation water was applied when depleting water table inside the pipe reached a certain level.

The first treatment ($T_1$) was continuous submergence (1 to 5cm standing water) and the remaining three ($T_2$, $T_3$ and $T_4$) stood for an application of 5cm irrigation water when water level in the pipe fell 15, 20 and 25cm below the G.L., respectively. Continuous standing water (5cm) was maintained in all the plots up to 28 days after transplantation (DAT) to avoid pre-apprehended weed infestation that could be awesome during crop establishment stage. A bowl of 1.5liters was used to irrigate the plots from the buffer zones by throwing water in. The seedlings were transplanted maintaining hill to hill distance of 15cm and row to row distance of 25cm. The first and the last hills were kept at 7.5cm away from their nearest levees resulting in 25 hills along the length and 10 hills along the width. Since the grains of BRRI hybrid 2 got ripened earlier than the BRRI*dhan* 29, the former was harvested (01 May 2008) two weeks earlier than the harvesting date (May 14, 2008) of the latter. Matured plants inside 1m square of land were harvested for subsequent analysis. Moisture content of the grains, however, was adjusted to 14% equivalent moisture content after measuring through digital grain moisture meter for subsequent analysis. Quantitative information related to yield and all the yield contributing characters *viz.,* plant height, effective tillers, length of the panicle, no. of spike lets per panicle, no. of filled and unfilled grains per panicle, 1000 grain weight, grain yield, straw yield, harvest index and water use efficiency of the two varieties (BRRI*dhan* 29 and BRRI hybrid2) were analyzed to obtain the effect for AWDI on rice production.

## 3. Results and Discussions

Significant consequences of AWDI on the production of *Boro* rice were observed as given in Tables 1. The analysis showed that varietal effect on plant height was statistically significant at 1% probability level. The tallest plant (107.00cm) was found in BRRI hybrid2 ($V_2$). The shortest plant (101.95cm) was found in BRRI *dhan29* ($V_1$). Variation in plant height might be due to the differences in the genetic make–up of the varieties. The result is in consistent with findings of [8] who also reported a variable plant height existed among the varieties. The highest number of total tillers hill$^{-1}$(10.96) was found in BRRI hybrid2 and the lowest number of total tiller was found (10.63) in BRRI *dhan*29. The variation in number of total tillers hill$^{-1}$ might be due to varietal characteristics. The highest number of effective tillers (9.11) was found in BRRI hybrid2 and the lowest number of effective tiller hill$^{-1}$(8.68) was found in BRRI *dhan29*. The highest number of non-effective tillers (1.95) was found in BRRI *dhan*29 and the lowest number of non-effective tiller hill$^{-1}$ (1.85) was found in BRRI hybrid2. The highest length of panicle (22.92cm) was found in BRRI *dhan*29. The lowest length of panicle was (22.80cm) in BRRI hybrid2. The results showed that the highest grain yield (137.64) was achieved from BRRI hybrid2. The lowest grain yield (118.45) was achieved from BRRI *dhan*29. The highest number of unfilled grains panicle$^{-1}$ (22.64) was found in BRRI*dhan* 29. The lowest number of unfilled grains panicle$^{-1}$ was found from BRRI hybrid2. The result showed that the highest weight of 1000- grain (23.65g) was obtained from BRRI hybrid2. The lowest weight of 1000-grain (23.35g) was obtained from BRRI*dhan*29. Grain yield was statistically significant at 1% level of probability. The highest grain yield (5.64 t/ha) was achieved from BRRI hybrid2. The lowest grain yield (4.93t/ha) was achieved from BRRI*dhan* 29. These differences occurred due to variations of genetic make-up among the varieties. The result shows that the highest straw yield (6.70tha$^{-1}$) was found from BRRI hybrid2. The lowest straw yield (5.83tha$^{-1}$) was found from BRRI *dhan*29. The highest yield occurred due to higher plant height, higher total tiller hill$^{-1}$ and lower number non-effective tiller hill$^{-1}$. These results are consistent with those obtained by [9] who reported differences in straw yield among varieties.

The highest biological yield (12.34tha$^{-1}$) was obtained from BRRI hybrid2 and the lowest one (10.76tha$^{-1}$) was obtained from BRRI*dhan* 29. Maximum harvest index (45.73%) was obtained from BRRI*dhan* 29 and the minimum harvest index (45.65 %) was obtained from BRRI hybrid2.

The experiment aimed in exploring the possible effects of different irrigation treatments on the production and production related parameters. Different yield contributing characters *viz*., plant height (cm), number of effective tillers per hill, panicle length (cm), total number of filled grains per panicle, number of unfilled grains per panicle; 1000 seed weight (gm), grain yield (t/ha) and straw yield (t/ha) for each of the varieties were analyzed. Statistical relationships of the effect of four treatments on the individual yield contributing parameters are given with their detail statistical analysis in Table 2.

***Table 1.*** *Varietal (BRRI dhan29 and BRRI hybrid2) Effect on the Yield and Yield Contributing Characters.*

| Variety | Plant height (cm) | Total tiller Hill$^{-1}$ | Effective tiller Hill$^{-1}$ | Non-effective tiller Hill$^{-1}$ | Panicle length (cm) | Filled grain panicle$^{-1}$ | Unfilled grain panicle$^{-1}$ | 1000-GW | Grain yield (t/ha$^{-1}$) | Straw yield (t/ ha$^{-1}$) | Biological yield | Harvest Index |
|---|---|---|---|---|---|---|---|---|---|---|---|---|
| $V_1$ | 101.95 | 10.63 | 8.68 | 1.95 | 22.92 | 118.45 | 22.64 | 23.35 | 4.93 | 5.83 | 10.76 | 45.73 |
| $V_2$ | 107.00 | 10.96 | 9.11 | 1.85 | 22.80 | 137.64 | 20.49 | 23.65 | 5.64 | 6.70 | 12.34 | 45.65 |
| LSD | 0.369 | 0.106 | 0.072 | 0.094 | 0.348 | 0.926 | 0.354 | 0.176 | 0.047 | 0.047 | 0.066 | 0.319 |
| Level of sig | ** | NS | ** | NS | NS | ** | ** | NS | ** | ** | ** | NS |

In a column figures with same letter or without letter do not differ significantly whereas figures with dissimilar letter differ significantly (as per DMRT), $V_1$ = BRRI*dhan* 29, $V_2$= BRRI hybrid 2, NS = Not Significant, * = Significant at 5% level of probability, ** = Significant at 1% level of probability.

***Table 2.*** *Effect of Different Irrigation Treatments on the Yield and Yield Contributing Characters.*

| Treatment | Plant height (cm) | Total tiller Hill$^{-1}$ | Effective tiller Hill$^{-1}$ | Non-effective tiller Hill$^{-1}$ | Panicle length (cm) | Filled grain panicle$^{-1}$ | Unfilled grain panicle$^{-1}$ | 1000-GW | Grain yield (t/ha$^{-1}$) | Straw yield (t/ha$^{-1}$) | Biological yield | Harvest Index |
|---|---|---|---|---|---|---|---|---|---|---|---|---|
| $T_1$ | 103.43b | 10.40b | 8.06c | 2.34a | 23.41 | 119.32c | 26.80a | 23.07b | 4.71c | 6.12b | 10.83c | 43.61b |
| $T_2$ | 105.25ab | 12.14a | 11.06a | 1.08b | 22.89 | 141.94a | 14.07c | 24.48a | 5.69a | 6.57a | 12.26a | 46.33a |
| $T_3$ | 103.45b | 10.86b | 8.67b | 2.19a | 21.72 | 127.38b | 23.04b | 22.91b | 5.45b | 6.16b | 11.61b | 46.96a |
| $T_4$ | 105.78a | 9.78c | 7.78c | 2.00a | 23.43 | 123.57bc | 22.36b | 23.53b | 5.27b | 6.22b | 11.49b | 45.86a |
| LSD | 0.738 | 0.212 | 0.145 | 0.187 | 0.696 | 1.852 | 0.707 | 0.352 | 0.095 | 0.093 | 0.132 | 0.637 |
| Level of sig | * | ** | ** | ** | NS | ** | ** | ** | ** | ** | ** | ** |

In a column figures with same letter or without letter do not differ significantly whereas figures with dissimilar letter differ significantly (as per DMRT), $T_1$ = Continuous Standing water, $T_2$= Irrigation when water is 15cm below from the soil surface , $T_3$ = Irrigation when water is 20cm below from the soil surface, $T_4$ = Irrigation when water is 25 cm below from the soil surface, NS = Not Significant, * = Significant at 5% level of probability, ** = Significant at 1% level of probability

The analysis showed that the irrigation treatments had significant effect on plant height at 5 % level of probability. The highest plant height (105.78cm) was obtained in treatment $T_4$ (irrigation when water is 25cm below from the soil surface) and the lowest (103.45cm) in Treatment $T_1$ (continuous flooding). This result is in agreement with the findings of [10] who reported that treatment having continuous flooding could not improved plant height. The highest number of effective tillers per hill (11.06) was found in treatment $T_2$ followed by treatment $T_3$ (8.67) and treatment $T_1$ (8.06). The lowest number of effective tiller per hill (7.78) was found in treatment $T_4$.

It was found that the highest number of filled grains (141.94) per panicle was obtained in treatment $T_2$ (irrigation when water is below 15cm from the soil surface) followed by treatments $T_3$ (Irrigation when water is 20cm below from the soil surface) and $T_4$ (Irrigation when water is 25cm below from the soil surface). The lowest number of filled grains per panicle (119.32) was found for treatment $T_1$. Thus the result showed that applying irrigation water in rice field when water level goes 15 to 25cm below G.L does not really reduce the total number of filled grains compared to that nursed with 5cm standing water. However, treatment $T_1$ (continuous standing water) decreased the number of filled grains. The highest grain yield (5.69t/ha) was obtained from treatment $T_2$ (irrigation when water is below 15cm from the soil surface) and the lowest yield grain (4.71t/ha) was obtained from treatment $T_1$ (continuous standing water). The results shows that the grain yield did not decreased when plants suffered little water stress. The second highest yield grain (5.45 t/ha) was found in the treatment $T_3$ (when irrigation is 20cm below from the soil surface). The maximum straw yield (6.57t/ha) was found from the treatment $T_1$. The minimum straw yield (6.12t/ha) was found from treatment $T_1$. The irrigation treatments and of the experiment did not have any significant effect on the harvest index either at 1% or 5% level of probability. The highest value of harvest index (46.96 %) was found for the treatment $T_3$ and the minimum for the $T_1$ (43.61%).

Effect of the interaction between the varieties and the treatments was also found to be statistically significant at 1% level of probability table 3. The tallest plant height (109.17cm) was found for the interaction $V_2 \times T_2$ ($V_2$=BRRI hybrid2, $T_2$= Irrigation when water is 15cm below from the soil surface). The interaction effect (variety × Irrigation) had any effect on the panicle length of the varieties.

The cause of the non significant output of the panicle length might have occurred due to insufficient photosynthesis from the less vigorous crop canopy and reduced leaf area of BRRI*dhan* 29 and BRRI hybrid2. The interaction effect of the varieties and treatments also came significant at 5% level of probability. The highest number of filled grains (145.65) was, however, marked for the interaction ($V_2 \times T_2$) and the lowest number of filled grains (95.50) was obtained from $V_1T_1$. The highest 1000 grain weight (24.80) was obtained for the interactions ($V_2 \times T_2$) and the lowest 1000 grain weight (22.80) was obtained for the $V_2 \times T_3$. The study raveled that the varieties $V_1$ and $V_2$ and interaction effect between variety × treatments produced statistically insignificant variation in 1000 grain weight among themselves. Thus, it was clear from the interaction effect that AWDI method of irrigation treatments did not reduced the 1000 grain weight as irrigation delayed. The interaction between the varieties and treatments also produced significant results for grain yield at the 5% level of probability. The highest grain yield of BRRI hybrid2 (6.28 t/ha) was obtained for the interaction ($V_2 \times T_2$) and the lowest grain yield (4.18t/ha) was obtained from the interaction ($V_1 \times T_1$). The interaction between the varieties and treatments also produced significant straw yield at the 1% level of probability. The highest straw yield (7.06 t/ha) was obtained for the interaction ($V_2 \times T_2$) and lowest straw yield (5.26t/ha) was obtained for the interaction ($V_1 \times T_1$). Interaction effect of the variety and the treatments were found insignificant either at the 1% or 5 % level of significant. The highest harvest index (47.08 %) for the treatment ($V_2 \times T_2$). The lowest harvest index (44.28 %) for the treatment ($V_1 \times T_1$). Interaction effect of the variety and the treatments were found insignificant either at the 1% or 5 % level of significant.

***Table 3.** Mean Effect of the interaction Between Varieties and Irrigation Treatments on the Yield and Yield Contributing Characters of BRRI dhan29 ($V_1$) and BRRI hybrid2 ($V_2$)*

| | Plant height (cm) | Total tiller $Hill^{-1}$ | Effective tiller $Hill^{-1}$ | Non-effective tiller $Hill^{-1}$ | Panicle length (cm) | Filled grain $panicle^{-1}$ | Unfilled grain $panicle^{-1}$ | 1000-GW | Grain yield (t/ $ha^{-1}$) | Straw yield (t/ $ha^{-1}$) | Biological yield | Harvest Index |
|---|---|---|---|---|---|---|---|---|---|---|---|---|
| $V_1T_1$ | 99.85c | 10.23 | 7.86 | 2.37 | 22.98 | 95.50e | 27.31a | 22.83 | 4.18d | 5.26e | 9.44e | 44.28 |
| $V_1T_2$ | 101.32c | 11.95 | 11.04 | 0.91 | 23.92 | 138.22b | 17.88d | 24.16 | 5.10c | 6.08c | 11.18cd | 45.59 |
| $V_1T_3$ | 99.93c | 10.47 | 8.37 | 2.10 | 21.57 | 118.17d | 22.06c | 23.02 | 5.08c | 5.75d | 10.83d | 46.93 |
| $V_1T_4$ | 106.72ab | 9.86 | 7.44 | 2.42 | 23.23 | 121.93cd | 23.32c | 23.38 | 5.33c | 6.23c | 11.57c | 46.11 |
| $V_2T_1$ | 107.02ab | 10.56 | 8.27 | 2.30 | 23.85 | 143.13ab | 26.29ab | 23.31 | 5.24c | 6.97a | 12.21b | 42.93 |
| $V_2T_2$ | 109.17a | 12.32 | 11.07 | 1.25 | 21.87 | 145.65a | 10.27e | 24.80 | 6.28a | 7.06a | 13.33a | 47.08 |
| $V_2T_3$ | 106.98ab | 11.24 | 8.97 | 2.27 | 21.87 | 136.58b | 24.02bc | 22.80 | 5.82b | 6.57b | 12.39b | 46.99 |
| $V_2T_4$ | 104.84b | 9.70 | 8.12 | 1.58 | 23.63 | 125.20c | 21.39c | 23.67 | 5.21c | 6.21c | 11.41c | 45.62 |
| LSD | 1.48 | 0.42 | 0.29 | 0.37 | 1.39 | 3.70 | 1.41 | 0.70 | 0.19 | 0.19 | 0.26 | 1.27 |
| Level of sig | ** | NS | NS | NS | NS | ** | ** | NS | ** | ** | ** | NS |

In a column figures with same letter or without letter do not differ significantly whereas figures with dissimilar letter differ significantly (as per DMRT), * = Significant at 5% level of probability, ** = Significant at 1% level of probability, NS = Not significant

### *3.1. Irrigation Treatments (AWDI)*

Irrigation treatments were applied at different stages of the growing period depending on the depletion of the water level in the perforated pipe. The very first treatment stated at the end of the fourth week after transplantation. During this time 5cm standing water was kept to avoid weed infestation in the plots. Table 4 shows that the highest number of irrigation (14 nos.) was given to the plots with treatment $T_1$ (continuous flooding) for BRRI *dhan*29. The other three treatments *viz.*, $T_2$, $T_3$ and $T_4$ received a total of 9, 8 and 7 nos. of irrigation for BRRI *dhan*29 while 12, 9, 8 and 7 number of irrigation for BRRI hybrid2, respectively. Water required for crop establishment and water received from the rainfall was estimated to be 53.3cm during the growing period for each of the treatments. For BRRI hybrid2, maximum amount of water (112.20cm) was required for $T_1$, while, second maximum (91.20cm) for $T_2$ was followed by other two treatments, $T_3$ (87.20cm) and $T_4$ (81.20cm). For BRRI *dhan*29 the treatments $T_1$, $T_2$, $T_3$ and $T_4$ required 122.2, 97.2, 92.2 and 87.2cm of water, respectively.

***Table 4.*** *Total Number of Irrigation Required for Different Irrigation Treatments.*

| Treatment | *No. of Irrigation | | Rainfall+ water for crop established (cm) | Total water required (cm) | |
|---|---|---|---|---|---|
| | BRRI *dhan*29 | BRRI hybrid2 | | BRRI*dhan*29 | BRRI hybrid2 |
| $T_1$ | 14 | 12 | 55.3 | 122.2 | 112.20 |
| $T_2$ | 9 | 9 | 55.3 | 97.20 | 91.20 |
| $T_3$ | 8 | 8 | 55.3 | 92.20 | 87.20 |
| $T_4$ | 7 | 7 | 55.3 | 87.20 | 81.20 |

*One irrigation means application of 5cm irrigation water

The graphical representation of water usage by different treatments after transplantation is shown in Fig.1

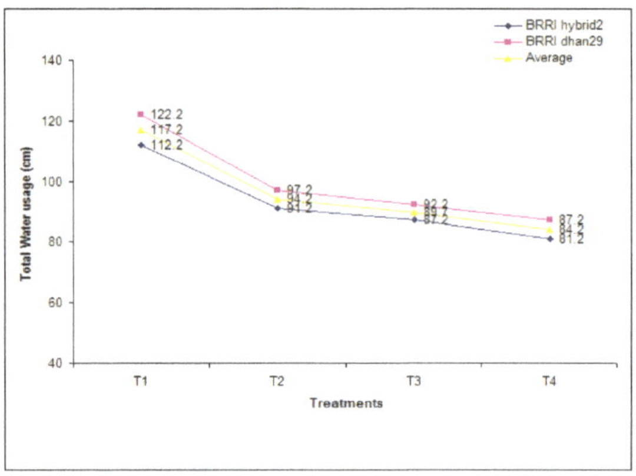


***Fig 1.*** *Water usage of different treatments for the production of BRRIdahn29 and BRRI hybrid2.*

### *3.2. Water Use Efficiency*

Water use efficiencies for the individual effect of different treatments were derived along with the values of WUE for the 8 interactions between treatments and varieties (Table 5). The highest water use efficiency, WUE was found to be 87.38kg/ha/cm ($V_1 \times T_2$). All the highest water use efficiencies were found in the combinations having variety $V_1$ (BRRI*dhan*29). The lowest WUE was obtained in the treatment $T_1$ for $V_2$. In case of BRRI*dhan*29 ($V_1$) the highest WU was found to be 87.38 kg/ha/cm of water and the lowest was found to be 86.11 kg/ha/cm. The second highest WUE highest WUE (87.38 kg/ha/cm) was found in the treatment $T_2$ though it gave poor yield (5.10t/ha). Treatment $T_2$ gave high yield with high water use efficiency (85.55kg/ha/cm) among the others (Table 5).

***Table 5.*** *Water use Efficiency for Different Treatments and Interactions.*

| Interactions | Total water required (cm) | Water applied (cm) | Grain Yield (t/ha) | Water use efficiency (kg/ha/cm) | Treatments | Average total water required (cm) | Average grain yield (t/ha) | Water use efficiency kg/ha/cm) | Water productivity (t/ha/cm) |
|---|---|---|---|---|---|---|---|---|---|
| $V_1T_1$ | 118.69 | 137.83 | 4.18 | 86.11 | $T_1$ | 116.25 | 4.71 | 84.34 | 0.029 |
| $V_1T_2$ | 116.08 | 132.83 | 5.10 | 87.38 | | | | | 0.037 |
| $V_1T_3$ | 115.55 | 133.33 | 5.08 | 86.66 | $T_2$ | 113.64 | 5.69 | 85.55 | 0.037 |
| $V_1T_4$ | 115.55 | 130.33 | 5.33 | 86.66 | | | | | 0.039 |
| $V_2T_1$ | 113.82 | 137.83 | 5.24 | 82.57 | $T_3$ | 113.11 | 5.45 | 84.83 | 0.037 |
| $V_2T_2$ | 111.21 | 131.83 | 6.28 | 84.31 | | | | | 0.046 |
| $V_2T_3$ | 110.68 | 131.33 | 5.82 | 84.27 | $T_4$ | 113.11 | 5.27 | 85.79 | 0.043 |
| $V_2T_4$ | 110.68 | 131.33 | 5.21 | 84.27 | | | | | 0.038 |

Water productivity was found to be the highest (0.046 t/ha/cm) in treatment $T_2$ (irrigation when water is 15cm below the soil surface) followed by treatment $T_3$ (0.043t/ha/cm) (irrigation when water is 20cm below the soil surface) and a minimum of 0.029 t/ha/cm treatment $T_1$ (continuous flooding). From these results, it can be seen that the water productivity decreased with the increase of irrigation water.

The observed effects of AWDI on rice cultivation are in agreement with Prior research. Numerous studies focused on manipulating the depth and interval of irrigation water have reported that continuous standing water of rice during cultivation is not essential for obtaining high yields. [11], [12], [13] reported that maintaining a very thin water layer, at saturated soil condition, or alternate wetting and drying can reduce water applied to the field by about 40–70 percent compared with the traditional practice of continuous shallow submergence, without a significant yield loss. A similar result was obtained by [14] concluded that a standing depth of water throughout the season is not needed for high rice yields. They added that about 40–45 percent of the water normally used in irrigating the rice crop in the dry season was saved by applying water in small quantities only to keep the soil saturated throughout the growing season, without sacrificing rice yields. [15] reported increased water productivity (1.26kg/$m^3$) in AWDI plot at 9cm ponding depth compared to continuous flooding ($0.96 kg/m^3$). [16], [17] concluded that in Southern China, AWDI for rice should be more widely used because of its potential to conserve water (20-35% reduction in water use), increase water productivity (from $0.65–0.82 kg/m^3$ to $1.18–1.50 kg/m^3$ after the application of AWDI), increase rice yield (15-28%), and potential to improve the root environment (*i.e.*, soil oxygen content increased by 120-200%). [18], [19] also reported a reduced irrigation water requirement for non-flooded rice by 20–50 percent than for flooded rice, with the difference strongly dependent on soil type, rainfall, and water management practices. [19], however, reported a decrease in rice yields under non-flooded conditions that was proportional to the level of water stress experienced by the plants. There is a concern that the AWDI method of water management promotes greater weed populations, thus requiring more labor for weed management. Association-[20], [21] reported that SRI methodology required approximately 38–54% more labor than conventional methods. According to [22], 62% of the extra labor was needed for weed management while 17% was for transplanting. Alternate drying and wetting of the fields allows for good aeration of the soil and better root growth thereby increasing rice yield and water use efficiency [22]. However, the efficient use of water is the most controversial component in rice farming and also one of the most difficult aspects for farmers to master. In order to achieve the necessary control of water levels, farmers must have a level field and a functioning irrigation system that allows for the precise control of the inflow and outflow of any individual field.

Experiments and field survey of the AWDI method of cultivating rice from different parts of the world have demonstrated the utility of AWDI for water saving in irrigated rice cultivation. This experiment also indicated that Water Productivity Index increased and that land productivity (*i.e.*, yield per unit of land) did not differ from conventionally standing water. This field experiment confirms that AWDI is a sustainable method in irrigated rice cultivation with benefits on water saving and maintaining the productivity comparable to conventional standing water.

The increased productivity of water and its resource saving aspects are likely to be the vital factors that will make farmers and other stakeholders adopt AWDI in water-scarce areas. However, it is difficult to make general conclusions as AWDI methods adopted in a certain area may not transfer to other areas because of variability in topography, soil, and climatic conditions across the rice agro-ecological zones. Therefore, it is important that comparative studies be conducted in different environments to verify this practice as a way to save water under conditions of water scarcity while maintaining, or increasing, crop yields.

# 4. Recommendation

In order to achieve a large-scale spread and adoption of AWD, at least in regions where water scarcity poses a threat to sustain and further improve rice cultivation, a number of constraints and issues at national, regional and local levels have to be overcome as suggested by the findings.

Since the further spread of AWD at this stage depends to a great extent on the actions taken and efforts made at the organizational level to improve and institutionalize the dissemination process in Bangladesh.

Lastly, the study offers some general recommendations and lessons for disseminating natural resource management technologies, based on experiences in Bangladesh, which are specific to the dissemination of AWD technology. AWD dissemination should become a priority/vital issue on the agenda of the National Agricultural Technology Coordination Committee (NATCC), formulate a national Plan of Action of AWD dissemination in Bangladesh, develop strategic partnerships for disseminating AWD, including local government representatives in local processes of AWD dissemination will help to further promote the technology among farmers, design training to fit AWD use in command areas of irrigation systems, improve monitoring and evaluation of AWD dissemination.

# Acknowledgements

The authors greatly acknowledge Asian Development Bank-Japan Scholarships Program (ADB-JSP) for funding support. Sincere thanks are due to the laboratory members of agro-environmental engineering (AEE) and collaborating farmer in Kashiwa, Chiba, Japan for their active support and cooperation throughout the field experiment.

# References

[1] S.I. Bhuiyan, "Water management in relation to crop production: Case study on rice" Outlook on Agriculture. 1992. vol. 21, pp293-299.

[2] P.K. Sharma, "Effect of Period Moisture Stress on Water-use Efficiency in Wetland Rice" Oryza, 1989, vol. 26, pp252-257.

[3] W.D. Ruckelshaus, "Toward a sustainable world" Scientific American, 1989 vol. 261,3, pp.114 - 120

[4] Government of Western Australia, "Hope for the Future: The Western Australian State Sustainability Strategy". Department of the Premier and Cabinet, Perth, Western Australia, 2003.

[5] P. Rogers, K.F., Jalal and J.A. Boyd, "An Introduction to Sustainable Development, 2nd Ed., 2000, Earthscan, London, UK.

[6] L. Feng, B.A.M. Bouman, T.P. Tuong, R.J. Cabangon, Y. Li, G.. Lu and Y. Feng "Exploring options to grow rice using less water in northern China using a modelling approach: I. Field experiments and model evaluation" Agricultural Water Management, 2007, Vol. 88, Issues 1–3, 16 March 2007, pp.1–13

[7] B.A.M. Bouman, L. Feng and T.P. Tuong, G.L.H. Wang and Y. Feng, "Exploring options to grow rice under water-short conditions in northern China using a modeling approach. II: Quantifying yield, water balance components, and water productivity" Agricultural Water Management 2007, vol. 88, pp.23-33.

[8] A.M. Shamsuddin, M.A. Islam and A. Hossain, "Comparative Studies on the Yield and Agronomic characters of nine Cultivars of Aus Rice" Bangladesh J. Agril. Sci., 1988 vol. 15,1 pp. 121-124.

[9] M.J.U. Chowdhury, A.U. Sarker, M.A.R. Sarker and M.A. Kashem, "Effect of variety and number of seedling hill$^{-1}$ on the yield and its components on late transplanted aman rice" Bangladesh J. Agric. Sci., 1993, vol. 20, pp. 311–316.

[10] M.A. Hassan, M.A. Salam and M.R. Arsan, "Influence of certain factors on severity of stalk rot disease of grain sorghum in Upper Egypt. Aust J Agric Sci. 1996; vol. 27, 179–189.

[11] S. Hatta, Water consumption in paddy field and water saving rice culture in the tropical zone. Japan Journal of Tropical Agriculture, 1967, vol.11, pp. 106-112.

[12] D.F. Tabbal, R.M. Lampayan, and S.I. Bhuiyan, "Water-efficient irrigation technique for rice. In V. V. N. Murty, & K. Koga (Eds.), Soil and water engineering for paddy field management" Proceedings of the International Workshop on Soil and Water Engineering for Paddy Field Management, 1992, 28-30 January. Bangkok, Thailand: Asian Institute of Technology.

[13] C.B. Singh, T.S. Aujla, B.S. Sandhu, and K.L. Khera, "Effects of transplanting data and irrigation regime on growth, yield and water use in rice (*Oryza sativa*) in northern India" Indian Journal of Agricultural Sciences, 1996, vol. 66, pp.137-141.

[14] S.I. Bhuiyan and T.P. Tuong, "Water use in rice production: Issues, research opportunities and policy implications. Paper presented at the Inter-Center Water Management Workshop", 29-30 September 1995.Colombo, Sri Lanka: International Irrigation Management Institute. Geneva: 1996, World Health Organization.

[15] V. Anbumozhi, E. Yamaji and T. Tabuchi, "Rice crop growth and yield as influenced by changes in ponding water depth, water regime and fertigation level" Agricultural Water Management, 1998, 37, 241- 253. doi:10.1016/S0378-3774(98)00041-9, http://dx.doi.org/10.1016/S0378-3774(98)00041-9

[16] Z. Mao, "Principle and technique of water saving irrigation for rice" Hubei Province, People's Republic of China: 1993 Wuhan University of Hydraulic and Electric Engineering.

[17] Z. Mao, "Environmental impact of water-saving irrigation for rice" In Irrigation scheduling: From theory to practice. Proceedings of the ICID/FAO Workshop on Irrigation Scheduling, Rome, Italy, 12-13 September 1995. Rome: FAO, 1996.

[18] S. Keisuke, E. Yamaji, S. Sato, P.S. Budhiharto and M. Mizoguchi, "Sustainability of System of Rice Intensification: Benefits of SRI focusing on effects of intermittent irrigation on yield increase and water savings" In: Proceeding of PAWEES 2007 6th International Conference on Sustainable Rural Development and Management 18 October 2007. South Korea: Seoul National University, pp. 25-37.

[19] G. Davids "Use of the Capital Intensive Irrigation Systems in Rice" Davids Engineering, 1998 Inc. California, USA.

[20] Association-Tefy-Saina, "Couts de Revient Dans les Systmes de Riziculture Divers" Association Tefy Saina, 1995, National Workshop on SRI, Antananarivo.

[21] H.W. Rakotomalala, "Comparison entre la Rizicullture Traditionnelle et le Systeme de Riziculture Intensive dans La Region de Ranomafana Science Agronomique" Antananarivo: Universite d'Antananarivo. September 2011 ISSN 1916-9752 E-ISSN 1916-9760, Journal of Agricultural Science, 1997, .vol. 3, 3.

[22] N. Uphoff, "The System of Rice Intensification (SRI) as a methodology for reducing water requirements in irrigated rice production. Paper for International Dialogue on Rice and Water: Exploring Options for Food Security and Sustainable Environments" held at IRRI, 7-8 March 2006, Los Baños, Philippines.

# Evapotranspiration and water stress estimation from TIR and SWIR bands

**Girolimetto Daniela, Venturini Virginia**

Centro de Estudios Hidro Ambientales, Facultad de Ingeniería y Ciencias Hídricas, Universidad Nacional del Litoral, C. C. 217, Santa Fe, (3000) Argentina

**Email address:**
dgirolimetto@fich.unl.edu.ar (G. Daniela), vventurini@fich.unl.edu.ar (V. Virginia)

**Abstract:** The World agriculture depends on water availability; thus, a successful water management system would assure food for the World. For several decades, the scientific community has developed methods to support water management. These models include the estimates of the main water loss in the system, i.e. the evapotranspiration (ET). In turn, the satellite technology encouraged the development of new models to monitor large regions. In this work, we present a modified ET estimation adapting the F parameter introduced by Venturini et al., in 2008. Additionally, a new simple index to estimate water stress (WS) for different types of surfaces, is also presented. The relative evaporation represented by F is derived from the soil moisture condition following the formulation of Barton and computed from the surface reflectance in the shortwave infrared bands (SWIR). The new ET and WS equations are applicable, with different satellite datasets, to any remote region since they are based on universal relationships. The preliminary results show errors of about 11% in ET. In general, the new WS index would have values of approximately 0.8 for a dry surface and 0.4 for a wet surface.

**Keywords:** Evapotranspiration, Water Stress, Remote Sensing, TIR, SWIR

## 1. Introduction

The evapotranspiration (ET) is the most significant water withdrawal in the water balance. Therefore, in recent years, the scientific community has given special attention to the ET impact on global circulation models and has devoted efforts to estimate how much water is globally evaporating [1]. Even today, a precise ET calculation is a real challenge for water resources practitioners. Moreover, the need to monitor large regions has motivated the development of new methods for calculating ET from remote sensing data. Thus, diverse ET models have been developed and widely applied with varying results [e.g. 2-8]. The importance of ET in the global water balance was remarked by the satellite mission Aquarius/SAC-D, a cooperation program between the NASA and Argentina. This mission seeks to determine how the ocean responds to the combined effects of evaporation, precipitation, ice melt and river runoff, in order to analyze their impact on the global water distribution as well as the global freshwater availability (http://aquarius.nasa.gov/).

The world agriculture depends on the availability of water, being the water management one of the main components of the success or failure of modern agriculture. Thus, indices to determine water stress (WS) of vegetation have been widely used to assist farmers in maximizing production. Many of them have been developed using remotely sensed data, allowing the WS estimation in large regions to be made [9-13]. For instance, Moran et al., [9] published a definition of WS in terms of latent heat fluxes, which can be expressed as $WS=1-(ET/E_{pot})$, where $E_{pot}$ is the potential evapotranspiration and the ratio $ET/E_{pot}$ is the relative evaporation. This formulation emphasizes the importance of an accurate ET calculation, not only for water management systems and risks but also for the agriculture and related food industries.

Even though there is a large set of ET models, we focussed on Venturini et al.'s method [8] because it presents interesting advantages. The authors presented a new formulation to derive ET maps from remotely sensed data without applying auxiliary or site-specific relationships. The method was based on Granger's complementary relationship and Priestley–Taylor equation to eliminate the wind function and the resistance parameters, by including the relative evaporation concept (F). The method was applied in different environments with errors of about 15% from the mean ET and proved to be better than similar ET models [14].

In this paper, we propose a simple methodology that relates the F concept to the surface moisture availability concept defined by Barton [15]. This new calculation exploits the water absorption properties of the SWIR region. The SWIR energy is absorbed by water; therefore, these bands are sensitive to the surface moisture content variations [16] and had been related to the concept of relative humidity of a surface. This new F concept is used in Venturini et al.'s method to compute ET and to derive a new WS index.

## 2. Methodologies

In this session, Venturini et al.'s method is briefly presented before the rationale behind the F parameter, and the new WS index is introduced.

### *2.1. Venturini et al.'s Method*

Venturini et al., modified the Priestley–Taylor (P-T) model [17] using Granger's complementary relationship [18] and the relative evaporation, defined as the ratio between ET and $E_{pot}$. The relative evaporation proposed by Granger and Gray [3] was used for defining a coefficient F, assuming that the wind function affects ET and $E_{pot}$ in a similar way. The authors also assumed that saturation and actual vapor pressures can be calculated from the saturation vapor pressure curve (SVP) with temperature data. Thus, $F=ET/E_{pot}$ can be expressed as

$$F = \frac{ET}{E_{pot}} = \frac{e_s - e_a}{e_s^* - e_a} \cong \frac{(T_u - T_d)\Delta_1}{(T_s - T_d)\Delta_2} \quad (1)$$

where $e_s$ is the surface actual water vapor pressure, $e^*_s$ is the surface saturation water vapor pressure, $e_a$ is the air actual water vapor pressure, $T_s$ is the surface temperature and $T_u$ is the temperature of the surface if it is brought to saturation without changing the actual surface vapor pressure, which is analogous to the dew point temperature ($T_d$) definition [8]. Therefore, $E_{pot}$= ET/F (see equation 1) combined with Granger´s complementary equation renders ET as a function of the wet environment evapotranspiration ($E_w$), as follows,

$$ET = E_w \frac{F(\Delta + \gamma)}{F\Delta + \gamma} \quad (2)$$

where $\gamma$ is the psychrometric constant, $\Delta$ is the slope of the SVP curve, and $E_w$ is the wet environment evapotranspiration.

The Priestley and Taylor equation [17] was used to compute $E_w$. Consequently, by combining the Priestley–Taylor expression with equations (2) and (1), the following model (hereinafter named as $ET_{V-V}$) was derived

$$ET_{V-V} = \alpha \frac{F\Delta}{F\Delta + \gamma}(R_n - G) \quad (3)$$

where $\alpha$ is a P-T parameter, assumed to be equal to 1.26 for saturated surfaces, $R_n$ is the net radiation, and G is the soil heat flux.

The new variable introduced in this formulation, $T_u$, is necessary to determine $e_s$ and then F (see equation 1). The authors determined $T_u$ from the SVP curve using $T_s$ and $T_d$ (see Figure 1); however, they recognize that the $T_u$ calculation did not follow the physics of the problem [19]. A full discussion regarding the definition and calculation of $T_u$ can be found in [8] and [19].

Once $T_u$ is estimated, $e_s$ can be computed, and then F and ET can be calculated.

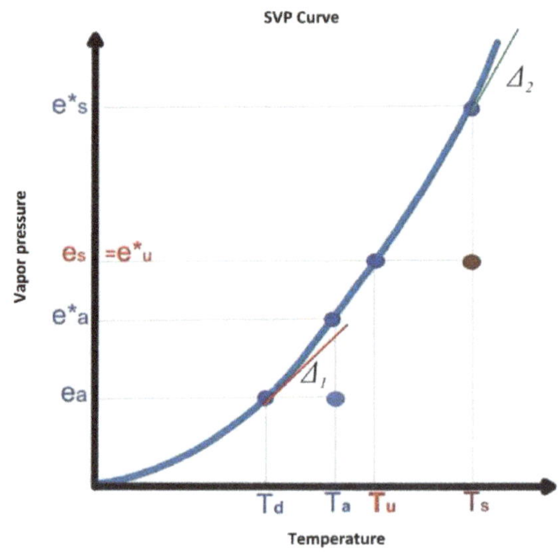


*Figure 1. SVP curve $T_u$, $T_s$, $e_s$, $e^*_s$ relationship in de $T_s$ context.*

### *2.2. The New Method to Estimate $e_s$*

The $e_s$ estimation could be enhanced by using a surface variable, such as the surface moisture content. The new $e_s$ calculation involves the surface water availability proposed by Barton [15], as follows:

$$\sigma = \frac{e_s}{e_s^*} \quad (4)$$

where σ is an indicator of the near surface moisture availability [15].

Barton analyzed the ground evaporation data in terms of σ that is directly related to the soil moisture (SM) content. Hence, the author empirically related σ to the bare SM data for the Deniliquin (Australia) region that are calculated with data from active microwave sensors. The author proposed the following relationship:

$$\sigma = \frac{1.8SM}{SM + 30} \quad SM < 37.5 \quad (5)$$

$$\sigma = 1 \quad SM > 37.5 \quad (6)$$

where SM is the soil moisture content (in percentage).

Deniliquin is a semi-arid region where the bare soil surface saturates with moisture contents higher than 37.5%. The soil is clayey with a patchy vegetation pattern dominated by mesophytic species forming a relatively dense sward. In this region, the mean annual rainfall and evaporation amount to 40 cm and 161 cm respectively and the mean temperature varies from 40 °C in July and 24 °C in January. Therefore, equations (5) and (6) were derived from site-specific relationships and were difficult to extrapolate to other areas. Barton derived σ for bare soil surfaces and it is not evident how it varies for mixed soil-vegetation. Nevertheless, Barton stated that this parameterization is convenient because it can be simply incorporated in an ET model and that the formulation would not need the determination of either the surface temperature or the nature of the surface.

From the remote sensing point of view, the SWIR energy is absorbed by water; therefore, these bands are sensitive to the surface moisture content variations [16]. Many authors combined the NIR and SWIR reflectance to study changes in foliar water content [20-27, 16, 28-29]. Fensholt and Sandholt in 2003 [24] explained that even though the atmosphere, the leaf/surface internal structure and the dry matter content might affect the vegetation SWIR reflectance, their effects are negligible compared with water absorption. Simulations of the leaf reflectance of SWIR bands showed that over 50% of the changes in reflectance of the SWIR region are due to the absorption caused by the water content in the vegetation [30].

The strong water absorption occurring at wavelengths > 1.0 μm makes the mix surface moisture the main cause for the SWIR variation. Thus, the SWIR reflectances of the soil and vegetation are negatively related to the moisture [26, 31]. Consequently, considering the surface as a vegetation-soil complex and assuming that the decrease of the reflectance in the SWIR is essentially due to the water content of the surface, we approximate σ as:

$$\sigma = \frac{Rsat}{Ri} \tag{7}$$

where Rsat is the SWIR reflectance of a saturated surface and Ri is the SWIR reflectance of any $i^{th}$-pixel.

Rsat is easily obtainable from the relation between Ri and SM, where ground data are available. Otherwise, it can be calculated from the mean Ri of the pixels identified as water in an image. In any case, the values of Ri close to zero would be associated to saturated pixels and can be considered as Rsat. This point is further explained in session 4.

If Rsat is known, it is possible to obtain σ and apply equation (4) to estimate $e_s$ ($e_s=\sigma e^*_s$). Introducing $e_s$ in equation (1),

$$F = \frac{\sigma e_s^* - e_a}{e_s^* - e_a} \tag{8}$$

Thus, ET is estimated from equation (3) with the new F that incorporates the physically consistent $T_u$ estimation.

### *2.3. New Water Stress Index WSIF*

ET is commonly used to index the vegetation stress, as well as to estimate the volume of water needed for irrigation. We used the new ET model in a water stress index, following Moran et al.'s concepts [9].

Moran et al. discussed the validity of the CWSI theory for partially vegetated areas. They derived an index based on the interpretation of the trapezoidal shaped IV-($T_s$-$T_a$) plot, where $T_a$ is the air temperature.

The WDI ponders two important assumptions associated to the relationship between IV and the difference $T_s$-$T_a$. The authors assume that the difference $T_s$-$T_a$ is linearly related to the percentage of vegetated area and to the canopy and soil temperatures. Another important statement made by Moran et al. is that, given a certain net energy ($R_n$), the temperatures of the foliage and soil are linearly related to the transpiration and evaporation respectively. Therefore, the variations in $T_s$-$T_a$ would be associated to ET and (for a partially vegetated area) WDI is:

$$WDI = 1 - \frac{ET}{E_{pot}} \tag{9}$$

where ($ET/E_{pot}$) is the relative evaporation.

Thus, it is possible to rewrite Moran et al.'s formulation replacing $ET/E_{pot}$ by F (see equation 1), as follows:

$$WSI_F = 1 - F = 1 - \frac{(e_s - e_a)}{(e_s^* - e_a)} = \frac{(e_s^* - e_s)}{(e_s^* - e_a)} \tag{10}$$

Knowing that $e_s = \sigma e^*_s$, with σ determined from the SWIR reflectance (see equation 4), $WSI_F$ can be re-written as follows,

Alpha

$$WSI_F = \frac{(e_s^* - e_s)}{(e_s^* - e_a)} = \frac{(e_s^* - \sigma e_s^*)}{(e_s^* - e_a)} \tag{11}$$

The new index tends to zero when σ tends to 1, i.e. for a saturated area, $e_s$ tends to the saturation vapor pressure ($e^*_s$) [see Figure 1 and equation (11)]; therefore, $WSI_F$ would tend to zero indicating the non-stress condition. On the contrary, when σ tends to zero, $e_s=\sigma * e^*_s$ tends to $e_a$ [see Figure 1 and equation (11)] and $WSI_F$ would be close to 1, reflecting a full stress condition.

## 3. Study Area and Data

### *3.1. Study Area*

The Southern Great Plain (SGP) region extends over the State of Oklahoma and southern part of Kansas, running from longitude 95.5° W to 99.5° W and from latitude 34.5° N to 38.5° N. It is a flat terrain, with heterogeneous land cover and a wide variety of weather throughout the year. Winters are very cold and summers are very hot and humid. The prairies support abundant wildlife.

This region has a relatively extensive and well distributed coverage of ground Energy Fluxes and Bowen stations, maintained by the Atmospheric Radiation Measurement (ARM) program (http://www.arm.gov). The stations are widely distributed over the whole domain (Figure 2). The sites are named according to the instruments located in each facility and numbered sequentially as each new center is opened; for example, "E5" refers to the extended facility site number 5. Extended facility instruments include an Energy Balance Bowen Ratio system (EBBR), which produces 30 minute estimates of the vertical fluxes of sensible and latent heat at the local surface.

In this work, we used data from the following stations: station E8 and E22, located in a grazed rangeland region; E4, located in an ungrazed rangeland area; E13, located in a region with pasture and wheat; E7, E9, E15, E20 and E27, located in pastures; E18 and E19 located in an ungrazed pasture area; E12, located in a native prairie; E10 located in an alfalfa region; E16 located in a wheat region; and E2 located in a grass region.

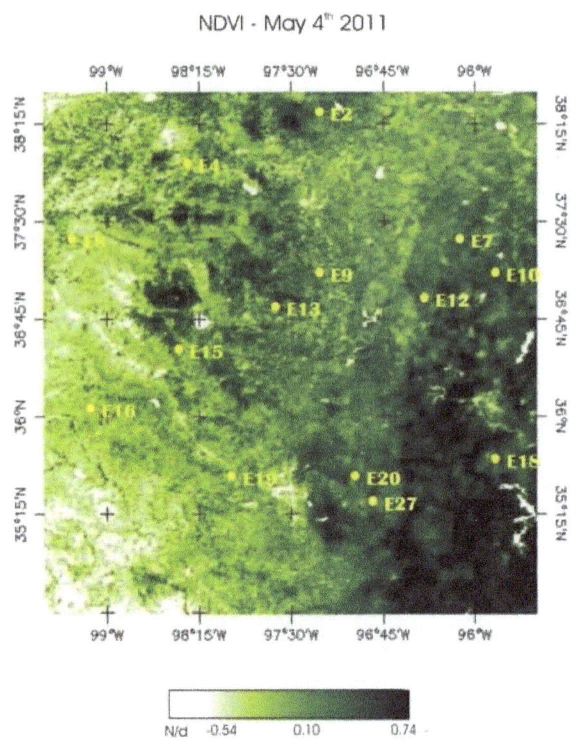


*Figure 2. Southern Great Plains Region and ground station locations.*

### *3.2. Data and Image*

The Energy Balance Bowen Ratio (EBBR) system computes 30-min estimates of the sensible and latent heat vertical fluxes at the local scale. The fluxes from the EBBR stations are calculated with observations of $R_n$, G, and the vertical gradients of temperature and relative humidity (RH). The instruments and measurement applications are well established and have been used for validation purposes in many studies [32-33, 8, 34]. Further information about the ARM EBBR data and methodology is available at http://www.arm.gov

The imageries used in this work were registered by the MODIS sensor. MODIS is one of the instruments on board EOS-Terra and EOS-Aqua satellites [http://modis.gsfc.nasa.gov/, 35-36].

Daytime MODIS-Aqua images for nine days during spring and summer, with at least 80% of the study area free of clouds, were selected. Table 1 summarizes the image information, including date, day of the year, satellite overpass time and image quality. MYD02, MYD07 and MYD11 were the three products needed for this application.

The MYD02 has corrected radiance, reflectance and geolocations for 36 bands at 1x1 km. The MODIS Atmospheric profile (MOD07) offers several parameters. Air temperature ($T_a$) and $T_d$ profiles were used in the current study. The spatial resolution of this daily product is 5×5 $km^2$ at 20 vertical atmospheric pressure levels [37]. This product was already used in a similar work by Bisht et al.'s [38] and Venturini et al.'s [8, 19]. Bisht et al.'s analyzed the agreement between MOD07 and $T_a$ and observed $T_a$ in the SGP, concluding that MOD07 is a good surrogate of field observations of $T_a$. The MYD11 product supplies $T_s$ images on a daily basis [39, 40, 19].

*Table 1. Date, day of the year, overpass time and image quality of the nine study days*

| Date | Day of the year (DOY) | Overpass time (UTC) | Image quality (% clouds) |
|---|---|---|---|
| April 6th 2011 | 96 | 19:30 | 15 |
| May 4th 2011 | 124 | 19:55 | 1 |
| May 26th 2011 | 146 | 19:20 | 2 |
| May 29th 2011 | 149 | 19:50 | 1 |
| June 5th 2011 | 156 | 19:55 | 14 |
| April 10th 2010 | 100 | 19:40 | 5 |
| June 4th 2010 | 155 | 19:45 | 18 |

## 4. Results

### *4.1. Preprocessing*

The MODIS images were georeferenced from the latitude and longitude provided in each product. $T_a$ and $T_d$ images corresponded to the vertical pressure level of 950 hPa, the profile closer to the surface in MYD07. It is worth mentioning that $T_a$ and $T_d$ are assumed to be homogenous over the area of 5 $km^2$ [37], while $T_s$ from the MYD11 product is provided in pixels of 1 $km^2$. The study area was pulled out of each image and projected on a grid of 445 columns by 445 rows, with a resolution of approximately 1 km per pixel.

The images of the SWIR band at 2130 nm (R7) were used to approximate $\sigma$. R7 was previously corrected for atmospheric effects following the dark object method [41].

$e^*_s$ and $e_a$ were estimated with Buck's equation [42], using the $T_s$ and $T_d$ images respectively. Finally, ET and $WSI_F$ were computed for the SGP region.

### 4.2. *Analysis of σ*

The Rsat parameter must be estimated to obtain σ from equation (4). In this work, Rsat was attained from R7 and SM ground data, in gravimetric percentage.

EBBR-SM data at 5 cm were selected because they were the nearest to the surface. Figure 3 shows the relationship between SM data and the R7 of the pixel where the EBRR station is located. These data indicate that R7 turns out to be asymptotic at 0.06 for SM values greater than 25%, suggesting the saturation condition of the surface for this region.

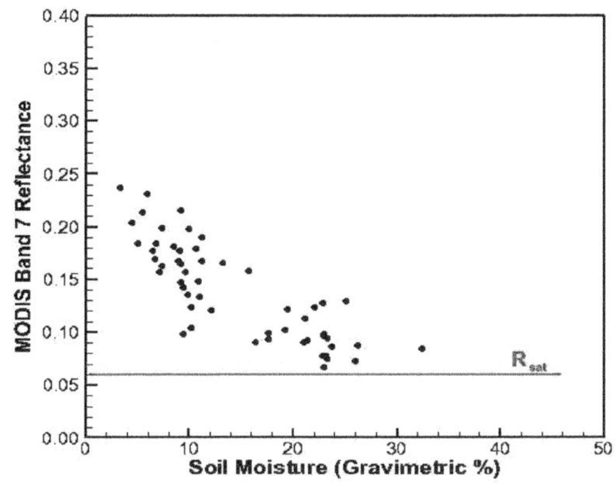


***Figure 3.*** *Ri versus soil moisture (gravimetric %) plot where Rsat limit is indicated.*

This value is consistent with the regional water reflectance (RWR) in the SWIR band (Table 2) obtained as $RWR=(\sum R_d^{water})/n$, where $R_d^{water}$ is the mean reflectance of the pixels identified as water (NDVI <0) in the SWIR band on a given day, and n is the total of the days analyzed.

***Table 2.*** *Water reflectance in the SWIR band for each day analyzed and RWR.*

| Date | $R_d^{water}$ |
|---|---|
| April 6th 2011 | 0.049 |
| May 4th 2011 | 0.044 |
| May 26th 2011 | 0.082 |
| May 29th 2011 | 0.034 |
| June 5th 2011 | 0.095 |
| April 10th 2010 | 0.052 |
| June 4th 2010 | 0.065 |
| $RWR=(\sum R_d^{water})/n$ | 0.059 |

***Table 3.*** *Regional Minimum, Mean and Standard deviation of σ.*

| Date | Min | Mean | Standard deviation |
|---|---|---|---|
| April 6th 2011 | 0.15 | 0.40 | 0.077 |
| May 4th 2011 | 0.16 | 0.37 | 0.098 |
| May 26th 2011 | 0.20 | 0.55 | 0.178 |
| May 29th 2011 | 0.16 | 0.40 | 0.132 |
| June 5th 2011 | 0.14 | 0.38 | 0.140 |
| April 10th 2010 | 0.17 | 0.41 | 0.076 |
| June 4th 2010 | 0.17 | 0.44 | 0.112 |
| June 5th 2009 | 0.22 | 0.58 | 0.150 |
| August 22th 2009 | 0.18 | 0.59 | 0.171 |

Then, σ was obtained for each of the days tested with Rsat and R7 of each pixel for each selected day.

The regional statistics of σ (means, minimums and standard deviation) are shown in Table 3.

The maximum σ value is approximately equal to 1 for all the days studied, which suggests that there were saturated pixels during the studied period. The minimum σ value varies from 0.14 to 0.22, indicating that there may be pixels with little SM. In general, the σ regional mean value is larger than 0.4, indicating that the region is drying; the standard deviations vary from 0.076 to 0.178, which is consistent with different degrees of dispersion around the mean.

Barton [15] did not reported values of σ for mixed soil-vegetation surfaces because σ is difficult to assess in mixed surfaces. Thus, we had to establish a criterium for the saturation condition of a soil+vegetation pixel. Hence, we assumed that a value of SM greater than 25% represents the saturation limit for a mixed surface and that σ values greater than 0.70 would indicate a saturated surface. It is worth noting that these limits may vary with the soil and vegetation type and that the index range interpretation may vary from region to region.

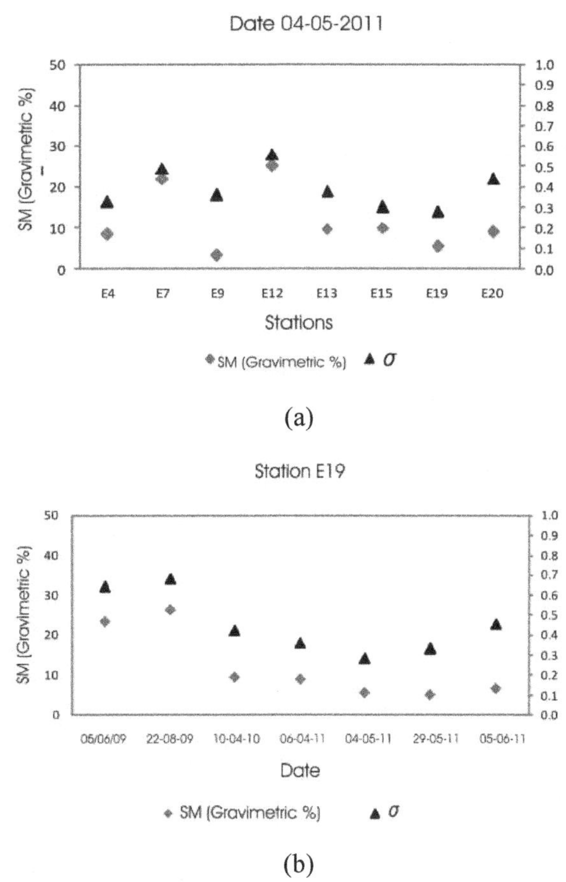


***Figure 4.*** *(a)* σ versus *soil moisture (SM) for day 04/05/2011 and (b)* σ versus *SM for station E19.*

The results obtained on 04/05/2011, with only 1% of cloud cover, were selected to exemplify the general regional variations. Figure 4(a) shows the areal distribution of SM and $\sigma$ for the day 04/05/2011, where a good general agreement is observed. Station E19 was selected to exemplify the time variation because the data were available in 80% of the days analyzed. Figure 4(b) displays a temporal variation of SM and $\sigma$ for station E19. Both variables ($\sigma$ and SM) seem to follow the same variation over time, although SM is a point measurement and $\sigma$ is the response of a mixed pixel of approximately 1 $km^2$.

### 4.3. Validation of ET Modified from $\sigma$ ($ET_{G-V}$)

The ET estimates ($ET_{modeled}$) computed from Venturini et al.'s original and modified equations were contrasted with observed ET ($ET_{ground}$). The validation procedure presented in this session was widely used in others works [4, 43, 5, 38, 44, 8, 19].

The regional maximum, minimum and average ET values are shown in Table 4; clearly, $ET_{V-V}$ and $ET_{G-V}$ methods yielded similar results. The mean ET ground data differs from the mean ET models in 13.10 $Wm^{-2}$ and -14.69 $Wm^{-2}$ for $ET_{V-V}$ and $ET_{G-V}$ respectively. While $ET_{V-V}$ overestimates the ET, $ET_{G-V}$ underestimates the regional observed ET from field measurements. However, these differences are comparable to those presented by other authors [43-44, 8].

***Table 4.*** *Overall Maximum, Minimum and Mean* $ET_{V-V}$, $ET_{G-V}$ *and* $ET_{ground}$ *for 54 sample pixels.*

| ET model | Max ($Wm^{-2}$) | Min ($Wm^{-2}$) | Mean ($Wm^{-2}$) |
|---|---|---|---|
| $ET_{ground}$ | 544.63 | 180.18 | 350.62 |
| $ET_{V-V}$ | 478.97 | 196.63 | 363.72 |
| $ET_{G-V}$ | 492.91 | 185.48 | 335.93 |

The bias, the root mean square error (RMSE) and the correlation coefficient (r) were used to analyze the soundness of the new model. The bias was calculated as $[\sum(ET_{ground}-ET_{modeled})]/n_{obs}$ and the RMSE as $[\sum(ET_{ground}-ET_{modeled})^2/n_{obs}]^{0.5}$, where $n_{obs}$ is the number of observations (54 points). The $ET_{V-V}$ method presented a RMSE of approximately 19% from the observed ET values. The bias was 4%. The $ET_{G-V}$ methodology yielded a RMSE of 39.92 $Wm^{-2}$ and -14.69 $Wm^{-2}$ for the bias, which represent 11% and 4% from the observed mean ET respectively. The correlations were r =0.92 for $ET_{G-V}$ and r=0.74 for $ET_{V-V}$.

The $ET_{G-V}$ model would imply a significant improvement in ET estimates with remotely sensed data. For instance, Venturini et al.'s in 2008 [8] published values for the RMSE and the bias of about 18% and 15% from the observed mean ET for the same region. Kalma et al.'s [45] conducted a thorough analysis where results from about 30 published ET validation studies were summarized. These authors reported RMSE values of about 50$Wm^{-2}$ and relative errors of about 15 to 30%. Long and Singh [46] recently published RMSEs of 45.6 $Wm^{-2}$ and 63.1 $Wm^{-2}$ using Landsat TM/ETM+ and ASTER images respectively.

Table 5 shows the RMSE and the bias for each analyzed day for $ET_{V-V}$ and $ET_{G-V}$ models. In general, the RMSEs for the $ET_{G-V}$ model are lower than 13% from the mean ET of each day; the biases are up to 8% from the observed mean ET. Only the day 04/06/2010 presented a RMSE of 18% with a bias of the same order. The quality of the image of day 04/06/2010 is not very good and few stations are free of clouds, which may cause uncertainties in the imagery. The $ET_{V-V}$ method presents RMSE maximum values of approximately 30% from the observed ET mean.

These preliminary results suggest that the $ET_{G-V}$ model correlates better with the measurements than the $ET_{V-V}$ model.

The state of the surface moisture introduced by $\sigma$ is consistent with the $T_u$ and $e_s$ definition; consequently, $\sigma$ improves the results of the $ET_{V-V}$ model. The new $ET_{G-V}$ method yields precise ET estimates while it is simple to apply with remotely sensed data and easily coded for routinary applications without the user supervision.

***Table 5.*** *ET ($Wm^{-2}$) comparison between observations and proposed method estimates for each day analyzed and the global RMSE and bias during the study period.*

| Date | # of obs. | $ET_{V-V}$ | | $ET_{G-V}$ | |
|---|---|---|---|---|---|
| | | RMSE | Bias | RMSE | Bias |
| April 6th 2011 | 6 | 96.52 | 28.74 | 44.34 | 23.07 |
| May 4th 2011 | 6 | 98.84 | 70.95 | 36.03 | -23.83 |
| May 26th 2011 | 4 | 20.95 | 10.84 | 16.97 | 7.01 |
| May 29th 2011 | 6 | 70.63 | 3.54 | 39.15 | -23.77 |
| June 5th 2011 | 8 | 55.42 | 24.16 | 36.78 | -18.82 |
| April 10th 2010 | 5 | 44.67 | 42.42 | 29.63 | -8.75 |
| June 4th 2010 | 4 | 51.69 | -19.97 | 84.05 | -74.97 |
| June 5th 2009 | 8 | 65.95 | -24.67 | 31.53 | -19.95 |
| August 22th 2009 | 7 | 34.24 | -11.94 | 25.85 | -2.92 |
| Global RMSE and bias | 54 | 65.89 | -13.10 | 39.92 | 14.69 |

### 4.4. $WSI_F$ Results

Knowing that $ET_{G-V}$ improves the ET estimates, we then tested the new $WSI_F$ presented in equation (11).

The $WSI_F$ regional average values are greater than 0.50 in all dates analyzed, suggesting that the region is within a moderately stressed period. These results are comparable with those found by Girolimetto and Venturini in 2013 [13], who estimated water stress from the NDVI-$T_s$ plot in the same region and days analyzed. The $WSI_F$ standard deviation value varies between 0.097 and 0.234, which is consistent with different degrees of dispersion around the mean and different regional surface moisture distributions.

Since WS is not directly measurable, a validation strategy was needed. Thus, ground values, $X_{obs}$, were obtained as $1-(ET_{obs}/E_{pot\_Samani})$ and $1-(ET_{obs}/E_{pot\_Makkink})$. The RMSE was calculated as $[\Sigma(X_{obs}-WSI_F)^2/n]^{0.5}$. Samani [47] and Makkink's equations [48] were chosen for their simplicity and little field data requirement. The difference between $1-(ET_{obs}/E_{pot\_Samani})$ and $WSI_F$ showed a bias of -0.046 and a RMSE of 0.094, which represents 19% from the mean value of WS. Comparison between $1-(ET_{obs}/E_{pot\_Makkink})$ and $WSI_F$ showed a bias and a RMSE of -0.083 and 0.120 respectively, consistent with Colaizzi et al.'s results [10]. In all cases, the RMSEs are of the same order, indicating that errors of the methods are about 25-30% from the average WS.

$WSI_F$ was also contrasted with observations of the SM, soil temperature ($T_{soil}$), precipitation of the previous 5 days (PP) and $ET_{obs}$ to analyze the sensitivity of the index to capture different surface conditions. The relationship between the $WSI_F$ and SM observations (at 5 cm from the surface) is shown in Figure 5. The relationship presented a correlation coefficient of 0.80 with the set of data available. This value is consistent with those published by other authors [49,24].

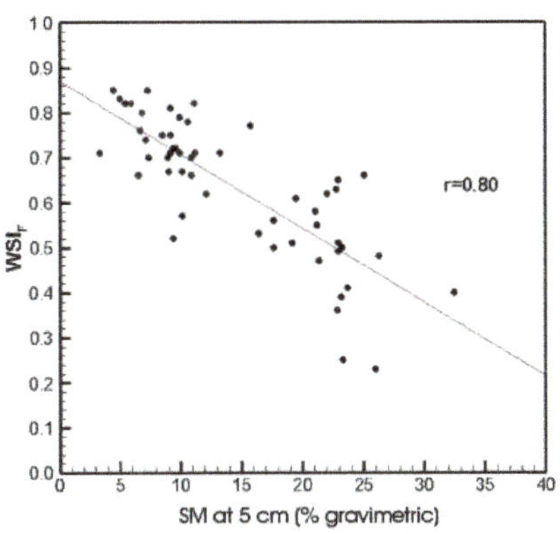


***Figure 5.*** *$WSI_F$ and SM ground observations relationship.*

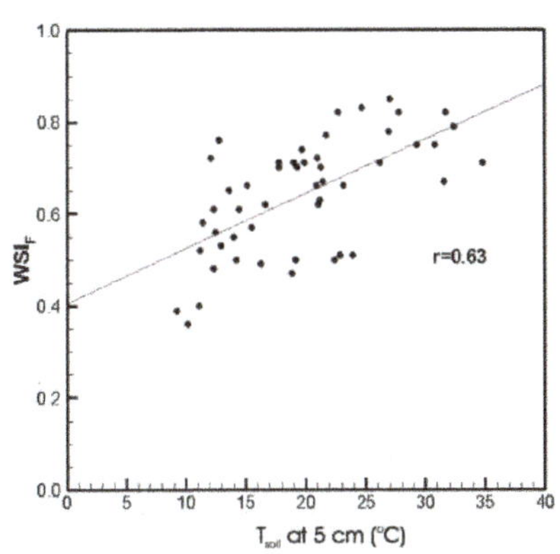


***Figure 6.*** *$WSI_F$ and SM and $T_{soil}$ observations relationship.*

The relation between $WSI_F$ and $T_{soil}$ at 5 cm (Figure 6) yielded r=0.63. The soil temperature and WS vary with the soil water content, i.e. when the water content in soil decreases, WS and $T_{soil}$ increase.

The relationship between PP and $WSI_F$ shown in Figure 7 presents an r of 0.84, indicating that $WSI_F$ is sensitive to variations in PP.

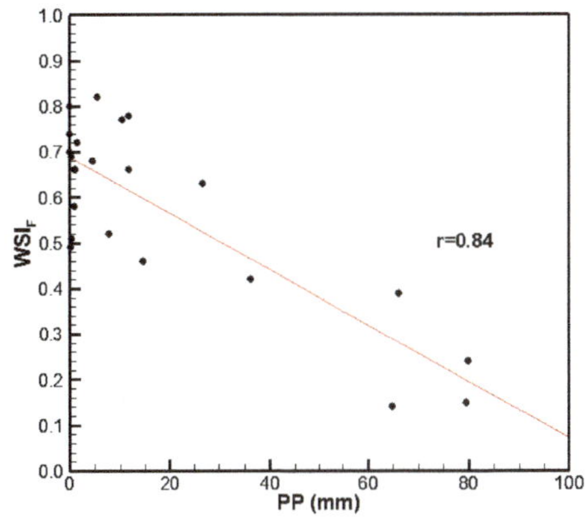


***Figure 7.*** *$WSI_F$ and PP observations relationship.*

Finally, the relationship between $WSI_F$ and $ET_{obs}$ presented in Figure 8 yielded a correlation of 0.71. The new index has a well-defined inverse relationship with $ET_{obs}$ in the study area. $WSI_F$ values of about 0.8 are related to values of ET of about 200 $Wm^{-2}$, while unstressed surfaces ($WSI_F \sim 0.3$) evapotranspire at a rate of 500-600 $Wm^{-2}$.

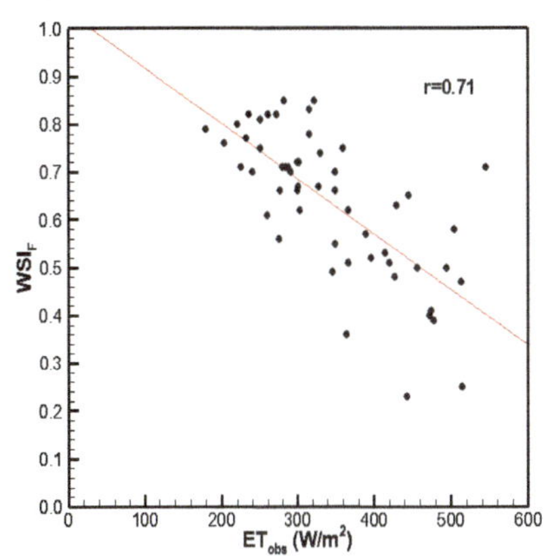


***Figure 8.*** *$WSI_F$ and $ET_{obs}$ relationship*

# 5. Summary and Conclusion

In this paper, we propose a new simple methodology to obtain the relative evaporation parameter (F) introduced by Venturini et al. in 2008. This new F is derived from the soil moisture σ parameter defined by Barton [15] as $e_s/e^*_s$, which in turn can be obtained from the surface reflectance in the SWIR bands. SWIR energy is absorbed by water; therefore, these bands are basically sensitive to variations in moisture content of a surface, although there are other factors that might affect the band response. Physically speaking, $e_s$ and the reflectance are inversely proportional to the surface moisture; accordingly, we propose to obtain σ as the relationship between the reflectance in the SWIR of the saturated pixel (Rsat) and the reflectance of the pixel i (Ri). Rsat can be estimated from the Ri *vs* SM curve or from the average value of the pixels identified with NDVI < 0. In any case, Rsat would tend to the minimum minimonum of R7.

The spatial and temporal analysis showed that the σ variation represents the SM data despite the different scales of observation and the pixel size. SM observations are valid for a few centimeters, the moisture being measured below the soil surface, and σ is the response of a mixed pixel of approximately 1 $km^2$.

As mentioned, σ enhanced the methodology to obtain the main parameter of Venturini et al.'s method. $ET_{G-V}$ was compared with ET observed values and $ET_{V-V}$ results. Even though this comparison is not exhaustive, results indicate that $ET_{V-V}$ overestimates the $ET_{obs}$, while $ET_{G-V}$ underestimates the $ET_{obs}$..

The $WSI_F$ derived from σ is based on the water vapor pressure gradient, combining the effects of surface moisture and temperature with those of the air, reflecting the stress caused by atmospheric conditions and surface parameters. In fact, considering the soil-plant-atmosphere complex system as a continuous, the soil face should be considered as a reservoir that supplies water to the system. Thus, any deficit that occurs in the reservoir adversely affects the system performance. In such a case, the physical stress process involves surface and atmospheric demands.

The $WSI_F$ contrast with different WS indicators yields results similar to those published in other works. $WSI_F$ is sensitive to SM and PP, and represents the rate of ET. We emphasize that SM, ET and PP observations are completely independent from the $WSI_F$.

The relationship between $WSI_F$ and SM ($R^2$=0.8) provided information about the index limits. For instance, a soil with 5% of moisture matches a $WSI_F$ of about 0.8 and a SM of 30% agrees with a $WSI_F$ of about 0.3. The $WSI_F$ would get close to 0 for surfaces with 25-30% of water, which represents a saturation point for most of the soils.

The main advantage of $ET_{G-V}$ and $WSI_F$ is that they are based on universal relationships. Although Rsat is a local parameter that would depend on the soil and vegetation type, it can be obtained from the reflectance of SWIR bands of each analyzed scene. This advantage renders the ET and WSIF equations applicable with different satellite missions and any remote region. $ET_{G-V}$ and $WSI_F$ required little processing of satellite images, resulting convenient for end users who require quick and easy to apply methods.

## References


[1] R.K. Vinukollu, E.F. Wood, C.R. Ferguson and J.B. Fisher, "Global estimates of evapotranspiration for climate studies using multi-sensor remote sensing data: Evaluation of three process-based approaches", Remote Sensing of Environment, 2011, vol. 111, pp. 801-823.

[2] R.D. Jackson, R.G. Reginato and S.B. Idso, "Wheat canopy temperature: A practical tool for evaluating water requirements", Water Resources Research, 1977, vol. 13, pp. 651-656.

[3] R.J. Granger, and D.M. Gray, "Evaporation from natural nonsaturated surfaces", Journal of Hydrology, 1989, vol. 111, pp. 21-29.

[4] L. Jiang and S. Islam, "Estimation of surface evaporation map over southern Great Plains using remote sensing data", Water Resources Research, 2001, vol. 37, Issue 2, pp. 329-340.

[5] K. Nishida, R.R. Nemani, S.W. Running and J.M Glassy, "An operational remote sensing algorithm of land evaporation", Journal of Geophysical Research, 2003, vol. 108, No. D9, pp.4270.

[6] J.M. Norman, M.C. Anderson, W.P. Kustas, A.N. French, J. Mecikalski and R. Torn, "Remote sensing of surface energy fluxes at 101-m pixel resolutions", Water Resources Research, 2003, vol. 39, Issue 8, pp. 1221-1232.

[7] R. Rivas and V. Caselles, "A simplified equation to estimate spatial reference evaporation from remote sensing-based surface temperature and local meteorological data", 2004, Remote Sensing of Environment, vol. 83, pp. 68-76.

[8] V. Venturini, S. Islam and L. Rodríguez, "Estimation of evaporative fraction and evapotranspiration from MODIS products using a complementary based model", Remote Sensing of Environment, 2008, vol. 112, pp. 132-141.

[9] M.S. Moran, T.R. Clarke, Y. Inoue and A. Vidal, "Estimating crop water deficit using the relation between surface-air temperature and spectral vegetation index", Remote Sensing of Environment, 1994, vol. 49, Issue 3, pp. 246-263.

[10] P.D. Colaizzi, E.M., Barnes, T.R. Clarke, C.Y. Choi, M. Peter, P.M. Waller, J. Haberland and M. Kostrzews, "Water Stress Detection Under High Frequency Sprinkler Irrigation with Water Deficit Index", Journal of Irrigation and Drainage Engineering. JANUARY/FEBRUARY 2003.

[11] L.E. Pertovt, R. Rivas, J. Schirmbeck, W.O.G. Coelho and L. Vives, "Análisis de condicionantes ambientales del estrés hídrico de la vegetación en el sur de Brasil mediante imágenes NOAA – AVHRR, Boletín Geológico y Minero, 2008, vol. 119 Issue 1, pp. 119-124.

[12] L.A. Mendez-Barroso, J. Garatuza-Payan and E.R. Vivoni E.R., "Quantifying water stress on wheat using remote sensing in the Yaqui Valley, Sonora, Mexico", Agricultural Water Management, 2008, vol. 95, Issue 6, pp. 725-736.

[13] D. Girolimetto and V. Venturini, "Water stress estimation from NDVI-Ts plot and the wet environment evapotranspiration", Advances in Remote Sensing, 2013, vol. 2, Issue 4, pp. 283-291.

[14] Z. Sun, Q. Wang, Z. Ouyang and Y.Yang, "Evaluation of a complementary based model for mapping land surface evapotranspiration", Hydrology and Earth System Sciences, 2012, vol. 9, pp. 3029-3062

[15] I.J. Barton, "A parameterization of the evaporation from nonsatured surfaces", Journal of Applied Meteorology, 1979, vol. 18, pp. 43-47.

[16] D. Chen, J. Huang and T. Jackson, "Vegetation water content estimation for corn and soybeans using spectral indices derived from MODIS near- and short-wave infrared bands", Remote Sensing of Environment, 2005, vol. 98, pp. 225-236

[17] C.H.B. Priestley and R.J. Taylor, "On the Assessment of Surface Heat Flux and Evaporation Using Large-Scale Parameters", Monthly Weather Review, 1972, vol. 100, pp. 81–92.

[18] R.J. Granger, "A complementary relationship approach for evaporation from nonsaturated surfaces", Journal of Hydrology, 1989, vol. 111, pp. 31-38.

[19] V. Venturini, L. Rodríguez and G. Bisht, "A comparison among different modified Priestley and Taylor's equations to calculate actual evapotranspiration with MODIS data", International Journal of Remote Sensing, 2011, vol. 32, pp. 1319-1338.

[20] E.R. Hunt, B.N. Rock and P.S. Nobel, "Measurement of leaf relative water content by infrared reflectance", Remote Sensing of Environment, 1987, vol. 22, pp. 429-435.

[21] B.C. Gao, "NDWI-A normalized difference water index for remote sensing of vegetation liquid water from space", Remote Sensing of Environment, 1996, vol. 58, pp. 257-266.

[22] J. Penuelas, J. Piñol, R. Ogaya and I. Filella, "Estimation of plant water concentration by the reflectance water index (R900/R970)", International Journal of Remote Sensing, 1997, vol. 18, pp. 2869-2875.

[23] P. Ceccato, S. Flasse, S. Tarantola, S. Jacquemoud, J.M. Gregoire, "Detecting vegetation leaf water content using reflectance in the optical domain", Remote Sensing of Environment, 2001, vol. 77, pp. 22-33.

[24] R. Fensholt and I. Sandholt, "Derivation of a shortwave infrared water stress index from MODIS near- and shortwave infrared data in a semiarid environment", Remote Sensing of Environment, 2003, vol. 87, pp. 111-121.

[25] D.A. Sims and J.A. Gamon, "Estimation of vegetation water content and photosynthetic tissue area from spectral reflectance: a comparison of indices based on liquid water and chlorophyll absorption", Remote Sensing of Environment, 2003, vol. 84, pp. 526-537.

[26] P.J. Zarco-Tejada, C.A Rueda and S.L. Ustin, "Water content estimation in vegetation with MODIS reflectance data and model inversion methods", Remote Sensing of Environment, 2003, vol. 85, pp. 109-124.

[27] M. Maki, M. Ishiahra and M., Tamura, "Estimation of leaf water status to monitor the risk of forest fires by using remotely sensed data", Remote Sensing of Environment, 2004, vol. 90, pp. 441-450.

[28] Y.B. Cheng, P.J. Zarco-Tejada, D. Riano, C.A Rueda and S.L. Ustin, "Estimating vegetation water content with hyperspectral data for different canopy scenarios: Relationships between AVIRIS and MODIS indexes", Remote Sensing of Environment, 2006, vol. 105, pp.354-366.

[29] M. Trombetti, D. Riano, M.A. Rubio, Y.B. Cheng and S.L. Ustin, "Multi-temporal vegetation canopy water content retrieval and interpretation using artificial neural networks for the continental USA", Remote Sensing of Environment, 2008, vol. 112, pp. 203-215.

[30] P. Ceccato, N. Gobron, S. Flasse, B. Pinty and S. Tarantola, "Designing a spectral index to estimate vegetation water content from remote sensing data: Part 1. Theoretical approach", Remote Sensing of Environment, 2002a, vol. 82, pp. 188-197.

[31] M.T. Yilmaz, Jr. E.R. Hunt, L.D. Goins, S.L Ustin, V.C. Vanderbilt and T.J. Jackson, "Vegetation water content during SMEX04 from ground data and Landsat 5 Thematic Mapper imagery", Remote Sensing of Environment, 2008, vol. 112, pp. 350-362.

[32] W.J. Shuttleworth, "Evaporation models in hydrology. In: Schmugge, T.J., André, J.C. (Eds.). Land Surface Evaporation: Measurement and Parameterization". Springer, New York, 1991, pp. 93-120.

[33] J.M. Lewis, "The story behind the Bowen ratio", Bulletin of the American Meteorological Society, 1995, vol. 76, pp. 2433-2443.

[34] V. Venturini, L. Rodríguez and G. Bisht, "A comparison among different modified Priestley and Taylor's equations to calculate actual evapotranspiration with MODIS data", International Journal of Remote Sensing, 2010, vol. 32, Issue 5, pp. 1319-1338.

[35] C.O. Justice, J.R.G. Townshend, E.F. Vermote, E. Masuoka, R.E. Wolfe, N. Saleous, D.P. Roy and J.T. Morisette, "An overview of MODIS Land data processing and product status", Remote Sensing of Environment, 2002, vol. 83, Issue 2-Jan, 3-15.

[36] E.F. Vermote, N.Z. Saleous and C.O. Justice, "Atmospheric correction of MODIS data in the visible to middle infrared: first results", Remote Sensing of Environment, 2002, vol. 83, Issue 2-Jan, pp. 97-111.

[37] W.P. Menzel, S.W. Seemann, J. Li and L.E. Gumley, "MODIS Atmospheric Profile Retrieval Algorithm Theoretical Basis Document", 2002, Version 6, Reference Number: ATBD-MOD-07. http://modis.gsfc.nasa.gov/data/atbd/atbd_mod07.pdf

[38] G. Bisht, V. Venturini, L. Jiang and S. Islam, "Estimation of net radiation using MODIS (Moderate Resolution Imaging Spectroradiometer) Terra Data for clear sky days", Remote Sensing of Environment, 2005, vol. 97, pp. 52-67.

[39] Z. Wan and J.A. Dozier, "A generalized split-window algorithm for retrieving land-surface temperature from space", IEEE Transactions on Geoscience and Remote Sensing, 1996, vol. 34, Issue 4, pp. 892-905.

[40] V. Venturini, G. Bisht, S. Islam and L. Jiang, "Comparison of evaporative fractions estimated from AVHRR and MODIS sensors over South Florida", Remote Sensing of Environment, 2004, vol. 93, pp. 77-86.

[41] P. Chavez, "Image based atmospheric corrections-revisited and improved", Photogrammetric Engineering and Remote Sensing, 1996, vol. 62, pp. 1025-1036.

[42] A.L. Buck, "New equations for computing vapor pressure and enhancement factor", Journal of Applied Meteorology, 1981, vol. 20, pp. 1527-1532.

[43] W.P. Kustas, J.M. Norman, M.C. Anderson and A.N. French, "Estimating sub-pixel surface temperature and energy fluxes from the vegetation index-radiometric temperature relationship", Remote Sensing of Environment, 2003, vol. 85, pp. 429-440.

[44] N. Batra, S. Islam, V. Venturini, G. Bisht and L. Jiang, "Estimation and Comparison of Evapotranspiration from MODIS and AVHRR sensors for Clear Sky Days over the southern Great Plains", 2006, Remote Sensing of Environment, vol. 103, pp. 1-15.

[45] J.D. Kalma, T.R. McVicar and M.F. McCabe, "Estimating Land Surface Evaporation: A Review of Methods Using Remotely Sensed Surface Temperature Data", Surveys in Geophysics, 2008, Vol. 29, Numbers 4-5, pp. 421-469

[46] D. Long and V.P. Singh, "A Two-source Trapezoid Model for Evapotranspiration (TTME) from satellite imagery", Remote Sensing of Environment, 2012, vol. 121, pp. 370-388.

[47] G.H. Hargreaves and Z.A. Samani, "Estimating of potential evapotranspiration", Journal of Irrigation and Drainage Engineering, 1982, E–ASCE 108, pp. 223-230.

[48] G.F. Makkink, "Testing the Penman formula by means of lysimeters", Journal of the Institution of Water Engineers, 1957, vol. 11, pp. 277-288.

[49] N.R. Patel, A.N. Mehta and A.M. Shekh, "Canopy temperature and water stress quantification in rainfed pigeonpea (Cajanus cajan (L.) Millsp.)", Agricultural and Forest Meteorology, 2001, vol. 109, pp. 223-232.

# 9

# Local soil knowledge and its use in crop allocation: Implications for landscape-scale agricultural production and conservation efforts in Talamanca, Costa Rica

**Leigh A. Winowiecki[1,*], Matthew P. Whelan[2], Paul A. McDaniel[3], Marilyn Villalobos[2], Eduardo Somarriba[2]**

[1]Soils Research Area, International Center for Tropical Agriculture (CIAT), Nairobi, Kenya
[2]Tropical Agricultural Research and Higher Education Centre (CATIE), Turriabla, Costa Rica
[3]Plant, Soil and Entomological Sciences Department, University of Idaho, Moscow, Idaho, USA

**Email address:**
l.a.winowiecki@cgiar.org (L. A. Winowiecki)

**Abstract:** Smallholder farmers have developed knowledge about soil and crop management through long-term, on-farm observations. Yet, food insecurity and land degradation continue to threaten livelihoods, globally. Part of the solution lies in linking local and scientific knowledge to develop improved agricultural strategies. We argue that incorporating farmer knowledge in the design and/or technology transfer processes of land management and restoration projects will improve the likelihood of success. The objectives of this study were to understand how farmers in the remote Cabécar communities in southeastern Costa Rica use soil knowledge to allocate and manage crops and to explore how to combine local and scientific knowledge to achieve wider adoption of sustainable agricultural techniques. Cabécar farmers in the foothills of the Talamanca Mountains practice both no-input subsistence and cash-crop farming. They cultivate both annual and perennial crops on their multi-parceled farms. Twenty-three stratified random heads of households were selected and interviewed. A variety of participatory methods were used including, participant observation (1.5 years), semi-structured interviews, farm mapping exercises, farm-transect walks and a participatory community workshop. Results indicate that farmers identified three distinct soil types using primarily soil color, texture and landscape position. Most interestingly, farmers' understanding of soil properties was tied directly to site suitability for specific crops. While farmers identified limitations of certain soil types' capacity to sustain a particular crop, farmers did not embark on restoration efforts or employ nutrient management techniques to improve the productivity of the soil. Furthermore, farmers had limited knowledge on processes of soil development and nutrient cycling. Instead of amending the soils to increase productivity, farmers allocated crops according to existing perceived suitability of the soil. These insights illustrate the barriers to farmer's adoption of new agricultural strategies. These data and experiences will be used to guide future agricultural research in the region, specifically projects that promote the restoration of unproductive soils by providing alternatives to burning and incorporating organic amendments.

**Keywords:** Local Soil Knowledge, Agroforestry, Costa Rica

## 1. Introduction

Farmers develop soil knowledge by conducting long-term observational experiments on their farm. Utilizing these experience-based insights and incorporating the knowledge of local people, whose livelihoods depend on the land, can improve the communication between researchers and farmers and potentially improve the success of extension projects. Local knowledge develops through experience over time and is unique because it seamlessly incorporates both historical and cultural context (Winklerprins, 1999; Oudwater and Martin, 2003), and is embedded in land-management practices (Krogh and Paarup-Laursen, 1997; Sillitoe, 1998). Previous studies have highlighted several benefits of integrating local soil knowledge into projects including: addressing the immediate needs of the local people (Winklerprins, 1999); increased success of sustainable development and conservation projects

(Winklerprins, 2001; Barrios and Trejo, 2003, Boreux *et al.*, 2009); expanding available datasets by including historical accounts (Huntington, 2000); and the development of GIS soil databases that incorporate both local and scientific soil classifications to create land-use suitability indices (Gobin *et al.*, 1998, 2000; Barrera-Bassols *et al.*, 2006). However, it is also important to acknowledge that farming communities do not always have the opportunity to farm the most productive lands (Ryder, 2003) and that local soil knowledge may also include recent agricultural extension efforts (Winklerprins and Barrera-Bassols, 2004). Recently, studies of local soil knowledge have shifted from documenting local soil classification systems to incorporating farmers' knowledge in the development of holistic soil management approaches (Talawar and Rhoades, 1998; Winklerprins and Barrera-Bassols, 2004; Barrios *et al.*, 2006). For example, national ecosystem assessments are combining qualitative local knowledge with quantitative soil data to better understand land degradation in the USA (Herrick *et al.*, 2010). Due to the complex nature of local knowledge, some authors suggest research should focus more on problem-oriented knowledge integration processes, rather than merely how to integrate the knowledge systems (Raymond *et. al.*, 2010). Methods that capture and incorporate farmer knowledge in order to prioritize local needs are still much needed.

Many biophysical, cultural, and socio-economic factors influence farmers' selection of crops and development of land-management practices over time (Dove, 1985; Schusky, 1989). These factors include local, regional, and international economies and markets, cultural knowledge, site suitability, climate, native soil fertility and soil characteristics, availability of labor and land, the influence of extension agents and the sharing of knowledge among farmers. All of these factors are often considered simultaneously and incorporated into local soil knowledge decision-making processes.

The Cabécar indigenous territory in southeastern Costa Rica provides an excellent study area to investigate local soil knowledge and demonstrate its application for incorporation into improved land management strategies. The indigenous territory borders two national parks and encompasses large tracts of forests, with smallholder subsistence and cash crop farmers living within this matrix. The territory lies within the northern foothills of the Talamanca Mountains and is part of the proposed Meso-American Biological Corridor, which aims to conserve and connect coastal and montane ecosystems. Due to its location and high conservation value to preserve structural connectivity within the corridor, the indigenous territory has been the center of several development projects promoting agricultural diversity (Dahlquist et al., 2007) and biodiversity conservation in cacao agroforestry systems (Andrade and Detlefsen, 2003; Somarriba et al., 2003). Yet, despite more than 20 years of extension efforts in the area, little is known about the local ecological knowledge of the Cabécar indigenous people (Whelan, 2005). The objectives of this research were threefold: 1) to assess if and how farmers distinguished between different soil types on the landscape; 2) to understand how farmers expressed soil knowledge through crop allocation; and 3) to suggest how to incorporate these data into development projects and land-management strategies.

## 2. Materials and Methods

### *2.1. Study Area*

The Talamanca region in southeastern Costa Rica borders Panama and is home to the Cabécar people (Figure 1). The Cabécar indigenous territory was established in 1977 by the Costa Rican government. It is 23,000 ha in size and supports a population of 3,500 (Andrade and Detlefsen, 2003). Cultural remains indicate that the Cabécar and other indigenous peoples have existed in the Talamanca region for over 3000 years (Borge and Castillo, 1997). Yet, they battled with Spanish conquistadors, Chiriquí Land Company (later know as United Fruit Company (UFC)), oil exploration initiatives and hydroelectric companies to maintain their presence in the region (Somarriba, 1993; Villalobos and Borge, 1998). For example, communities were relocated to accommodate the expansion of banana in Talamanca for UFC, but returned between 1940 and 1970 to reclaim their land and livelihoods in the Talamanca valley and foothills. Currently, the indigenous Cabécar smallholder farmers inhabit the Talamanca Mountain foothills and small portions of the alluvial valley. They practice both subsistence agriculture and cash-crop production, including organic banana (*Musa* AAA) and cacao (*Theobroma cacao*) agroforestry systems, shifting cultivation (rotation of basic grain crops and fallow) systems, as well as chemical-intensive plantain (*Musa* AAB) production (Somarriba and Harvey, 2003). Talamanca is the poorest canton in Costa Rica (Municipalidad of Talamanca, 2003) and farmers continue to face marginalization due to limited infrastructure and limited access to markets and health care (Gomez, 2001).

Geomorphologically, the Atlantic Slope of the Talamanca Mountains contains three distinct regions: the Talamanca Valley, which covers about 12,000 ha (18% of the territories) and contains 80% of the population within the Bribri and Cabécar indigenous territories (Borge and Castillo, 1997); the foothills, ranging in elevation from 100-600 m; and mountainous terrain above 600 m. Average annual temperature of the study region is 25 °C and annual precipitation is between 2200 - 3100 mm (Kapp, 1989).

No official soil survey has been conducted in Talamanca, but in general, the dominant soil type in the alluvial floodplains is classified as an Entisol and the foothill slopes are dominated by Inceptisol and Ultisol soil orders (Borge and Castillo, 1997, Polidoro *et al.*, 2008). The Talamanca Mountains and associated foothills are rugged, remote, and prone to natural disasters such as landslides and flooding.

Yet, farmers cultivate the land throughout the entire region. Due to its difficulty to access, the Talamanca region lacks well-documented biophysical and agricultural data and thus provides an excellent example to utilize local knowledge to better understand indigenous smallholder farming techniques.

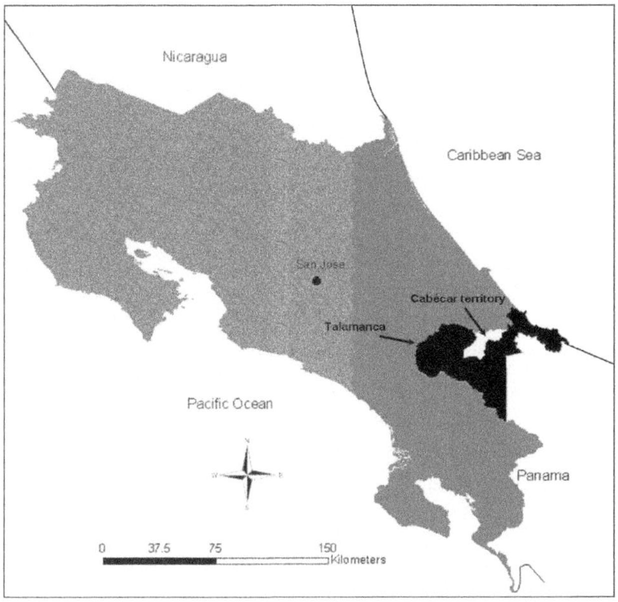


***Figure 1.*** *Location map of the Cabécar Indigenous Territory in southeastern Costa Rica.*

### *2.2. Selection of Communities and Households*

Communities were selected based on the following criteria: 1) ample, available land resources to practice subsistence, rotation, and cash-crop agriculture; 2) close proximity (~7 km) to each other with similar landforms; 3) homogenous ethnicity; and 4) willingness to accept us into their community. Three Cabécar communities (Sibuju, San Miguel, and San Vicente) were selected for the study, as these were representative of communities residing in the foothill region.

Twenty-three households were selected using a stratified random sampling design. Combined, the three Cabécar communities have approximately 75 households, which were stratified into three categories based on age of head of household: 20-40-, 40-60- and over 60-years old. Random samples were drawn from each category until roughly 30% of community was reached.

### *2.3. Participant Observation and Open-Ended, Semi-Structured Interviews*

Communities, farms and farmers were visited repeatedly over the course of 1.5 years (June 2005 to December 2006) to observe farming practices, soil properties and management strategies and to discuss farming techniques with farmers. Five preliminary interviews were conducted with farmers during this time to help design future interview questions.

Subsequently, semi-structured, open-ended interviews with the head of the household were conducted between January and March 2006 at the farmer's farm by the two lead authors. The main objectives of the interviews were to determine if farmers distinguished between different soil types in the region, how farmers described soil, if crops were allocated according to soil type and what management strategies farmers utilized for each crop and soil type. Basic information (size of household, land area farmed, etc.) about each household was also gathered. Average length of the interview was ~3 hours. Several family members often accompanied the head of household during the interview.

### *2.4. Participatory Mapping Exercise: Current and Historical Land Use*

The household members drew maps of their farm, including the location of each parcel within the farm, indicating the size of each land use and the crops planted. Landscape attributes and land-use history of each parcel was also recorded. For example, if a parcel currently was in fallow, we asked how long it has been in fallow and what the cropping cycles have been, including how the plot was managed, and if the land was burned for each rotation. Areas of different soils, as recognized by the farmers, were indicated on the farm map.

### *2.5.Farm Transect Walks*

Following the mapping exercise, farm transect walks were conducted to provide another opportunity to discuss soil types and crop suitability in the field as well as to validate the information collected during the interviews. After each interview, farmers selected transects across their farm to show the different soil types they described earlier. We excavated the soil to examine and describe it. Observations and discussion about crop allocation, suitability and productivity occurred at each soil observation point. Locations were revisited at a later time to dig soil pedons and collect soil samples (Section 2.8).

### *2.6. Participatory Community Workshop*

In addition to participant observation and farm transect walks, a community workshop was held after all these data were compiled in order to triangulate the results. The workshop was held at a local community center on September 29, 2007. Data were presented orally as well as in written form. All community members were invited, including farmers who were not interviewed. Additional questions were asked at the workshop. The objectives of this workshop were to: 1) assess if we accurately interpreted the farmers' knowledge from the interviews; 2) determine if farmers agreed on the different soil types in the region and the crop allocation patterns on these soil types; and 3) provide an additional forum (semi-formal community meeting) for farmers to share their soil knowledge with community members and researchers.

### *2.7. Qualitative Data Analysis*

Extensive notes were taken during all interviews and the

workshop. All notes were transcribed and typed. Data were compiled, analyzed and interpreted according to basic qualitative procedures (Creswell, 2003). Data (responses) were organized into categories and themes were identified. Responses under each theme were quantified using descriptive statistics.

### 2.8. Soil Sampling

Soil pedons were sampled on the same farms where the interviews were conducted (n=9 farms), and according to the soil types identified by the farmers. Since farmers associated soil types with specific landscape positions, three pedons were excavated on the three different landscape positions (alluvial flood plain, sloping lands and low lying areas (e.g. depressions)). Samples were taken by genetic horizon and transported to the University of Idaho Pedology Laboratory. Soil samples were air dried, gently crushed, and sieved to < 2mm. Sand, silt, and clay contents were determined using wet sieving, sonification, and centrifugation (Gee and Bauder, 1986). pH was measured on a 1:1 (soil:deionized water) slurry with standard pH electrode (Thomas, 1982). Exchangeable cations (Ca, Mg, and K) were extracted using $NH_4OAc$ and leachate was analyzed using inductively coupled plasma (ICP) spectroscopy (Sumner and Miller, 1996). Soils were classified using both Soil Taxonomy (Soil Survey Staff, 2006) and World Reference Base for Soil Resources (IUSS Working Group WRB, 2006).

## 3. Results

### 3.1. Demographic Information

Twenty-three households were interviewed in three Cabécar indigenous communities (Sibuju, San Vicente and San Miguel), representing 30% of the total number of households. Ninety-one percent of the interviews were conducted with men as head of the household and two interviews were conducted with widowed women. Head of household age ranged from 26 to 75 years old, with six households in the 20-40-year-old category, 11 in the 40-60-year-old category and six in the over 60-year-old category. Average number of household members for all three communities was five. Seventy percent of the Cabécar farmers interviewed were not born in the community where they currently live. Length of time the head of household has lived in the current community ranged from four to 45 years, with a mean of 27 years.

Principal crops identified by farmers include rice (*Oryza sativa*), maize (*Zea mays*), beans (*Phaseolus vulgaris*), cacao (*Theobroma cacao*), banana (*Musa* AAA) and plantain (*Musa* AAB).

### 3.2. How Famers Describe Soil

Talamanca foothill farmers used primarily color and texture to distinguish between soil types. Farmers were asked what types of soil exist on their farm. Ninety-six percent of farmers identified two distinct soil types and called them *Red soil* ("tierra colorada") and *Black soil* ("tierra negra") (Table 1). The remaining four percent did not identify a *Red* or *Black soil*. When asked what soils exist in the larger region, seventy percent of farmers described a *Sandy soil* ("tierra arenosa") located in the alluvial floodplain. It was often challenging for a farmer to describe the soil without relating it to a crop or landscape position. Yet, 74% of the farmers described *Red soil* as being hard. Terms such as dry, sticky, clayey, sterile and occurring on sloping lands or ridgetops were also used (Table 1). Fifty-two percent of the farmers described *Black soil* as smooth. Other descriptors included loose, moist, sandy and found in low-lying areas of the landscape (Table 1). All participants who identified *Sandy soil* described it as sandy, best for plantain production and occurring in the alluvial floodplain. Since many of the households moved from and farmed on the opposite side of the Talamanca Mountains, farmers were asked to describe the soil there. Pacific-slope soils were described as dry, sterile, and less productive than soils of the Atlantic Slope.

Only half of the households recognized soil depth as characteristic of the soil. Talamanca farmers do not plow or use mechanization on the farm, and basic grain crops do not require deep planting. Therefore they do not regularly examine the soil below 30-cm depth. Exposed riverbanks, landslides, tree falls and scarce road cuts provided opportunities for farmers to observe subsurface soil.

***Table 1.*** *Descriptive words (in Spanish with English translation) used by farmers to describe the three soil types (n= 23 interviews).*

| Soil Type (% of Households who Identified this) | Farmer Description (Spanish) | English Translation | % of Respondents using this Description |
|---|---|---|---|
| *Red Soil* (96) | Duro | Hard | 74 |
| | Pegajoso | Sticky | 48 |
| | Arcilloso | Clayey | 35 |
| | Seco | Dry | 22 |
| | Estéril | Sterile | 9 |
| | Altas, montaña | Steep, sloping land | 61 |
| *Black Soil* (96) | Suave | Smooth | 52 |
| | Suelto | Loose | 39 |
| | Arenoso | Sandy | 35 |
| | Húmedo | Moist | 17 |
| | Bajuras, bajos | Low-lying areas | 57 |
| *Sandy Soil* (70) | Arenoso | Sandy | 70 |
| | Suelto | loose | 17 |
| | Arena mesclado con tierra | Sand mixed with soil | 13 |
| | La isla | floodplain | 70 |

In general, Talamanca farmers use basic visual and textural observations to describe the soils in their region. Similarly, smallholder farmers in southwestern Nigeria used texture and color, as well as visual perception of drainage and soil density to classify soil (Osunade, 1988). Soil classification of rural farmers in Mexico and Burikna Faso was based on soil color, texture, consistency and moisture retention (Williams and Ortiz-Solorio, 1981; Dialla, 1993).

These studies report that farmers designated different soil types using soil characteristics that were visually observable or physically discernable, referring mostly to surface horizons.

### *3.3. Crop Suitability and Implications for Agricultural Productivity*

Farmers commonly discussed each soil type within the context of crop suitability. Through experimentation, farmers have learned that pasture, upland rice (*Oryza sativa*), peach palm (pejibaye) (*Bactris gasipaes*), coffee (*Coffea robusta*), pineapple (*Ananas comosus*) and oranges (*Citrus aurantium*) were suitable crops for *Red soil* (Table 2). *Red soil* was stated as being suitable for houses and other structures. In contrast, the *Black* and *Sandy soil* types were viewed as highly productive and fertile soils that supported the cultivation of most local crops including beans (*Phaseolus vulgaris*), maize (*Zea mays*), banana (*Musa* AAA), plantain (*Musa* AAB), cacao (*Theobroma cacao*), and cassava (*Manihot esculenta*) (Table 2).

Farmers also associated crop success with each soil type. While a specific crop may grow on the *Red soil*, its yield was reported to be too low to continue cultivation in the future. Yet, *Red soil* is the dominant soil type in the foothill region. When asked how farmers rehabilitate the *Red soil* to make it more suitable for cultivation, farmers stated that the soil needed to rest (e.g., as fallow) and/or they converted the land into pasture. Soil conservation measures were not mentioned or observed.

Talamanca farmers are not alone in establishing soil types that are strongly correlated to crop suitability. Farmers in Burkina Faso associated each soil type with site suitability for specific crops (Dialla, 1993, Ericksen and Ardon, 2003). Habarurema and Steiner (1997) noted that farmers in Rwanda classified soil based on agricultural suitability, and little correlation was found between farmers' soil classification scheme and Soil Taxonomy (Habarurema and Steiner, 1997). This type of knowledge is extremely useful for planning and executing agricultural extension programs.

***Table 2.*** *Suitable crops and associated management/conservation techniques identified by the interviewed households, listed for each soil type (AFS= agroforestry system).*

| Soil Type | Farmer-identified Suitable Crops | Management Techniques | Limitations Identified |
|---|---|---|---|
| *Red Soil* | Citrus | No input/manual weeding/pruning | --- |
| | Coffee | No input/manual weeding/pruning | Landslides/low yield |
| | Pasture | No input/low density | Hardening of soil |
| | Peach Palm | No input | --- |
| | Pineapple | No input/low density | --- |
| | Upland Rice | No input/manual weeding/burning | Landslides/high weed density |
| *Black Soil* | Beans | No input/manual weeding/burning | Landslides/low yield |
| | Maize | No input/manual weeding/burning | Landslides/high weed density |
| | Cacao | Organic AFS | Low yields/disease |
| | Banana | Organic AFS | Pests and disease |
| | Upland Rice | No input/manual weeding/burning | Landslides/high weed density |
| | Cassava | No input/manual weeding | --- |
| *Sandy Soil* | Plantain | Monoculture/inorganic fertilizer/pesticides/nematicides | Flooding and disease |
| | Banana | Organic AFS | Flooding and disease |
| | Beans | No input/manual weeding/burning | Landslides/low yield |

### *3.4. Landscape Position*

The Talamanca foothill region is an undulating and diverse landscape. In fact, farmers, often discouraged, described the landscape as broken "quebrada", uneven, or hilly. Sixty-one percent of farmers referred to the *Red soil* as the soil that is on the ridgetops. Fifty-seven percent of farmers described the *Black soil* as existing in the low-lying lands or upland depressions. Seventy percent of farmers specifically described the *Sandy soil* as existing in the alluvial floodplain in the valley. This information is invaluable when conducting a reconnaissance soil survey in the area.

A few farmers explained their understanding of this landscape-soil relationship by illustrating how rain washes ("se lava") the *Black soil* away from the ridgetops and slopes and deposits it in these low-lying areas ("bajuras"). Similarly, farmers in Burkina Faso described four different soil types in their region that were linked to directly to different landscape positions (Gray and Morant, 2003). It is important to understand how farmers view the soil and its formation in order to better communicate soil processes and soil-management strategies that influence these processes.

### *3.5. Land-use Trajectory*

Talamanca foothill farmers managed between one and five farms with an average of 2.7 farms per head of household with an average combined farm size of 47 ha. Farmers often managed a farm near their home and other farms located in a different community. This results in a patchwork of land ownership throughout the landscape and provides a variety of landforms for each individual farmer to cultivate and manage. In addition to having several farms, Talamanca farmers divided their farms into different parcels. Foothill farmers managed between 4 and 13 parcels within their farms, with each parcel potentially having a variety of land-management strategies, history or soil type. Figure 2 illustrates the land-use trajectory of a typical foothill farmer who has a topographically diverse farm and a number of land uses within the farm. Historically, all parcels were managed similarly for the cultivation of basic grains. For example, a forested area within the farm was selected for conversion to basic grain production (0.25-1 ha), which included slash and burning. After one cropping season, the parcel was allowed to go into a fallow period and a new parcel of forested land was selected for conversion for the next season. In summary, this 52-ha farm was slowly slashed and burned, one hectare at time, until after several years, the entire farm had been subjected to cultivation. A shift in this land-use pattern occurred when perennial crops were introduced on this farm. Site selection of perennial crops is an important decision for the farmer, as the crop will likely exist on the site for 5 to 50 years.

Farmers reported using landscape position, observation of soil characteristics and experience from past production of the basic grains to help decide where to plant the perennial crop. For example, beans were considered a more sensitive crop than upland rice and required more fertile and less-steep land. Therefore, if beans produced well on a particular parcel, the farmer assumed that perennial crops would also produce well and this location was selected for the introduction of agroforestry systems.

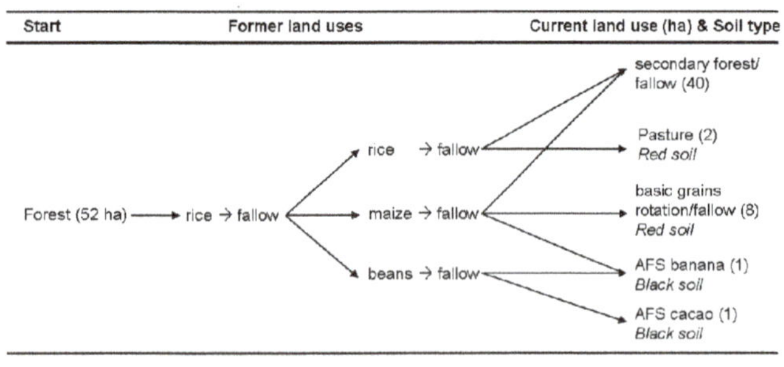


*Figure 2. Land-use trajectory of a typical farm in the foothills of Talamanca. The farm was divided into five parcels, all of which were managed similarly until the inclusion of perennial crops and pasture (AFS is agroforestry system).*

### 3.6. Soil Chemical and Physical Characteristics

Morphological, physical, and chemical differences existed between the three soil types identified by farmers. As the names suggest, *Red* and *Black soil* have moist soil colors that are considerably different (Figure 3). Mean clay concentrations in each soil type were significantly different from each other ($p<0.05$), with the *Sandy soil* having the lowest mean clay concentration (22%), followed by the *Black soil* (45%) and the *Red soil* (54%) (Figure 3). Mean pH values ranged from 4.5 for the *Red soil* to 6.9 for the *Sandy soil* (Table 3). Apparent effective cation exchange capacity (ECEC) is a measure of the clay fraction's contribution to the cation exchange and storage capacities of the soil. A low apparent ECEC value of 12 $cmol_c$ per kg of clay or less indicates a soil dominated by low-activity, kaolinitic clay (Soil Survey Staff, 2006). *Sandy soil* had the highest mean apparent ECEC, indicating high-activity clays present, while the *Red soil*, had relatively low ECEC (Table 3).

*Table 3. Mean pH values and apparent effective cation exchange capacities (ECEC) per kg of clay for soil horizons in each of the three soil types identified by farmers.*

| | pH | | | ECEC ($cmol_c$ kg $clay^{-1}$) | | |
|---|---|---|---|---|---|---|
| Soil Type | n | Mean | Range | n | Mean | range |
| Red | 18 | 4.5 | 4.2-4.7 | 11 | 30 | 17-58 |
| Black | 15 | 5.0 | 4.4-6.3 | 11 | 49 | 26-72 |
| Sandy | 18 | 6.9 | 5.6-7.7 | 8 | 169 | 109-295 |

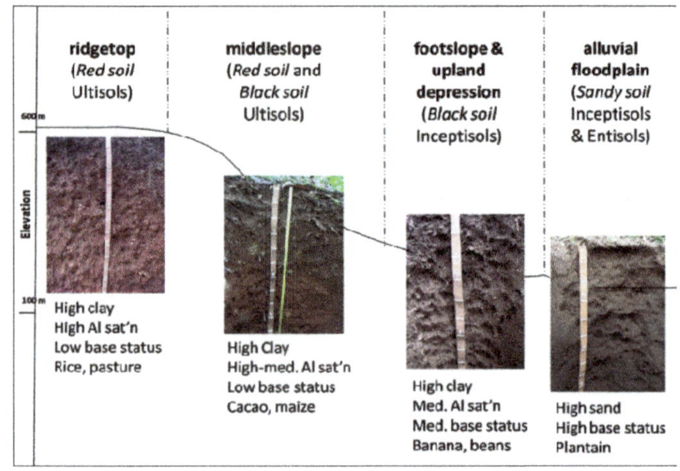


*Figure 3. Landscape position, farmer classifications, US Taxonomy classification, soil characteristics and crop suitability of soil of the Talamanca foothills.*

### 3.7. Relating Soil Chemical Indicators and Local Soil Knowledge

Soil chemical analyses have important implications for soil fertility and availability of nutrients. For example, due to the low storage and exchange capacities of the *Red soil*, management techniques that encourage efficient biocycling and minimize losses due to leaching are needed in order to continue to farm on the *Red soil*. Soil chemical characteristics are also reflected in crop productivity. Farmers state that neither banana, plantain, nor maize produce well in the *Red soil*. This could be due to the high clay content inhibiting growth of fine roots, the low base status unable to satisfy the nutrient requirements of the crop, or potentially high Al saturation inhibiting root growth and nutrient uptake. While farmers have no way to chemically measure soil fertility, their observations of crop productivity gave an accurate perception of the fertility of the soil. Farmers perceived *Black soil* as better for crop production than the *Red soil*. Though pH values of *Black soil* are still considered acidic and clay content is comparable to the *Red soil*, the higher ECEC of the *Black soil* makes it more suitable for most crops compared to the *Red soil*.

The three soils types identified by farmers are also classified differently using Soil Taxonomy (Figure 3). Due to their low base status and clay-rich subsoil, *Red soils* were classified in Soil Taxonomy as Ultisols (Udults). *Black soils* were classified as Inceptisols (Udepts). *Sandy soils* were classified as Entisols (Fluvents) due to their high base status and that they formed in alluvial sediments.

# 4. Conclusions and Discussion

### *4.1. Changing Soils*

Farmers commented on the current and historic soil characteristics of their farm, noting if the soil had changed since they first started cultivating it less than 40 years ago. Seventy percent of households reported that the soil had changed negatively over this time period. Thirty-nine percent of these farmers indicated that landslides had a large impact on the land. In addition to landslides, 56 % commented that soil had changed due to the continual practice of slash-and-burn-agriculture. Farmers also commented that burning led to soil sterilization and low productivity due to overuse. In addition, all of the farmers in the over 60-year-old category referred to the soil on Pacific Slope as sterile due to over-burning. Despite this, 96% of the farmers interviewed burned regularly. During the community workshop, farmers actively debated the necessity of burning. The most common argument for burning was weed control. Yet, other farmers suggested a labor-intensive alternative of manually mulching the fallow vegetation and planting the crop within this mulch layer. It is important to note that while farmers acknowledge that the soil had changed over time, they did not see themselves as agents to make a positive change toward rehabilitating the soil. This is an important barrier to overcome for a successful implementation of soil improvement programs in the area.

### *4.2. Soil Management and Restoration Efforts*

Despite the limited fertility of some foothill soils and in contrast to many other regions around the world, Talamanca farmers do very little to no active land manipulation to improve soil conditions for crop production. This may, in part, be due to the large farm sizes within the region. Talamanca foothill farmers designated overused soil (often *Red soil*) as tired ("*cansado*") and converted the land use to pasture or fallow. Pasture was seen as a final land use and farmers did not report converting pasture to any other land use. As population continues to increase, management practices that put these *Red Soils* back into production will be needed. Extension projects also need to address the disconnect Talamanca farmers have concerning their possible role in soil degradation and their ability to manipulate the land to create favorable soil conditions.

Farmers mentioned that trees provide several benefits including timber, shade, fruit and protection of natural springs (water sources). However, no active tree planting to promote these benefits was observed. Most farmers kept shade trees on their farm for their timber value or fruit production. In similar studies conducted in Chiapas, Mexico, organic coffee farmers remarked that shade trees provided litterfall, which contributes to soil formation, yet farmers did not recognize the role of trees in maintaining soil moisture or enhancing nutrient uptake (Grossman, 2003).

There are several potential explanations for the lack of soil management techniques employed in Talamanca. For example, no extension efforts have promoted the application of organic amendments; market values for crops are too low to encourage applying or paying for amendments; and only inorganic fertilizers are readily available on the local market, which are not suitable for organic cacao and banana production. Despite these potential obstacles, soil-management techniques are needed, especially on the *Red soil*. Identifying these obstacles and how farmers view the soil will aid in communicating and developing sustainable land-management techniques.

### *4.3. Combining Knowledge Systems*

The fusion of these learning systems has the potential to play an important role in improving land management strategies and enhancing conservation efforts in the Talamanca Indigenous Reserves. For example, addressing the disconnect in farmers' knowledge regarding their role and ability to improve (or degrade) the soil will help improve dissemination of new technologies (e.g., techniques that enhance nutrient cycling). Tree planting initiatives will need to incorporate local knowledge in dissemination to improve communication on the benefits of trees. For example, communicating how trees can rehabilitate degraded lands by reducing susceptibility of landslides and improving soil fertility, while contributing to infrastructure and protecting watersheds. Previous initiatives promoted trees because they enhanced structural connectivity and biodiversity.

Once knowledge gaps are identified, utilizing and acknowledging a common vocabulary is necessary to create effective dialogues and changed behavior (Eigenbrode et al., 2007). Specifically, Talamanca farmers use the words "*cansada o esteril*" tired or sterile to describe unproductive soil. They also use the term "*vitamin*", not nutrient, to describe nutrients needed by crops. It is important to use the local agricultural vocabulary to better communicate soil data and research results to the farmers.

As smallholder farmers continue to abort diverse agricultural systems to adopt seemingly lucrative (often monoculture) cash crops, effective dialogue between farmers, researchers and extension scientists are needed to develop innovative agricultural solutions that increase production and improve economic livelihoods. Understanding how farmers view the land and soil is an initial step to encourage effective communication and trust between involved parties to accomplish these goals. Subsistence farmers have developed knowledge of soil and its site suitability because it directly impacts their survival. In Talamanca, farmers communicated their knowledge, asked questions, and shared their ideas and concerns. Through these interactions, future project themes and workshops were identified and include: nutrient (*vitamin*) cycling; effects of fire on soil; and rehabilitation of degraded lands. Talamanca farmers were concerned about the productivity of the land within indigenous territories and about how their children and grandchildren will survive financially amidst the current socio-economic conditions. Successful extension and research projects must acknowledge farmers reliance on productive soil, incorporate the local agricultural vocabulary and beliefs, and address the current concerns of the farmers (e.g., soil degradation and low crop productivity) in order to develop integrated programs that improve both human livelihoods and soil conditions.

## Acknowledgements

The authors would like to thank the Cabécar indigenous communities of Sibuju, San Vicente and San Miguel, and the indigenous associations of ADITICA, ADITIBRI, and ACOMUITA. The authors would especially like to thank Celia Lopéz and Margarita Alvarado for their logistical support and all members of the Talamanca Team. Thank you to Craig Watt for offering a Social Scientist perspective and thank you to Jodi Johnson-Maynard for your thoughtful review of this paper. This work was supported by NSF grant 0114304 and the University of Idaho College of Graduate Studies.

## References

[1] Andrade, H., Detlefsen, G., 2003. Principales actores de Talamanca. Agroforestería en las Américas 10, 6-11.

[2] Bailgar, V.C., Fageria, N.K., 2005. Soil aluminum effects on growth and nutrition of cacao. Soil Science and Plant Nutrition 51, 709-713.

[3] Barrera-Bassols, N., Zinck, J.A., Van Ranst, E., 2006. Local soil classification and comparison of indigenous and technical soil maps in a Mesoamerican community using spatial analysis. Geoderma 135, 140-162.

[4] Barrios, E., Delve, R.J., Bekunda, M., Mowo, J., Agunda, J., Ramisch, J., Trejo, M.T., Thomas, R.J., 2006. Indicators of soil quality: A South-South development of a methodological guide for linking local and technical knowledge. Geoderma 135, 248-259.

[5] Barrios, E., Trejo, M.T., 2003. Implications of local soil knowledge for integrated soil management in Latin America. Geoderma 111, 217-231.

[6] Bertsch, P.M., Bloom, P.R., 1996. Aluminum. In: Sparks, D.L. (Ed.), Methods of Soil Analysis. Part 3 Chemical Methods. Soil Science Society of America and American Society of Agronomy, Madison, WI, pp. 517-543.

[7] Borge, C., Castillo, R., 1997. Cultura y conservación en la Talamanca Indígena. EUNED, San Jose.

[8] Creswell, J.W., 2003. Research Design: qualitative, quantitative, and mixed method approaches. Sage Publications, Thousand Oaks.

[9] Dahlquist, R.M., Whelan, M.P., Winowiecki, L., Polidoro, B., Candela, S., Harvey, C.A., Wulfhorst, J.D., McDaniel, P.A., Bosque-Perez, N.A., 2007. Incorporating livelihoods in biodiversity conservation: a case study of cacao agroforestry systems in Talamanca, Costa Rica. Biodiversity and Conservation 16, 2311-2333.

[10] Dialla, B.E., 1993. The Mossi indigenous soil classification in Burkina Faso. Indigenous Knowledge and Development Monitor 1, 17-20.

[11] Dove, M.R., 1985. Swidden Agriculture in Indonesia: Subsistence strategies of the Kalimantan Kantu. Mounton Publishers, Berlin.

[12] Eigenbrode, S.D., O'Rourke, M., Wulfhorst, J.D., Altoff, D.M., Goldberg, C.S., Merril, K., Morse, W., Nielsen-Pincus, M., Stephens, J., Winowiecki, L., Bosque-Perez, N.A., 2007. Employing a Philosophical Dialogue in Collaborative Science. BioScience 57, 55-64.

[13] Ericksen, P., Ardon, M., 2003. Similarities and differences between farmer and scientist views on soil quality issues in central Honduras. Geoderma 111, 233-248.

[14] Gee, G.W., Bauder, J.W., 1986. Particle size analysis. In: Klute, A. (Ed.), Methods of Soil Analysis. Part I. . Soil Science Society of America and American Society of Agronomy, Madison, pp. 383-441.

[15] Gobin, A., Campling, P., Deckers, J., Feyen, J., 1998. Integrated toposequence analysis at the confluence zone of the River Ebonyi; headwater catchment, southeastern Nigeria. Catena 32, 173-192.

[16] Gobin, A., Campling, P., Deckers, J., Feyen, J., 2000. Integrated toposequence analysis to combine local and scientific knowledge systems. Geoderma 97, 103-123.

[17] Gomez, V.F., 2001. Analisis de seleccion de mejoras en produccion sostenible y conservacion de la biodiversidad en fincas indıgenas de cacao en Talamanca, Costa Rica. CATIE, Turrialba, Costa Rica, p. 171.

[18] Gray, L.C., Morant, P., 2003. Reconciling indigenous knowledge with scientific assessment of soil fertility changes in southwestern Burkina Faso. Geoderma 111, 425-437.

[19] Grossman, J.M., 2003. Exploring farmer knowledge of soil processes in organic coffee systems of Chiapas, Mexico. Geoderma 111, 267-287.

[20] Habarurema, E., Steiner, K.G., 1997. Soil suitability classification by farmers in southern Rwanda. Geoderma 79, 75-87.

[21] IUSS Working Group WRB, 2006. World reference base for soil resources 2006: World Soil Resources Reports No. 103. FAO, Rome.

[22] Kapp, G.B., 1989. Perfil ambiental de la zona Baja de Talamanca, Costa Rica. CATIE, Turrialba, Costa Rica, p. 97.

[23] Krogh, L., Paarup-Laursen, B., 1997. Indigenous soil knowledge among the Fulani of northern Burkina Faso: linking soil science and anthropology in analysis of natural resource management. GeoJournal 43, 189-197.

[24] Municipalidad of Talamanca, 2003. Plan local de desarrollo 2003-2013. Municipality of Talamanca, Bribri, Costa Rica, p. 77.

[25] Nelson, D.W., Sommers, L.E., 1996. Total carbon, organic carbon, and organic matter. In: Sparks, D.L. (Ed.), Methods of Soil Analysis. Part 3 Chemical Methods. Soil Science Society of America and American Society of Agronomy, Madison, WI, pp. 961-1010.

[26] Osunade, M.A.A., 1988. Soil suitability classification of small farmers. Professional Geographer 40, 194-201.

[27] Oudwater, N., Martin, A., 2003. Methods and issues in exploring local knowledge of soils. Geoderma 111, 387-401.

[28] Polidoro, B., 2007. Ecological risk assessment of current-use pesticides in the Sixaola watershed, Costa Rica. Plant, Soil and Entomological Science. University of Idaho, Moscow, USA, p. 212.

[29] Polidoro, B., Winowiecki, L., Johnson-Maynard, J.L., McDaniel, P.A., Morra, M., 2008. Suelos del valle y del piedemonte en Talamanca: un paisaje dinámico para el almacenamiento de carbono. Agroforestería en lasAméricas 46, 40-44.

[30] Ryder, R., 2003. Local soil knowledge and site suitability evaluation in the Dominican Republic. Geoderma 111, 289-305.

[31] Schusky, E.L., 1989. Culture and agriculture: An introduction to traditional and modern farming systems. Bergin and Garvey Publishers, New York.

[32] Sillitoe, P., 1998. Knowing the land: soil and land resource evaluation and indigenous knowledge. Soil Use and Management 14, 188-193.

[33] Soil Survey Staff, 2006. Keys to soil taxonomy. USDA-NRCS, Washington D.C.

[34] Somarriba, E., 1993. Allocation of farm area to crops in an unstable Costa Rican agricultural community. University of Michigan, Ann Arbor, USA, p. 165.

[35] Somarriba, E., Harvey, C., 2003. ¿Cómo integrar producción sostenible y conservación de biodiversidad en cacaotales orgánicos indígenas? Agroforestería en las Américas 10, 12-17.

[36] Somarriba, E., Trivelato, M., Villalobos, M., Súarez, A., Benavides, P., Moran, K., Orozco, L., López, A., 2003. Diagnóstico agroforestal de pequeñas fincas cacaoteras orgánicas de indígenas Bribri y Cabécar de Talamanca, Costa Rica. Agroforestería en lasAméricas 10, 24-30.

[37] Steiner, K.G., 1998. Using farmers' knowledge of soils in making research results more relevant to field practice: Experience from Rwanda. Agriculture Ecosystems and Environment 69, 191-200.

[38] Sumner, M.E., Miller, W.P., 1996. Cation exchange capacity and exchange coefficients. In: Sparks, D.L. (Ed.), Methods of Soil Analysis. Part 3 Chemical Methods. Soil Science Society of America and American Society of Agronomy, Madison, pp. 1201-1229.

[39] Talawar, S., Rhoades, R., 1998. Scientific and local classification and management of soils. Agriculture and Human Values 15, 3-14.

[40] Thomas, G.W., 1982. Exchangeable cations. In: Page, A.L. (Ed.), Methods of soil analysis. Part 2. 2nd ed. Agronomy Monograph 9. ASA and SSSA, Madison, pp. 159-165.

[41] Villalobos, M., Borge, C., 1998. Talamanca en la encrucijada. EUNED, San Jose.

[42] Whelan, M.P., 2005. Reading the Talamanca Landscape: land use and livelihoods in the Bribri and Cabécar indigenous territories. CATIE, Turrialba, p. 123.

[43] Williams, B., Ortiz-Solorio, C.A., 1981. Middle American soil folk soil taxonomy. Annals of the Association of American Geographers 71, 335-358.

[44] Winklerprins, A.M.G.A., 1999. Local soil knowledge: A tool for sustainable land management. Society and Natural Resources 12, 151-161.

[45] Winklerprins, A.M.G.A., McGrath, D.G., 2000. Smallholder agriculture along the lower Amazon floodplain, Brazil. PLEC News and Views 16, 34-42.

[46] Winklerprins, A.M.G.A., 2001. Why context matters: Local soil knowledge and management among an indigenous peasantry on the Lower Amazon Floodplain, Brazil. Etnoecologica 5, 6-20.

[47] Winklerprins, A.M.G.A., Barrera-Bassols, N., 2004. Latin American ethnopedology: A vision of its past, present, and future. Agriculture and Human Values 21, 139-1554.

# Effects of mineral nitrogen and phosphorus fertilizers on yield and nutrient utilization of bread wheat (*Tritcum aestivum*) on the sandy soils of Hawzen District, Northern Ethiopia

**Bereket Haileselassie[1, *], Dawit Habte[2], Mehretab Haileselassie[1], Gebremedhin Gebremeskel[3]**

[1]Mekelle Soil Research Center, Tigray Agricultural Research Institute, Mekelle, Ethiopia
[2]Kulumsa Agricultural Research Center, Ethiopian Agricultural Research Institute, Assela, Ethiopia
[3]Mekelle Agricultural Research Center, Tigray Agricultural Research Institute, Mekelle, Ethiopia

**Email address:**
yalem03@yahoomail.com (B. Haileselassie), dawithabte99@yahoo.com (D. Habte), mehretabhs@yahoo.com (M. Haileselassie), medkel17@gmail.com (G. Gebremeskel)

**Abstract:** On-farm field experiments were conducted to assess the effects of nitrogen (N) and phosphorus (P) on yield, yield components, nutrients uptake and use efficiencies, protein content of bread wheat, residual soil nitrogen and phosphorus availability and economic return of fertilizer application. The locations of the experiments were in two nearby farmers' field at Hawzen district, Northern Ethiopia. Factorial combinations of five N levels (0, 46, 69, 92 and 138 kg N/ha) and four levels of P (0, 46, 69 and 92 kg $P_2O_5$/ha) were laid out in a randomized complete block design with three replications. Experimental Field 1 was relatively better in its soil fertility than experimental Field 2 especially in its phosphorus availability and cation exchange capacity (CEC). However, both fields were low in their total soil nitrogen content. Grain yield of wheat significantly increased by 46% and 15% in Field 1 and Field 2, respectively at nitrogen application rate of 46 kg N/ha than the control. Nitrogen fertilization increased straw yield significantly only in experimental Field 1. Phosphorus application at a rate of 46 kg $P_2O_5$/ha increased significantly grain and straw yields by 38 % and 46 %, respectively in Field 2 than control. It is also economical to apply phosphorus at this rate for such farmers' field because the grain yield marginal rate of return was 179 %. Consistent results were observed on wheat nutrient uptakes and nutrient use efficiencies. Grain and straw yields of wheat were not significantly affected by the main effect of phosphorus application in Field 1. These results are consistent with the soil analysis result before planting where the soil phosphorus was found to be low in Field 1 and very low in Field 2. Previous management by the farmers may have influenced the soil fertility status. There were no interaction effects of nitrogen and phosphorus to affect wheat productivity and nutrient utilization in both fields. Future approach to soil fertility management should consider the past management history and should also be site specific based on soil fertility assessment. As a result, a combined application of 46 kg N/ha and 46 kg $P_2O_5$/ha are recommended to achieve sustainable bread wheat production on the sandy soils of Hawzen district. But in fields with higher initial soil phosphorus level, there is no need of applying phosphate fertilizer.

**Keywords:** Bread Wheat, Marginal Rate of Return, Nitrogen, Nutrient Use Efficiency, Phosphorus, Protein Content

## 1. Introduction

Wheat is an important large-acre crop which is grown on more acres globally than any other and provides a major share of the nutritional requirements for the growing world population [1]. It is a staple food for about one third of the world's population [2], being cultivated in Ethiopia on about 1.51 million hectares and delivering about 3.3 million tons of grain yields which makes Ethiopia the largest wheat producer in sub Saharan Africa [3]. Ethiopia is the second largest wheat producer in sub-Saharan Africa, after South Africa [4]. However the Ethiopian government is forced to import wheat every year because of higher demand than supply. Wheat is the single most important staple crop imported from abroad and most of the humanitarian food aid and commercial import takes in the form of wheat [4].To feed the growing human population and fill the yield gaps between wheat consumption and production in Ethiopia, increasing production of wheat is paramount important. Besides wheat straw is an integral component of livestock feed and commonly used as a roof thatching material both in the rural and urban areas in Ethiopia. Increasing wheat production in Ethiopia can be possible by increasing productivity of small holder producers in the mid and highlands and by bringing more area into wheat production in the lowlands. In the mid and highlands, wheat production is constrained by both biotic and abiotic factors such as diseases and pests, poor soil fertility and moisture stress. Poor agronomic and soil management, inadequate level of technology generation and dissemination are the most significant constraints to increased wheat production in the highlands and mid highlands of Ethiopia. Low soil fertility and slow progress in developing wheat cultivars with durable resistance to disease are considered the most important constraints limiting wheat production in Ethiopia [4, 5].

With fertilizers becoming important inputs to maximize crop yields, and nitrogen fertilizer being determinant in wheat production in Ethiopia, several N fertilizer trials have been undertaken to determine the optimum level. Results indicated that N rates had significant positive effect on yield up to 230 kg $ha^{-1}$[6]. Crop response to nitrogen fertilizer is influenced by factors such as nitrogen fertilizer management, soil type, crop sequence and supply of residual and mineralized nitrogen [7]. A good supply of nitrogen to the plant stimulates root growth and development as well as uptake of other nutrients [8,9]. The total wheat N uptake increased with increasing N application [10]. Nitrogen rate significantly influenced grain yield, protein content, N uptake efficiency, N biomass production efficiency, N utilization efficiency, N use efficiency for grain and N use efficiency for protein yield of wheat [11]. Increasing levels of nitrogen increases grain and dry matter yields, number of kernels per head and plant height of wheat [2, 10, 12). Nitrogen application increased the number of tillers at late tillering stage of crop growth of wheat [10, 13]. Nitrogen nutrition stimulates tillering probably due to its effect on cytokinin synthesis [14]. There are contradictory evidences on the effect of nitrogen on thousands seed weight of wheat. A report by [15] and [10] indicated reduction in thousands grain weight of wheat by N fertilizer application, despite increase in yields. Furthermore, [15] concluded that N fertilizer applied at rates to optimize yield response would not necessarily give a comparable improvement in 1000-grain weight. On the other hand, [2, 16) indicated nitrogen application increased thousands seed weight of wheat. Adequate nitrogen supply improves protein content of vegetative organs as well as storage tissues [17] and manufactures protein from carbohydrates [18]. Nitrogen fertilization increased protein content of wheat and the highest grain protein content (13.09%) was obtained at N rate of 120 kg N/ha [16]. A report from [10] indicated residual soil total N tended to remain almost the same irrespective of different levels of N applications.

Phosphorus is the second most essential element of crop production and it is deficient in most soils around the world to achieve maximum yields [9]. Adequate phosphorus enhances many aspects of plant physiology like fundamental process of photosynthesis, flowering, seed formation and maturation [9, 19]. Fertilizer P application significantly and positively influenced grain yield and number of tiller of wheat [10]. A report of [17] indicated that cereal crops requires about 20 kg P/ha for normal production. Moreover, residual soil available P content increased with increasing rates of applied P up to 20 kg P/ha after sowing [10].

Nitrogen and phosphorus are considered as the most deficient nutrients in soils of Ethiopia including Hawzen [20, 21]. This indicates that nitrogen and phosphorus are the most yield limiting factors of cereals including wheat production in Ethiopia. Soil fertility status also varies within adjacent farms or plots mainly due to preceding individual farmer's soil management practices. Though there were information on the effects of nitrogen and phosphorus on wheat productivity in Ethiopia, a comprehensive work is scanty in sandy and degraded soils like the case of Hawzen. This study investigates effects of N and P fertilizers and their factorial combinations on wheat production based on indicative parameters.

# 2. Materials and Methods

## *2.1. Area Description*

On-farm field experiments were conducted in 2013 growing season at Hawzen district, Northern Ethiopia in two nearby farmers' fields represented as Field 1 and Field 2 (located at 39° 27' 2" E and 13° 15' 16" N, altitude of 2263 meter above sea level for Field 1 and at 39° 27' 30" E and 13° 59' 9" N, altitude of 2275 meter above sea level for Field 2). Hawzen district is located in tepid to cool sub moist mountains plateau agro-ecological zone [22]. Soils in Hawzen district are one of the most degraded soils in the Tigray region (Northern Ethiopia) which are believed to be very poor in soil organic matter. Wheat is grown during the main season from June to September (Figure 1). The wheat growing season of 2013 had relatively enough rainfall compared to the long term average. Annual mean maximum temperature is 24°C and mean minimum is 7.7 °C. The rainfall data for the district was summarized from the nearest district data of Hawzen Office of Agriculture and Rural Development [23]. Temperature and potential evapo-transpiration data were estimated using the LocClim 1.0 software [24].

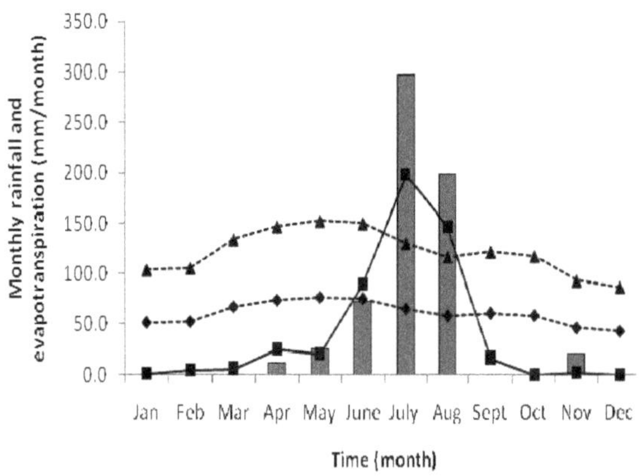


***Figure 1.*** *Average monthly rainfall (solid line and squares), actual monthly rainfall in 2013 (bars) and estimated potential evapo-transpiration (dashed line and triangles) and estimated half potential evapo-transpiration (dashed line and diamonds).*

### *2.2. Experimental Design and Procedures*

The treatments were five levels of nitrogen (0, 46, 69, 92 and 138 N kg/ha) and four levels of phosphorus (0, 46, 69 and 92 $P_2O_5$ kg/ha). The treatments were replicated three times in a Randomized Complete Block Design in a plot size of 3 m by 4 m. Nitrogen and phosphorus were applied in the form of urea and Triple Super Phosphate (TSP), respectively. Phosphorus and half of the nitrogen rates were incorporated into the soil at planting. The remaining half of the nitrogen rates were applied during tillering.

The initial experimental fields' soils were analyzed for texture, organic carbon, total nitrogen, Cation Exchange Capacity (CEC), available P, exchangeable Ca and Mg, and pH. Soil samples at flowering and after harvest were also analyzed for total nitrogen and available phosphorus. The methods used for physico-chemical analysis were: pH [25], organic carbon (modified Walkley and Black method [25]), texture (hydrometer method procedure of Bouyoucos [26]) available phosphorus [27], total nitrogen [Kjeldehal method [28], CEC and exchangeable Ca and Mg (Neutral ammonium acetate method [29]). Grain and straw samples were analyzed for nitrogen and phosphorus. Plant total nitrogen was analyzed using (Kjeldehal method [28]) and phosphorus using (dry ashing).

The bread wheat variety used in this experiment was Pica flor (*kakaba*) planted on 10 July 2013 for Field 1 and 11 July 2013 for Field 2. Harvesting was conducted on 30 October for Field 1 and 23 October for Field 2. Data on number of plant height and spike length were taken at maturity. Above ground biomass from sample quadrants were sun-air dried before weighing. The spikes were threshed and cleaned and grain yield was weighed. The straw yield was calculated by subtracting grain yield from the above ground biomass. Grain and straw yields were adjusted to 12.5 % moisture content.

Nutrient recovery and uptake, agronomic and physiological nutrient use efficiency were calculated from establised formulas and facts as follows:

1. The total uptake of nitrogen or phosphorus were calculated by multiplying the grain and straw yield (kg/ha) with the nitrogen or phosphorus concentration in (%) of each treatment as follows:

   a) N or P uptake of grain or straw (kg/ha)
   = Yield of grain or straw (kg/ha) x nutrient concentration of grain or straw (%) x $10^{-2}$ Eq [1]
   b) Total N or P uptake = N or P uptake of grain + N or P uptake of straw Eq [2]

2. The efficiency of N or P applied in the different treatments was estimated by calculating the N or P recovered in the above ground biomass of wheat (straw and grain) from the N or P applied in the treatments using the formula described by [30]
   Apparent N or P recovery (%) =
   (N or P uptake of treatment in kg/ha – N or P uptake of control in kg/ha) x 100 /(N or P added in kg/ha) Eq [3]

3. Agronomic and physiological N or P use efficiencies were calculated using the formula described by [31].
   a) Agronomic N or P Use Efficiency = (Grain yield of treatment – Grain yield of control) /(N or P added in kg/ha) Eq [4]

   b) Physiological N or P Use efficiency = (Grain yield of treatment – Grain yield of control)/(N or P uptake of treatment – N or P uptake of control) Eq [5]
   Where yield and N or P added and uptakes were in kg/ha

4. Grain protein content was calculated as % protein content = % N grain content × 5.7 Eq [6] [32].

### *2.3. Data Analysis*

Generated data were subjected to analysis of variance. All analysis were performed with GenStat $12^{th}$ edition statistical software package (VSN international).

## 3. Results and Discussion

### *3.1. Soil Properties before Planting*

The chemical and physical properties of the soils of the experimental fields are indicated in Table 1. The soil pH for Field 1 is neutral and Field 2 is slightly acidic. Both fields are very low in organic carbon content (<2 %), medium in CEC (in between 13-25 meq/100 g soil), low in total nitrogen (in between 0.05-0.125 %), very low in exchangeable calcium (<2 cmol(+)/kg soil), high in exchangeable magnesium (>3 cmol (+)/kg soil) and low in available phosphorus (< 10 mg/kg soil) in Field 1 and very low ( <5 mg/kg soil) in Field 2 [33]. The soil textural classes for both experimental fields are sandy loam. From the results of the soil analysis it can be depicted both nitrogen and phosphorus may be yield limiting for wheat production in the area. Experimental Field 2 was relatively low in its soil fertility than experimental Field 1 especially in its phosphorus availability and CEC.

*Table 1. Surface (0-20 cm) soil properties of experimental fields*

| Soil properties | Field 1 | Field 2 |
|---|---|---|
| $pH_{water}$(1:2.5) | 6.83 | 5.94 |
| Organic carbon (%) | 0.67 | 0.61 |
| P-Olsen (mg $kg^{-1}$) | 9.48 | 4.45 |
| Total-N (%) | 0.098 | 0.102 |
| CEC (meq/100 g soil) | 20.4 | 14.16 |
| Exchangeable Ca (cmol(+)/kg soil) | 1.2 | 1.4 |
| Exchangeable Mg (cmol(+)/kg soil) | 4.8 | 5.2 |
| Clay (%) | 14 | 14 |
| Silt (%) | 17 | 7 |
| Sand (%) | 69 | 79 |

### 3.2. Yield Components

Number of effective tillers per plant of wheat significantly increased due to the main effect of nitrogen ($P < 0.001$) and phosphorus ($P = 0.038$) fertilizers application in experimental Field 1 (Table 2). Phosphorus application increased significantly number of effective tillers per plant in Field 2. However, nitrogen fertilization had no significant effect on number of tillers per plant in Field 2. These results are supported by the findings of [13] and [10]. Plant height and head length were significantly increased due to main effect of nitrogen and phosphorus fertilization in both fields (Table 2). A report of [2] also shows nitrogen fertilization improved plant height of wheat. A significant decrease in thousands seed weight of wheat was observed due to nitrogen application. The current result is supported by [15] and [10] findings. However, the trend was not observed due to phosphorus application and thousands seed weight of wheat was not significantly affected by phosphorus application. All the measured yield components were not affected significantly by the interaction effect of nitrogen and phosphorus fertilization (Table 2).

### 3.3. Grain and Straw Yields

Grain yield of wheat significantly increased due to the main effect of nitrogen fertilization in both the experimental fields (Table 3). The result is supported by [6, 12]. Highest grain yields were obtained at the highest nitrogen rate of 138 kg N/ha in both fields but was not statistically different than the rate of 69 kg N/ha in Field 1 and 46 Kg N/ha in Field 2. Straw yield of wheat significantly increased due to nitrogen fertilization in Field 1. The highest straw yield was recorded at the highest nitrogen rate though it was not statistically different than the rate 69 kg N/ha. Straw yield was not affected significantly due to main effect of nitrogen in Field 2. This reveals that nitrogen is more yield limiting factor in Field 1 than Field 2 as both grain and straw yields of wheat responded to nitrogen application while only grain yield of wheat responded to nitrogen in Field 2.

Grain and straw yields of wheat were significantly increased due to the main effect of phosphorus fertilization in Field 2 (Table 3). Highest grain yield was obtained at a rate of 69 kg $P_2O_5$/ha though it was not statistically different than the preceding lower rate (46 kg $P_2O_5$/ha).This is supported by [17]. Highest straw yield was obtained at the highest phosphorus rate (92 kg $P_2O_5$/ha) though it was not statistically different than the preceding lower rate (69 kg $P_2O_5$/ha). Grain and straw yields of wheat were not significantly affected by the main effect of phosphorus in Field 1. Though it was not statistically significant, 8% grain yield and 11% straw yield increase were observed in Field 1 at the lowest phosphorus fertilization rate (46 kg $P_2O_5$/ha). The results indicated that phosphorus is yield limiting factor of wheat in Field 2 but not in Field 1. This is supported by the soil analysis result prior to planting where the available soil phosphorus content in Field 2 is very low while the available soil phosphorus in Field 1 is low in its class but near the range of sufficiency of phosphorus in soil (> 10 mg/kg) (Table 1).

Grain and straw yields of wheat were not affected significantly due to the interaction effect of nitrogen and phosphorus fertilization in both experimental fields (Table 3). Earlier works also indicated non-significant interaction effect of nitrogen and phosphorus on wheat yield [10].

Average grain and straw yields of wheat were better in Field 1 than Field 2 indicating a better soil fertility status of Field 1 than Field 2 (Table1).

### 3.4. Nutrient Uptakes and Recoveries

Total nitrogen and phosphorus uptakes of wheat increased with an increasing rate of nitrogen and phosphorus, respectively in Field 1 (Figure 2). In Field 2 a trend of increase in nitrogen and phosphorus uptakes were observed till a rate of 69 kg N/ha and 69 kg $P_2O_5$/ha, respectively. However, a trend of decrease in both nutrients uptake were observed in the higher application rates of each nutrients.

Highest recovery fraction of nitrogen was obtained at a rate of 69 kg N/ha in both the experimental fields (Figure 3). Nitrogen recovery was higher in Field 1 than Field 2. Indicating nitrogen is yield limiting factor of wheat in Field 1.This is supported by yield response of wheat (Table 3). Highest recovery fraction of phosphorus was obtained at the highest rate of phosphorus (92 kg $P_2O_5$/ha) in Field 1 and at a rate of 46 kg $P_2O_5$/ha in Field 2. Highest recovery fraction at the highest phosphorus application rate in Field 1 is an indication that phosphorus is not yield limiting factor of wheat. Phosphorus recovery was higher in Field 2 than Field 1. Indicating phosphorus is yield limiting factor of wheat in Field 2. These are consistent with the soil phosphorus content and the yield response of wheat to phosphorus application (Table 1 and Table 3).

***Table 2.*** *Effect of N and P fertilizers on some agronomic parameters of wheat.*

| | Field 1 | | | | Field 2 | | | |
|---|---|---|---|---|---|---|---|---|
| | NETPP | PH (cm) | HL(cm) | TSW (gm) | NETPP | PH (cm) | HL (cm) | TSW (gm) |
| **N rate (kg/ha)** | | | | | | | | |
| 0 | 2.10 | 67.37 | 6.18 | 40.03 | 3.32 | 68.67 | 6.12 | 32.39 |
| 46 | 3.10 | 75.98 | 6.97 | 38.87 | 3.53 | 69.85 | 6.67 | 31.35 |
| 69 | 3.24 | 76.97 | 7.30 | 38.14 | 2.60 | 70.58 | 7.08 | 28.53 |
| 92 | 3.14 | 79.12 | 7.13 | 37.87 | 3.70 | 71.92 | 6.77 | 30.12 |
| 138 | 3.91 | 79.82 | 7.50 | 34.73 | 3.78 | 70.32 | 6.97 | 29.05 |
| LSD | 0.64 | 4.28 | 0.377 | 3.5 | NS | 2.038 | 0.34 | 1.88 |
| P-value | <0.001 | <0.001 | <0.001 | 0.05 | NS | 0.042 | <0.001 | <0.001 |
| **P rate (kg $P_2O_5$/ha)** | | | | | | | | |
| 0 | 2.63 | 73.07 | 7.00 | 38.43 | 2.34 | 60.84 | 5.77 | 31.36 |
| 46 | 3.09 | 74.96 | 6.60 | 38.32 | 3.57 | 73.04 | 7.07 | 29.94 |
| 69 | 3.48 | 77.49 | 7.25 | 37.74 | 3.96 | 74.56 | 6.92 | 30.45 |
| 92 | 3.21 | 77.88 | 7.20 | 37.19 | 3.68 | 72.64 | 7.12 | 29.40 |
| LSD | 0.57 | 3.8 | 0.337 | 3.13 | 0.78 | 1.823 | 0.305 | NS |
| P-value | 0.038 | 0.05 | 0.002 | NS | <0.001 | <0.001 | <0.001 | NS |
| N*P interaction | NS | NS | NS | NS | NS | NS | NS | NS |
| CV (%) | 25 | 6.8 | 6.5 | 11 | 31 | 3.1 | 7.5 | 7.5 |

LSD: Least Significant Difference, P-value: Probability value, CV: Coefficient of Variation, NETPP: number of effective tillers per plant, PH: plant height, HL: head length and TSW: thousands seed weight

***Table 3.*** *Effect of N and P fertilizers on grain and straw yields of wheat*

| | Grain yield (kg/ha) | | Straw yield (kg/ha) | |
|---|---|---|---|---|
| | Field 1 | Field 2 | Field 1 | Field 2 |
| **N rate (kg/ha)** | | | | |
| 0 | 2496 | 2309 | 5715 | 5239 |
| 46 | 3656 | 2662 | 6623 | 5499 |
| 69 | 4131 | 2447 | 7244 | 6243 |
| 92 | 4398 | 2663 | 7592 | 5804 |
| 138 | 4443 | 2914 | 7897 | 6256 |
| LSD | 577 | 408 | 966 | NS |
| P-value | <0.001 | 0.048 | <0.001 | NS |
| **P rate (kg $P_2O_5$/ha)** | | | | |
| 0 | 3551 | 1922 | 6357 | 4044 |
| 46 | 3837 | 2647 | 7056 | 5918 |
| 69 | 3988 | 2997 | 7044 | 6444 |
| 92 | 3922 | 2828 | 7285 | 6827 |
| LSD | NS | 402 | NS | 837 |
| P-value | NS | <0.001 | NS | <0.001 |
| N*P interaction | NS | NS | NS | NS |
| CV (%) | 18.2 | 19 | 16.7 | 19.5 |

LSD: Least Significant Difference, P-value: Probability value, CV: Coefficient of Variation

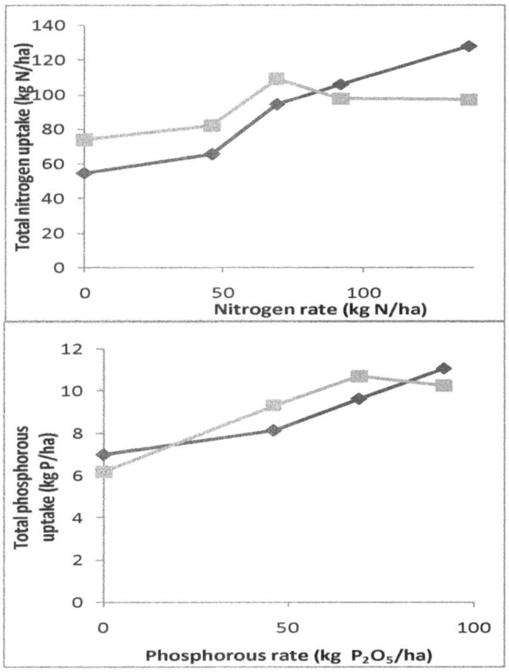


*Figure 2. Nitrogen and phosphorus uptake as influenced by nitrogen and phosphorus application for Field 1 (a line and a triangles) and for Field 2 (a line and squares).*

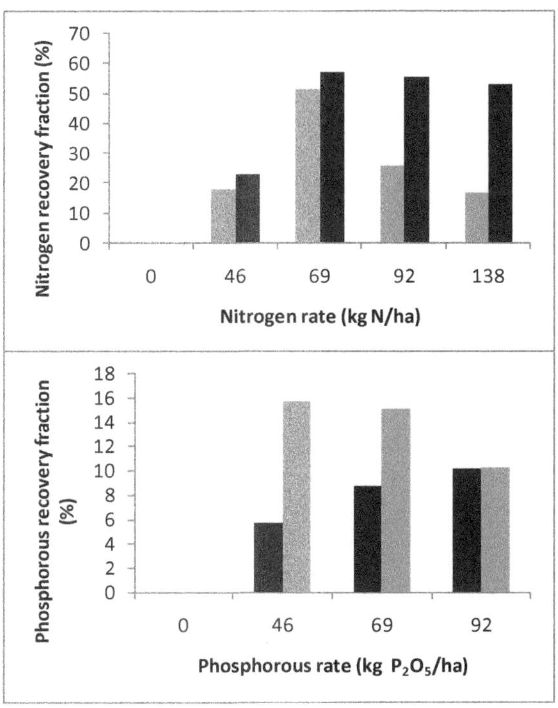


*Figure 3. Nitrogen and phosphorus recovery fractions for Field 1 (black bars) and for Field 2 (grey bars).*

### 3.5. Agronomic and Physiological Nutrient Use Efficiency

Highest agronomic use efficiency (AUE) of nitrogen was obtained at a rate of 46 kg N/ha in both the experimental fields (Figure 4). AUE of nitrogen was higher in Field 1 than Field 2. AUE of phosphorus was higher in Field 2 than Field 1 and the highest efficiencies were obtained at a rate of 46 kg $P_2O_5$/ha. The AUE of N and P showed a decreasing trend in both fields except at 69 kg N/ha for Field 2. The results on both N and P agronomic efficiency are supported by the effects of N or P fertilization on yields, nutrient recovery and the soil analysis result before planting (Table 1, Table 3 and Figure 3).

Physiological use efficiencies (PUE) of nitrogen in wheat were higher in Field 1 than Field 2 except at the highest nitrogen rate where both fields attained the same PUE (Figure 5). Highest PUE (109 kg grain yield/kg N total uptake) was attained at 46 kg N/ha in Field 1. However the highest PUE for Field 2 was 42 kg grain yield/kg N total uptake at the highest N level. Indicating nitrogen is more yields limiting in Field 1 than Field 2. Highest PUE of phosphorus (253 kg grain/kg P total uptake) was attained at 46 kg $P_2O_5$/ha in Field 1 with a decreasing trend in the higher phosphorus application rates (Figure 5). A comparable PUE ranged 222 to 237 kg grain/kg P total uptake were obtained in all the three phosphorus application rates in Field 2.

### 3.6. Grain Protein Content due to Nitrogen Fertilization

Grain protein contents of wheat were highest at 138 kg N/ha (11.9 %) and 69 kg N/ha (13.36%) application rates for Field 1 and Field 2, respectively indicating the required amount of nitrogen fertilizer to attain the higehst protein content was less in Field 2 than Field 1 (Figure 6). The result is consistent with the trend of nitrogen uptake (Figure 2). A report by [16] also indicated nitrogen fertilization increased protein content of wheat and the highest grain protein content (13.09%) was obtained at N rate of 120 kg N/ha. The lowest grain protein contents were recorded at nitrogen application rate of 46 kg N/ha (6.2%) and the control (9.5%) for Field 1 and Field 2, respectively. These showed the increases in yield of wheat (Table 3) due to nitrogen fertilization at a rate of 46 kg N/ha were attained decreasing the grain protein content of wheat in Field 1 and increasing the protein content of wheat in Field 2.These could be due to dilution effect because Field 1 was more responsive to nitrogen fertilzation than Field 2 and the percent change of increase in wheat grain yield from the control was higher in Field 1 than Field 2 at this rate (Table 3). This indicates the target of wheat nitrogen fertilization in highly responsive soil should not be yield only but it should put into account grain protein content. Because wheat provides more protein than any other cereal crops globally [16]

### 3.7. Soil Nitrogen and Phosporous Availabilty due to Nitrogen and Phosporus Fertilization

Residual soil total nitrogen content didn't show a trend either due to phosphorus and nitrogen fertilization (data not shown). A report of [10] also shows residual total N tended to remain almost the same irrespective of different levels of applications. However residual soil phosphorus availability showed a trend of increase starting at a rate of 69 kg N/ha in Field 1 and a decrease at the same rate in Field 2 due to nitrogen fertilization both at heading and at harvest (Figure 7). This might be pH difference of the fields (Table 1).

Phosphorus could be fixed in Field 2 and becomes available in Field 1 due to acidifing effect of the nitrogen.

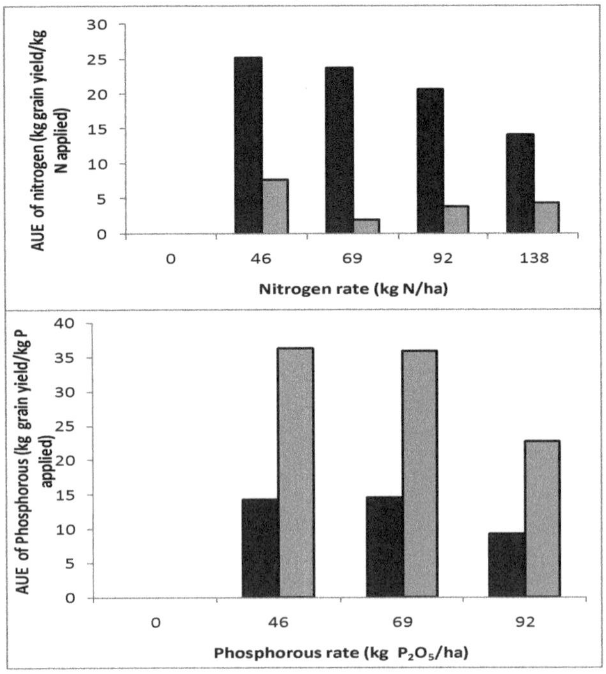


***Figure 4.*** *Agronomic use efficiencies of nitrogen and phosphorus for Field 1 (black bars) and for Field 2 (grey bars).*

An increasing trends of residual soil phosphorus availability were observed with an increasing application of phosphorus fertilizer rates at flowering in both fields and at harvest in Field 1(Figure 8).The incremental trend was observed up to 69 kg $P_2O_5$/ha at flowering in Field 2. A report of [10] also indicated that the soil available P content increased with increasing rates of applied P up to 20 kg P/ha after sowing.

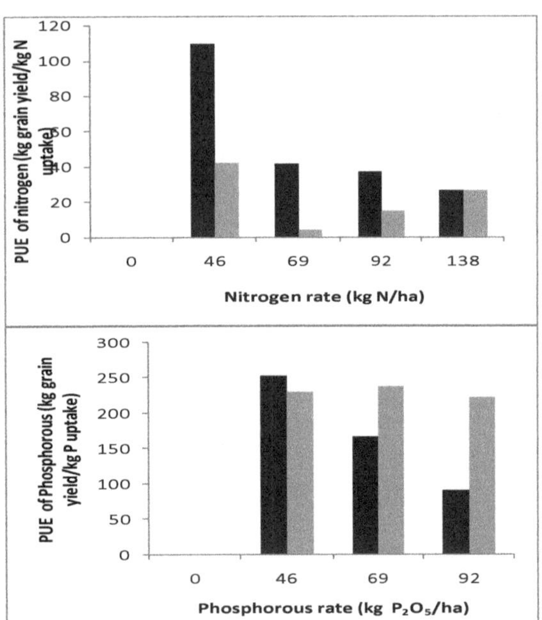


***Figure 5.*** *Physiological use efficiencies of nitrogen and phosphorus for Field 1 (black bars) and for Field 2 (grey bars).*

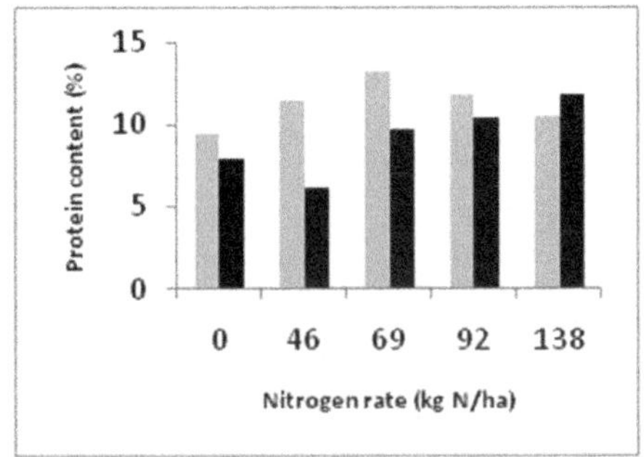


***Figure 6.*** *Grain protein content of wheat as influenced by nitrogen ferilization for field1 (black bars) and for field2 (grey bars).*

### *3.8. Partial Budget Analysis*

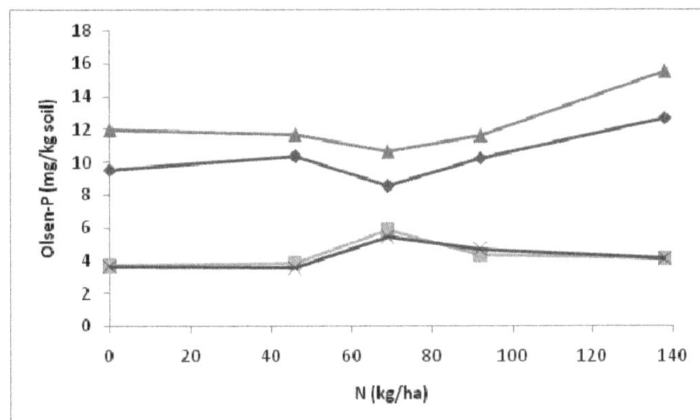


***Figure 7.*** *Soil P availability collected at two growth stages as influenced by N fertilization triangles and a line Field 1 at harvest, diamonds and a line Field 1 at flowering, rectangles and a line Field 2 at flowering, and crosses and a line Field 2 at harvest*

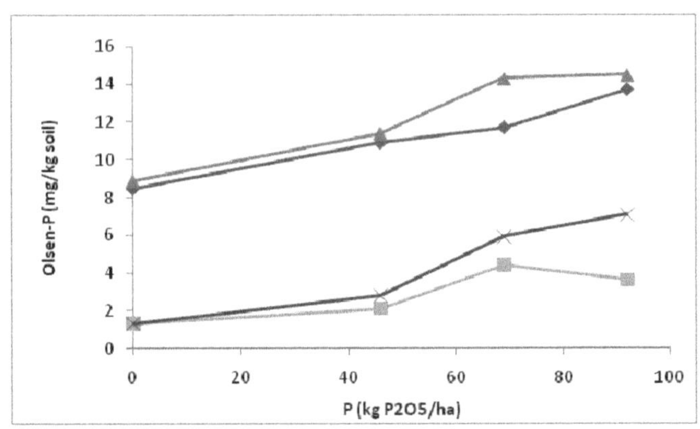


***Figure 8.*** *Soil P availability collected at two growth stages as influenced by P fertilization triangles and a line Field 1 at harvest, diamonds and a line Field 1 at flowering, rectangles and a line Field 2 at flowering, and crosses and a line Field 2 at harvest*

An increase in output will always rise profit as long as the marginal rate of return is higher than the minimum rate of return i.e. 50 to 100% [34]. Data in Table 4 shows that the marginal rate of return at the nitrogen application rates of 46, 69 and 92 kg N/ha were greater than 100% in Field 1 showing that it is economical to apply up to 92 kg N/ha. But Field 2 showed an economically feasible application rate at 46 kg N/ha since it has greater than 50% marginal rate of return. The other treatments were eliminated by the concept of dominance analysis [34] since the net benefit incurred decreased as the cost increased. Wheat straw yield is also important as of grain yield to the small holder farmers because wheat straw is an integral livestock feed. Considering straw in the partial analysis would improve the economic return of nitrogen fertilization in both fields.

Since the marginal rate of return in Field 1 due to phosphorus application is less than 50 to 100%, application of phosphate fertilizer is not economical (Table 5). This is in line with the initial soil P availability (Table 1). But in Field 2, it is economically profitable to apply up to 69 kg $P_2O_5$/ha. Considering straw yield in the partial analysis would also improve the economic return of phosphorus fertilization in Field 2.

***Table 4.*** *Partial budget analysis of bread wheat produces by applying nitrogen fertilizer*

| Fertilizer Rate (kg N/ha | Fertilizer Cost (Birr) | Fertilizer Application & Transport Cost [Birr] | Total Variable Cost (TVC) [Birr] | Grain Yield (kg/ha) | Total Revenue (TR) [Grain yield*11] | Net Revenue [TR-TVC] | Marginal Rate of Return (ratio) | Marginal Rate of Return (%) |
|---|---|---|---|---|---|---|---|---|
| | | | | **Field 1** | | | | |
| 0 | 0 | 0 | 0 | 2496 | 17971.2 | 17971.2 | | |
| | | | | | | | 4.59 | 459 |
| 46 | 1252 | 240 | 1492 | 3656 | 26323.2 | 24831.2 | | |
| | | | | | | | 3.98 | 398 |
| 69 | 1878 | 300 | 2178 | 4131 | 29743.2 | 27565.2 | | |
| | | | | | | | 1.8 | 180 |
| 92 | 2504 | 360 | 2864 | 4398 | 31665.6 | 28801.6 | | |
| | | | | **Field 2** | | | | |
| 0 | 0 | 0 | 0 | 2309 | 16624.8 | 16624.8 | | |
| | | | | | | | 0.70 | |
| 46 | 1252 | 240 | 1492 | 2662 | 19166.4 | 17674.4 | | |

***Table 5.*** *Partial budget analysis of bread wheat produce by applying phosphate fertilizer*

| Fertilizer Rate (kg $P_2O_5$/ha | Fertilizer Cost (Birr) | Fertilizer Application & Transport Cost [Birr] | Total Variable Cost (TVC) [Birr] | Grain Yield (kg/ha) | Total Revenue (TR) [Grain yield*11] | Net Revenue [TR-TVC] | Marginal Rate of Return (ratio) | Marginal Rate of Return (%) |
|---|---|---|---|---|---|---|---|---|
| | | | | **Field 1** | | | | |
| 0 | 0 | 0 | 0 | 3551 | 25567.2 | 25567.2 | | |
| | | | | | | | 0.10 | 100 |
| 46 | 1626 | 240 | 1866 | 3837 | 27626.4 | 25760.4 | | |
| | | | | | | | 0.24 | 24 |
| 69 | 2439 | 300 | 2739 | 3988 | 28713.6 | 25974.6 | | |
| | | | | **Field 2** | | | | |
| 0 | 0 | 0 | 0 | 1922 | 13838.4 | 13838.4 | | |
| | | | | | | | 1.79 | 179 |
| 46 | 1626 | 240 | 1866 | 2647 | 19058.4 | 17192.4 | | |
| | | | | | | | 1.88 | 188 |
| 69 | 2439 | 300 | 2739 | 2997 | 21578.4 | 18839.4 | | |

# 4. Conclusions

Nitrogen and phosphorus are known as the most yield limiting nutrients constraining cereal productivity including wheat in Ethiopia. Though there were some evidences on the effects of nitrogen and phosphorus on wheat productivity, a comprehensive work is scanty in sandy and degraded soil like the case of Hawzen. The results of this experiment indicated increase in wheat grain and straw yields, nutrient uptakes, nutrient recoveries, agronomic and physiological use efficiencies with mineral nitrogen and phosphorus fertilization. Even though the experimental fields were near to each other, wheat responses varied to the different nutrients differently. Biologically a combined application (46 to 138 kg N/ha and 46 kg $P_2O_5$/ha) of NP fertilizers are required. But in fields with higher initial soil phosphorus level like Field 1, there is no need of applying phosphate fertilizer. The economic analysis also supports this since the marginal rate of return in Field 1 due to phosphorus application is less than 50 to 100% indicating application of phosphate fertilizer is not economical. This indicates phosphorus fertilization should follow a soil test based approach to get an economical yield. Generally it can be concluded that a combined application of 46 kg N/ha and 46 kg $P_2O_5$/ha are recommended to achieve sustainable bread wheat production on the sandy soils of Hawzen district.

## Acknowledgements

This research was funded by grants from Eastern Africa Agricultural Productivity Project-Wheat Regional Center of Excellence (EAAPP-WRCoE)). The support from Mekelle Soil Research Center and Mekelle Agricultural Research Center are highly acknowledged for the facilitation of field and laboratory works.

## References

[1] Y. Shapiro. Monsanto's plan and prospects for wheat improvement through breeding and biotechnology. In: Reynolds M. and Eaton D., eds. Book of Abstract. Complementary strategies to raise wheat yield potential. Workshop held at CIMMYT, E. Baton, Mexico, DF., CIMMYT. 2009.

[2] M. I. Hussain. and S. H. Shah. Growth, yield and quality response of three wheat (*Triticum aestivum L.*) varieties to different levels of N, P and K. *Int. J. of Agri. and Bio.*, 4(3): 362-364. 2002.

[3] Ethiopian Central Statistical Agency (CSA). The Federal Democratic Republic of Ethiopia Central Statistical Agency. Agricultural sample survey, report on area and production of major crops, Addis Ababa, Ethiopia. 2013.

[4] Demeke. M. and F. Di Marcantonio. Analysis of incentives and disincentives for wheat in Ethiopia. Technical notes series, MAFAP, FAO, Rome. 2013.

[5] Teklu. E. and Hailemariam. T. Agronomic and Economic Efficiency of Manure and Urea Fertilizers Use on Vertisols in Ethiopian Highlands. *Agricultural Science in China*. 8(3): 352-360. 2009.

[6] Debreziet Agricultural Research Center (DZARC). Annual Research Report of 1989.DZARC, Debre-Zeit, Ethiopia. Pp. 64- 69. 1989.

[7] J. A. Lory. G. W. Randall. and M. P. Russelle. A classification system for Factors affecting crop response to nitrogen fertilization. *Agron. J.* 87: 869-76. 1995.

[8] Food and Agricultural Organization (FAO). Fertilizers and their use 4th ed. International fertilizer industry association, FAO, Rome, Italy. 2000.

[9] N. C. Brady. and Weil, R.R. The Nature and Properties of Soils ($13^{th}$ ed.). Pearson Education Ltd., USA. 2002.

[10] Damene. D. D. Yield response of bread wheat (*Triticum aestivum* L.) to applied levels of N and P fertilizers on Nitosol of Dawro Zone, South Western Ethiopia, a thesis (M.Sc.), The School of Graduate Studies, Alemaya University, Ethiopia. 2003.

[11] Haile. D. Nigussie. D. and Ayana. A. Nitrogen use efficiency of bread wheat: Effects of nitrogen rate and time of application. *J. Soil Sci. and Plant Nutr.* 12 (3): 389-409. 2012.

[12] Minale. L. Alemayehu A. Tilahun. T. and Abreham. M. Response of Bread Wheat to Nitrogen and Phosphorus Fertilizers at Different Agroecologies of Northwestern Ethiopia. In: M.G. Kinyua. J. Kamwaga. J.O. Owuoche. A. C. Ndiema. P.N. Njau. D. K. Friesen. D. Ouya. (Eds). Proceedings of the 12th Regional Wheat Workshop for Eastern, Central, and Southern Africa. Nakuru, Kenya, 22–26 November 2004. Mexico, D.F.; CIMMYT and KARI. 2006.

[13] M. Tahir. Ali. M. Anjum. S. Iqbal. and M.Yamin. Evaluation of the Effect of Use of N.P. Fertilizer in Different Ratios on the Yield of Wheat (*Triticum aestivum*) Crop. *Pak. J. Life Soc. Sci.* 2(2): 145-147. 2004.

[14] M. A. Botella. A. C. Cerda. and S. H. Lips. Dry matter production, yield and allocation of carbon-14 assimilate by wheat as affected by nitrogen source and salinity. *Agron. J.* 85 (5): 1044 – 1049. 1993.

[15] M. J. Gooding. and W. P. Davies. Wheat Production and Utilization, Systems, Quality, and Environment. CAB International, USA. 1997.

[16] H. Iqtidar. A. K. Muhammad, and A. K. Ejaz. Bread wheat varieties as influenced by different nitrogen levels. *J. Zhejiang Univ Sci.* 7(1): 70-78. 2006.

[17] K. Mengel. and E. A. Kirkby. Principles of Plant Nutrition. Panimo Publishing Corporation, New Delhi, India. 1996.

[18] S. L. Tisdale. W. L. Nelson. J. D. Beaton. and J. L. Havlin. Soil Fertility and Fertilizers (5th ed.). Macmillan Publishing Company, USA. 1993.

[19] F. N. Salisbury. and C. W. Ross. Plant Physiology. $4^{th}$ ed. Wads Worth Publishing Company, Belmant, California.1992.

[20] Asnakew. W. Tarekegn. M. Mengesha. B. and Tefera. A. Soil fertility management studies and wheat in Ethiopia. In: Hailu. G. D. G. Tanner and Mengistu H. (eds.). Wheat Research in Ethiopia. A Historical Perspective.IAR, CIMMYT, Addis Ababa, Ethiopia. Pp 112-144.1991.

[21] Bereket H. and Yirgalem W. Status Survey of Total Plant and Total and DTPA Extractable Soil Copper, Zinc, Iron and Manganese in Different Agro Ecological Zones, Soils and Districts of Tigray, Ethiopia. Tigray Agricultural Research Institute, Mekelle, Ethiopia. 2012.

[22] Anonymous. Agro Ecological Zonation of Ethiopia. Ministry of Agriculture, Addis Ababa, Ethiopia. http://www.fao.org/ag/AGP/AGPC/doc/counprof/ethiopia/ethiopia.htm (January, 2014). 2000.

[23] Hawzen Office of Agriculture and Rural Development, unpublished metrological data, Hawzen, Tigray, Ethiopia. 2013.

[24] Food and Agricultural Organization (FAO). Web LocClim, Local Climate Monthly Estimator. http://www.fao.org/sd/locclim/srv/locclim.home (February, 2014). 2002.

[25] M.L. Jackson. Soil Chemical Analysis. Prentice Hall of India Pvt. Ltd., New Delhi. 1967.

[26] P.R. Day. Particle fractionation and particle-size analysis. In: C.A. Black. D.D. Evans. J.L. White. L.E. Ensminger. F.E. Clark. (eds). Method of Soil Analysis, Part 2. ASA, Inc. Madison, Wis, U.S.A. 1965.

[27] S.R. Olsen. C.V. Cole. F.S. Watanabe. and L.A. Dean. Estimation of available phosphorus in soil by extraction with sodium bicarbonate. USDA Circular 939:1-19. Gov. Printing Office Washington D.C. 1954.

[28] J.M. Bremner. and C.S. Mulvaney. Total nitrogen. In A.L. Page. R.H. Miller. and D.R. Keeney. (eds). Methods of Soil Analysis, Part 2. Argon. Mongr. 9. 2 eds. ASA and SSSA, Madison, WI. 1982.

[29] C.A. Black. D.D. Evans. J.L. White. L.E. Ensminger and F.E. Clark. Method of Soil Analysis, Part 2. ASA, Inc. Madison, Wis, USA. 1965.

[30] G. R. Nziguheba. C.A. Merckx. Palm. and P. Mutuo. Combined Use of *Tithonia diversifolia* and Inorganic Fertilizers for Improving Maize Production in a Phosphorus Deficient Soil in Western Kenya. In: A. Bationo. (eds). Managing Nutrient Cycles to Sustain Soil Fertility in Sub Saharan Africa. Academy Science Publishers. Nairobi, Kenya. 2004.

[31] J.K. Syers. A.E. Johnston. and D. Curtin. Efficiency of Soil and Fertilizer Phosphorus Use: Reconciling changing concepts of soil phosphorus behavior with agronomic information. FAO Fertilizer and Plant Nutrition Bulletin 18. Food and Agricultural Organization of the United Nations. Rome. 2008.

[32] American Association Cereal Chemists (AACC). Approved Methods of the American Association Cereal Chemists. American Association of Cereal Chemists, Inc., St. Paul, Minnesota. 2000.

[33] J.R. Landon. Booker Tropical Soil Manual: A handbook for soil survey and agricultural land evaluation in the tropics and sub tropics (ed.). Longman Scientific & Technical. Booker Tate. UK. 1991.

[34] CIMMYT. From Agronomic data to farmers recommendations: An economic training manual. Completely revised edition, Mexico, D.F. 1998.

# 11

# Farmers' perception on agricultural technologies a case of some improved crop varieties in Ghana

**Emmanuel Asiedu – Darko**

Council for Scientific and Industrial Research-Plant Genetic Resources Research Institute, Bunso Eastern Region, Ghana

**Email address:**
easiedudarko@yahoo.com

**Abstract:** The survey was carried out to find out farmers perception on improved crop varieties from March to June 2012 in the Ashanti, Eastern, and Northern regions of Ghana. Adoption of farming technologies is very crucial to agricultural development. In Ghana, a greater percentage (70%) of the people is in the agricultural sector. Improvement in agriculture will have direct positive impact on the livelihood of the people. Farmer perception on agricultural technology influences their decision to adopt the technology or not. The study revealed that farmers perceived the improved crop varieties with particular reference to Maize (*Zea maize*), Cassava (*Manihot esculentus*) and Oil Palm (*Elaies guineensis*) as lacking some good characteristics of the landraces and also expensive to adopt. The need for farmers to be actively involved in the development of improved crop varieties was also highlighted. The study recommends that all stakeholders (Plant Breeders, Agronomists, Post Harvest Technologists, Ministry of Food and Agriculture (MOFA) and Farmers) should be actively involved in the development of farming technologies.

**Keywords:** Perception, Improved Crop Variety, Agricultural Technologies, Farmer

## 1. Introduction

Agricultural research activities have given birth to several improved crop varieties in Ghana. The agricultural institutions under the Council for Scientific and Industrial Research (CSIR) notably Crops Research Institute (CRI) and Savannah Agricultural Research Institute (SARI) have released several improved crop varieties to improve agricultural activities in the country. Some of these improved varieties include the following: Aziga, Golden Jubilee, Etubi (Maize varieties), Agbelifia, Esam bankye, bankye hemaa, Nyeri-kogba (Cassava varieties), Edorpo-Munikpa, Kpaneli, (Groundnut varieties) and Zaayura and Songotra (Cowpea varieties). These improved crop varieties are introduced to farmers by Extension Department of Ministry of Food and Agriculture (MOFA).

The number of people working in the agricultural sector worldwide remains substantial. According to Food and Agriculture Organization (FAO, 2000) about 96.5% of these people live in developing countries. The active agricultural population had appreciated by nearly 60% in developing countries over the last 25 years, regardless of increasing urbanization (FAO, 2000)

The FAO report for the year 2000 revealed that in spite of this increase, availability and affordability of food is a major problem in many developing countries. Half of the world's population is still underfed and affected by some form of malnutrition and nutrient deficiency diseases which often have tragic health consequences.

Agriculture dominates Ghana's economy as the single largest contributor to Gross Domestic Product (GDP) while majority of the Ghanaian population, especially rural folks depend on it as the main source of livelihood. In 2008 agriculture's percentage contribution to GDP and Foreign Exchange was 33.6% and 37.9% respectively, (ISSER, 2008).

It is also a source of employment, employing more than half the population on a formal and informal basis, and accounting for almost half of GDP and export earnings, thus a major foreign exchange earner for the country (Osabutey, 2009).

In terms of agricultural performance Ghana has not done badly at all but the question is will Ghana be able to sustain the recent impressive agricultural performance? The need to keep pace with current agricultural activities and not to derail the current trend should be aligned with technological changes in the agricultural sector. Increased agricultural productivity depends primarily on accepting cultural and technological changes at the rural farm level. Peasant farmers could improve their productivity if they adopt improved farming techniques; however, some new practices are sometimes complicated, making adoption difficult for non-literate farmers (Apantaku, *et al* 2008). Peasant farmers must understand these farming techniques before they can adopt and successfully use them and this requires effective teaching by agricultural extension service (agents) ( Kesley and Hearne, 1995). According to Farinde and Jibowo (1996), the adoption and use of any extension teaching method depends on characteristics of the method, the type of audience to be reached by extension workers and the type of message (agricultural innovation) to be disseminated.

Farmers continuous use of landraces traditional (low yielding varieties) will make it difficult for the government vision to achieve agricultural growth rate of 4% as envisaged in the country's Vision 2020 a mirage.

One of the key factors that influence the farmers decision to adopt or not to adopt farming technology is the perception associated with the technology which are sometimes backed by cultural and traditional considerations.

Perception as defined by Van de Ban and Hawkin (1988) is the process by which information or stimuli is received and transformed into psychological awareness. Norton and Mumford posited that on the basis perception the farmer weighs the benefits to be derived from adopting the technology before a decision was reached

The objective of the study was to ascertain farmers perception on improved crop varieties and seeks answers on the following:

1. The reasons behind why farmers have certain perception on some improved crop varieties.
2. How these perceptions influence farmers decisions to adopt improve crop varieties.

# 2. Materials and Methods

The research adopted the descriptive and explanatory survey research design. This is because the study was aimed at describing how farmers perception affect adoption of farming technologies and explain its effects on agricultural productivity Data was collected from farmers within the operational areas of the Council for Scientific and Industrial Research namely Ejura, (Lat. $0.7^0$ 40'N) Nyapkala(Lat. $0.9^0$ 28'N) and Kusi (Lat.$0.5^0$ 54' N The target population for the study was farmers in these areas

Three hundred farmers made up of 225 males and 75 females were sampled randomly from the population of 446 farmers who grow maize, cassava, oil palm, cowpea and millet. A list of farmers groups in the operational areas which was obtained from extension agents at the operational areas was used for the sampling of the respondents. Questionnaire was used to solicit responses from the farmers. In addition, four different focus group discussions involving 15 farmers in each group were also conducted to collect qualitative data for the study. Farmers who are direct beneficiaries of technologies developed by CSIR scientists were targeted for focus group discussion.

# 3. Results and Discussions

The main crops cultivated by the respondents are cereal (maize, millet sorghum beans), root and tuber (cassava) and oil palm. These crops are cultivated on subsistence level with the surplus offered for sale to generate income.

Even though some of the farmers had adopted some improved crop varieties, majority of the farmers grow the improved and the traditional varieties. However substantial number of the farmers were still holding on to the traditional crop varieties thus these farmers had not adopted any kind of improved crop varieties (Table 1).

***Table 1.*** *Technology used by farmers*

| Technology used | Frequency | Percentage |
|---|---|---|
| Improved crop varieties only | 123 | 41 |
| Traditional crops only | 84 | 28 |
| Both | 93 | 31 |
| Total | 300 | 100 |

Farmers who grow both traditional and improved crop varieties argued that some traditional crops still have desirable traits such as good taste and ease of preservation which the improved crop varieties lack.

The respondents who have not adopted any technology (improved crop varieties) were of the view that it is expensive to adopt new technologies. The farmers thus explained that using the improved crop varieties required inputs such as fertilizers and other chemicals, which they could not afford. Some of the farmers interviewed maintained that at times farmers who cultivated the improved varieties had bumper harvest but found it difficult to market the produce. On the more serious note the respondents stated that they found the improved crop varieties lacking some desirable traits such as good taste and ease of preservation as indicated on Table 2.

***Table 2.*** *Reasons cited by farmers for non adoption of improved crop varieties*

| Reason | Frequency of responses | Percentage (%) |
|---|---|---|
| Taste | 46 | 24.47 |
| Ease of Preservation | 62 | 32.98 |
| Ready Market | 42 | 22.34 |
| Cost of Production | 38 | 20.21 |
| Total | 188 | 100 |

Majority of respondents (72%) indicated that that they had adopted improved crop varieties whilst 28% said they had not adopted any improved crop varieties. The reasons cited by these farmers for non-adoption included; ease of preservation, taste, ready market and cost of production.

They also complained about their marginalization in the planning and execution of the dissemination activities.

The knowledge and experience of the farmers play a vital role in the dissemination and adoption of new technologies. Cohen and Levinthal (2000) stated in the theory of absorptive capacity that, there should the willingness by the scientists to incorporate farmer's local innovations into their research activities and also the local farmers should be willing to share their local knowledge. This would help make dissemination of technologies very easy which would increase adoption rate.

To bring a greater cooperation between farmers and researchers there was the need to incorporate farmers knowledge in agricultural research. In conducting research in Spain, Ashby et al (2000) coined CIAL a Spanish acronym for *Comite de Investigation Agricola Local* or Local Agriculture Research Committee. CIALs are a form of local agricultural research group belonging to, managed by and providing a service to a rural community. The research team was made up of volunteer farmers, chosen because of their experimentation aptitude and, supported by a facilitator. The CIAL aims to link farmer-researchers with formal research systems, thus increasing local capacity to exert demands on the formal systems and access potentially useful skills, information and research. The CIALs created an opportunity for participants to systematically assess research processes and results and to then translate these reflections into adjustment of the research and management activities. In essence, the CIAL process is one of joint experimentation and learning.

All the 123 farmers who were growing only improved crop varieties (Table 1) stressed the need for them to be part of the planning of the dissemination process. Their reason was that they could offer useful suggestions which would improve the methods of dissemination.

When a Plant Breeder from CSIR was interviewed, he indicated that at the beginning of every year, a planning session is organized for stakeholders (Plant Breeders, Agronomist, Post Harvest Technologist, MOFA and Farmers) in agriculture activities. He said such meetings offer the opportunity for to raise issues concerning the technology, the mode of dissemination and how best the technologies could be adopted. He cited an example that farmers complained of poor husk of *obatanpa* an improved maize variety which maize breeders addressed by improving the husk cover so that insects cannot easily attack the crop on the field.

Farmers find it extremely difficult to do away with traditional varieties because they maintained that they find them tastier and easier to preserve as compared to the improved varieties. These traits of the traditional varieties have motivated them to continuously cultivate them regardless of the fact that yields were low. The ease of preservation brings to the fore the question of how the dissemination of improved varieties were packaged. The package does not include post- harvest technology to help the farmers to be able to preserve their bumper harvest. In a situation whereby the farmers find it difficult to preserve their produce, it would be a disincentive for them to adopt the improved crop varieties.

Farmers who have adopted wholly or partially improved crop varieties stated that high yield and resistance to drought were the key element that motivated them to adopt the technologies. The farmers contended that in the past poor agronomic practices coupled with use of landraces gave low yields. With the introduction of improved varieties and improved methods of farming yields have improved remarkably.

On what could be done to make adoption of technology easy for the farmers, the farmers mentioned provision of inputs such as fertilizers, chemicals (insecticides and weedicides) as a way of helping farmers to adopt the technologies . There was also the need to create and guaranteed prices to minimize post-harvest losses and ensure guaranteed income for farmers.

In explaining the adoption problem, or lack thereof, farmers stated that the problem does not wholly lie in dysfunctional extension systems or in the poverty of the people in the region. In their view, the lack of market for their excess produce upon adoption of new technologies is the main problem. Hence, many technologies exist but unexploited. Farmers were also unable to adopt them because of resource constraints (World Bank 2003).

Igben (1987) identified non availability of demonstration plots as one of the factors hampering the adoption of farming technologies. Very few farmers advocated more and effective education as the key to improve adoption.

The focus group discussions organized for maize, cassava and oil palm farmers revealed some pertinent issues which need to be addressed. The focus on maize and cassava was that these crops are staple foods which are consumed by people all over the country. Again maize is one of the food security crops under the national buffer stock system which is currently in operation in the northern Ghana.

According to the farmers *obatanpa* an improved maize variety is good but its storage was a problem. The high weevil infestation makes it difficult to store for a long period. A farmer who has had some knowledge in *obatanpa* explained that the protein content is high and as such when it is allowed on the field for a long time it attracts the weevil. To solve the problem according to the farmer the crop should be harvested early.

Another issue raised by the farmers was that it cost five hundred Ghana cedis (GH500.00) to cultivate one acre of maize, and ideally adhering to all the agronomical practices, the farmer is expected to get 8 to 9 bags of maize. As a result of insects' infestation and the inability to follow some agronomical practices, one acre maize farm could fetches only 5 bags which when sold at the current market price of the commodity usually make it unprofitable.

The farmers also complained of the cost of seeds of the improved varieties with particular reference to maize. This compels them to select seeds from the previous harvested maize from the field; a situation according to agricultural scientists' is not a good practice. The scientist explained further that planting the improved variety near the traditional variety result in contamination of the improved variety. The need for government to subsidize the price of improved variety seeds was stressed by the farmers.

Concerning cassava there is an improved variety known as *afisiafi* which is good for *garri* processing aside being equally good for other uses of cassava like *fufu* preparation. The peel of the variety was a source of worry to the farmers. The peel colour is not the same as the traditional cassava and as such marketing it for other purposes apart from *garri* processing becomes a problem.

The problem with oil palm according to the farmers is that the seed nuts are expensive making it unaffordable to most of them. The farmers are of the view that some of the improved oil palm varieties were not suitable for the preparation of palm nut soup.

The study revealed that some of the farmers had the perception that growing some of the improved variety was affront to their certain cultural beliefs. One of these cultural beliefs mentioned especially in the Northern Ghana was this festival which had to be celebrated before harvesting of crop (millet) can be done. However in the case where the date of this festival varies, farmers who cultivate early maturing millet stand to lose the crop if it matures before the celebration of the said festival.

Other cultural beliefs were that in certain communities certain crops traits tag the crop a taboo under certain conditions. For example maize cobs with grains of varied colours or twin-bunched plantain are taboos in certain communities.

## Recommendation

The need for government to assists farmers in the acquisition of the planting materials of the improved crop varieties is very crucial to agricultural development of the country. The policy of seed production unit under Ghana Cocoa Board (COCOBOD) to produce improved variety cocoa seeds to sell to farmers at highly subsidized prices is laudable and must be extended to crops in Ghana.

The study also recommends that all stakeholders (Plant Breeders, Agronomists, Post Harvest Technologists, Ministry of Food and Agriculture (MOFA) and Farmers) should be actively involved in the development farming technologies.

## References


[1] Apantaku SO, Olufade O, Adebayo K, Oyekanmi, A. (2008). Analysis of effects of extension teaching methods on farmer's level of cassava and maize production in Ogun State, Nigeria. *Ghana Journal of Agriculture Science, 41,* Part1.

[2] Ashby JA, Braun AR, Gracia T, Guerrero MP, Hernandez LA, Quiros CA.(2000) *committees in Latin America.* Cali, Colombia. Centro International de Agricultural Tropical, Working paper No. 5 Washington D.C.

[3] Cohen WM, Levinthal DA. (1990). Absorptive capacity: A new perspective on learning and innovation. *Administrative Science Quarterly. 35,* 128-152.Inquiry: Rethinking human organizational towards a positive theory of change, StipesPublishing, Chicago

[4] FAO, and World Bank (2000). *Agricultural knowledge and information systems for rural development (AKTS/RD): Strategic vision and guiding principles.* FAO and World Bank, Rome.

[5] Farinde AJ, Jibowo AA. (1996). Factors associated with effectiveness of extension Teaching methods used in the transfer of agricultural innovation in Lagos State, Nigeria. Paper presented at the 8th Annual Conference of Nigeria Rural Sociological

[6] Association NRSA on Sustainable Rural Development in Nigeria: An Agenda for the Third Republic, March, 1994.

[7] Gabre-Madin EZ, Haggblade S. (2004). Successes in African agriculture: Results of an expert survey. *World Development, 32,* 745-766.

[8] Heong KL, Escalada MM, Mai V (1994). An analysis of insecticides use in rice: case studies in the Philipines Vietnam International Journal of Pest Management 1994 40(2)173-178.

[9] Igben MS. (1987). Issues and problems in the administration of the Ministry of Agriculture based extension services in Nigeria. *Agricultural Administration and Extension,* 27, 215-230.

[10] Institute of Statistical Social and Economic Research (2009). *A Report on the State of Ghanaian Economy.* Accra, ISSER.

[11] Kelsey D, Hearne C. (1995). *Co-operative extension work.* New York: Comstock Publishing Associates.

[12] Kenyon L, Fowler M. (2000). Factors affecting the uptake and adoption of outputs of crop protection research on yams in Ghana. In S.D. Hainsworth, & S. J. Eden-Green (Eds.), Sustaining change: proceedings of a workshop on the factors affecting uptake and adoption of outputs of crop protection research in yam. Department for International

[13] Development (DFID) Crop Protection Programme (CPP) research outputs (pp.15-25). Kent, UK.

[14] Norton GA. Mumford JD (1983). Decision making in pest control In: Coaker, T. H.( Ed)Applied Biology Vol 8 Academic Press, New York.

[15] Osabutey PD (2009). Salvaging the agricultural industry through a national fund. *The Ghanaian Chronicle.*

[16] Rogers EM (1995). *Diffusion of innovation,* (4th ed.). New York: The Free Press.

[17] Röling N, (1995).What to Think of Extension?: A comparison of three models of extension

[18] practice. *An Article for Francophone Issue of the AERDD Bulletin.* Reading, UK: University of Reading.

[19] Van de Ban AW Hawkin, H. (1988). *Agricultural extension.* NY: John Wiley and Sons.World Bank. (2000). Attacking Poverty. *World Development Report* Washington, D. C.: World Bank.

# 12

# Mercury stress resistances in *Nauclea orientalis* seedlings inoculated with arbuscular mycorrhizal fungi

**Hanna Artuti Ekamawanti[1, 2], Yadi Setiadi[3], Didy Sopandie[4], Dwi Andreas Santosa[5]**

[1]Tropical Silviculture Major, Post-graduate School of Bogor Agricultural University, Bogor, Indonesia
[2]Faculty of Forestry, Tanjungpura University, Pontianak, Indonesia
[3]Laboratory of Forest Biotechnology, Research Center for Biological Resources and Biotechnology, Bogor Agricultural University, Bogor, Indonesia
[4]Department of Agronomy and Horticulture, Bogor Agricultural University, Bogor, Indonesia
[5]Magister Study Program on Soil and Environmental Biotechnology, Department of Soil and Land Resources, Bogor Agricultural University, Bogor, Indonesia

**Email address:**

hanna.artuti@gmail.com (H. A. Ekamawanti), ysetiad55@gmail.com (Y. Setiadi), didysopandie@gmail.com (D. Sopandie), dsantosa@indo.net.id (D. A. Santosa)

**Abstract:** The effect of inoculation with the arbuscular mycorrhizal (AM) fungi on the resistance of mercury (Hg) in lonkida (*Nauclea orientalis*) seedlings were investigated using an artificially contaminated river sand media at the concentration of 0, 375 and 750 $\mu$M Hg at screen house. Mercury accumulation was lower in mycorrhizal roots than in non-mycorrhizal roots when Hg was added at the rates of 750 $\mu$M. However, Hg accumulation and its translocation to the leaves were very low, either in mycorrhizal or non-mycorrhizal seedlings. The amount of Hg retained in roots was higher than in leaves, irrespective of applying different Hg concentration. *N. orientalis* seedlings have a high tolerance to Hg (> 70%). Under these experimental conditions, *N. orientalis* shows a high resistance and capacity to retain Hg in roots. Mycorrhizal symbiosis showed a significant effect of the resistance to Hg of *N. orientalis* plants grown in Hg-polluted and Hg-unpolluted media. Furthermore, the beneficial effects of the AM fungi observed in this study aroused an interest in considering the role of AM fungi in plant-based strategies of remediation of highly Hg-polluted soils.

**Keywords:** *Nauclea Orientalis*, Mercury, Resistance, Arbuscular Mycorrhizal Fungi, Remediation

## 1. Introduction

Mercury is ranked third on The Agency for Toxic Subtances and Disease Registry (ATSDR) 2011 Substance Priority List - a listing algorithm prioritizes substances based on frequency of occurrence at National Priorities List (NPL) sites - toxicity and potential for human exposure to the substances found at NPL sites [1]. Anthropogenic activities and the agricultural soils because of the annual import of toxic Hg [2]. $Hg^{2+}$ causes toxicity symptoms in higher plants through adverse effects on biochemical and physiological process [2], reduces chlorophyll levels in leaves [3] and disturbs photosynthetic activity [4]. Change in uptake and translocation of mineral nutrients have been reported in the presence of Hg [5]. In addition, Hg triggers oxidative stress by inducing production of $H_20_2$, lipid peroxide, and reactive oxygen species (ROS) in many plants, such as tomato, cucumber and alfalfa [6,7].

The use of plants and soil microbes for remediation of Hg-polluted is an emerging area of interest because it provides an ecologically sound and safe method. It has been proposed that plants can be used to remediate soils polluted with such element, either by stabilization, extraction, and volatilization [8-10]. Plants species selection is a key aspect for the sucess of remediation soils polluted and long-term reclamation of degraded lands. Hg accumulation in several plants has been studied, such as white clover [11], alfalfa [12], lentil, chickpea [13] and common vetch [14]. However, very few studies have been conducted on the Hg accumulation in tropical trees, such as *Nauclea orientalis*. *N. orientalis* is a pioneer species, a hardy species with dry land reclamation potential, and gives some services to the environment, such as to controls soil loss on riparian areas (erosion control), and soil improver because leaf litter on decay replenishes soil fertility [15]. If this tree species is proposed for phytoremediation on Hg-polluted soil, plant Hg uptake and resistance response should be evaluated under controlled conditions, prior to field establishment a screen to identify suitable candidate species.

An alternative solution is to combine the advantages of microbe plant symbiosis in the plant rhizosphere into an effective clean-up technology. Arbuscular mycorrhizal (AM) fungi are ubiquitous soil, forming symbiotic associations with roots of the majority of plant species [16]. It has been addressed that mycorrhizal inoculation can affect the accumulation of metals such as Cu, Cd, Zn, and As by plants and enhance the tolerance of host plants to contamination of these metals in soil [17-19]. However, the behavior of Hg in mycorrhizal plants is still lack and the potential of AM fungi has not yet been fully explored with respect to its Hg phytoremediation potentials. The first report about the effects of mycorrhizal inoculation on Hg behavior in soil-plant system is done [20] and it has been shown that Hg uptake was lower by mycorrhizal roots of maize than by non-mycorrhizal roots.

The interaction between AM fungi, tree species (such as *N. orientalis*) and Hg was the subject of this study because of the possibility of the beneficial effect of mycorrhizae in improving the resistance of plants against Hg toxicity. The study was carried out to determine the resistance to Hg stress in *N. orientalis* seedlings inoculated with AM fungi grown on artificial Hg-polluted river sands media in pot experiment. Such knowledge would help clarify the potential of tree species (*N. orientalis*) and AM fungi as phytoremediation agents of Hg-polluted soils.

## 2. Materials and Methods

### 2.1. Plant Culture in River Sands at Different Rates of Hg

Eleven-months old seedlings of non-mycorrhizal and mycorrhizal *N. orientalis* were obtained from the nursery. Each plant was grown in a PVC pot (10 cm in diameter and 30 cm in height), containing 5 kg coarse river sands (1-2 mm) as substrate. Mycorrhizal inoculum (25 g) (Laboratory of Forest Biotechnology, Research Center for Biological Resources and Biotechnology, Bogor Agricultural University collection) was inoculated to mycorhizal seedlings for second inoculation and sterilized mycorrhizal inoculum (25 g) and its filtrate (10 mL each pot) was applied to non-mycorrhizal seedlings. All seedlings were maintained by adding the nutrient solution for 8 weeks. The composition of the nutrient solution was as follows (mmol $L^{-1}$) [21]: $Ca(NO_3)_2$ 1.5; $KNO_3$ 4.0; $KH_2PO_4$ 1.5 and $MgSO_4$ 1.0. Micronutrients were supplied ($\mu$mol $L^{-1}$): Fe-EDDHA 36; $MnSO_4.H_2O$ 33; $ZnSO_4.7H_2O$ 1.6; $CuSO_4.5H_2O$ 1.6; $H_3BO_3$ 46 and $(NH_4)_6Mo_7O_{24}.4H_2O$ 0.1. The pH of nutrient solution ranged for 5.5 to 6.0. A factorial experiment with completely randomized design with three replicates for each treatment was used. Hg treatments were 0, 375 and 750 $\mu$M, supplied as $HgCl_2$ salt and applied to *N. orientalis* seedlings, either with or without AM fungi inoculation treatments. The experiment was carried out in screen house of Forest Microbiology Laboratory, Research Center of Rehabilitation and Forest Conservation, Bogor, under the following conditions: 28°C-33°C temperature and 80% - 95% humidity.

### 2.2. Harvesting

Plants were harvested after 30 days to Hg exposure and divided into roots, stems and leaves. All plant material was thoroughly washed with tap water followed by a subsequent rinse in 20 mM EDTA solution for 1 minute and then in deionized water for 2 minutes. Total fresh weight of each tissue was determined, then a representative plants were oven-dried to constant weight at 70°C for 3 days prior to preparation for measuring dry weights and nutrients analysis, and at 50°C before preparation for Hg analysis. The rest of the material plants was frozen in liquid $N_2$ and stored at -20°C before chlorophyll and malondialdehyde (MDA) concentration analyzing. Water content was determined based on the difference of fresh weight and dry weight.

### 2.3. Analytical Determination

Dried plant sample was acid digested using 25 mL of deionized water, 3.5 mL of concentrated $HNO_3$ and 0.5 mL of $HClO_4$, added to 0.5 g dry weight (DW) in a test tube. The digested solution was analyzed for mercury concentration using inductively coupled plasma-optical emission spectrometry (ICP-OES, Spectro Genesis) excitation, and calibration was performed by using standard Hg solution [22]. Total Hg and nutrient (P, Ca and Mg) concentration were determined by ICP-OES. Chlorophylls was extracted and estimated according to Arnon in [23]. Lipid peroxidation in plant tissues was based on an estimate of malondialdehyde (MDA) concentration, as described by Heath and Packer in [12]. Mycorrhizal colonization was determined following [24].

### 2.4. Calculation and Statistical Analysis

Some variables were calculated to study the resistance to mercury stress of *N. orientalis* seedlings [7, 25]:

$$\text{Hg uptake (nmol Hg g}^{-1}\text{ DW)} = \frac{\text{nmol Hg plant}^{-1}}{\text{g DW}}$$

where DW is the entire-plant dry weight.

$$\text{Hg translocation (nmol Hg g}^{-1}\text{ DW)} = \frac{\text{nmol Hg leaves}^{-1}}{\text{g DW}}$$

where DW is the entire-plant dry weight.

Concerning uptake and translocation, the same procedure

was used for P, Ca, and Mg [25]. Tolerance index (TI) of mycorrhizal and non-mycorrhizal seedlings to Hg treatments were determined based on roots dry weight [26], as:

$$TI\ (\%) = \frac{\text{DW of roots at Hg treated}}{\text{DW of roots at non Hg treated of the same treatments}} \times 100$$

A statistical analysis for means comparison was carried out using ANOVA and least significant difference (LSD) test with CoStat 6.400.

# 3. Results and Discussions

## *3.1. Mercury Accumulation by* N. Orientalis *Seedlings*

Hg accumulation in roots of *N. orientalis* seedlings was significantly increased consistently with increasing of Hg rates (Table 1), either within non-mycorrhizal or mycorrhizal seedlings. Hg concentration in the roots was 53- to 1,633-fold and Hg content in the roots was 45- to 1,513-fold higher than the control (without Hg treatment). In line with this, Hg accumulation in roots was 4- to 10-fold the concentration found in shoots of several plant species such as *B. chinensis* [7], *Brassica napus* [27] and *Zea mays* [28].

However, at the 750 $\mu$M of Hg treatment, Hg accumulation in roots of mycorrhizal seedlings was significantly decreased by 87% of Hg concentration and 84% of Hg content when compared to control (non-mycorrhizal plants). This result shows that at the highest concentration of Hg, the AM fungi could limit Hg store in roots. This is in agreement with [20] result that mycorrhizal inoculation significantly decreased Hg concentration in maize roots with Hg at the rates of 2.0 and 4.0 mg $kg^{-1}$. Hyphae of AM fungi, which contain free amino, hydroxyl, carboxyl, and other groups [16], colonize outside and inside the roots. Outside the roots, the hypahe and root surface could adsorp Hg so that Hg translocation into roots could be inhibited, and inside the roots, it change cell wall components of plant, hence possibly enhancing the sequestration of Hg [20]. In line with this, buffering heavy metal-stress had been assigned, at least partly, to selective immobilization of heavy metals in those root tissues that contain fungal structures [29] or to the high metal sorption capacity of the extraradical mycelium of AMF [30].

This finding supports results from numerous studies reporting that AM fungi often protect plants against high accumulation of toxic elements in the shoots, as it was reported for Cu [17], Al [31], Cd [17,32], Zn [33], Pb [17,34], U [35] and As [36]. In addition, glomalin, an insoluble glycoprotein which is contained in hyphae and spores of AM fungi, had a high binding capacity for Cu, Cd and Pb [17], hence it could act as a biostabilizer for the remediation of polluted soils. Glomalin could be considered for Hg sequestration in hyphae and spores of AM fungi as well. This presumption need to be investigated in further research.

***Table 1****. Hg concentration and Hg content in roots and leaves of N. orientalis seedlings grown for 30 days in river sands treated with different Hg concentrations and mycorrhizal inoculation.*

| Hg ($\mu$M) | Arbuscular mycorrhizal inoculation | | | |
|---|---|---|---|---|
| | Without | With | Without | With |
| | Hg concentration (ng Hg $g^{-1}$ DW) | | | |
| | *Roots* | | *Leaves* | |
| 0 | 0.42 ± 0.13 d | 2.06 ± 0.87c | nd | nd |
| 375 | 423.43 ± 74.90 a | 383.37 ± 121.10 ab | nd | 0.13 ± 0.13 |
| 750 | 685.99 ± 16.82 a | 109.09 ± 23.79 b | 2.10 ± 1.34 | 8.00 ± 0.96 |
| | Hg content (ng Hg) | | | |
| | *Roots* | | *Leaves* | |
| 0 | 3.14 ± 1.02 d | 14.62 ± 5.03c | nd | nd |
| 375 | 2900.04 ± 1112.77 a | 2307.52 ± 991.22 ab | nd | 0.53 ± 0.53 |
| 750 | 4749.53 ± 1707.50 a | 651.13 ± 161.07 b | 12.44 ± 7.46 | 51.92 ± 10.35 |

Significant differences among Hg treatments are indicated by different letter (mean ± SE, n = 3; LSD's test, p<0.05). nd = not detected

The very small of Hg concentration and content in leaves of *N. orientalis*, either in non-mycorrhizal or mycorrhizal seedlings, showed that Hg accumulation in leaves was inhibited (Table 1). However, there is a tendency that Hg concentration and Hg content were higher in leaves of mycorrhizal seedlings than non-mycorrhizal seedlings when the highest Hg was applied. This result indicates that AM fungi play an important role in accumulation of Hg in plant.

Hg uptake by seedlings and Hg translocation of root to shoot were very limited as well (Table 2). These results are in agreement with previous studies [37], that roots were the main organs for Hg retention in *Halimione portulacoides*. Various studies have indicated that only a very small amount of Hg is translocated to plant shoot after root uptake and Hg in shoots mainly comes from the uptake of air Hg [7,11,38,39]. Metal retention in roots can be a strategy for protecting the more sensitive aerial parts from the deleterious effects induced by metal stress [40] and it has been reported that in most plant species, metal ions are preferably retained in root tissues and only small portions are translocated to leaves. Hg accumulated by mycorrhizal and non-mycorrhizal seedlings was mostly distributed in root tissues, suggesting that an exclusion strategy for Hg tolerance widely exists in them. In addition, evaporation of Hg from soil/growth medium to atmosphere was promoted by mycorrhizal treatment [20].

### *3.2. Effects on Plant growth and Nutrient Distribution*

Early and direct phytotoxic symptoms of heavy metals, such as Hg, are the reduction in cell proliferation and inhibition of growth, and are frequently used as a phytotoxic index, followed by analyzing several physiological parameters [41,42]. Hg treatment caused no significant differences in biomass (fresh and dry weight of plants) and water content in non-mycorrhizal and mycorrhizal seedlings of *N. orientalis*, when compared to the control (without Hg treatment) (data not shown). It means that there was not a reduction of plant growth caused by Hg application up to 750 $\mu$M during 30 days of Hg exposure. In line with observations on growth, the non-mycorrhizal and mycorrhizal seedlings have a high tolerance to Hg (>70%) (Table 3). There were not toxicity symptoms in *N. orientalis* seedlings grown for 30 days after Hg exposure.

***Table 2.*** *Hg uptake and translocation in N. orientalis seedlings grown for 30 days in river sands treated with different Hg concentrations and mycorrhizal inoculation.*

| Hg ($\mu$M) | Arbuscular mycorrhizal inoculation | | | |
|---|---|---|---|---|
| | Without | With | Without | With |
| | Hg uptake* (nmol g$^{-1}$ DW) | | Hg translocation* (nmol g$^{-1}$ DW) | |
| 0 | nd | nd | nd | nd |
| 375 | nd | nd | nd | nd |
| 750 | 0.016 ± 0.006 | 0.040 ± 0.005 | 0.003 ± 0.002 | 0.013 ± 0.001 |

* Mean ± SE, n = 3; nd = not detected

Roots of inoculated plants were extensively colonized by AM fungi, while non-inoculated controls remained non-mycorrhizal (Table 3). The percentage of AM colonization was slightly reduced in the presence of Hg in river sands substrate. This result indicates that the concentrations of Hg in the substrate was slightly harm to AM fungi. Reference [26] reported that sensitivity of AM symbionts to heavy metal contaminated soil expressed as a reduction in root colonization. In contrast, [20] reported that addition of Hg in soil did not significantly influence root colonization rate in maize.

***Table 3.*** *Tolerance index and AM colonization of N. orientalis seedlings grown for 30 days in river sands treated with different Hg concentrations and mycorrhizal inoculation.*

| Hg ($\mu$M) | Arbuscular mycorrhizal inoculation | | | |
|---|---|---|---|---|
| | Without | With | Without | With |
| | Tolerance index* (%) | | AM colonization (%) | |
| 0 | 100.0 | 100.0 | 0 | 78.1 |
| 375 | 89.1 | 71.1 | 0 | 66.0 |
| 750 | 95.9 | 77.2 | 0 | 58.5 |

* The tolerance index was based on root dry weight

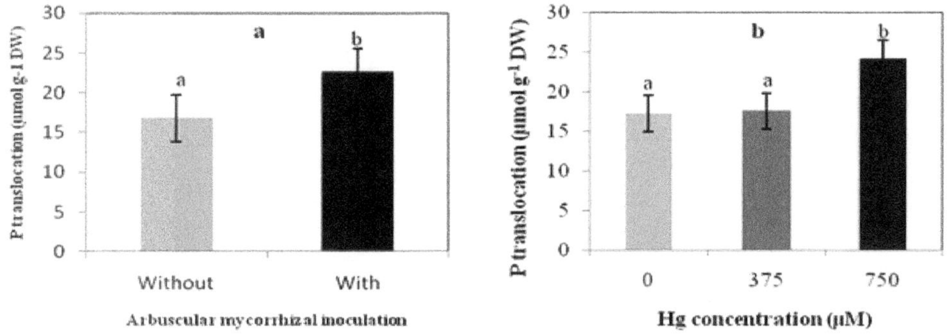


***Figure 1.*** *P translocation in seedlings of N. orientalis treated with arbuscular mycorrhizal inoculation (a) (mean ± SE, n = 9) and different Hg concentration (b) (mean ± SE, n = 6). Significant difference among Hg treatments and AM inoculation are indicated by different letter (LSD's test, p<0.05).*

Another interesting result in Table 3 was that the presence of AM fungi could increased 8.6% of tolerance index of seedlings treated with 750 $\mu$M Hg compared with seedlings treated with 375 $\mu$M. This result emphasizes that AM fungi could be potentially effective in protecting seedlings exposed to high levels of Hg concentration. The AM fungi ability to alleviate heavy metals stress of plants grown in heavy metal contaminated soil was previously proved by [31,43]. The highly AM colonization in seedlings treated with Hg indicated that AM fungi still function in Hg substrate.

In addition, AM fungi still affected uptake and

translocation of nutrients (P, Ca and Mg) as well. P translocation was increased significantly in seedlings treated with AM inoculation as single factor (Fig. 1A). AM fungi increased P translocation in plant 35% higher than control (without AM inoculation). Previous research reported that AM fungus (*Glomus mosseae*) increased the absorption of P for plant growth on red clover (*Trifolium pratense*) treated with different Zn rate levels [44].

On the other hand, P translocation to leaf was significantly increased 40% in seedlings treated with the highest of Hg application (Fig. 1B), higher than control (without Hg). This result indicates that increasing P translocation to leaf was induced by the highest of Hg application. In addition, AM inoculation caused significantly increased in Ca and Mg uptake and translocation of *N. orientalis* treated with the highest of Hg application (Table 4). Nutrient uptake and translocation to shoot depend on water movement in xylem, and a reduction of water absorption could also disturb micronutrients distribution [5]. Water content in each tissue of seedlings did not decrease during Hg exposure, indirectly, showed that water movement within the seedlings was not disturbed by Hg application. Therefore, nutrients distribution (especially P, Ca and Mg) in *N. orientalis* seedlings were not inhibited. In reference [5] stated that the increase of some nutrients in response to trace toxic elements could be a strategy to avoid toxicity in plants, although a contribution of concentration effects after Hg application cannot be ruled out. The role of AM fungi in the increase of plant nutrients could also be an important strategy to enhance the resistance to Hg stress.

***Table 4.*** *Ca and Mg uptake and translocation in* N. orientalis *seedlings grown for 30 days in river sands treated with different Hg concentrations and mycorrhizal inoculation.*

| Hg (μM) | Arbuscular mycorrhizal inoculation | | | |
|---|---|---|---|---|
| | Without | With | Without | With |
| | Ca uptake (nmol Hg g$^{-1}$ DW) | | Ca translocation (nmol Hg g$^{-1}$ DW) | |
| 0 | 55.3 ± 2.2 b | 53.4 ± 3.7 bc | 94.8 ± 4.5 b | 96.7 ± 13.1 b |
| 375 | 42.0 ± 2.5 d | 44.5 ± 0.9 cd | 80.4 ± 7.7 b | 90.0 ± 3.3 b |
| 750 | 41.8 ± 5.0 d | 67.5 ± 1.5 a | 79.3 ± 9.4 b | 136.5 ± 12.6 a |
| | Mg uptake (nmol Hg g$^{-1}$ DW) | | Mg translocation (nmol Hg g$^{-1}$ DW) | |
| 0 | 16.7 ± 1.3 b | 16.2 ± 0.6 b | 78.0 ± 7.1 b | 79.4 ± 8.1 b |
| 375 | 16.5 ± 1.5 b | 17.7 ± 0.2 b | 85.1 ± 3.0 b | 97.2 ± 3.0 b |
| 750 | 15.9 ± 0.4 b | 24.1 ± 1.2 a | 82.7 ± 7.6 b | 131.9 ± 10.3 a |

Significant differences among Hg treatments are indicated by different letter (mean ± SE, n = 3; LSD's test, p<0.05)

### 3.3. *Effect of Mercury and Arbuscular Mycorrhizal Fungi on Chlorophyll and Lipid Peroxidation*

Chlorophylls and MDA have been used as biomarkers, which can be used in the early diagnosis of metal toxicity [45]. Mercury inhibits enzymes involved in chlorophyll synthesis [46] and consequently, reduced chlorophyll levels have been observed [4]. In this research, chlorophyll (a, b and total) concentration in *N. orientalis* seedlings remained unchanged during exposure to Hg treatment (data not shown). This means that the process of photosynthesis was not impaired by Hg up to 750 μM on river sands media. This result was contrasts with research as in [5] which chlorophyll a concentration in *R. induratus* and *M. vulgare* were decreased significantly, while chlorophyll b levels remained unchanged by 5, 10, and 50 μM Hg rates when cultivated after Hg exposure for 2 months. There were not toxicity symptoms in leaves of *N. orientalis* seedlings during 30 days of Hg exposure. Enhancement of chlorophyll b concentration was observed in *N. orientalis* seedlings inoculated with AM fungi (Fig. 2). AM fungi resulted in significant increased chlorophyll b concentration by 15% when compared to the control (non-mycorrhiza). Increased chlorophyll b concentration could increase the accessory pigment that collects the energy to pass on to chlorophyll a.

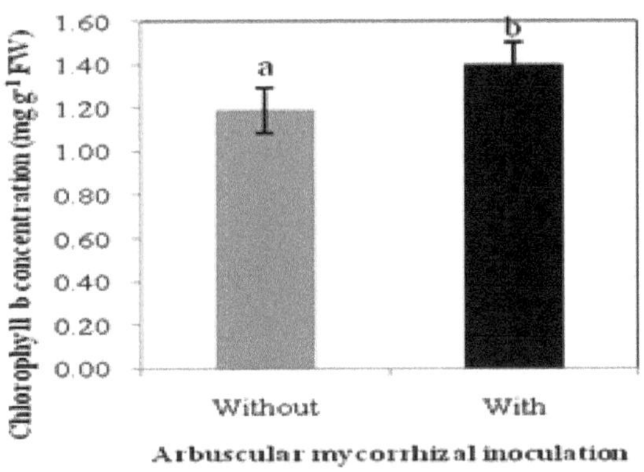


***Figure 2.*** *Chlorophyll b concentration of seedlings inoculated with arbuscular mycorrhiza (mean ± SE, n = 9). Significant difference for each AM inoculation treatments are indicated by different letter (LSD's test, p<0.05).*

Exposure of plants to non-redox-reactive heavy metal, such as Hg, resulted in oxidative stress as indicated by lipid peroxidation, $H_2O_2$ accumulation, and an oxidative burst [47]. Lipid peroxidation reflected the toxic effects of Hg on the cell membrane in tissues [7]. In this research, only malondialdehyde (MDA) was measured as a variable of lipid peroxidation. Roots of *N. orientalis* seedlings did not show any significantly increased in lipid peroxidation when Hg treatment is applied as a single factor (data not shown). It showed that oxidative stress was not observed in roots with Hg application, even though Hg was accumulated in roots. On the other hand, MDA concentration in roots of mycorrhizal seedlings was significantly decreased by 76% when it was compared to the control (non-mycorrhizal) (Fig. 3). It means that lipid peroxidation was reduced by AM fungi, consequently, the oxidative stress in roots could be minimized.

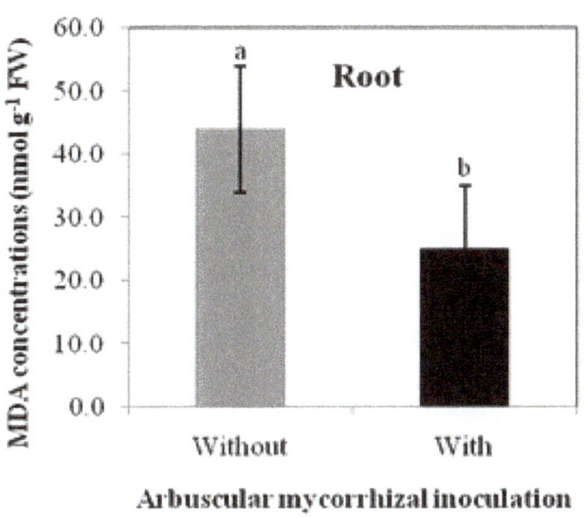


***Figure 3.*** *MDA concentration in roots of seedlings inoculated with arbuscular mycorrhiza (mean ± SE, n = 9). Significant difference among AM inoculation treatments are indicated by different letter (LSD's test, p<0.05).*

In contrast to MDA concentration in roots, MDA concentration in leaves of *N. orientalis* seedlings was significantly affected by combined effects of Hg application and AM inoculation (Fig. 4). At the 0 $\mu$M and 375 $\mu$M Hg concentration, MDA concentration witin leaves of mycorrhizal seedlings were higher than within non-mycorrhizal seedlings, but they did not differ significantly with the MDA concentration in leaves of mycorrhizal seedlings at the 750 $\mu$M Hg concentration.

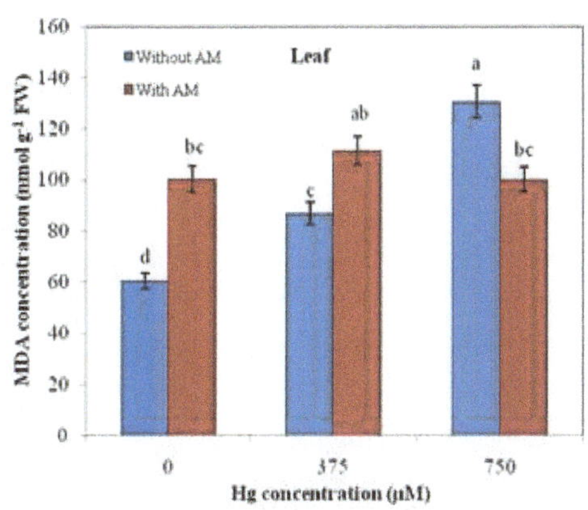


***Figure 4.*** *MDA concentration in leaves of N. orientalis seedlings with different Hg concentration (mean ± SE, n = 3). Significant difference among Hg treatments are indicated by different letter (LSD's test, p<0.05).*

It indicates that the inoculation of AM fungi to *N. orientalis* seedlings minimized the increased of lipid peroxidation when Hg concentration was increased. However, in non-mycorrhizal seedlings, MDA concentration in leaves increased with increased Hg concentration. At the highest concentration of Hg, MDA concentration in leaves of non-mycorrhizal plants was increased significantly by 116% compared to the mycorrhizal plants. In line with this, MDA levels in tomato plant tissues increase in response to Hg [4] as a consequence of higher lipid peroxidation. Although there was the significant increase of MDA concentration, it did not cause toxicity in non-mycorrhizal seedlings. It reveal that *N. orientalis* is included Hg tolerance species.

## 4. Conclusions

A major part of Hg accumulated by *N. orientalis* seedlings was in root. Translocation of Hg to the aerial part of the plant (leaves) was limited so that the contents were relatively constant even though the Hg concentration in the growth media was higher. Mycorrhizal symbiosis showed a significant effect of the resistance to Hg stress of *N. orientalis* plants grown in Hg treated and non-treated Hg media. *N. orientalis* seedlings have a high tolerance to Hg exposure (>70%). The recently results indicate the potential of *N. orientalis* and AM fungi to be used in the long-term land reclamation strategy, including land use purposes beneficial in areas Hg-polluted.

## Acknowledgements

We thanks to the Southeast Asian Regional Centre for Tropical Biology (SEAMEO BIOTROP), thesis research grants for 2012, financially supported this research.

## References

[1] [ATSDR] Agency for Toxic Substances and Disease Registry, [Internet], The Priority List of Hazardous Subtances that Will Be the Subject of Toxicological Profiles. Atlanta, Georgia (US): Department of Health and Human Services, 2013. [cited 2013 August 30]. Available from http://www.atsdr.cdc.gov/SPL/index.html.

[2] [M Patra, A Sharma, Mercury toxicity in plants, Bot Rev, Vol. 66, pp. 379-422, 2000.

[3] J. Sobrino-Plata, C. Ortega-Villasante, M.L. Flores-Cáceres, C. Escobar, F.F. Del Campo, L.E. Hernández, Differential alterations of antioxidant defenses as bioindicators of mercury and cadmium toxicity in alfalfa, Chemosphere, Vol. 77, pp. 946–954, 2009.

[4] U.H. Cho, J.O. Park, Mercury-induced oxidative stress in tomato seedlings, Plant Sci, Vol. 156, pp. 1-9, 2000.

[5] E. Moreno-Jiménez, J.M. Penalosa, E. Esteban, R.O. Carpena, Mercury accumulation and resistance to mercury stress in Rumex induratus and Marrubium vulgare grown on perlite, J Plant Nutr Soil Sci, Vol. 170, pp. 485-494, 2007.

[6] S. Shiyab, J. Chen J, F.X. Han, D.L. Monts, F.B. Matta, M. Gu, Y. Su, Phytotoxicity of mercury in Indian mustard (Brassica juncea L.), Ecotoxicology and Environmental Safety, Vol. 72, pp. 619-625, 2009. DOI:10.1016/j.ecoenv.2008.06.002.

[7] J. Chen, S. Shiyab, F.X. Han, D.L. Monts, C.A. Waggoner, Z.M. Yang, Y. Su, Bioaccumulation and physiological effects of mercury in Pteris vittata and Nephrolepsis exaltata. Ecotoxocology, Vol. 18(1), pp. 110-121, 2009. DOI 10.1007/s10646-008-0264-3.

[8] D.E. Salt, M. Blaylock, N.P.B.A. Kumar, V. Dushenkov, B.D. Ensley, I. Chet I, I. Raskin, Phytoremediation: a novel strategy for the removal of toxic metals total carotenoids, using various solvents with spectrophotometers of different from the environment using plants, Biotechnology, Vol. 13, pp. 468-474, 1995.

[9] B.H. Robinson, R. Schulin, B. Nowack, S. Roulier, M. Menon, B.E. Clothier, S.R. Green, T.M. Mills, Phytoremediation for the management of metal flux in contaminated sites, Forest Snow Landscape Res, Vol. 80(2), pp. 221-234, 2006.

[10] I.D. Pulford, N.M. Dickinson, "Phytoremediation Technologies Using Trees", in Trace elements in the environment, M.N.V. Prasad, K.S. Saiwan, R. Naidu, Eds. New York: Taylor and Francis, CRC Press, pp. 375-395, 2005.

[11] M. Greger, Y. Wang, C. Neuschútz, Absence of Hg transpiration by shoot after Hg uptake by roots of six terrestrial plant species, Environmental Pollution, Vol. 134, pp. 201-208, 2005.

[12] C. Ortega-Villasante, R. Rellán-Alvarez, F.F. Del Campo, R.O. Carpena Ruiz, L.E. Hernández, Cellular damage induced by cadmium and mercury in Medicago sativa, Journal of Experimental Botany, Vol. 56, pp. 2239-2251, 2005.

[13] L. Rodríguez, J. Rincón, I. Asencio, L. Rodríguez-Castellanos, Capability of selected crop plants for shoot mercury accumulation from polluted soils: Phytoremediation perspectives, International Journal of Phytoremediation, Vol. 9, pp. 1-13, 2007.

[14] M.J. Sierra, R. Millán, E. Esteban, A.I. Cardona, T. Schmid, Evaluation of mercury uptake and distribution in Vicia sativa L. applying two different study scales: Greenhouse conditions and lysimeter experiments, Journal of Geochemical Exploration, Vol. 96, pp. 203-209, 2008.

[15] World Agroforestry Centre, A tree species reference and selection guide. AgroForestryTree Database, PROSEA,http://www.worldagroforestrycentre.org/sea/products/afdbases/af/asp/SpeciesInfo.asp?SpID=17928. [cited March 13, 2012]

[16] S.E. Smith, D.J. Read, Mycorrhial Symbiosis. 3rd ed., Great Britain: Academic Press, 2008.

[17] M.C. Gonzalez-Chavez, S.F. Wright, K.A. Nichols, The role of glomalin, a protein produced by arbuscular mycorrhizal fungi, in sequestering potentially toxic elements, Environmental Pollution, Vol. 130, pp. 317-323, 2004.

[18] M. Janoušková, D. Pavlíková, M. Vosátka, Potential contribution of arbuscular mycorrhiza to cadmium immobilisation in soil, Chemosphere, Vol. 65, pp. 1959-1965, 2006.

[19] A.P.G.C. Marques, R.S. Oliveira, A.O.S.S. Rangel, P.M.L. Castro, Zinc accumulation in Solanum nigrum is enhanced by different arbuscular mycorrhizal fungi, Chemosphere, Vol. 65, pp. 1256-1263, 2006.

[20] Y. Yu, S. Zhang, H. Huang, Behavior of mercury in a soil–plant system as affected by inoculation with the arbuscular mycorrhizal fungus Glomus mosseae, Mycorrhiza, Vol. 20, 407-414, 2010.

[21] P. Zornoza, R. Millan, M.J. Sierra, A. Seco, E. Esteban, Efficiency of white lupin in removal of mercury from contaminated soils: Soil and hydroponic experiments, Journal of Environmental Sciences, Vol. 22(3), pp. 421-427, 2010.

[22] F.X.X. Han, A. Banin, Y. Su, D.L. Monts, M.J. Plodinec, W.L. Kingery, G.E. Triplett, Industrial age anthropogenic inputs of heavy metals into the pedosphere, Naturwissenschaften, Vol. 89: 497-504, 2002.

[23] D.I. Arnon, Copper enzymes in isolated chloroplast. Polyphenoloxidase in Beta vulgaris. Plant Physiol, Vol. 24, pp. 1-15, 1949.

[24] M. Brundrett, N. Bougher, B. Dell, T. Grove, N. Malajczuk, Working with Mycorrhizas in Forestry and Agriculture. ACIAR Monograph 32. 374 + x p., 1996.

[25] Y. Wang, M. Greger, Plant and environment interactions: Clonal differences in mercury tolerance, accumulation, and distribution in willow, J Environ Qual, Vol. 33, pp. 1779-1785, 2004.

[26] G.H. Rabie, Contribution of arbuscular mycorrhizal fungus to red kidney and wheat plants tolerance grown in heavy metal-polluted soil. African Journal of Biotechnology, Vol. 4(4), pp. 332-345, 2005.

[27] S. Iglesia-Turiño, A. Febrero, O. Jauregui, C. Caldelas, J.L. Araus, J. Bort , Detection and quantification of unbound phytochelatin 2 in plant extracts of Brassica napus grown with different levels of mercury, Plant Physiology, Vol. 142, pp. 742-749, 2006.

[28] R. Rellán-Álvarez, C. Ortega-Villasante, A. Álvarez-Fernández, F.F. del Campo, L. E. Hernández, Stress responses of Zea mays to cadmium and mercury, Plant and Soil, Vol. 279, pp. 41-50, 2006.

[29] M. Kaldorf, A. J. Kuhn, W.H. Schrőder, U. Hildebrandt, H. Bothe, Selective element deposits in maize colonized by a heavy metal tolerance conferring arbuscular mycorrhizal fungus, J Plant Physiol, Vol 154, pp. 718-728, 1999.

[30] E.J. Joner, R. Briones, C. Leyval, Metal-binding capacity of arbuscular mycorrhizal mycelium, Plant Soil, Vol. 226, pp. 227-234, 2000.

[31] G. Rufyikiri, S. Declerck, J.E. Dufey, Arbuscular mycorrhizal fungi might alleviate aluminium toxicity in banana plants, New Phytol, Vol. 148(2), pp. 343-352, 2000.

[32] X. Yu, J. Cheng, M.H. Wong, Earthworm-mycorrhiza interaction on Cd uptake and growth of ryegrass, Soil Biol Biochem, Vol. 37, pp. 1-7, 2004.

[33] S.H. Burleigh, B.K. Kristensen, I.E. Bechmann, A plasma membrane zinc transporter from Medicago truncatula is upregulated in roots by Zn fertilization, yet down - regulated by arbuscular mycorrhizal colonization, Plant Mol Biol, Vol. 52(5), pp. 1077-1088, 2003.

[34] R. Malcova, M. Vosatka, M. Gryndler, Effects of inoculation with Glomus intraradices on lead uptake by Zea mays L. and Agrostis capillaris L., Appl Soil Ecol, Vol. 23(1), pp. 255-267, 2003.

[35] G. Rufyikiri, L. Huysmans, J. Wannijn, M. van Hees, C. Leyval, I. Jakobsen, Arbuscular mycorrhizal fungi can decrease the uptake of uranium by subterranean clover grown at high levels of uranium in soil, Environ Pollut, Vol. 130, pp. 427-436, 2004.

[36] Y. Yu, S.Z. Zhang, H. L. Huang, L. Luo, B. Wen, Arsenic accumulation and speciation in maize as affected by inoculation with arbuscular mycorrhizal fungus Glomus mosseae, J Agr Food Chem, Vol. 57, pp. 3695-3701, 2009.

[37] M. Válega, A. I. G. Lima, E. M. A. P. Figueira, E. Pereira, M.A. Pardal, A.C. Duarte, Mercury intracellular partitioning and chelation in a salt marsh plant, Halimione portulacoides (L.) Aellen: strategies underlying tolerance in environmental exposure, Chemosphere, Vol. 74, pp. 530-536, 2009.

[38] J.A. Ericksen, M.S. Gustin, Foliar exchange of mercury as a function of soil and air mercury concentrations, Sci Total Environ, Vol. 324, pp. 271-279, 2004.

[39] L. Fay, M. Gustin, Assessing the influence of different atmospheric and soil mercury concentrations on foliar mercury concentrations in a controlled environment, Water Air Soil Poll, Vol. 181, pp. 373-384, 2007.

[40] E. Lozano-Rodríguez, L. E. Hernández, P. Bonay, Distribution of cadmium in shoots and root tissues of maize and pea plants: physiological disturbances, Journal Experimental Botany, Vol. 48, pp. 123-128, 1997.

[41] A. Schützendübel, P. Schwanz, T. Teichmann, K. Gross, R. Langenfeld-Heyser, D.L. Goldbold, A. Polle, Cadmium-induced changes in antioxidant systems, hydrogen peroxide content and differentiation in Scot pine roots, Plant Physiol, Vol. 127, pp. 887-898, 2001.

[42] L. E. Hernández, C. Ortega-Villasante, M. B. Montero-Palmero, "Heavy Metal Perception in a Microscale Environment: A Model System Using High Doses of Pollutants", in: Metal Toxicity in Plants: Perception, Signaling and Remediation, D. K. Gupta, L. M. Sandalio, Eds. Berlin Heidelberg: Springer-Verlag, 2012. DOI 10.1007/978-3-642-22081-4_2.

[43] U. Hildebrandt, M. Regvar, H. Bothe, Arbuscular mycorrhiza and heavy metal tolerance, Phytochemistry, Vol. 68, pp. 139-146, 2007.

[44] S. Cheng, Effects of Heavy Metals on Plants and Resistance Mechanisms. A State-of-the-Art Report with Special Reference to Literature Published in Chinese Journals. ESPR- Environ Sci & Pollut Res, Vol. 10(4), pp. 256-264, 2003. DOI http://dx.doi.or~10.1065/espr2002.11.141.

[45] M. N. V. Prasad, "Biomarkers", in Heavy Metal Stress in Plants. From Molecules to Ecosystem, 2nd ed., M. N. V. Prasad, J. Hagemeyer, Eds. Berlin: Springer-Verlag, pp. 445-448, 2003.

[46] D. D. K. Prasad, A. R. K. Prasad, Altered d-aminolevulinic acid metabolism by lead and mercury in germinating seedlings of bajra (Pennisetum typhoideum), J Plant Physiol, Vol. 127, pp. 241-249, 1987.

[47] A. Schützendübel, A. Polle, Plant responses to abiotic stresses: heavy metal-induced oxidative stress and protection by mycorrhization, J Exp Bot, Vol. 53, pp. 1351-1365, 2002.

# 13

# Predicting growth and yield models for *Eucalyptus* species in Aek Nauli, North Sumatera, Indonesia

**Siti Latifah[1], Teodoro Reyes Villanueva[2], Myrna Gregorio Carandang[2], Nathaniel Cena Bantayan[2], Leonardo M. Florece[3]**

[1]Forestry Program Study, Sumatera Utara University, Medan, Indonesia
[2]Forest Resources Management, University of the Philippines Los Banos, Los Banos, Philippines
[3]Environmental Science, University of the Philippines Los Banos, Los Banos, Philippines

**Email address:**
sitilatifah 164@yahoo.co.id (S. Latifah), teodoro_villanueva@yahoo.com (TR. Villanueva), mgc405@yahoo.com (M. G. Carandang), ncbantayan@gmail.com (NC. Bantayan)

**Abstract:** This study was conducted primarily to develop a yield prediction model for *Eucalyptus* spp plantations in Aek Nauli, North Sumatera, Indonesia as a contribution to sustain development and management of forest plantations. Data for growth and yield model were collected from the inventory and permanent sample plots (PSPs). The data in this study used 650 rhombic plots consisting of 106 PSPs and 544 inventory plots with several variations of plot size. Stands' features referred to diameter, height, merchantable volume, age, species, spacing, site index, basal area, and density of *Eucalyptus* species. Models using initial age, specifically, model 2 was found consistently to be the best model in most *Eucalyptus* plantations. Among the models using initial and projection age, model 4 was the better one. Model 2 using original ages looks better than model 4 because of it is being more reliable and its sigmoid growth curve. Nonetheless, significant differences were noticed between different models for predicting the merchantable volume of *Eucalyptus* spp. Plantations. Growth and yield models can be used to identify the best growing species of *Eucalyptus* spp. *E.* hybrid is recommended for plantation in this study area because it had the highest of merchantable volume.

**Keywords:** Eucalyptus, Growth, Yield, Merchantable Volume, Forest Plantation

## 1. Introduction

Indonesia is the world's largest exporter of tropical timber, generating income of US$ 5 billion annually, and more than 48 million hectares (55 per cent of its remaining forests) are earmarked for logging.. During 1990- 2005, the country lost more than 28 million ha of forest indicating annual change in forest cover of -1,871,400 ha or annual deforestation rate of 2.0 per cent. Total forest loss since 1990 is around 24.1% (Mongabay, 2006).

One potential method for reducing tropical forest destruction is the creation of environmentally sustainable plantations that can supply wood and paper pulp product in order to reduce demand for wood from virgin forest (Hartono, 2002).

The Government of Indonesia is developing timber estates to increase forest sector wood products. Wood industries are increasing rapidly; they need more than 40 million $m^3$ of wood production every year. This cannot be supplied from the country's natural forests alone and Indonesia, thus, need developing forest plantations. These programs are executed by private companies with forest concession areas (e.g. Industrial Timber Estate or *HTI_ Hutan Tanaman Industry* ) or State Forest Corporations (e.g. Perhutani and Inhutani) (Pramono and Agung, 1996).

In this case, Indonesian government policy will help rapid increase in Eucalypt plantations for industrial wood use. *Eucalyptus* is favored for its rapid growth, indigenous origin and multipurpose use. Its major uses in Indonesia are for pulp (PT. TPL, 2008). Sumatera However, in Indonesia, the estimation of yield and growth model for *Eucalyptus* spp has not been adequately investigated.

Growth modeling of plantation timber species is a vital tool for predicting yield from future harvesting and financial returns. One important criterion of sustainable forest management is long term maintenance of yield. These requirements have implications for model design, its implementation and commercial use. For robust predictions, reliable models and input data of high accuracy are a must. Sustained yield regulation is a prerequisite for sustainable management of forest resources. Its importance is recognized internationally in developing standards for sustainable forest management following ITTO guidelines' Principle 17. In order to ensure a sustained production of timber from each forest management unit, a reliable method for controlling timber yield should be adopted (ITTO, 1990).

One of the three essential elements of forest management is prognosis, that is, growth and yield prediction. Forest cannot be managed without proper information on the future availability of wood and/or-sub product ( Bantayan, 2006). Such information is obtained by growth and yield models. According to Vanclay (2003), there are many good reasons for building models. Building a model can help to analyze data from experiments and observation plots. It can help to focus research plans and to anticipate results, so that measurement programs and techniques can be planned accordingly. Modeling may be the most efficient way to examine experimental data, investigate implications and formulate optimal silvicultural guidelines

For these reasons, managing these forests has greatly increased the demand for accurate growth and yield information on forest management. Thus, this study essentially seeks to answer the questions: (1) what is the growth and yield performance of *Eucalyptus* plantations in Aek Nauli, North Sumatera Indonesia?; (2) What factors (as independent variables) are related to the stand yield?;(3) Which is the best growing species of *Eucalyptus* plantations ?

## 2. Material and Method

### *2.1. Study Site*

The study was conducted in Aek Nauli sector of the Industrial Timber Estate (*Hutan Tanaman Industri* - HTI) namely, PT. Toba Pulp Lestari (PT. TPL), from December 2008 – March 2009. It is in the territorial jurisdiction of Porsea subdistric, Simalungun regency, North Sumatera Province, Indonesia (Figure 1). The study area whose plots were used in the study belongs to climate A based on Schmith and Ferguson (1951). It has the highest annual rainfall in October and the lowest in August. Based on its topography, the study area has a rainfall range of 85 – 434 mm/month, it has an elevation range from 350 to 1400 msal and it has top soil depth ranges from 10 to 44 cm. The study area covered an area 189,975 ha with Aek Nauli sector has an area 21341.8 ha (Figure 1). Geographically, it is located between 02° 40'00" to 02° 50'00" north latitude and 98° 50'00" to 99° 10'00" east longitude.

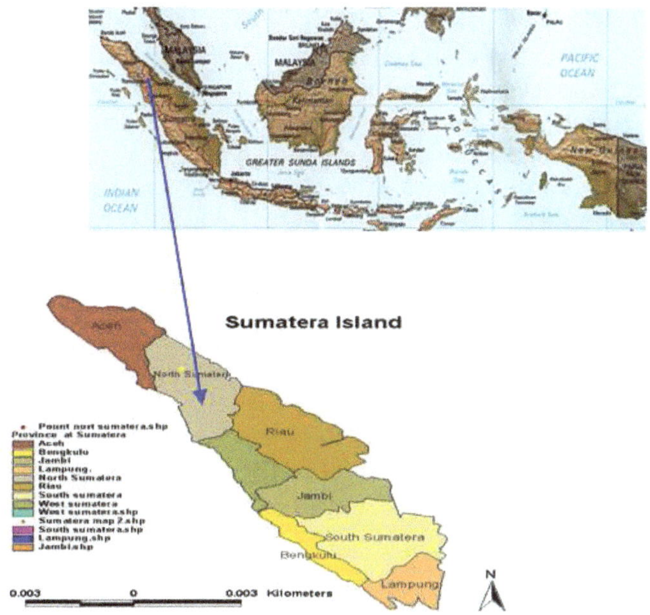


*Figure 1. Location of study site*

### *2.2. Materials*

The data in this study used 650 rhombic plots consisting of 106 permanent sample plots (PSPs ) and 544 inventory plots with several variations of plot size, from 0.02 to 0.04 ha. PSPs have an area of 3.44 ha. It consisted of 106 plot or 273 observations. Inventory plot has an area of 21.76 ha. It consisted of 544 plots ~~(Table 2)~~. Thus, the total area of the plots in this study site is 25.2 ha.

Each selected sample plot was located in an area where the stand was of even age, uniformly spaced, and disease free. Ten percent (10%) of total plots equal to sixty five (65) rhombic plots were taken with random samples from inside 650 plots which were used for validation of the yield model. Dependent variable of the yield stand was merchantable volume. Based on the analyses of factors affecting yield, growth and yield were expressed as a function of independent variables (Davis, 2001). Generally, yield (V) is considered to be a function of basal area, top height, age, site index, stand density, spacing as stand variables while top soil, rain fall, slope, elevation, and type of soil as geographic variables.

Species were grouped into *E.* hybrid, *E. grandis* (*E.gra*), *E. pellita, E. urophylla* (*E. uro*), mixed *Eucalyptus,* and all *Eucalyptus species.* The data sets for the model were derived from 470 *E.* hybrid, *94 E.grandis*, 23 *E. pellita*, 46 *E. urophylla* and 119 mixed *E.* observations. *E.* hybrid is breeding from E. *urophylla* vs *E. grandis* and *E. pellita* vs *E. grandis*. Mixed *Eucalyptus* observations consisted of *E.*hybrid, *E.grandis*, *E. pellita*, and *E. urophylla species.*

### *2.3. Data Analysis*

According to Davis et al. (2001), growth and yield can be measured in physical units such as volume, basal area, height, density and weight. Also, they can be measured in value, which is the measure of economic and social interest. In this study area, focus concentrated on physical measures of growth and yield namely, merchantable volume, because it forms the basis of value measures.

There have been many general forms of standard volume equations presented by many authors for a variety of conditions. There is no specific equation which is generally applicable and valid for all species. When tree volume is expressed as a function of both height and diameter, the relationship is called standard volume equation.

The regression equations used the logarithmic transformation models which were modified from the best model developed by some research such as that of Silva et al (2007), Pereira et al (2006), Yue et al ( 2007) ,Villar (2005), Aswandi (2000) Bi et al., ( 1994), Bennet (1970) as cited by Clutter (1983), Sullivan and Clutter (1972). For the purpose, the following equations were generated. The models tested are shown in Table 1.

***Table 1.*** *Regression models tested for estimating growth and yield for Eucalyptus spp*

| Model Number | Model |
|---|---|
| 1 | $\ln V = bo + b_1 \ln BA + b_2 \ln (Age^{-1}) + b_3 (SI) + b_4 (SD) + b_5 (SP) + b_6 (TS) + b_7 (R) + b_8 (SL) + b_9 (E) + b_{10} (DA) + b_{11} (DB) + b_{12} (DC) + b_{13} (DD)$ |
| 2 | $\ln V = bo + b_1 (\ln BA * \ln TH) + b_2 (Age^{-1}) + b_3 (SI^{-1}) + b_4 (SD) + b_5 (SP) + b_6 (TS) + b_7 (R) + b_8 (SL) + b_9 (E) + b_{10} (DA) + b_{11} (DB) + b_{12} (DC) + b_{13} (DD)$ |
| 3 | $\ln V = bo + b_1 (\ln BA * \ln Age) + b_2 (Age^{-1}) + b_3 SI/(Age^{-1}) + b_4 \ln (SD) + b_5 (SP) + b_6 (TS) + b_7 (R) + b_8 (SL) + b_9 (E) + b_{10} (DA) + b_{11} (DB) + b_{12} (DC) + b_{13} (DD)$ |
| 4 | $\ln V_2 = bo + b_1 \ln (V_1) + b_2 (\ln BA * (Age_1/Age_2)) + b_3 (1/Age_2) + + b_4 (1 - Age_1/Age_2)) + b_5 (1/SI) + b_6 \ln (SD) + b_7 (SP) + b_8 (TS) + B_9 (R) + B_{10} (SL) + b_{11} (E) + b_{12} (DA) + b_{13} (DB) + B_{14} (DC) + B_{15} (DD)$ |
| 5 | $\ln V_2 = bo + b_1 \ln (V_1) + b_2 \ln BA + b_3 (1 - Age_1/Age_2)) + b_4 (1/Age_2) + B_5 (SI/Age_2) + B_6 \ln (SD) + b_7 (SP) + b_8 (TS) + b_9 (R) + b_{10} (SL) + B_{11} (E) + b_{12} (DA) + b_{13} (DB) + b_{14} (DC) + b_{15} (DD)$ |
| 6 | $\ln V_2 = bo + b_1 \ln (V_1) + b_2 (\ln BA * (Age_1/Age_2)) + b_3 (1/Age_2) + b_4 (1 - Age_1/Age_2)) + B_5 (SI/Age_2) + B_6 \ln (SD) + b_7 (SP) + b_8 (TS) + b_9 (R) + b_{10} (SL) + B_{11} (E) + b_{12} (DA) + b_{13} (DB) + b_{14} (DC) + b_{15} (DD)$ |

where :

V= merchantable volume in cubic meter per hectare

V1= merchantable volume in initial ages in cubic meter per hectare

V2= merchantable volume in projection ages in cubic meter per hectare

BA= basal area in square meters per hectare

TH= top height in meter

Age= original age in year

Age1 and Age 2 = initial and projection ages in year

SI= site index in meter

SD= stand density (tree/ha)

SP= original spacing in square meters;

TS= depth of top soil in centimeter (cm)

R= rain fall monthly (mm)

SL= slope in percent (%)

E= elevation in masl

DA, DB, DC and DD= dummy variable for group of soil

ln= natural logarithm or logarithm to the base 2.718281828

$b_0, b_1, \ldots bn$ = regression coefficient

1,2,3= model using original age

4,5,6 = model using initial and projection ages

One approach to simplifying multiple regression equations is the stepwise regression. The procedure is probably the most widely used method for subset regression analysis (ANALYSTSOFT, 2008).

All equations were tested to determine which one is the best predictor of growth and yield for *Eucalyptus* species. The merchantable volume models were tested and respective fits of each the models were mainly assessed using criteria as follows:

1. The best model is the one which gives the largest coefficients of determination ($R^2$) and coefficients of correlation (r), smallest mean square error (MSE) and values of Mallow's C(p) and Variance Inflation Factor (VIF)
2. Significant F–values as computed in the analysis of variance (ANOVA) that test the overall regression given the intercept term;
3. Consider the significance of each the independent variables;
4. Randomness of the residual as shown in the graph of the residual error values versus fitted merchantable volume values;
5. Considers the parsimony of parameters. It is a subset of the independent variables which have to be small enough to manage cost and analysis facilitated but large enough for adequate description, control or prediction. That means, when all the leading equations appear to fit the data equally well, the simplest equation should be chosen as the best model.

The criteria that determine the best model is weighted with each criterion in the same scale as follows:

Score of weighted = [ ($R^2$) + (r) - (MSE) – (Cp) + (Σ independent variables are significant / Σ independent variables in the model ) + (Σ independent variables are not multicollinerity / Σ independent variables in the model )]

The model selected should provide the highest accuracy as the models are verified against a separate set of yield data. The validation of the models developed in this study was done by using the chi-square test to determine the accuracy of the prediction models. This method enables one to determine whether the models follow the real world, can be examined using chi-square test of goodness of fit. This method enables one to determine whether the models follow the real world, can be examined using chi-square test of goodness of fit. Specifically, the form of the chi-square is as follows:

$$\chi 2_{C(\alpha, n-1)} = \sum_{i-1}^{n} \frac{(Oi - Ei)^2}{Ei}$$

where :

$\chi 2$ C = is the chi-square value calculated
α = level of significance
n = number of observations
Oi = observed value
Ei = expected (true) value

# 3. Results and Discussion

### 3.1. Selecting the Best Model

The regression equations were tested using several candidate regressions from non-linear models. When the relationship is non –linear, the model should be based on the nature of the curve representing given relationship. Two general types of equations are commonly used in regression analysis involving variables that exhibit non-linear relationship namely polynomial and exponential equations. These equations can be transformed into linear equations thru polynomial or logarithmic transformation.

Forest growth models in this study involving the logarithm of merchantable volume are the dependent variable. The use of the logarithm of yield as dependent variable is a convenient way to mathematically express the interaction of the independent variables in their effect on yield. Stand age, in most yield analysis, has been expressed as a reciprocal to allow for the 'leveling off'( asymptotic) effect of yield with increasing age. Site index is not often transformed prior to fitting, but sometimes logarithmic or reciprocal transformations are employed. In some models, height of dominant stand has been used in conjunction with age and the variable site index eliminated. The measure of stand density is commonly subjected to logarithmic transformation, particularly in models employing basal area, but the exact form in which density is included is quite variable, especially for plantation models which utilize number of trees per unit area as an independent variable. (Avery and Burkhart, 1994 )

The regression equations used two groups of independent variables. One group ( model 1, 2 and 3) used original age, the basal area, height, site index, stand density, spacing, top soil, rain fall, slope, elevation and type of soil as independent variable. The other group used initial and projection age, volume, the basal area, height, site index, stand density, spacing, top soil, rain fall, slope, elevation and type of soil in initial age as independent variable.

Among the models using initial age, model 2 of the stepwise procedure was found consistently to be the best model because it was accepted in almost *Eucalyptus* species groups and also the model had the highest of the coefficients of determination ($R^2$) and coefficients of correlation (r), and the lowest mean square error (MSE) and Mallow's C(p), almost all independent variables are significant and also almost the independent variables is not multicollinerity. Similarly, among the models using initial and using initial and projection age, model 4 was the best (Table 2)

### 3.2. Validation

Model validation is possibly the most important step in the model building sequence. It is also one of the most overlooked. A good representation of how the effectiveness a yield model as a predictive tool is its behavior in a real situation. Figure 2 and 3 show the comparison of graph between the actual data and predicted data by the best models for all species group. The angle between Y-axis and a line on the graph is nearly $45^0$ . Likewise is the angle between X- axis and a line on the graph.

***Table 2.*** *Coefficients of determination ($R^2$), coefficients of correlation (r), mean square error (MSE), and Mallow's C(p), P-value and VIF among the best models*

| Species Group | Model | Model Adequate | | Prescision | | Test of Hyptheses on β's | Multi Collinear (MC) |
|---|---|---|---|---|---|---|---|
| | | $R^2$ | R | MSE | C(P) | P-Value | VIF |
| *E.* hybrid | 2 | 0.919 | 0.957 | 0.7105 | 2.4271 | 4/6 significant | 6/6 not mc |
| | 4 | 0.910 | 0.954 | 0.2532 | 3.1891 | 8/9 significant | 9/9 not mc |
| *E.grandis* | 1 | 0.983 | 0.991 | 0.0341 | 2.5881 | 4/5 significant | 3/5 not mc |
| | 4 | 0.953 | 0.976 | 0.0341 | 3.2016 | 4/6 significant | 6/6 not mc |
| *E. pellita* | 2 | 0.995 | 0.997 | 0.0055 | 2.9787 | 4/4 significant | 2/4 not mc |
| | 5 | 0.982 | 0.991 | 0.0152 | 3.8920 | 5/5 significant | 5/5 not mc |
| *E. urophylla* | 2 | 0.925 | 0.962 | 0.5840 | 0.0174 | 4/4 significant | 4/4 not mc |
| | 4 | 0.990 | 0.995 | 0.0259 | 18.8819 | 6/6 significant | 4/6 not mc |
| Mixed *Eucalyptus* | 3 | 0.913 | 0.956 | 0.3977 | 3.4095 | 5/6 significant | 6/6 not mc |
| All species | 2 | 0.903 | 0.950 | 0.7763 | 2.2983 | 8/8 significant | 7/8 not mc |
| | 4 | 0.830 | 0.911 | 0.3775 | 3.6742 | 6/6 significant | 6/6 not mc |

mc : multicollinerity

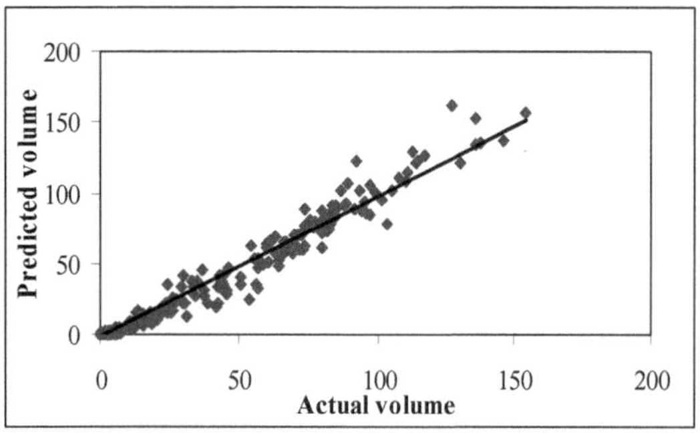


*Figure 2. Actual and predicted volume all species group using original age*

The capability of any prediction model to give accurate estimates is the real test for any such model. It is not necessary to establish that a model is "perfect", which is probably impossible because all models are abstractions of reality. But it is important to establish that the model is correctly implemented and free of errors. To test the accuracy of the yield model whether the model particularly follows the real world or not, two sets of observations were used in testing the goodness of fit of the yield model using chi-square ($\chi 2$) test.

Based on the results of computation, it can be concluded that there is no significant difference between the actual and the predicted value of merchantable volume. This was because p-value of the chi-square ($\chi 2$) of the model using original ages for E. *hybrid, E.grandis*, *E. pellita*, *E. urophylla*, *E. mix* and all species observations, respectively is 0.1601, 0.7587, 0.9952, 0.6191, 0.9999 and 0.0557 while p-value of the chi-square ($\chi 2$) of the model using initial and projection ages for E. *hybrid, E.grandis*, *E. pellita*, *E. urophylla*, and all species observations, respectively is 0.0505, 0.707, 0.3279, 0.2934 and 0.0538. The p-value of the chi-square ($\chi 2$) of both group models is greater than the corresponding 0.05 level of significance

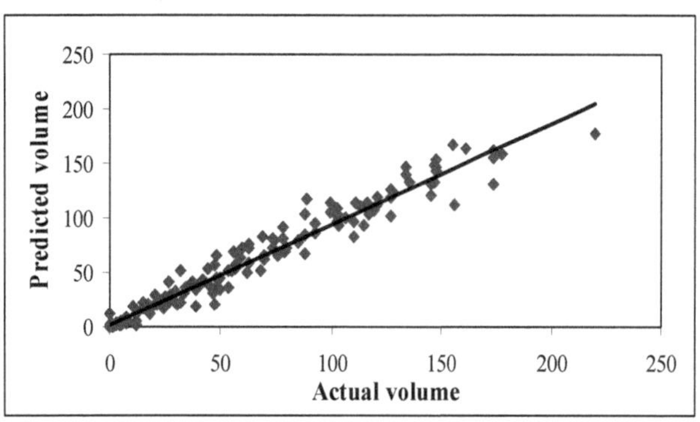


*Figure 3. Actual and predicted volume all species group using initial and projection age*

The residual plot is the best model shown which offers a good example of what a problem-free plot should look like (Figure 4.). The residual plot is the best model shown which offers a good example of what a problem-free plot should look like. There are no curved trends in the plot, the average of the residuals is zero, and the points are equally represented about the x-axis. This residual represents the difference between the observed response merchantable volume and the value predicted by the regression line. If the regression model represents the data correctly, the residuals should be randomly distributed around the line of error = 0 with zero mean.

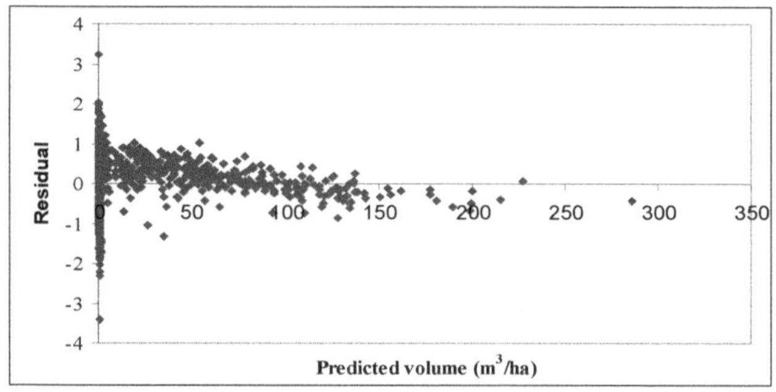


*Figure 4. Example of residuals plotted of all Eucalyptus species for the best model (model 2) using original ages*

Both the best models using original ages and using initial and projection age are precise enough since there no significant differences in results compared to those using the chi-square method. Both models are biologically realistic, because all independent variables of both models follow the logical factors affecting yield. Models using original age are better than models using initial and projection age because it had the highest total weighted score for those criteria and also it is more reliable because the growth of stands follows a sigmoid shape of the curve. (Figure 5 and 6.)

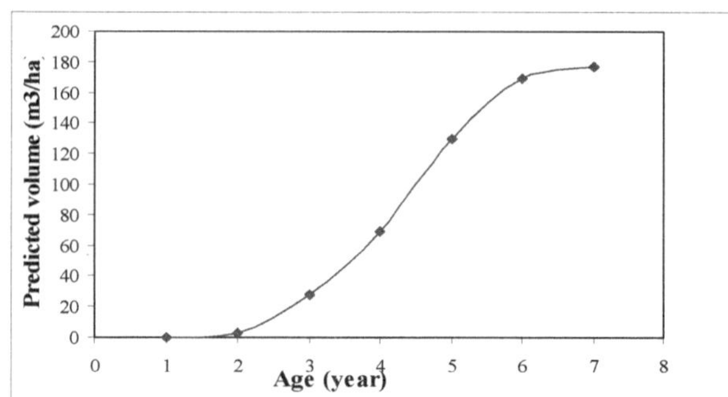


*Figure 5. Yield curve for all Eucalyptus species using original ages*

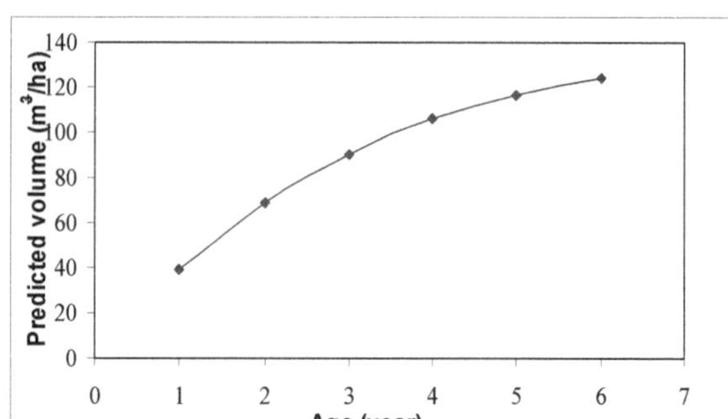


*Figure 6. Yield curve for all Eucalyptus species using initial and projection age.*

### *3.3. The Application of the Best Model to Get the Best Growing Species*

The timber properties of grown *Eucalyptus* in the past have proven to have problems, but with the current increased interest in biomass for fiber and energy production, Eucalyptus species may now have a commercially viable endues. Various *Eucalyptus* species have been planted in Aek Nauli sector with varying degrees of success. As mentioned in the beginning of this chapter, in this study area, the species are grouped into *E. hybrid* (E.hyb), *E. grandis* (*E.gra*), *E pellita E. urophylla* (*E. uro*), E. Mix *and E all species*. The problem is in deciding which species to plant that have high productivity. The author would like to find out if these six group species produce different average merchantable volumes. These yields were obtained by the prediction value of merchantable volume.

Models allow us to forecast future trends thereby helping us make more effective decisions (Villanueva, 2005) .Models using initial age was selected as the best model to predict growth and yield of *Eucalyptus* species, with a variant for each species. The final models for each eucalyptus species are:

For *E. hybrid*

$$V = \exp [\, 0.12043 + 0.46844 (\ln BA * \ln TH) - 1.84144 (Age^{-1}) + 0.03701 (TS) + 0.00129 (R) + 0.14059 (DC) \,]$$

For *E. grandis*

$$V = \exp [0.30190 + 1.38912 \ln BA - 0.40322 \ln (Age^{-1}) + 0.02682 (SI) - 0.00054 (N)]$$

For *E. pellita*

$$V = \exp [1.65117 + 0.42590 (\ln BA * \ln TH) - 0.93169 (Age^{-1}) - 0.00023 (E)]$$

For *E. urophylla*

$$V = \exp [-3.22314 + 0.79323 (\ln BA * \ln TH) + 0.00351 (R) + 0.00094771 (E)]$$

For mixed *Eucalyptus*

$$V = \exp [3.98198 + 0.81678 (\ln BA * \ln Age) - 2.16647 (Age^{-1}) + 0.07019\, SI/(Age^{-1}) + -0.90034 \ln (N) + 0.13791 (TS)]$$

For E. all species

$$V = \exp [(-1.07511) + 0.50996 (\ln BA * \ln TH) - 1.39383 (Age^{-1}) - 0.51483 (SI^{-1}) + 0.11307 (SP) + 0.04181 (TS) + 0.00103 (R) - 0.22406 (DA)]$$

Pereira et al (2006) reported that the prediction equation for *Eucalyptus grandis* in central Western region of Minas Gerais Brazil, including age, site index and basal area were fitted. These models fitted were also found to be good enough to predict growth and yield in *E. grandis* plantations in North Sumatera, Indonesia (Aswandi, 2000).

***Table 3.*** *Analyses models of variance (ANOVA) of the different yield prediction for the E. hybrid, E.grandis, E. pellita, E. urophylla, mixed Eucalyptus and all species*

| SV | SS | DF | MS | F | P-Value | F Crit |
|---|---|---|---|---|---|---|
| Between Groups | 1240.726 | 5 | 248.145 | 42.782 | 2.64E-41 | 2.220 |
| Within Groups | 8688.8041 | 1498 | 5.800 | | | |
| Total | 9929.530 | 1503 | | | | |

Based on the result of the analysis of variance using the six models , it can be concluded that there is significant difference between models among them to predict the merchantable volume of eucalyptus species. This is because the value of test statistic F is 42.782 is greater than the corresponding tabular F value of 2.22 at 0.05 level of significance ( Tabel 3). Consequently, each Eucalyptus species has the specific model to predict the yield.

When the result of ANOVA indicates significant difference between models among them, we need to look which group of means are different or not. Based on the result of Duncan test, there is no significance difference between models for mixed *Eucalyptus, E. pellita, E. urophylla* and all species while there is no significance difference between models for *E. pellita, E. urophylla, E. hybrid, E. grandis, E. hybrid* and all eucalyptus species. It is because p-values are 0.078 and 0.124, respectively which are greater than 0.05 of level significance .

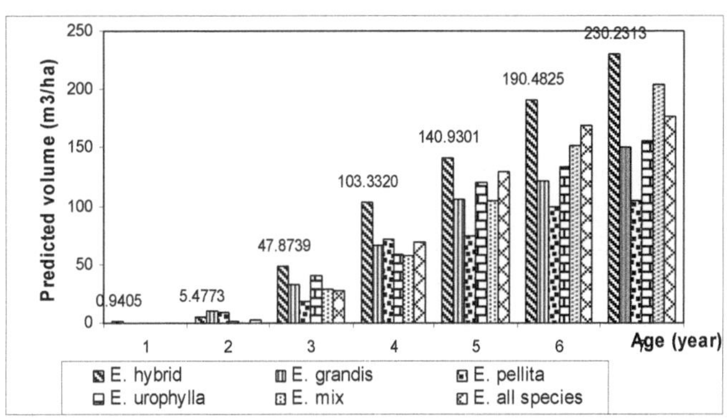


***Figure 7.*** *Predicted volume for different Eucalyptus species*

Based on the specific model for every Eucalyptus spp. to predict the yield, Figure 7 shows the graph for each *Eucalyptus* species. This figure indicates that *E. hybrid* observations has the highest (230 $m^3$ $ha^{-1}$) of merchantable volume in year 7 and *E. pellita* observations has the lowest (104 $m^3$ $ha^{-1}$) of merchantable volume in year 7.

*E.* hybrid has the highest (230 $m^3ha^{-1}$) of predicted volume in year 7 with final model V = exp [ 0.12043 + 0.46844 ( ln BA*ln TH) -1.84144 ($Age^{-1}$) + 0.03701 ( TS) + 0.00129 ( R) + 0.14059 (DC) ]. It means that there is an increase in volume by 0.46844 per unit increase in multiple basal area with top height holding other independent variables constant. There is a decrease in volume by 1.84144 per unit increase in reciprocal age holding other independent variables constant. There is an increase in volume by 0.03701 per unit increase in depth of top soil holding other independent variables constant. There is an increase in volume by 0.0129 per unit increase in rainfall holding another independent variables constant and last there is a increase in volume by 0.14059 per unit increase in dummy variable for group of soil holding another independent variables constant.

The wrong choice of the species on a given site, wrong site selection for a given species and lack of tending operations were the principal factors for the low yield. It is assumed that the highest of merchantable volume is *E. hybrid;* the recommended plantation for this study area is *E. hybrid.* This species is suitable, as suggested by International development agencies, because the breeding new crops is important for ensuring sustainable forest management by developing new varieties that are higher-yielding, resistant to pests and diseases, drought-resistant or regionally adapted to different environments and growing conditions.

## 4. Conclusions

In relation to the objectives of the study, there are several conclusions reached as follows:

1. Based on six criteria used in this study, Model 2 using original ages can be applied to accurately growth and yield at study site. Each *Eucalyptus* species have the specific model to predict growth and yield for the study site.
2. There are eight independent variables affecting the yield of *Eucalyptus* namely basal area, height, age, site index, spacing, depth of top soil, rainfall and group of soil.
3. The quality of predictions obtained from any type of growth and yield model will depend on the availability of suitable data to calibrate and to initialize the model before each simulation.
4. With the highest value of averages merchantable volume at 7 year, *Eucalyptus* hybrid is the best growing species to be planted at the study site.

## Acknowledgement

This research is part of PhD thesis of the first author funded by SEAMEO-BIOTROP DIPA 2008. The authors extend deeply acknowledgment to the Ministry of Education and Culture – Republics Indonesia for scholarship and support to accomplish this paper. Sincerely appreciation is also extended to anonymous reviewer for correction and comments.

## References

[1] ANALYST SOFT. 2008. Stat Plus. http://www.analystsoft.com/en/products/statplus

[2] AVERY, T.E and H.E BURKHART. 1994. Forest Measurement. Fourth Edition. New York: McGraw-Hill, Inc. 408 pp.

[3] ASWANDI, 2000. Growth and Yield of *Eucalyptus grandis* Hill ex Maiden at Aek Nauli Simalungun. North Sumatera.

[4] BANTAYAN, N.C. 2006. Gis in the Philippines : Principles and Applications in Forestry and Natural Resources. First Edition. Published by PARRFI and AKECU. Philippines

[5] BENNET, F. A. 1970. Yield and Stand Structural Patterns for Old Field Plantations on Slash Pine. U. S. D. A. For. Serv., Res. Paper SE-60.

[6] BENNET, F. A. 1970. Yield and stand structural patterns for old-field plantations of slash pine. U.S.D.A. For. Serv., res. Paper SE-60

[7] BI, H., O. CHIKUMBO, and V. JURSKIS. 1994. Yield Equations for Regrowth Forests Regenerated From Fire on The Southeast Coast of New South Wales.http://www.daff.gov.au/__data/assets/pdf_file/0017/50462/nsw_ed_frm5.pdf

[8] CLUTTER, JL, J.C. FORTSON, L.V. PIENAAR, G.H. BRISTER, and R.L. BAILEY. 1983. Timber Management: A quantitative approach. Canada: John Willey & Sons, Inc. 332 p

[9] DAVIS, L.S., K.N. JOHNSON, P.S. BETTINGER and T.E. HOWARD. 2001. Forest Management to Sustain Ecological, Economic and Social Values. McGraw-Hill Companies, Inc. 1221 Avenue of the Americas, New York. NY 10020. 804 p.

[10] HARTONO, B.T. 2002. Can Forest Plantations Alleviate Pressure on natural Forest? an Efficiency Analysis in Indonesia. Research Report no. 2002- RRI in Economy and Environment Program for Southeast Asia.

[11] ITTO. 2001. Annual Review and Assessment of The World Timber Situation.

[12] ITTO. 1990. ITTO Guidelines for The Sustainable Management of Natural Tropical Forest. ITTO Policy Document Series, I., Policy Development Series 1. ITTO, Yokohama, 18 pgs.

[13] MONGABAY. 2006. INDONESIA. http://rainforests.mongabay.com/20indonesia.htm

[14] PEREIRA AJ, HELIO G.L, JOAO C.C.C, AND AGOSTINHO L.S. 2006. Use Data from Permanent Sampling Point in Growth and Yield Modeling. Http://www.scielo.br/scielo.php

[15] PERSEROAN TERBATAS TOBA PULP LESTARI (PT. TPL). 2008. Site Productivity Mapping.

[16] PERSEROAN TERBATAS TOBA PULP LESTARI (PT. TPL). 2008. Summary of the Annual Work Plan Efforts Timber Forest Products Utilization in the Industrial Timber Plantation Forests.

[17] PRAMONO, I.B. and A. PUDJIHARTA. 1996 Research Experiences on Eucalyptus in Indonesia. http://www.fao.org/docrep/005/AC772E/ac772e09.htm.

[18] SILVA, A. L., JOAO, C. C. C., HELIO, G. L. AGUSTINHO, L. D. S., and PABLO, F. L. 2007. Growth and Yield Prediction Using The Modified Buckman Model. Revista. Árvore vol.30 no.5 Viçosa Sept./Oct. 2006.

[19] SULLIVAN, A. D., and J. L.. CLUTTER. 1972. A simultaneous growth and yield model for loblolly pine. For. Sci. 18: 76-86.

[20] VANCLAY, J.K. 2003. Growth Modeling and Yield Prediction For Sustainable Forest Management. Paper presented at the ITTO Workshop on Growth and Yield, 24-28th June 2002. Kuala Lumpur, Malaysia. http://jkv.50megs.com/R083_mf.pdf

[21] VILLAR, R. G. 2005. Optimization Model Based on Geographic Information System for a Forest Production Unit in Bukidnon, Philippines. PhD Dissertation Institute of Renewable Natural Resources. University o the Philippines Los Banos, Laguna, Philippines

[22] VILLANUEVA, T. R. 2005. Upland Ecosystem Management. Second Edition. University of the Philippines. Open University, Laguna, Philippines.

[23] WEI, X, and NUNO M.G.B. 2000. Genetic Gains and Levels of Relatedness From Best Linier Unbiased Prediction Of *Eucalyptus Europhylla* for Pulp Production In Southeastern China. Canadian Journal Of Forestry Research. 30 (10): 1601-1607

[24] YUE, W.,VALERIE, M., BAKER, and THOMAS ,G. 2007. Modeling and Prediction of Dominant Height and Site Index of *Eucalyptus globulus* Plantations Using a Nonlinear Mixed-Effects Model Approach. Canadian Journal of Forest Research, Volume 37, Number 8, 1 August 2007 , pp. 1390-1403(14)

14

# Study of phosphorus adsorption and its relationship with soil properties, analyzed with Langmuir and Freundlich models

**Fisseha Hadgu[1, *], Heluf Gebrekidan[1], Kibebew Kibret[1], BirruYitaferu[2]**

[1]School of Natural Resource Management and Environmental Science, Haramaya University, Haramaya, Ethiopia
[2]Amhara Regional Agricultural Research Institute, Amhara National Regional State, Ethiopia

**Email address:**
habenfhb@gmail.com (Fisseha H.), helufgebrekidan@yahoo.com (Heluf G.), kibebewkibrett @yahoo.com (Kibebew K.), birru_yitaferu2002@yahoo.com (Birru Y.)

**Abstract:** Phosphorus (P) adsorption capacity and its relationship with some properties of five different soil types (Fluvisols, Cambisols, Vertisols, Arenosols and Nitosols) of Tigray Region, northern Ethiopia were studied during the main cropping season in 2012. Besides, fractionation of the total soil P in to Ca-P, Fe-P and Al-P was carried out and their relationships with P adsorption characteristics of the soil type under study were also studied. Adsorption data for the different soil types were obtained by equilibrating the respective soil samples for 24 hours at room temperature with 25 ml of 0.01 M $CaCl_2$, containing 0, 60, 120, 180, 240, 300, 360, 420, 480 and 540 mg $kg^{-1}$ of applied external P as $KH_2PO_4$. The results of this study showed that Ca-P was highest followed by Fe-P in all the soils except in the Arenosols. Testing the Langmuir and Freundlich adsorption isotherms both showed well fitted with the correlation coefficients (r) that ranged from 0.951 to 0.999 in all the soils. Comparing the two adsorption models, the Freundlich model was slightly better in Vertisols, Arenosols and Nitosols while the Langmuir model was better in the Fluvisols and Cambisols in predicting P adsorption. The adsorption maximum (b) obtained from the Langmuir isotherm ranged from 296.6 mg P $kg^{-1}$ soil in Fluvisols to 820.4 mg P $kg^{-1}$ soil in Vertisols. The constant related to bonding energy calculated from the Langmuir adsorption model varied from 0.0151 to 0.0447 L $mg^{-1}$. The Freundlich adsorption parameters also ranged from 7.28 to 30.15 mg $kg^{-1}$ for the Freundlich isotherm constant ($K_f$) and from 0.4874 to 1.012 for slope of the Freundlich isotherm (1/n). Among the soil properties ($CaCO_3$, CEC, organic matter, EC, pH, and sand, silt, and clay contents), clay content, CEC and $CaCO_3$ were positively correlated with the Langmuir adsorption maximum. The adsorption maximum significantly ($P \leq 0.05$) correlated (r = -0.902 but decreased with increasing sand content. The constant ($K_f$) positively and significantly ($P \leq 0.05$) correlated (r = 0.953) with EC of the soil. The remaining soil properties studied were not significantly correlated with the adsorption parameters.

**Keywords:** Phosphorus Content, Isotherms, Freundlich Model, Langmuir Model, Soil Characterization, Ethiopian Soils

## 1. Introduction

Adsorption which is the net accumulation of chemical species at the interface between a solid phase and an aqueous solution phase determine availability of native soil P and the rate of P applied to soils in fertilizers. When soluble P compounds are added to the soil, they undergo a series of complex reactions which lead to the decline of the availability of the added P. The compounds react rapidly with soil minerals by precipitation reactions and adsorption onto the soil's solid particle surfaces. Adsorption reaction is one of the principal processes involved in the retention of P (1).

The physico-chemical characteristics of soils largely influence the parameters of adsorption process and this eventually governs the availability of nutrient ions to the growing plants. Among the soil properties affecting the P adsorption capacity are soil texture (2), organic matter content (3), soil pH (4) and $CaCO_3$ content (5) of the soil. The surface charge of clay minerals (and oxides) is partly pH dependent, so that anion exchange capacity increases as pH decreases (6). According to (6) possible major factors affecting P adsorption of soils can be ascribed as physicochemical properties of soils such as clay type and clay content. They reported higher P adsorption by Entisols which had satisfactory level of organic matter and medium level available P than that of other soils in Sudan.

In addition to adsorption, the release of P through natural processes is very scanty, whereas the sinks for P especially at the ground of ocean are huge (7). Therefore the amount of P in soil solution is low. Phosphorus deficiency in crop plants is therefore a widespread nutritional disorder. Its deficiency in soils usually occur due to too low concentration of orthophosphate in the soil solution rather than from inadequate total P content which is partly due to P adsorption in the soil. Many researchers have established that P is essential in plants in energy transfer, being constituent part of adenosine Tri phosphate. Phosphorus is required particularly by the growing tips of plants, hence its importance for root growth. Any shortage, especially in the very early stages, reduces root growth and nutrient uptake and this can adversely affect the growth of crops for the rest of a growing season. Phosphorus fertilization and its management is a means of improving soil P for crop production in the cropping system. Better management of P fertilizer can be achieved by studying the P adsorption behavior of the soil. Phosphorus adsorption curve helps in predicting the fertilizer P needed to replenish the soil solution P to a level optimum for a particular crop (8). The Langmuir adsorption equation was popular among soil chemists for monitoring P adsorption and calculating the crop P requirements since 1957. The Freundlich and Langmuir adsorption isotherms are usually used to describe adsorption of different compounds by soils.

The Ethiopian soils, similar to the other agricultural soils of the tropics, are generally low in available P. Several authors have reported independently that 70-75% of some Ethiopian agricultural soils are deficient in plant available P (9, 10 and 11). In Tigray, northern Ethiopia, most of the farmers usually apply P fertilizers without judging the P status of the soils for which P adsorption study have significant role to play. Fluvisols, Cambisols, Arenosols, Vertisols and Nitosols are the common types of soils in the whole Tigray as well as in the study areas. Vertisols in Adigudom, Adwa, Wukro, Shire and Michew areas of Tigray Regional National State are low in P and micronutrients (12). Many local experimental results also indicated that these soils are low in these plant nutrients. However, very little detailed work has been done on the P adsorption of Ethiopian soils (13; 14; 15 and 11) in general and no P adsorption study has been conducted in Tigray in particular and most of the studies conducted in Ethiopia as well as in the Region so far on these soils have mainly been on fertilizer experiments. The objectives of this study were therefore to quantify P adsorption capacity, bonding energy and other adsorption parameters of the Langmuir and Freundlich models on major soils; to examine the relationship of these adsorption parameters with dominant soil properties and to determine the contents of different forms of P (Ca-P, Fe-P and Al-P) and correlate these forms with adsorption parameters.

# 2. Materials and Methods

## *2.1. Description of the Study Area*

The study was conducted in 2012 at the soil laboratory of the Mekelle Agricultural Research Center, northern Ethiopia. Representative soil samples including Fluvisols from Birki site (latitude 13°42'N; longitude 39° 39' E and altitude of 2065 masl); Cambisols from Agulae site (latitude 13° 41' N; longitude: 39° 35' E and altitude 1991 masl); Vertisols from Wukro (latitude 13° 45' N; longitude 39° 36' E and altitude 1988 masl); Arenosols from Wukro (latitude 13° 48' N; longitude: 39° 35' E and altitude 1986 masl) in Eastern Zone and Nitosols from Mailaham (latitude 14° 9' 16"; N:longitude 38° 18' 10"E and altitude 1940 masl) in North Western Zone of Tigray National Regional State were collected for this study. The four soil samples were classified as Eutric Fluvisols (Agulae series), Eutric Cambisols (Agulae series), chromic Vertisols (Gormodo series) and Cambic Arenosols (Abraha series) by (16). All the soils are found in semi-arid agro-ecology of Tigray. The topography of the Tigray Region including the surrounding of the sampling sites is mainly the extension of the central highlands which comprises mountains, highland plateaus and gorges. The samples were collected from flat terrain with slopes seldom exceeding 2%, representing farms supplemented by irrigation. According to (17), the mean rainfall for the District of Wukro in eastern Tigray (the nearest place with meteorological data for Eastern Zone soil sampling sites) for the years 1963 to 1997 was 650 mm. The mean annual rainfall for the four sampling sites from Birki, Agulae and Wukro areas is about 590 mm and the maximum and minimum temperatures range from 23-28 and from 9-14 °C, respectively (National Meteorological Service Agency, Wukro Station). The mean annual rainfall for the Nitosols which is located in Shire area is 987 mm having five months LGP (length of growing period) which is longer than the LGP at Wukro that has three months (12).

## *2.2. Soil Sampling and Characterization*

Surface soil samples from five soil types, which were selected based on preliminary survey as well as on secondary information, were collected from 20 sub sampling sites per composite sample at a depth of 0-15 cm using auger before sowing. The composite soil samples for the laboratory analysis were air dried, crushed using pestle and mortar and passed through a 2 mm diameter sieve for analysis of most of the soil chemical properties. A portion of the disturbed soil sample was taken and sieved using 0.5 mm diameter for the determinations of organic matter.

One composite sample was prepared by mixing the different soil subsamples for the laboratory analyses of soil properties, P adsorption and iron (Fe), calcium (Ca) and aluminium (Al) P as well as for determination of soil texture. The samples were transported to Mekelle for laboratory analysis.

Particle size distribution was determined by the hydrometer method (18). Once the sand, silt, and clay separates were calculated in percent, the soil was assigned to a textural class based on the soil textural triangle using International Soil Science Society (ISSS) system (19).

Soil pH in water was determined by the glass electrode pH meter (20) at 1:2.5 soils to water ratios. The electrical conductivity (EC) of the soils was measured according to the method described by (20). The cation exchange capacity (CEC) was determined by the method described by (21). Exchangeable potassium (K) and Exchangeable sodium (Na) were determined using flame photometer as described by (22), while calcium (Ca) and magnesium (Mg) were determined by atomic absorption spectrophotometer method (23). Percent base saturation (PBS) was calculated from exchangeable K ion, Ca ion, Mg ion, Na ion and CEC. Calcium carbonate ($CaCO_3$) was determined by titration according to (24). Soil available P was analyzed using Olsen method (39) modified by (25). To determine organic carbon, the (26) method was employed. Finally, the organic matter content of the soil was calculated by multiplying the organic carbon percentage by 1.724.

Inorganic P was fractionated by the method of (27) as modified by (28) to obtain Fe-P, Al-P and Ca-P contents of the soils. Phosphorus in the soil was determined colorimetrically by the molybdenum blue color method of (29).

### *2.3. Phosphorus Adsorption Studies*

Phosphorus adsorption was studied using the method of (30). A duplicate of 3 g portion of each soil sample containing levels of 0, 60, 120, 180, 240, 300, 360, 420, 480 and 540 mg P $kg^{-1}$ as $KH_2PO_4$ in 25 ml of 0.01 M $CaCl_2$ was shaken on an end-to-end shaker at 150 oscillations per minute for 24 hours at room temperature. After equilibrium achieved, sod solution was centrifuged in 50-ml centrifuge tube for 10 minutes at 3000 revolution per minute and the supernatant was decanted. The decanted part then was filtrated with a filter paper (Whatman number 40). Phosphorus in equilibrium solution was determined by Olsen method.

Amount of P adsorbed by the soils was determined by the difference between initial and final amounts of P concentrations in the equilibrium solution. The data were used to fit the linearized Langmuir and Freundlich equations, hence to determine the model parameters; and for model validation. The Freundlich and Langmuir models were used to describe P adsorption:

Langmuir equation:

$$\{C/(x/m)\} = 1/Kb + C/b$$

Where, C = Concentration of P in soil solution at equilibrium (mg P $L^{-1}$), x/m = Amount of P adsorbed (mg $kg^{-1}$ soil), b = Adsorption maximum (mg P $kg^{-1}$ soil), 1/K b = Y - Intercept and K = Constant, i.e. adsorption affinity (L $mg^{-1}$ P).

Freundlich equation:

$$x/m = K_f C^{1/n}$$

Where, x/m = P adsorbed (mg P $kg^{-1}$ soil), C = Concentration in soil solution at equilibrium (mg P $L^{-1}$), $K_f$ = Proportionality constant for Freundlich equation, 1/n = Slope of the curve (log C vs. log x/m) and Log $K_f$ = Y-intercept.

### *2.4. Data Analysis*

Regression techniques were used for model validation and to determine the functional relationship between the model parameters and selected soil properties including P adsorption maximum (b) values with soil properties including percent sand silt and clay, pH, EC, $CaCO_3$, organic matter content, CEC as well as Fe-P, Ca-P and Al-P contents. In addition,. Mean square error (MSE) and correlation coefficients were used to compare the performance of two parameterized adsorption isotherms.

## 3. Results and Discussion

### *3.1. Soil Properties*

Soils of the study sites varied in texture and chemical characteristics. Some important soil properties are presented in Table 1. The textural classes of the soils under investigation are sandy loam for the Fluvisols, Cambisols and Arenosols, whereas it is clay for the Vertisols and Nitosols based on the soil textural triangle of the ISSS system (19). As per rating suggested by (31), the soil organic matter content was very low for Nitosols, low for the Fluvisols and Arenosols and moderate for Cambisols and Vertisols. The very low amount of organic matter in Shire area may be due to its faster decomposition because of high temperature of the area. The data further revealed that the Arenosols and Nitosols soil samples were neutral and the Fluvisols, Cambisols and Vertisols were mildly alkaline. On the basis of EC limit purposed by (32), the soils under investigation fall in the category of very low EC for Arenosols and Nitosols, low for the Fluvisols, medium for Vertisols and high for Cambisols. The Fluvisols and Arenosols have low; Cambisols and Nitosols had moderate and Vertisols had high CEC values, on the basis of CEC rating by (33). As per rating of (34) $CaCO_3$ content was moderate in all the soils. The PBS calculated from these cations was moderate for the Nitosols, high for the Arenosols and very high for the other three soils according to rating by (33). Based on the rating set by (35), the available phosphorus in the plow layer of all the soils is found low (Table 1).

*Table 1. Physical and chemical properties of Fluvisols, Cambisols, Vertisols, Arenosols and Nitosols in the study areas*

| Soil property* | Fluvisols | Cambisols | Vertisols | Arenosols | Nitosols |
|---|---|---|---|---|---|
| Sand (%) | 74 | 69 | 42 | 75 | 32 |
| Silt (%) | 16 | 17 | 12 | 5 | 26 |
| Clay (%) | 10 | 14 | 46 | 20 | 42 |
| Textural class | SL | SL | Clay | SL | Clay |
| pH 1:2.5 ($H_2O$) | 7.50 | 7.60 | 7.70 | 7.00 | 6.70 |
| EC (dS $m^{-1}$) | 0.11 | 0.62 | 0.18 | 0.05 | 0.04 |
| $CaCO_3$ (%) | 6.0 | 7.5 | 9.0 | 7.0 | 8.0 |
| OM (%) | 1.30 | 2.63 | 2.35 | 1.32 | 0.94 |
| Available P (mg $kg^{-1}$) | 5.00 | 11.00 | 3.60 | 3.60 | 5.80 |
| Exchangeable Ca (cmol(+) $kg^{-1}$) | 5.90 | 12.40 | 11.40 | 6.90 | 8.10 |
| Exchangeable Mg (cmol(+) $kg^{-1}$) | 1.00 | 1.90 | 8.20 | 0.90 | 1.20 |
| Exchangeable Na (cmol(+) $kg^{-1}$) | 0.24 | 0.10 | 0.20 | 0.10 | 0.13 |
| Exchangeable K (cmol(+) $kg^{-1}$) | 0.82 | 0.86 | 0.90 | 0.70 | 0.79 |
| CEC (cmol(+) $kg^{-1}$) | 8.80 | 17.40 | 22.80 | 11.00 | 20.80 |
| Base saturation (%) | 90.45 | 87.90 | 90.80 | 86.55 | 63.03 |

OM = Organic matter, CEC = Cation exchange capacity, SL = Sandy loam.

### *3.2. Relationship between Soil Properties and P Adsorption Parameters*

In all the soils, P adsorption characteristics were influenced by some soil characteristics at varying levels. Adsorption maximum was significantly ($P \leq 0.01$ for clay and $P \leq 0.05$ for CaCO3 and CEC) and positively correlated with clay (r = 0.975), CaCO3 (r = 0.912) and CEC (r = 0.885) (Table 2). Sand content was significantly ($P \leq 0.05$) and negatively correlated (r = -0.902) with Langmuir adsorption maxima (b). No significant correlations were observed between Langmuir adsorption parameters and the other soil characteristics studied (Table 2). According to (36), most soil properties used in predicting soil P are interrelated and this makes it difficult to identify the principal components contributing to P-availability in soils. Multiple correlation coefficient of K with both clay and CaCO3 contents was also non significant (R2 = 0.6179). Similarly (37) reported that clay contents were major contributors to P adsorption in Australian soils. The positive relationship of P adsorption with clay content and negative correlation with sand content may be related with larger surface area of clay as compared to sand. It could also be related to the relatively large number of positive charges that can react and strongly bind the negatively charged phosphate ions in soil solution. A negative non significant correlation was observed between the soil Olsen-P and P adsorption maxima (Table 2). According to (38) a positive correlation between soil Olsen-P and soil P adsorption was found in Iran. No significant correlations were observed between Langmuir parameters and soil organic matter (Table 2). Phosphorus adsorption at low concentrations of added P negatively correlates with organic carbon suggesting that organic matter may compete with P for adsorption sites. (39) and (40) showed that coefficient for soil OM to predict maximum P adsorption had positive sign and the soils with high OM had poorly crystalline iron oxides suggesting inhibiting effect on iron oxide crystallization through complexation increasing sorption capacity for phosphate.

*Table 2 Correlation between Langmuir adsorption parameters and selected soil characteristics of the five soils in the study areas*

| Soil property vs. Langmuir adsorption parameter* | Regression equation | Correlation coefficient(r) | Coefficient of determination ($R^2$) |
|---|---|---|---|
| Sand vs. b | b = 1184.44-10.21 (sand) | -0.902* | 0.8136 |
| Silt vs. b | b = 475.21+ 7.38 (silt) | 0.245 | 0.0600 |
| Clay vs. b | b = 234.78 + 13.35 (clay) | 0.975** | 0.8084 |
| pH vs. b | b = 1757.19-160.26 (pH) | -0.305 | 0.0929 |
| $CaCO_3$ vs. b | b = -798.98 +184.44 ($CaCO_3$) | 0.912* | 0.8316 |
| CEC vs. b | b = 81.17 + 31.63 CEC) | 0.885* | 0.7840 |
| EC vs. b | b = 634.82 -237.45 (EC) | -0.254 | 0.0643 |
| OM vs. b | b = 579.85 + 4.37 (OM) | 0.014 | 0.0002 |
| Olsen-P | b = 707.2-20.67 (Olsen-P) | -0.279 | 0.0781 |
| Sand vs. K | K = 0.0008 +0.0004 (sand) | 0.668 | 0.4457 |
| Silt vs. K | K = 0.025 - 00005 (silt) | -0.015 | 0.0002 |

| Soil property vs. Langmuir adsorption parameter* | Regression equation | Correlation coefficient(r) | Coefficient of determination ($R^2$) |
|---|---|---|---|
| Clay vs. K | K = 0.0399 -0.0006 (clay) | -0.804 | 0.6408 |
| pH vs. K | K =-0.044 +0094 (pH) | 0.033 | 0.1110 |
| $CaCO_3$ vs. K | K=- 0.0946 - 0.0094 ($CaCO_3$) | -0.864 | 0.7426 |
| CEC vs. K | K = 0.0.0490- 0.0015 (CEC) | -0.803 | 0.6449 |
| EC vs. K | K= 0.0237 - 0.0035 (EC) | 0.070 | 0.0001 |
| OM vs. K | K = 0.0284 - 0.0024 (OM) | -0.143 | 0.0001 |

** = Significantly correlated at P = 0.01; * = significantly correlated at P = 0.05; CEC = Cation exchange capacity; EC = Electrical conductivity; b = Langmuir adsorption maximum; K = Langmuir constant related to bonding energy; OM = Organic matter.

No significant correlations were observed between Langmuir adsorption parameters and soil organic matter (Table 2). (41) also reported that P adsorption was significantly correlated with clay and CEC. Positive and negative relationships of organic matter and P adsorption had also been reported by several researchers. (42) recorded on highly weathered soils that the presence of organic matter reduces P adsorption capacity due to direct result of competition for adsorption sites between phosphate and organic ligands. There was also suggestion that the possibility of organic matter to reduce positive charge on variable charge surfaces by lowering pH (7) and this decreases the attraction of P to the soil surface. On the contrary, positive relationships between the organic matter contents of soils and P adsorption was reported (43). They found that the role of organic matter in increasing the ability of soils to adsorb P that is attributed to its association with cations such as Fe, Al and Ca.

### *3.3. Phosphorus Fractionation*

***Table 3.*** *Phosphorus contents in the form of Al-P, Ca-P and Fe-P of the soils*

| Soil type | Al-P (mg $kg^{-1}$) | Fe-P (mg $kg^{-1}$) | Ca-P (mg $kg^{-1}$) |
|---|---|---|---|
| Fluvisols | 28.6 | 90.0 | 300.0 |
| Cambisols | 31.6 | 120.0 | 410.0 |
| Vertisols | 21.7 | 165.0 | 320.0 |
| Arenosols | 8.3 | 385.0 | 310.0 |
| Nitosols | 10.0 | 375.0 | 530.0 |

The amount of Ca-P was higher in all the soils except in the Arenosols in which Fe-P was higher compared to Ca-P and Al-P. This may be due to the location of the soils in semiarid agro-ecology in which Ca ion is not leached down from the upper soil layers to the subsurface soil horizons and reacts with P to form Ca-P. The Fe-P was next to Ca-P in all the other soils (Table 3). Similar results of P fractionation on Vertisols in India reported by (44) showed that Ca-P was the dominant form followed by Fe-P with very low amount of Al-P.

### *3.4. Langmuir and Freundlich Equations for the Studied Soils*

#### *3.4.1. Correlation between Adsorption Parameters*

Langmuir adsorption maximum (b) showed higher opposite correlation with Langmuir adsorption parameter (K) but all other correlations were not significant (Table 4). Significant correlation between Langmuir adsorption maximum (b) and slope of the Freundlich model (1/n) makes calculation of the value of P adsorption maximum from the Freundlich model possible with the help of this relationship. It can help if the Freundlich model showed superiority over the Langmuir model because it is impossible to calculate adsorption maximum directly from the Freundlich model (45). However, no significant correlation was observed between b and 1/n in this study (Table 4).

Significant ($P \leq 0.05$) correlation was observed between EC and $K_f$ of the Freundlich adsorption model. No significant correlation was observed between Freundlich adsorption parameters and other soil characteristics (Table 6). Multiple correlation coefficient of 1/n with both sand and clay contents was no significant ($R^2$ = 0.7045). Multiple correlation coefficient of $K_f$ with both pH and OM content was also non significant ($R^2$ = 0.6087).

***Table 4*** *Correlation between the Langmuir and/or the Freundlich equation (s)*

| Correlation between | Regression equation | Correlation Coefficient ( r) | Coefficient of Determination ($R^2$) |
|---|---|---|---|
| K and b | b = 10001.08 – 16943.5 (K) | -0.907* | 0.8226 |
| b and K | K = 0.0529- 0.00049 (b) | -0.907* | 0.8226 |
| b and 1/n | 1/n = 0.4127 + 0.007 (b) | 0.728 | 0.5300 |
| b and $K_f$ | Kf= 24.74 - 0.0197 (b) | -0.461 | 0.2125 |
| 1/n and b | b = -35.56 + 755.45 (1/n) | 0.728 | 0.5300 |
| $K_f$ and b | b = 729.01-10.775 ($K_f$) | -0.461 | 0.2125 |

*Correlation is significant at 0.05 level; = b = Langmuir adsorption maximum; K = Langmuir constant related to bonding energy; $K_f$ = Freundlich constant related to extent of adsorption; 1/n = Slope of the Freundlich isotherm.

***Table 5*** *Adsorption equations for Freundlich and Langmuir models for the five soils.*

| Soil type | Freundlich model | Correlation coefficient ( r) | LSE | Langmuir model | Correlation coefficient ( r) | LSE |
|---|---|---|---|---|---|---|
| Fluvisols | $x/m = 11.72C^{0.7467}$ | 0.985** | 19.45 | x/m = 13.244C/1 + 0.0447C | 0.987** | 14.39 |
| Cambisols | $x/m = 30.15C^{0.4874}$ | 0.994** | 75.14 | x/m = 12.114C/1 + 0.0257C | 0.974** | 19.02 |
| Vertisols | $x/m = 7.29C^{1.0054}$ | 0.999** | 14.96 | x/m = 12.356C/1 + 0.0151C | 0.951** | 32.37 |
| Arenosols | $x/m = 9.32C^{0.8707}$ | 0.999** | 5.83 | x/m = 11.109C/1 + 0.0207C | 0.974** | 27.48 |
| Nitosols | $x/m = 7.28C^{1.0124}$ | 0.999** | 19.29 | x/m = 12.906C/1 +0.0159C | 0.982** | 32.09 |
| Mean | - | 0.995** | | | 0.974** | |

** = Correlation is significant ate 0.01 level, n = 10; LSE = Least significant error

***Table 6*** *Correlation between Freundlich adsorption parameters and selected soil characteristics of the five soils in Tigray, northern Ethiopia*

| Soil property vs. Freundlich adsorption Parameter | Regression equation | Correlation coefficient(r) | Coefficient of determination ($R^2$) |
|---|---|---|---|
| Sand vs. $K_f$ | $K_f$= 0.5274 + 0.2161 (sand) | 0.446 | 0.1989 |
| Silt vs. $K_f$ | $K_f$ = 11.60 +0.1021 (silt) | 0.081 | 0.0066 |
| Clay vs. $K_f$ | $K_f$ = 22.08 – 0.338 (clay) | -0.578 | 0.3341 |
| pH vs. $K_f$ | $K_f$ = -57.33 + 9.65 (pH) | 0.429 | 0.1840 |
| $CaCO_3$ vs. $K_f$ | $K_f$ = 24.64 – 1.53 ($CaCO_3$) | -0.177 | 0.0313 |
| CEC vs. $K_f$ | $k_f$ = 14.51 - 0848 (CEC) | -0.056 | 0.0031 |
| EC vs. $K_f$ | $K_f$ = 5.51 + 38.20 (EC) | 0.953* | 0.9082 |
| OM vs. $K_f$ | $K_f$ = -1.41 + 8.53 (OM) | 0.649 | 0.4212 |
| Sand vs. 1/n | 1/n = 1.26 - 0075 (sand) | 0.686 | 0.4706 |
| Silt vs. 1/n | 1/n = 0.8044 + 0.0013 (silt) | 0.046 | 0.0021 |
| Clay vs. 1/n | 1/n = 0.5430 + 0.0107 (clay) | 0.809 | 0.6545 |
| pH vs. 1/n | 1/n = 2.51 - 0.2310 (pH) | -0.456 | 0.2080 |
| $CaCO_3$ vs. 1/n | 1/n= 0.1362 + 0.0918 ($CaCO_3$) | 0.471 | 0.2218 |
| CEC vs. 1/n | 1/n = 0.6197 + 0.0128 (CEC) | 0.372 | 0.1384 |
| EC vs. 1/n | 1/n= 0.9749 - 0.7519 (EC) | -0.833 | 0.6938 |
| OM vs. 1/n | 1/n = 1.08 + 0.1450 (OM) | -0.508 | 0.2581 |

* = Correlation is significant at P > 0.05 level; CEC = Cation exchange capacity; EC = Electrical conductivity; b = Langmuir adsorption maximum; K = Langmuir constant related to bonding energy; OM = Organic matter; 1/n = Slope of the Freundlich isotherm.

### *3.4.2. Model Validation*

The Langmuir and Freundlich generated equations are presented in Table 5 and graphs showing their validation are depicted in Figure 1 and 2. Generally, the choice between the models is often based on the goodness of fit (46). It is clear from Table 5 that both Freundlich and Langmuir models proved good based on average correlation coefficient values (r = 0.995 for Freundlich and ( r = 994 for Langmuir ) in all the soils under investigation. On individual soil basis, the Langmuir model and Freundlich model were good but they have slight differences. Taking the correlation coefficient and least significant error (LSE) into account, Langmuir model was better in Fluvisols and Cambisols which were also lower in P adsorption capacity (Table 5). However, on Vertisols, Arenosols and Nitosols which were relatively higher in P adsorption, Freundlich model showed superiority over the Langmuir model on the basis of correlation coefficient and LSE values (Table 5). The least square error calculated for each soil indicated lower error by Langmuir equations than Freundlich on Fluvisols and Cambisols but Freundlich model showed lower least square error for Vertisols, Arenosols and Nitosols which showed similar but opposite trend to the correlation coefficient. Different results had been reported on the effectiveness of the two adsorption models in predicting P adsorption. (46) and (47) reported better results of the Langmuir model than the Freundlich model. (48) and (49) found that ability of the Freundlich model to describe P adsorption curves was superior to the Langmuir model for calcareous soils. However, (45) had reported comparable results for prediction of P adsorption by both the Langmuir and the Freundlich models.

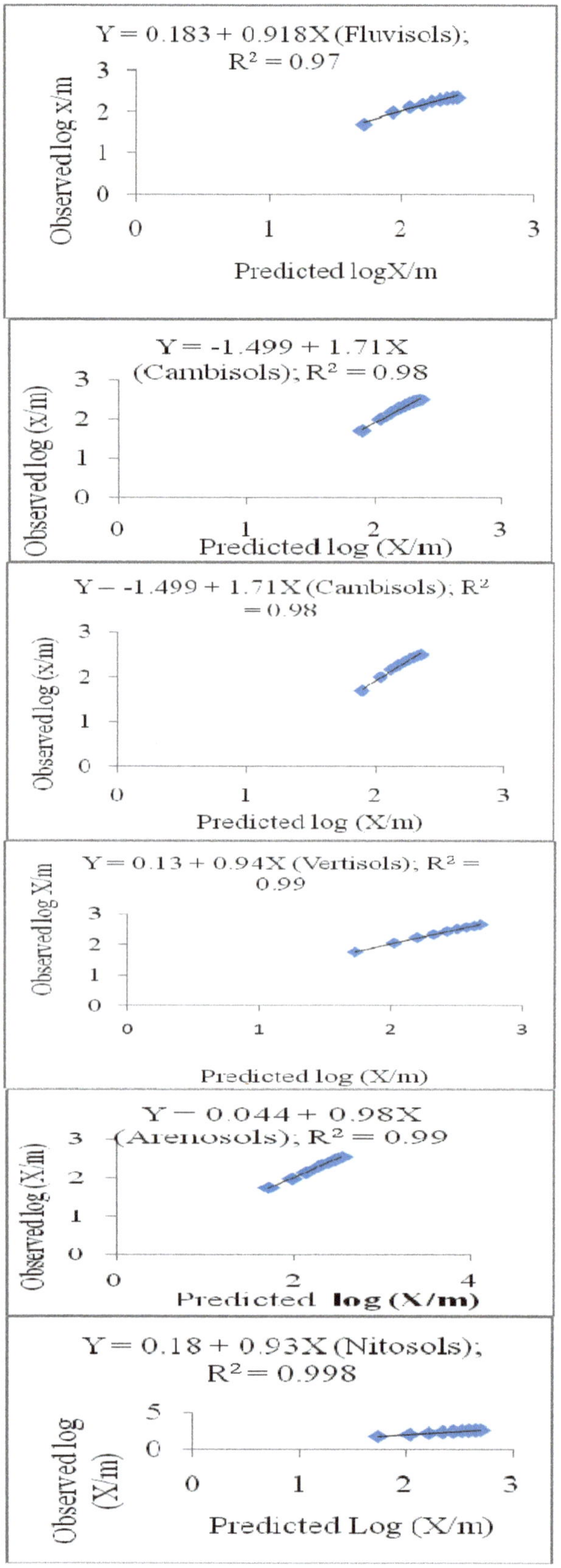


***Figure 1*** *Validation for the Freundlich adsorption isotherm of the five soils*

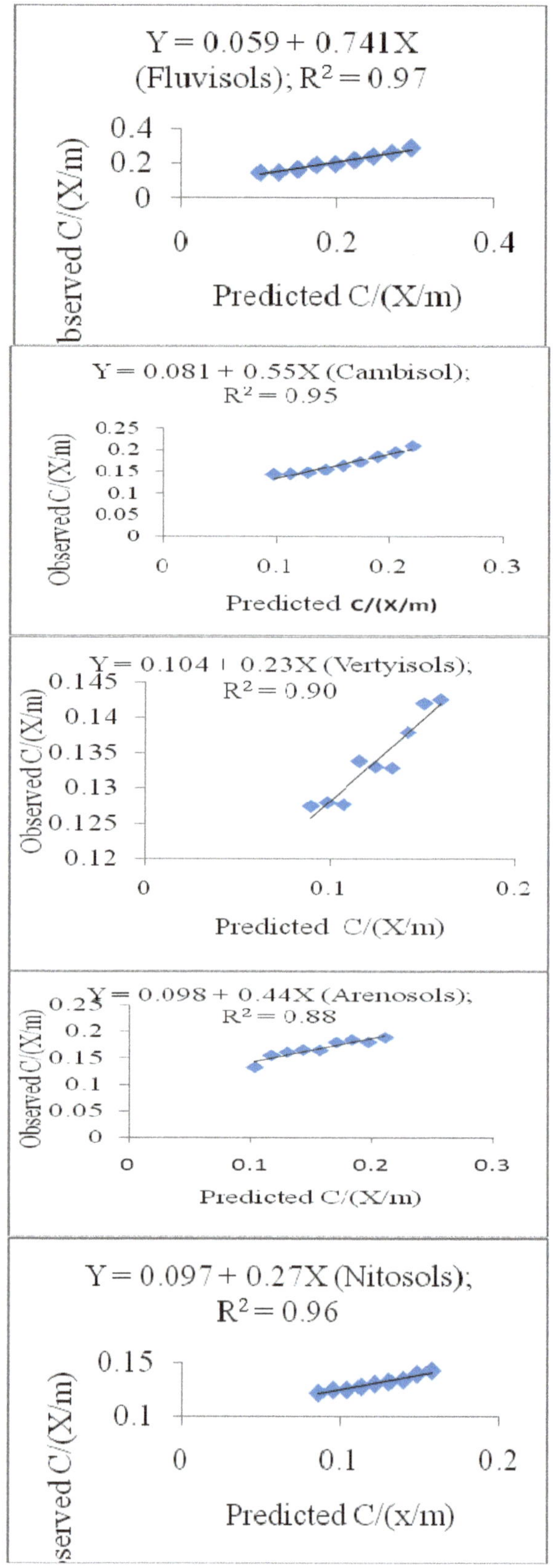


***Figure 2.*** *Validation for the Langmuir adsorption isotherm of the five soils.*

No significant ($P \leq 0.05$) correlation was observed between Langmuir and Freundlich adsorption parameters and Fe-P, Ca-P and Al-P of the soils. Multiple correlation coefficient of b with both Ca-P and Fe-P contents was also non significant ($P \leq 0.05$), (R2 = 0.5542). However, Fe-P, Ca-P and Al-P showed positive relations with most adsorption parameters (Table 7). Non significant ($P \leq 0.05$) correlation was obtained between Al-P and all adsorption parameters (Table 7). According to (50), hydrous iron and aluminium oxides and aluminosilicates react with phosphate solutions to produce an isomorphous series of iron and aluminium phosphates, the solubility criteria of which were believed to control the concentration of phosphate in the soil solution during 1950s. According to the report by (50), it is unlikely that the iron and aluminium phosphates persist for long to control dissolution and precipitation processes. They concluded that phosphate concentrations are largely independent of pH over the range of 4 to 8 and that the solution is invariably under saturated with resp.ect to all phosphate compounds. They stated based on studies done so far that iron and aluminum phosphates which were earlier assumed to occur in soil are more soluble than the corresponding metal oxides, which definitely do occur in soil and strongly adsorb phosphate ions. However, they stated that precipitation of calcium phosphates in soils above pH 6 = can control the phosphate concentration in the soils. The present results of P fractionation of the studied soils coincide with their report with respect to Fe-P and Al-P except Fe-P with K and Kf; Ca-P with K and Al-P with b.

***Table 7.*** *Correlation between Al-P, Ca-P and Fe-P with adsorption parameters of both Langmuir and Freundlich*

| Soil P forms vs adsorption parameter* | Regression equation | Correlation coefficient(r) | Coefficient of determination ($R^2$) |
|---|---|---|---|
| Fe-P vs. b | b = 410.80 + 0.7775 (Fe-P) | 0.489 | 0.2392 |
| Ca-P vs. b | b = 167.74 + 1.1219 (Ca-P) | 0.484 | 0.2346 |
| Al-P vs. b | b = 815.09 -11.37 (Al-P) | -0.532 | 0.2832 |
| Fe-P vs. K | K = 0.0362- 0.00005 (Fe-P) | -0.608 | 0.3694 |
| Ca-P vs. K | K =0.0445 - 0.00005 (Ca-P) | -0.432 | 0.1867 |
| Al-P vs. K | K = 0.0109 + 0.0067 (Al-P) | 0.589 | 0.3474 |
| Fe-P vs. $K_f$ | Kf = 20.82 - 0.0338 (Fe-P) | -0.497 | 0.2468 |
| Ca-P vs. $K_f$ | Kf = 9.58 + 0.0096 (Ca-P) | 0.096 | 0.0093 |
| Al-P vs. $K_f$ | Kf = 0.6972 + 0.6214 (Al-P) | 0.680 | 0.4621 |
| Fe-P vs. 1/n | 1/n = 0.6264 + 0.0009 (Fe-P) | 0.570 | 0.3246 |
| Ca-P vs. 1/n | 1/n = 0.7244 + 0.0003 Ca-P) | 0.120 | 0.0144 |
| Al-P vs. 1/n | 1/n = 0.6972 + 0.6214(Al-P) | 0.697 | 0.4621- |

b = Langmuir adsorption maximum; K = Langmuir constant related to bonding energy; $K_f$ (intercept) = Freundlich constant related to extent of adsorption; 1/n = Slope of the Freundlich isotherm.

### *3.5. Correlation between Langmuir and Freundlich Model Parameters and Selected Soil Properties*

### *3.6. Phosphorus Adsorption*

Comparing the highest amount of P adsorbed in the five soils it was found that the soils have different P adsorbing capacities (Table 8). Differences in the adsorption capacity observed between the different soil types could be due to the differences in soil physical and chemical properties. The highest adsorption was found at the highest applied rate of 540 mg P $kg^{-1}$ soil. The higher and the least average P adsorbed observed from the experiment were on Shire Nitosols and Birki Fluvisols, respectively.

***Table 8*** *Phosphorus adsorbed by different soils at 24 hours equilibration period*

| | Phosphorus adsorbed ( mg $kg^{-1}$) | | | | |
|---|---|---|---|---|---|
| Applied P ( mg $kg^{-1}$) | Brki Fluvisols | Agulae Cambisols | Wukro Vertisols | Wukro Arenosols | Shire Nitosols |
| 60 | 49.13 | 50.16 | 56.54 | 54.38 | 59.08 |
| 120 | 97.71 | 99.59 | 112.62 | 93.21 | 115.66 |
| 180 | 131.46 | 146.59 | 169.32 | 134.87 | 173.20 |
| 240 | 151.29 | 186.76 | 215.32 | 175.29 | 226.20 |
| 300 | 184.63 | 220.79 | 270.87 | 219.96 | 276.99 |
| 360 | 198.63 | 250.75 | 325.46 | 242.12 | 327.57 |
| 420 | 211.21 | 274.09 | 365.71 | 275.12 | 377.44 |
| 480 | 222.21 | 295.88 | 405.96 | 322.04 | 414.90 |
| 540 | 224.62 | 310.02 | 454.96 | 345.62 | 456.44 |
| Mean | 163.43 | 203.85 | 264.09 | 206.96 | 269.72 |

At low initial P addition (60 mg P $kg^{-1}$ soil), percent of P adsorption was maximum while at high initial P addition (540 mg P $kg^{-1}$), percent P adsorption was minimum. As increment of P addition, percent adsorption decreased. It can be concluded that soil P saturation can decrease adsorption, i.e. when the soil is saturated with P, rate of adsorption decreased. This reduction in percent P adsorption could be due to increasing concentration of applied P causing excess P on soil adsorption sites resulting in P release into solution. An inverse relationship between soil P saturation and adsorption was noticed by (45).

Linear Langmuir adsorption isotherm of the five soils showed different curves (Figure 3). The linear Freundlich adsorption isotherm curves also indicated differences among the soils (Figure 4). The Langmuir and Freundlich equations well fitted to the data on all the soils (Figures 3, 4 and Table 5). The two models showed similar curves for Wukro Vertisols and Shire Nitosols as well as for Agulae Cambisols and Wukro Arenosols among the soil types (Figures 3 and 6).Slope of the Freundlich model (1/n) is a measure of the heterogeneity of a system. A more homogeneous system will have (1/n) value approaching unity and a more heterogeneous system will have 1/n value approaching zero (49).

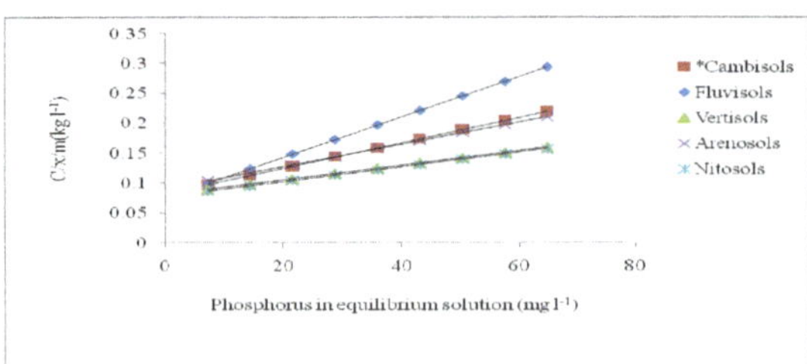


***Figure 3.*** *Isotherms for the adsorption of added P obtained using linear Langmuir model*

* = C/(x/m) = 0.077551 + 0.00337 (C) and $R^2$ = 0.97, for Fluvisols; C/x/m = 0.082551 + 0.002121 (C) and $R^2$ = 0.94, for Cambisols; C/x/m = 0.080935 + 0.001219 (C) and $R^2$ = 0.90, for Vertisols; C/x/m = 0.090021 + 0.00186 (C) and $R^2$ = 0.88, for Arenosols; C/x/m = 0.077483 + 0.001234 (C) and $R^2$ = 0.96, for Nitosols.

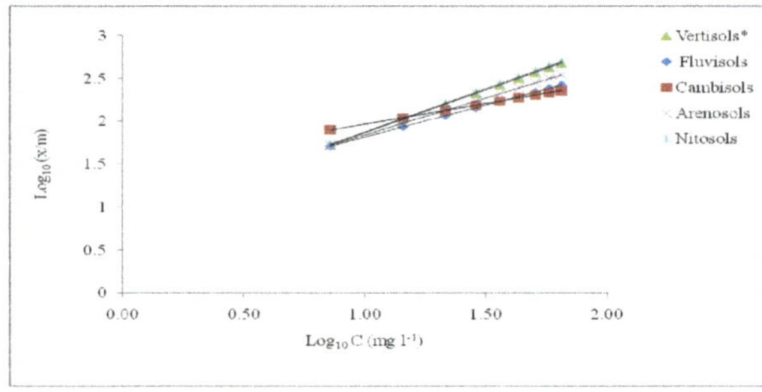


***Figure 4.*** *Isotherms for the adsorption of added P obtained using linear Freundlich model*

* = Log10 (x/m) = 1.0690 + 0.74072 × log10 (C) and $R^2$ = 0.97, for Fluvisols; log10(x/m) = 1.4793 + 0.48735 × log10(C) and $R^2$ = 98, for Cambisols; log10 (x/m) = 0.862582 + 1.00535 × log10 (C) and $R^2$ = 0.99, for Vertisols; log10 (x/m) = 0.96916 + 0.87073 × log10 (C) and $R^2$ = 0.99, for Arenosols; log10(x/m) = 0.862043 + 1.01240 × log10 (C) and $R^2$ = 0.99, for Nitosols.

Comparing the amount of P adsorbed in the five soils by estimating adsorbed P from Langmuir isotherm, it was observed that each soil has different capacities to adsorb P (168.69 -275.85 mg $kg^{-1}$). Similarly, P adsorption calculated from linear Freundlich adsorption isotherm showed differences in P adsorption of the soils with variations in magnitude from that of Langmuir (165.31 - 274.44 mg $kg^{-1}$). Similarly, it was reported that phosphate adsorption in many soils can be described well by the Langmuir or Freundlich adsorption isotherms (51).

The Langmuir adsorption maximum (b) for the five soils ranged from 296.6 mg P $kg^{-1}$ soil for Birki Fluvisols to 820.4 mg P $kg^{-1}$ soil for Wukro Vertisols (Table 9). The adsorption maximum in the Wukro Vertisols may be due to relatively higher clay, $CaCO_3$ and Ca-P contents in the soil. The constant related to bonding energy (K) was highest for Wukro Fluvisols [0.0447 L ($mg^{-1}$ P)] and lowest for Wukro Vertisols [0.0151 L $(mg\ P)^{-1}$]. Data plotted according to the linear form of the Freundlich model indicated exponent (1/n) values of less than or equal to one in all the soils (Table 9).

***Table 9.*** *Adsorption parameters of the Langmuir and the Freundlich models for the soils*

| Soil type | Langmuir parameters | | | | Freundlich parameters | | | | |
|---|---|---|---|---|---|---|---|---|---|
| | Slope = (1/b) | B (mg kg-1) | Y-intercept | Kb (mg kg-1) | K (L $mg^{-1}$ P) | Y-intercept = (log $K_f$) | $K_f$ (mg $kg^{1}$) | Slope = (1/n) | SPEC |
| Fluvisols | 0.00337 | 296.6 | 0.0755 | 13.24 | 0.0447 | 1.0690 | 11.72 | 0.7467 | 2.10 |
| Cambisols | 0.00212 | 471.6 | 0.0826 | 12.11 | 0.0257 | 1.4793 | 30.15 | 0.4874 | 9.82 |
| Vertisols | 0.00122 | 820.4 | 0.0809 | 12.36 | 0.0151 | 0.8623 | 7.29 | 1.0053 | 0.72 |
| Arenosols | 0.00181 | 537.5 | 0.0900 | 11.11 | 0.0207 | 0.9692 | 9.31 | 0.8707 | 1.25 |
| Nitosols | 0.00123 | 810.5 | 0.0775 | 12.91 | 0.0159 | 0.8620 | 7.28 | 1.0124 | 0.701 |

b = Langmuir adsorption maximum; 1/b = Slope of Langmuir model; K = Langmuir constant related to bonding energy; $K_f$ = Freundlich constant related to extent of adsorption; 1/n = Slope of the Freundlich isotherm and SPEC = Solution P equilibrium concentration.

Similar results were reported by (50). According to them, Freundlich model is empirical in nature and implies that adsorption decreases exponentially with increasing saturation of exchange sites with P.It is obvious from Table 9 that Cambisols had minimum (1/n) value, i.e. 0.4874 while Nitosols has maximum (1/n) value, i.e. 1.0124. Therefore, the Nitosols and Vertisols have relatively more homogeneous adsorption sites while Cambisols have more heterogeneous sites.

According to (52), $K_f$ is dependent on the solution concentration. (53) stated that $K_f$ is adsorption capacity while according to (54) it is extent of adsorption (mg $kg^{-1}$).

According to (45), solution P concentration equilibrium with adsorbed P was calculated by dividing Freundlich $K_f$ by antilog of the Freundlich slope [$K_f$/ antilog (1/n) = SEPC]. The values of SEPC calculated using these formulae for soils are indicated in Table 9.

# 4. Conclusions

The results of the study depicted that both Langmuir and Freundlich adsorption models are robust in predicting P adsorption in the soils of the Region. The results of this study also revealed differences in P adsorption among the soils studied. Nitosols and Vertisols showed relatively higher P adsorption maxima compared to the other soils. The differences in the P adsorption capacities of the soils indicated that the use of blanket P fertilizer rate recommendations without studying optimum rate for each crop, used in the study area, may fail to increase yield in the areas as it may lead to under-application or over-application of P. Phosphorus adsorption characteristics were influenced by some soil properties. The results identified that soil properties like clay, sand, CEC and $CaCO_3$ contents influence P adsorption in the study area. These further indicated that the properties of the different forms of soil P were in the order of Ca-P > Fe-P > Al-P in all the soils studied except in one soil. The result of this study could be used for further soil test based P fertilizer recommendations by considering the amount of P fixed in a growing season.

# Acknowledgment

The authors express their appreciation to the Ethiopian Institute of Agricultural Research for funding the research work.

# References

[1] G.M. Pierzynski, J.T. Sims and G.F., Vance. Soils and environmental quality. 3rd ed., Taylor and Francis, Boca Raton, USA. 2005.

[2] M.L. Leclerc M.C. Nolin, D. Cluis and R.R. Simard. Grouping soils of the Montreal lowlands (Quebec) according to fertility and P sorption and desorption characteristics. Can. J. Soil Sci., 81: 71-83. 2001.

[3] K. Daly, D. Jeffrey and H. Tunney. The effect of soil type on phosphorus sorption capacity and desorption dynamics in Irish grassland soils. Soil Use Manage, 17: 12-20. 2001.

[4] N.J. Barrow. Modeling the effects of pH on phosphate sorption by soils. J. Soil Sci., 35: 283-297. 1984.

[5] I. Bertrand, R.E. Holloway, R.D. Armstrong and M.J. Mclaughlin. Chemical characteristics of phosphorus in alkaline soils from Southern Australia. Aust. J. Soil Res., 41: 61-76. 2003.

[6] O.A Amelidris and H. S. Ahmed. Phosphorus sorption capacity as a guide for phosphorus availability of selected Sudanese soil series. African Crop Science Journal, 20-65. 2012.

[7] H. Moazed Y. Hoseini, A. Naseri and F. Abbasi. Determining phosphorus adsorption isotherm in soil and its relation to soil characteristics. International Journal of Soil Science, 5 (3): 131-139. 2010.

[8] Z.A. Mohammad, S. Hoque and K.T. Osman. A comparison of the Langmuir, Freundlich and Temkin equations to describe phosphate sorption characteristics of some representative soils of Bangladesh. International Journal of Soil Science, 7: 91-99. 2012.

[9] L. Pulschen. Soils and Plant Nutrition Section, Terminal report for the period May 1983-April 1987. Agricultural Research Center, Debre Zeit, Ethiopia. 1987.

[10] Desta B. Diagnosis of phosphorus deficiency in Ethiopian soils. 18p. Soil Science Bulletin, No: 3. Institute of Agricultural Research, Addis Ababa, Ethiopia. 1952.

[11] Tekalign M. and I. Haque. Phosphorus status of some Ethiopian soils. I. Sorption characteristics. Plant and Soil, 102: 261-266. 1987.

[12] Fassil, K. and C. Yamoah. Soil fertility status and Numass fertilizer recommendation of Typic Hapluusterts in the nrthern highlands of Ethiopia. World Applied Sciences Journal, 6(11): 1473-1480. 2009.

[13] C. Behn, M. Janssen, G.A. Yalemsew and B. Lennartz. Phosphorus contents and phosphorous sorption in soils of the Gilgel Gibe catchment, SW Ethiopia. The Smithsonian/NASA Astrophysics Data System, Geophysical Research, Abstracts. 2013.

[14] Birru Y., Heluf G.. and V.P. Gupta. Phosphorus sorption characteristics of some soils of the north-western highlands of Ethiopia. Ethiopian Journal of Natural Resources (EJNR). 5(1): 1-16. 2003

[15] D. Miressa and W.P. Robarge. Soil characteristics and management effects on phosphorus sorption by highland plateau soils of Ethiopia. Soil Science Society of America Journal, 63(5): 1455-1462. 1999

[16] HTSL (Hunting Technical Service Limited). Tigray rural development study, Annex 4. Land and water resources consultants in association with Sir M. Macdonald and Partners, London, UK. 1976.

[17] G.W. Meijerin. Alternative cropping practices in Ethiopia: A literature review, PIMEA Working Paper 2002-2005, Wageningen University, the Netherlands. 2002.

[18] P.R. Day. Particle fractionation and particle-size analysis. pp. 545-567. In: Black, CA, D.D. Evans, J.L. Ensminger and F.E. Clark (eds). Methods of soil analysis. Part I. ASA, WI, Madison, USA. 1965.

[19] D.L. Rowell. Soil science: Method and applications. Longman Scientific and Technical, Longman Group UK Limited Addison, Wesley, England. 350p. 1994.

[20] M. Peech. Hydrogen-ion activity. pp. 914-926. In: CA, Black, D.D. Evans, J.L. Ensminger and F.E. Clark (eds). Methods of soil analysis. Part II.ASA, WI, Madison, USA. 1965.

[21] H.D. Chapman. Cation exchange capacity. pp. 891-901. In: CA, Black, D.D. Evans, J.L. Ensminger and F.E. Clark (eds). Methods of soil analysis. Part II. ASA, WI, Madison, USA. 1965.

[22] P.R. Hesse. A text book of soil chemical analysis. Murray Publishers Limited, London, UK. 1971.

[23] FAO (Food and Agriculture Organization). The Euphrates Pilot Irrigation Project Methods of Soil analysis. Gadeb Soil laboratory (A laboratory Manual), Rome, Italy. 1974

[24] S.R. Olsen, C.V. Cole, L. Watanabe and A. Dean. Estimation of available P in soils by extraction with $NaHCO_3$. USDA Cir. 939, US Government Printing Office, Washington D. C., USA. 1954.

[25] F.S. Watanabe and S.R. Olsen. Test of an ascorbic acid method for determining phosphorus in water and NaHCO3, extracts from soil. Soil Science Society of America Proceeding, 29:677-678. 1965.

[26] A. Walkley and C.A. Black. An examination of the method for determining soils organic matter and a proposed modification of the chromic acid titration method. Soil Science, 37: 29-38. 1934.

[27] S.C. Chang and M. L. Jackson. Fractionation of soil phosphorus. Soil Science, 84: 133-144. 1957.

[28] G.W. Peterson and R.B. Corey. A modified Chang and Jackson procedure for routine fractionation of inorganic soil phosphates. Soil Science Society of America Proceedings, 30: 563-565. 1966.

[29] J. Murphy, and J.P. Riley. A modified single solution method for the determination of phosphate in natural waters. Analytica Chimica Acta, 27: 31-36. 1962.

[30] R.L. Fox. and E.J. Kamprath. Phosphate sorption isotherms for evaluating the phosphorous requirements of soils. Soil Sci. Soc. Am. Proc., 34: 902-907. 1970M. Ayaz, A. Saleem and M. Memon. Phosphorus adsorption parameters in relation to soil characteristics. Journal Chemical Society of Pakistan, 32 :129-139. 2010

[31] Charman P.E. and M.M. Roper. Soil organic matter. In: Charman P.E., Murphy B.W., eds. Soils-their properties and management, 3rd edition. Melbourne, Australia: Oxford University Press, pp. 276-285. 2007.

[32] R.J. Shaw. Soil salinity-electrical conductivity and chloride In: Peverill K.I., Sparrow L.A., and Reuter D.J., eds. Soil Analysis: An Interpretation Manual. pp. 129-145. CSIRO Publishing, Collingwood, Australia. 1999.

[33] A.J. Metson. Methods of chemical analysis for soil survey samples. Newzealand Department of Scientific and Industrial Research, Soil Bureau Bulletin. pp. 168-175. In: Hazelton, P.A, Murphy, B.W., eds. Interpreting soil test results: what do all the, numbers mean?. 2nd Edition. Government Printer,Wellington, New Zealand, New South Wales (NSW) Department of Natural Resources, CSIRO Publishing, Collingwood, Australia. 1961.

[34] F. Nachtergaele, H.V. Velthuizen, and L. Verelst. Harmonized world soil database. (version 1.1). FAO, Rome, Italy and Laxenburg, Austria.2009.

[35] J. Landon. Booker tropical soil manual: Handbook for soil survey and agricultural land evaluation in the tropics and subtropics. A John Wiley and Sons, New York. 1991.

[36] E.U. Onweremadu. Predicting phosphorus sorption characteristics in highly weathered soils of South-Eastern Nigeria. Research Journal of Environmental Sciences, 1: 47-55.2007.

[37] B. Singh and R.J. Gilkes. Phosphorus sorption in relation to soil properties for the major soil types of south-west Australia. Aust. J. Soil Res., 29: 602-618. 1991.

[38] Y. Hoseini and R. D. Taleshmikaiel 2013. Comparison of phosphorus adsorption isotherms in soil and its relation to soil properties. International Journal of Agriculture, 3 (1): 163-171. 2013.

[39] M. Ayaz, A. Saleem and M. Memon. Phosphorus adsorption parameters in relation to soil characteristics. Journal Chemical Society of Pakistan, 32 :129-139. 2010

[40] L. Zhuan-xi, Z. Bo, T. Jia-liang, and W. Tao. Ecological Engineering. pp 35-57. 2009.

[41] M. Arshed Rahmatulla M. Salim and M, Yousaf. Soil properties related to phosphorus sorption as described by modified Freundlich equations in some soils. International Journal of Agriculture and Biology, 4: 290-292. 2000.

[42] R. Burt., M.D. Mays, E.C. Benham and M.A. Wilson. Phosphorus characterization and correlation with properties of selected bench mark soil of the United States. Commun. Soil Sci. Plant Anal., 33: 117-142. 2002.

[43] S.K. Sanyal and S.K. De Datta. Chemistry of phosphorus transformations in soil. Adv. Soil Sci., 16: 1-120. 1991.

[44] ICRISAT (International Crops Research Institute for the Semi-Arid Tropics).. Farming Systems. Pp.255-259. In ICRISAT Annual Report 1984. ICRISAT, Patancheru, India. 1985.

[45] H. Amjad G.. Abdul and M. Ghulam. Use of models for phosphorus adsorption on some sodic soils of Punjab. International Journal of Agriculture and Biology, 2: 241-248. 2006.

[46] N.A. Polyzopoulos, V.Z. Keramidas and H. Kiosse. Phosphate sorption by some Alfisols of Greece as described by commonly used isotherms. Soil Sci. Soc. Am. J., 49: 81-4. 1985.

[47] M. Del Bubba, C.A. Arias and H. Brix. Phosphorus adsorption maximum of sands for use as media in subsurface flow constructed reed beds as measured by the Langmuir isotherm. Water Res., 37: 390-400. 2003.

[48] M.S. Bakheit, and A. Pakermanjie. Phosphate adsorption and desorption by calcareous soils of Syria. Commun. Soil Sci. Pl. *Anal.*, 24: 197-210. 1993.

[49] T. Gregory, L. K. Chelsey and K. D. Shimizu. A critical examination of the use of the Freundlich isotherm in characterizing molecularly imprinted polymers (MIPS). Analytica Chemica Acta., 528: 107-113. 2005.

[50] R.G. McLaren and K.C. Cameron.. Soil Science: An introduction to the properties and management of New Zealand soils. Oxford University Press, Auckland- New Zealand. 1990.

[51] S. Shai laja. Phosphate adsorption-desorption in selected Vertisols. An MSc Thesis Presented to Andhra Pradesh Agricultural University. 1989.

[52] S. Kuo, and E.G., Lotse. Kinetics of phosphate adsorption and desorption by lake sediments. Soil Sci. Soc. Am. Proc., 38: 50-54. 1974.

[53] M. Zhou and Y. Li. Phosphorus–sorption characteristics of calcareous soils and limestone from the southern Everglades and adjacent farmlands. Soil Sci. Soc. Am. J., 65: 1404-12. 2001.

[54] G.S. Bahl and G.S. Toor. Influence of poultry manure on phosphate availability and the standard phosphate requirement of crop estimated from quantity–intensity relationships in different soils. Bioresource Tech., 85: 317-322. 2002.

# Decomposition rate of pigs' manures and nutrient release pattern in wetland condition

**Bokossa Hervé Kouessivi Janvier[1], Saïdou Aliou[2, *], Fiogbé Emile Didier[1], Kossou Dansou[3]**

[1]Laboratory of Research on Wetlands (LRW), Department of Zoology, Faculty of Science and Technique, University of Abomey-Calavi, Benin
[2]Integrated Soil and Crop Management Unit (ISCM), Laboratory of Soil Sciences, Department of Crop Science, Faculty of Agronomic Sciences, University of Abomey-Calavi, Benin
[3]Laboratory of Plant Biology, Department of Crop Science, Faculty of Agronomic Sciences, University of Abomey-Calavi, Benin

**Email address:**
bokossakouessiv@yahoo.fr (Bokossa H. K. J.), saidoualiou@gmail.com (Saïdou A.), Edfiogbe@Yahoo.Fr (Fiogbe E. D.), kossoudansou@yahoo.com (Kossou D.)

**Abstract:** Recent studies reported the important contribution of animal dejections as organic manure for crop production but, little is known on the decomposition rate and nutrient release pattern in wetland of the manure of pigs nourished with diet enriched with *Azolla filiculoides*. A litter bag study was carried out under full control during 6 weeks in pots containing 25 liters of tap water and 15g of pigs' dejection in each decomposition bag. The experimental design was a completely randomized block design with three replications. The treatments consisted of dejections of pigs nourished with : T1 (recommended diet composition) consisted of 15% *Azolla* + 55% provender + 5% coconut copra + 5% oil palm + 5% soybean bran + 10% rice bran + 5% kitchen waste ; T2 (partially improved diet with *Azolla*) consisted of 30% *Azolla* + 65% rice bran + 5% oil palm ; T3 (improved diet with *Azolla)* consisted of 47.5% *Azolla* + 47.5% rice bran + 5% oil palm; and T4 (improved diet with cereal bran) consisted of 15% *Azolla* + 40% rice bran + 40% wheat bran + 5% oil palm). Four pigs per diet were considered leading to 16 white landrace pigs of six months age. Decomposition rates in water were significantly ($P < 0.05$) fast during the first week and became very slow during the following weeks. After the six weeks of experiment, 58.2 % of manure from treatments T1 and T4 were decomposed against 47.2 % for T2 and T3. However, 40.1, 53.3, 67.4 and 57.1% of total N content in treatments T1, T2, T3 and T4 respectively were released. As consequence, manure from improved diet with *Azolla* (T3 and T2) are suggested for integrated rice and fish production system. Nevertheless, decomposition and nutrients (P, K, Ca and Mg) release patterns had significantly ($P < 0.05$) increased in water in treatments T1 and T4 compared with the two treatments containing more *Azolla* in the diet (T2 and T3).

**Keywords:** Pig Manures, Pig Diets, *Azolla Filiculoides*, Mineralization Rate, Nutrient Release, Wetland

## 1. Introduction

During the recent decades, an increase attention was paid to soil and water pollution by chemical fertilizer and pesticide using in intensive agricultural system [1]. Much researches put emphasized on nitrate leaching in agricultural land due to an increasing concentration in ground and drinking water [2] as result of livestock breeding intensification. This practice is responsible of most of the environmental damage caused to the terrestrial and marine ecosystems. In fact, it is generally attributed to the supply of increase nitrogen fertilizer [1]. This suggests alternative input environment friendly such as the use of organic residues from agricultural activities to improve land productivity and also fish production system. Manure and animal dejection can be an opportunity to supply nutrients especially P and N for crop production [3, 4, 5]. N release from organic wastes mineralization has been successfully tested in agriculture [6, 7, 8, 9, 10]. These authors concluded that, the supply of manure in agricultural land is an excellent way to recycle nutrients and organic matter to improve crop production. Generally, farmyard manures or compost are suggested as an alternative method for the management of these organic residues [10, 11]. Knowledge on nutrients release from the decomposition and mineralization of organic manure is critical for the regulation of the ecosystem productivity, especially in nutrient-limited system.

However, [12] revealed that, rice–fish farming system constitutes an unique agro-landscape across the world, especially in tropical and sub-tropical area and the introduction of fish production to rice farming creates an integrated agro-ecological system. In that system, [13] found an increase of paddy rice yield compared with rice culture alone. It was also proved that, when the pigs were nourished with diet of high nutritive value such as *Azolla filliculoides*, it improves the quality of manure produced which could be used to feed fish and to supply nutrient to the crop in the wetland production system [14].

In such way, developing strategies to recycle nutrient from these residues especially those from livestock and pigs in the context of Benin agriculture is a challenge. However, in Benin, more than 518,600 pigs are breed with an increase of 3% the last four years [16, 17] leading to an important quantities of dejection produced which are source of environmental pollution. Furthermore, zooplankton production system based on the use of pigs' manure have been previously reported by [18] but, there is lack of knowledge on the decomposition and nutrient release pattern from animal manures especially pigs' dejection in the wetland for fish and rice production system.

This paper reports results of a study on the decomposition and nutrient release pattern of pigs' dejection in wetland condition using pots experiment. It aims to (i) study the decomposition rate in wetland condition of the manure of pigs nourished with diets of rice and wheat bran enriched with *Azolla filiculoides*, (ii) determine the amount of nutrients remain in the pigs' dejection during the decomposition process in the wetland condition and (iii) assess the amount of nutrient released in the water regarding the dejection of the pigs nourished with the different diets.

## 2. Materials and Methods

### *2.1. Experimental Site*

The experiment was carried out on the site of the Laboratory of Research on Wetlands (LRW), Department of Zoology, Faculty of Science and Technique, University of Abomey-Calavi, Benin. It is located between 6°42' N and 2°34' E and 17 m asl. The area is characterized by sub-equatorial climate with two rainy seasons (March to end of July and mid-September to November) and two dry seasons (August to mid-September and December to March). *Azolla filliculoides* an aquatic fern used to enrich the pigs' diet was produced during three months (May to July 2013) in ponds on the experimental site.

### *2.2. Pigs Feeding and Manure Collection*

Pigs feeding and manure collection according to the types of diet served to the animal were carried out on the experimental site. The manures were daily collected and mixed according to the diet served. They were collected from June to August 2013 and air dried at ambient temperature. About 50 kg of dejection were collected per types of diet given to the pigs. The diets were made based on results from experiments carried out by [17].

The diets consisted of recommended diet composition: 15% Azolla + 55% provender + 5% coconut copra + 5% oil palm + 5% soybean bran + 10% rice bran + 5% kitchen waste (T1); partially improved diet with *Azolla*: 30% *Azolla* + 65% rice bran + 5% oil palm (T2); improved diet with *Azolla* : 47.5% Azolla + 47.5% rice bran + 5% oil palm (T3) and improved diet with cereal bran: 15% *Azolla* + 40% rice bran + 40% wheat bran + 5% oil palm (T4). The feeding experiment was carried out from May to August 2013. Four pigs white landrace of six months age per type of diet were considered leading to 16 animals were followed for manure collection.

### *2.3. Experimental Design and Decomposition Process*

The experimental design was a completely randomized block design with three replications. The treatments consist of the different types of manure collected from the feeding experiment: T1 = recommended diet composition; T2 = partially improved diet with *Azolla*; T3 = improved diet with *Azolla* and T4 = improved diet with cereal bran.

Instead of plastic containers we used clay jars for the pot experiment as they create fresh condition like in the wetland. These clay jars were purchase in the village of Sè (South-West of Benin). The jars were filled with 25 liters of tap water [18]. Manure bags decomposition used in this experiment were constructed from nylon mosquito wire with a mesh size of 1 mm. The nylon was tied and suspended in the tap water and also adapted from litter-bags techniques [19].

For each treatment, 12 manure nylon bags containing 15 g oven dried manure materials were filled. Manure nylon bags were randomly placed in the 12 clay jars of one lot to be sampled each week. In total, 6 lots of 12 clay jars leading to 72 clay jars were used during the experiment. Manure nylon bags were sampled at 6, 12, 18, 24, 30 and 36 days, with one manure nylon bag randomly selected from each lot and transported to the laboratory for the analysis. During each sampling period, three manure nylon bags per treatment were collected. Furthermore, 15 water samples were also collected and sent to the laboratory for analysis.

Undecomposed manures were carefully put in aluminum crucible after removing the nylon bag and oven dried at 65°C for 72 h to determine manure mass remaining at the Laboratory of Wetland Research. The samples were then weighted to appreciate the weight losses. Samples were ground (mesh 0.1 mm) and sent to the Laboratory of Soil Science Water and Environment of Benin National Agricultural Research Institute (LSSEE/INRAB) for nutrient analysis. Water samples were analyzed at the Laboratory of Water Analysis of the General Directorate of Water (DGE, in charge of drinking water control in Benin) at Cotonou.

### *2.4. Manure and Water Analysis*

The analyses were carried out on the undecomposed pigs' manure and water during the decomposition process. Methods used for the analysis of the undecomposed manure were that described by NF EN 14082 norm used at the LSSEE/INRAB's laboratory. The analyses were performed on: organic carbon (dry ashing in a muffle furnace at 550°C for 24 h then, organic matter is determined by the difference between the treated sample and the mass of ash obtained assigned an empirical coefficient of 2). The ash obtained was used for determination of Ca, Mg, K and P content in the manure. The ash was gathering in 6N HCl involving a period of heating at 125°C then the residue was dissolved in 1N $HNO_3$. Ca, Mg and K were determined by atomic absorption spectrophotometry and P measured colorimetrically by ammonium molybdate with ascorbic acid at a wavelength of 660 nm. Total N was determined using Kjeldahl digestion in a mixture of $H_2SO_4$ selenium followed by distillation and titration.

Water sample collected were stored in a fridge at 4°C for assessing nutrient released during the mineralization process. Analyses included suspended solids, $NO_3^-$, $NO_2^-$ and $NH_4^+$ content. Suspended solids were measured using Standard NF EN 14082 method after water drying in an oven at 105°C for 24 hours followed by dry ashing of the residue in a muffle furnace at 550°C for 6 hours then, C content was determined. $NO_3^-$ using sodium salicylate method [20], nitrate ions in the presence of sodium salicylate form sodium paranitrosalicylate colored in yellow which is measured by colorimetry at a wavelength of 420 nm. $NO_2^-$ using reagent of Zambely method [20] where ammonium ions in the presence of sulfanilic acid and phenol form a yellow colored complex which is measured by colorimetry at a wavelength of 435 nm. $NH_4^+$ by NESSLER method [20], ammonia in the presence of this reagent forms reddish-brown color which is measured by colorimetry at a wavelength of 425 nm. $PO_4^{3-}$ using ammonium molybdate reagent the blue color was measured by colorimetry at a wavelength of 880nm.

$K^+$, $Ca^{2+}$ and $Mg^{2+}$ content in water were determined by atomic absorption spectrophotometry. Temperature and dissolved oxygen were measured in situ using an oxythermometer (OXI 197i-N27455), conductivity using a conductimeter (N27444-197i) and pH with a pH meter (197i-N27499).

### *2.5. Data Analysis*

Several models were tested using EXCEL software but it was found that, the single exponential equation provided a good fit of the different types of pigs' manures decomposition pattern in water over the complete time of observation (36 days) with coefficient of determination over 90%. The following single exponential model was fitted to the data:

$$Y = e^{-k*t}$$

Whereby, Y is the proportion of initial mass remaining at time *t*, and *k* is the decomposition factor. To calculate the *k* values, the formula was rewritten as:

$$\ln Y = -k * t$$

Hence the slope of the line, calculated by linear regression, is the *k* value.

The statistical analyses were performed using SAS 9.2 package. Analysis of variance was run on the decomposition data, nutrient remained in the manures and nutrient released in water considering time of decomposition and treatments (types of manure) as main factors, standard errors of the difference in means were calculated. The Student Newman-Keuls test was performed to compare differences in means among treatments. All significance levels were set at $P < 0.05$.

## 3. Results

### *3.1. Decomposition Pattern of the Different Types of Pigs' Manure in Water*

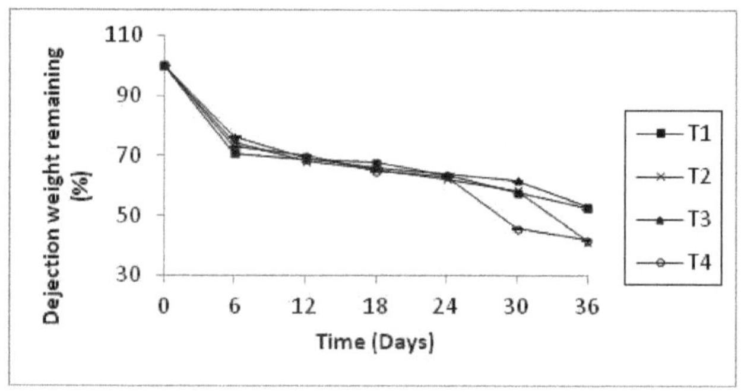


***Figure 1.*** *Decomposition pattern in water of manures from pigs nourished with different types of diet*

Note: T1 = Manure of pigs nourished with recommended diet composition; T2 = manure of pigs nourished with diet partially improved with Azolla; T3 = manure of pigs nourished with diet improved with Azolla ; T4 = manure of pigs nourished with diet improved with cereal bran.

The decomposition patterns *i.e.* mass remaining of the manure from pigs nourished with different types of diet are presented in Fig. 1. During the first week the graphs show similar and rapid decomposition of the manures in water. But, after two weeks, the decomposition trends were almost slow and do not vary significantly considering the manure types. Specifically, the decomposition pattern was steady from the second week until the fourth week. Then, the decrease of the mass remains were noticed with manure from pigs nourished with diet partially improved with *Azolla* (T2) and diet improved with cereal bran (T4) after five and six weeks of decomposition. The decomposition pattern of the manures from pigs nourished with diet improved with *Azolla* (T3) and manure of pigs nourished with recommended diet composition (T1) were almost similar after five and six weeks. Significant differences were noticed between treatments T2 and T4 in the one hand and treatments T1 and T3 in the other hand. Thus, at the end of the experiment, more than 41.8 % of the mass of the manure from treatments T4 and T2 were not decomposed against more than 52.8 % of the mass of the manure from treatments T1 and T3.

The decomposition constants k calculated for the different periods of experiment are presented in Table 1. The k values decrease fast during the first week and slowly during the following five weeks as previously shown.

### *3.2. Nutrient Remaining in the Manure*

The amount of nutrient remaining in the manure during the sampling periods is presented in Fig. 2. However, quantity of nutrient remained after the decomposition process is irregular over the time as shown with the decomposition rate. The losses of N, P, K, Ca and Mg were significantly ($P < 0.05$) fast during the first week of decomposition. From the second week to the subsequent weeks, nutrient concentration in the manure decreased slowly but, the amount remains relatively stable at the end of the experiment.

More than 50% of the N in the different manures was mineralized during the six weeks of the experiment with almost 50% of weight loss of the decomposed manure. This is much more important with manure of pigs nourished with diet improved with *Azolla* (T3 and T2). Considering the total amount of N, it is observed a significant release for treatment T3 and T2 compared with treatments T1 and T4. However, the amount of N release in treatments T1 and T4 are similar meaning that, the low level of nitrogen source in the diet of the pigs affects significantly the decomposition rate of the manure in the water.

The total P released in the manure of treatment T1 was significantly ($P < 0.05$) higher than that of treatments T2, T3, T4. About 2/3 of the initial contents of P in the different types of manure were released from the beginning to the end of the experiment. For the different types of manure, after two weeks of decomposition, almost the total amount of P is released leading to weak amount of P remained in the manure from the $4^{th}$ to the $6^{th}$ week of decomposition. In total, the amount of P that remain in the manure after six weeks of decomposition in treatment T1 is 1.3 time significantly ($P < 0.05$) higher than that of treatment T2 while it is almost similar with treatments T3 and T4.

The change in K, Mg and Ca concentration during the decomposition process follow the same trend as that of P with high amount released with treatment T1 compared with treatments T2, T3 and T4. It is also observed that, about 2/3 of the initial K contents were released over the six weeks. In total, the amount of K remaining in the manure after six weeks of decomposition in treatment T1 is 2.4, 2.6 and 1.9 times significantly ($P < 0.05$) higher than that of treatments T2, T3 and T4 respectively. However, the amount of Ca remaining in the manure after six weeks of decomposition in treatment T1 is 1.3, 1.2 and 1.4 times significantly ($P < 0.05$) higher than that of treatments T2, T3 and T4 respectively. Furthermore, the change with Mg content in the manure T1 which is the same with treatment T3 is 1.3, 1.2 times significantly ($P < 0.05$) higher than that of T2 and T4.

### *3.3. Effect on Water Quality*

Water chemical properties as affected by the manure decomposition process at the beginning and the end of the experiment are presented in Tables 2 and 3. The decomposition of manure has significantly affected water quality despite the relative short period of decomposition. However, the different types of manure have induced significant increase in N fractions ($NH_4^+$, $NO_3^-$, $NO_2^-$) at the end of experiment. The $K^+$, $Ca^{2+}$, $Mg^{2+}$ and phosphorus ($PO_4^{3-}$) content in water at the end of the decomposition process of all of the manure of pigs nourished with different diets have also significantly ($P < 0.05$) increased as induced by the large amount of nutrients released. This is much more important with treatment T1 compared with the other treatments.

We also observed an increase of the conductivity of the water induced by the decomposition of the manure as result of nutrient enrichment. However, manure from pigs nourished with diet improved with cereal bran (T4) presents the highest value of water conductivity compared with the other types of decomposed manure. The value of water conductivity in treatment T4 has significantly ($P < 0.05$) increased by 1.5, 1.14 and 1.2 compared with treatments T1, T2 and T3 respectively.

The pH values as affected by the decomposition process of the manures have significantly increased along the six weeks of experiment. The values are almost greater than 7 leading to alkaline area induced by the manure in decomposition which has released large amount of $NH_3$. The temperature fluctuates between 30 and 33°C reflecting the climatic condition during experimentation period. With regard to the suspended solids in the water, they present a remarkable increase between the beginning and the end of the experiment in all of the treatments. Manure from diet with high content of *Azolla filiculoides* (treatment T3 and secondarily T2) has release more suspended solids which were 1.4, and 1.3 times higher than treatments T4 and T1 respectively. Values of dissolved oxygen concentration increased significantly from the first day to the end of the experiment but, no significant ($P > 0.05$) difference was observed between treatments.

## 4. Discussion

### *4.1. Performance of the Decomposition Model Used*

The one component model $Y=e^{-k*t}$ used also applied by [21] has been used to determine the mineralization constant k. During the decomposition process a fast raise of the k values during the first week was observed then, the trends were progressively weak and constant until the end of the six weeks of observation. It was observed that, k is not constant during the time showing the weakness of the model used. However, [22] working in an aquatic environment indicated that, the rate of decomposition is not constant in the time and proposed the implication of several other constants and modified the formula developed by [23] suggesting the following equation: $Y=Y_o*e^{(ko/b)(e-bt-1)}$ which was not applicable in the case of our study. Our results show a fast mineralization of labile fraction contained in the manure during the first week then, during the following weeks, a slow mineralization process of the fraction rich in carbon (*i.e.* having high C/N ratio) until the end of the experiment. However, taking into account the various constituent of the organic residues, [19] studying the mineralization process of nitrogen in the organic residues suggested the following equation: $W_t = Ae^{-k1t} + (1 - A)e^{-k2t}$

where, Wt = percentage of weight loss at the time t, A the labile portion in relation to the total mass, 1-A the ratio of the resistant fraction in the total weight, $k_1$ the rate of decomposition for the labile fraction to fast deterioration within a year and $k_2$ for the rate of decomposition of the resistant fraction. As described, the equation suggested by [19] cannot be applied to short length experimentation. Nevertheless, the equation used in the present study has the merit to contribute in the raising of the mechanism of decomposition of the manure in water with the report of fast and fleeting mineralization process. This is very important when it comes to develop improving practice for fish and rice cultivation while limiting the risk of N volatilization. Furthermore, diets including rice bran, *Azolla filiculoides* and wheat bran were found to be an excellent practice which provides manure of good quality to be used in integrated fish-rice culture for saving resources.

### *4.2. Mineralization Process and Nutrient Release as Affected by the Pigs' Diet*

Our study shows large differences in the chemical composition, decomposition patterns and water chemical properties as result of decomposition process of the manures. N and P decreased fast at the first week and slowly for the last five weeks showing two phases of N mineralization. The works of [24] confirmed our result and revealed that, most of the N in green manure was mineralized shortly after application. These results were different from those found by [25] who found four phases of mineralization of litters.

This difference from our results can be due to the short time duration of the experiment and the decomposition environment. Beside the short duration of our study, one could mention watershed environment with probably different types of microorganism involved in the decomposition process compared with that of the soil.

Furthermore, [26] revealed that, higher microbial activity can be correlated with greater N release. In fact, low C: N ratio was found in all of the pig manures studied which could probably explain this fast mineralization trend observed during the first phase noticed. But in the second phase with the recalcitrant fraction of the manure the C: N ratio becomes high due to the rapid loss of N as consequence, slow N release is noticed during the succeeding weeks of decomposition.

***Table 1.*** *Decomposition constants (k.day$^{-1}$) calculated for different periods (t). Values were calculated with single exponential model ($Y = e^{-k*t}$), coefficient of determination ($R^2$) indicates goodness of fit.*

| Treatments | t = 6 days | | t = 12 days | | t = 18 days | | t = 24 days | | t = 30 days | | t = 36 days | |
|---|---|---|---|---|---|---|---|---|---|---|---|---|
| | k | $R^2$ | k | $R^2$ | k | $R^2$ | k | $R^2$ | k | $R^2$ | k | $R^2$ |
| T1 | 5.72 | 0.84 | 3.10 | 0.84 | 2.16 | 0.84 | 1.89 | 0.84 | 1.83 | 0.84 | 1.78 | 0.84 |
| T2 | 4.94 | 0.87 | 3.19 | 0.87 | 2.36 | 0.87 | 1.97 | 0.87 | 1.78 | 0.87 | 2.45 | 0.87 |
| T3 | 4.48 | 0.87 | 3.08 | 0.87 | 2.30 | 0.87 | 1.86 | 0.87 | 1.61 | 0.87 | 1.77 | 0.87 |
| T4 | 5.19 | 0.92 | 2.95 | 0.92 | 2.40 | 0.92 | 1.91 | 0.92 | 2.61 | 0.92 | 2.42 | 0.92 |

T1 = manure of pigs nourished with recommended diet composition; T2 = manure of pigs nourished with diet partially improved with Azolla; T3 = manure of pigs nourished with diet improved with Azolla ; T4 = manure of pigs nourished with diet improved with cereal bran.

***Table 2.*** *Average value (± standard errors) of water chemical (mg.L$^{-1}$) properties at the beginning and the end of the decomposition process of manure of pigs nourished with different types of diet*

| Treatments | $NO_2^-$ | | $NO_3^-$ | | $NH_4^+$ | | $PO_4^{3-}$ | |
|---|---|---|---|---|---|---|---|---|
| | Beginning | End | Beginning | End | Beginning | End | Beginning | End |
| T1 | 0.003±0.00a | 0.012±0.002a | 6.04± 0.00a | 12.41±1.19a | 0.04±0.00a | 0.36±0.05a | 0.14±0.00a | 2.80±0.2a |
| T2 | 0.003±0.00a | 0.007±0.001a | 6.04±0.00a | 9.00±0.20b | 0.04±0.00a | 0.23±0.06a | 0.14±0.00a | 2.02±0.32a |
| T3 | 0.003±0.00a | 0.009±0.002a | 6.04±0.00a | 8.60±0.72b | 0.04±0.00a | 0.35±0.07a | 0.14±0.00a | 2.41±0.26a |
| T4 | 0.003±0.00a | 0.005±0.000a | 6.04±0.00a | 7.30±0.14b | 0.04±0.00a | 0.31± 0.07a | 0.14±0.00a | 2.26±0.08a |

***Table 2.*** *continued*

| Treatments | $K^+$ | | $Ca^{2+}$ | | $Mg^{2+}$ | |
|---|---|---|---|---|---|---|
| | Beginning | End | Beginning | End | Beginning | End |
| T1 | 4.82± 0.00a | 13.18±0.52a | 1.20 ± 0.00a | 9.41± 0.30a | 1.21±0.00a | 3.51±1.5b |
| T2 | 4.82± 0.00a | 10.35±0.64ab | 1.20 ± 0.00a | 3.66± 0.56b | 1.21 ±0.00a | 4.06±0.6a |
| T3 | 4.82± 0.00a | 8.97±0.70bc | 1.20 ± 0.00a | 7.83 ± 0.21a | 1.21 ±0.00a | 3.61±1.5b |
| T4 | 4.82± 0.00a | 10.37±0.60ab | 1.20 ± 0.00a | 7.23± 1.01a | 1.21 ±0.00a | 4.28±0.9a |

***Table 3.*** *Average value (± standard errors) of water physical properties at the beginning and the end of the decomposition process of manure of pigs nourished with different types of diets*

| Treatments | Suspended solids (mg.l$^{-1}$) | | pH | | Temperature (°C) | |
|---|---|---|---|---|---|---|
| | Beginning | End | Beginning | End | Beginning | End |
| T1 | 110.00± 0.00a | 470.66±51.20a | 5.13 ±0.00a | 7.71± 0.27ab | 30.10 ± 0.00a | 33.77± 0.24a |
| T2 | 110.00± 0.00a | 571.00±60.79a | 5.13 ±0.00a | 7.69 ± 0.33ab | 30.10 ± 0.00a | 32.62 ± 0.16a |
| T3 | 110.00± 0.00a | 646.67±74.66a | 5.13 ±0.00a | 7.84 ± 0.32ab | 30.10 ± 0.00a | 32.33 ± 0.26a |
| T4 | 110.00± 0.00a | 453.33±42.20a | 5.13 ±0.00a | 8.14 ±0.23a | 30.10 ± 0.00 | 33.13± 0.46a |

***Table 3.*** *continued*

| Treatments | Dissolved $O_2$ (mg.l$^{-1}$) | | Conductivity (µS.cm$^{-1}$) | |
|---|---|---|---|---|
| | Beginning | End | Beginning | End |
| T1 | 8.70 ± 0.00a | 13.59 ± 1.44a | 99.00 ±0.00a | 199.19 ± 4.17a |
| T2 | 8.70 ± 0.00a | 13.48 ± 0.53a | 99.00 ±0.00a | 154.10±14.40a |
| T3 | 8.70 ± 0.00a | 14.17 ± 0.71a | 99.00 ±0.00a | 147.17±13.02a |
| T4 | 8.70 ± 0.00a | 13.12 ± 0.72a | 99.00 ±0.00a | 174.93±20.06a |

Note: Means followed by the same letter are not significantly different (P > 0.05) according to the Student Newman-Keuls test.

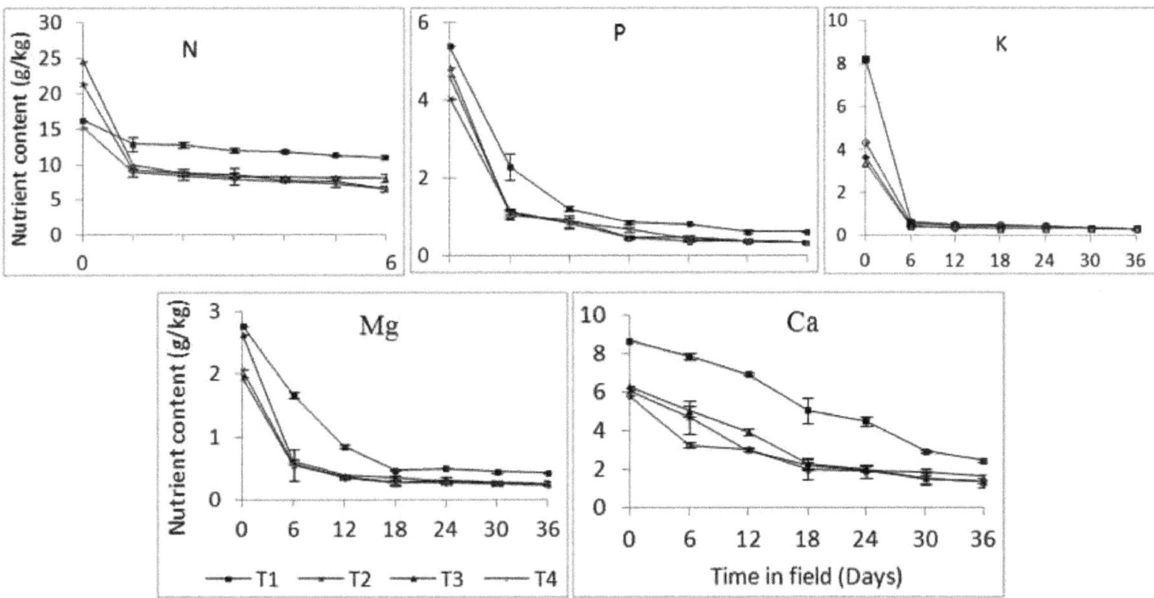


***Figure 2.*** *Changes in nutrient concentration (g kg$^{-1}$) during decomposition process of manures from diets: (T1 = Manure of pigs nourished with recommended diet composition; T2 = manure of pigs nourished with diet partially improved with Azolla; T3 = manure of pigs nourished with diet improved with Azolla ; T4 = manure of pigs nourished with diet improved with cereal bran.*

From our results, we observed that, only treatments T1 and T4 from diet enriched with few amount of *Azolla filiculoides* maintained a low C: N ratio during the experiment. However, *Azolla* contains 28.24 % of lignin [27, 28] as consequence it may slow down the decomposition process of the organic residue. From this value, [29] reported that, C and N content regulate the early phase of decay while, lignin become more important for explanation thereafter. With treatments T2 and T3 diet containing 30% and 47.5 % respectively of *Azolla filiculoides* the C: N ratios were high after a week. These results corroborate with [24] results studying litters decomposition in the soil and found that, the slow decomposition is mainly caused by the very highest C: N ratio due to the important lignin + polyphenol content in the residues. Furthermore, [19] reported also that, the amount of nutrient released can be correlated with C : N : P ratio. Manure from diet T3 has the highest value of N at the beginning and ongoing, these values become low due to the loss of N.

K, Ca and Mg values were high in treatment T1 but decreased over the weeks of experimentation. Our results indicate that, pig manure can supply all major nutrients (N, P, K, Ca, Mg) for rice production in wetland. At the same way, [30] prove that, organic materials such as farmyards have traditionally been used by rice farmers. It supplies all major nutrients (N, P, K, Ca, Mg, S,) necessary for plant growth [31]. In wetland especially inland valley conditions where it was noticed from our result nitrogen loss by volatilization during the decomposition process, it is suggested for rice-fish production system to split manure supply.

## 5. Conclusion

From the present study we can conclude that, manure

from diets T1 and T4 are an important source of nutrients P, K, Ca and Mg. However, manure from diets T3 and T2 contained more N due to the addition of *Azolla filliculoides* (47.5 % and 30 %) in the diet which had released an important quantity of N in the first week then after, leading to very weak concentration during the succeeding weeks of experiment. This important release of N could be harmful for rice production in wetland especially in the inland valley and having important nutrient demand during the growth period. Bases on these results, it is suggested for fish-rice production system pig manure from diet not too much rich in N in order to avoid N loss during the cropping period. When it comes to use this source of nutrient as strategy to improve crop production in the wetland or in the inland valley, it is suggested to split the amount of this organic manure to be applied according to the plant growth stage.

## Acknowledgements

The authors are grateful to the "Scientific Council of the University of Abomey-Calavi in Benin" for funding the present research in the framework of the project "*Optimization Integrated Agricultural Production System without Chemical Inputs* (OPASISI)".

## References

[1] F. Pilar, I. Castellar, and J. Navarro, "Nitrate leaching in pepper cultivation with organic manure and supplementary additions of mineral fertilizer". *Communications in Soil Science and Plant Analysis*, 36: 2889–2899, 2005.

[2] R.F. Follett, and J.A. Delgado,"Nitrogen fate and transport in agricultural systems." *Journal of Soil and Water Conservation,* 57: 402–408, 2002.

[3] M. Kaleem, A. Abbasi, H. Munazza, K. Abdul, and S. R. Khan, "Mineralization of three organic manures used as nitrogen source in a soil incubated under laboratory conditions." *Communications in Soil Science and Plant Analysis*, 38:13-14, 1691-1711. 2007.

[4] M.P. Bernal, J.A. Albuquerque, and R. Moral, "Composting of animal manures and chemical criteria for compost maturity assessment". A review. *Bioresource Technology* 100, 5444–5453, 2009.

[5] M. Inckel, S. Peter, T. Tersmette, and T. Veldskamp, "Fabrication et utilisation du compost." *Série AgroDoc.* N°8 : 71p., 2007.

[6] E. Vasconcelos, F. Cabral, and C.M.D.S. Cordovil, "Wheat yield and leachability of phosphorus and nitrogen in pig slurry amended soils." *Communications in Soil Science and Plant Analysis*, 30: 2245–2257, 1999.

[7] N. Vagstad, A. Broch-Due, and I. Lyngstad, "Direct and residual effects of pulp and paper mill sludge on crop yield and soil mineral N". *Soil Use Management,* 17: 173–178, 2001.

[8] G.W. Evers, "Ryegrass–Bermuda grass production and nutrient uptake when combining nitrogen fertilizer with broiler litter." *Agronomic Journal,* 94, 905–910, 2002.

[9] C.M.D.S. Cordovil, F. Cabral, and J. Coutinho, "Potential mineralization of nitrogen from organic wastes to ryegrass and wheat crops". *Bioresource Technology 98,* 3265–3268, 2007.

[10] L.G. Amadji, A. Saïdou and L. Chitou, "Recycling of organic residues in compost to improve coastal sandy soil properties and cabbage shoot in Benin." *International Journal of Biological and Chemical Sciences*, 3(2): 192 – 202, 2009.

[11] G.F. Huang, J.W.C. Wong, Q.T. Wu, and B.B. Nagar, "Effect of C/N on composting of pig manure with sawdust." *Waste Management,* 24: 805–813, 2004.

[12] L. Jianbo, and L. Xia, "Review of rice–fish-farming systems in China One of the Globally Important Ingenious Agricultural Heritage Systems (GIAHS)." *Aquaculture,* 260: 106–113, 2006.

[13] A.K.Y Haroon, and K.A. Pittman, "Rice-fish culture: feeding, growth and yield of two size classes of Puntius gonionotus Bleeker and Oreochromis spp. in Bangladesh." *Aquaculture,* 154: 26 l-281, 1997.

[14] A.Youssouf, A. Saidou, D. Mama, E.D Fiogbé and J-C. Micha, "Evaluation of nitrogen and phosphorus wastes produced by Nile Tilapia (Oreochromis niloticus L.) Fed Azolla-diets in earthen ponds." *Journals of Environmental Protection*, 3: 502-507, 2012.

[15] E.D. Fiogbé, and K.H. Gangbazo, "Production porcine avec Azolla." *Actes des 2$^{èmes}$ Journées Scientifiques Internationales des Universités Nationales du Bénin.* (13-16 avril 2004) ; 142-154, 2005.

[16] J-M.M. Accodji, E.D. Fiogbé, and K.H. Gangbazo, "Essai de valorisation d'*Azolla* (*Azolla microphylla* Kaulf) dans la production porcine en zone humide." *International Journal of Biological and Chemical Sciences,* 3(5) : 890 – 898, 2009.

[17] H. Agadjihouèdé, E. Montchowi, A. Chikou, and P.A Lalèyè, "Libération comparée de sels dans l'eau par la minéralisation de l'Azolla, la bouse de vache, la fiente de volaille et les sons de riz et de maïs utilisés en pisciculture." *International Journal of Biological and Chemical Sciences,* 5(5): 1883-1897, 2011.

[18] S. Qiu, A.J. Mc Comb, and R.W. Bell, "Leaf litter decomposition and nutrient dynamics in woodland and wetland conditions along a forest to wetland hill slope." *International Scholarly Research Network. ISRN Soil Science.* Vol 12, ID346850, 8 p., 2012.

[19] J. Rodier, "L'Analyse de l'eau. eaux naturelles, eaux résiduelles, eau de mer." 8$^{ème}$ édition DUNOD, 1383p., Juin 1996.

[20] E.A. Hartemink and J.N. O' Sullivan, "Leaf litter decomposition of *Piper aduncum*, *Gliricidia sepium*, and *Imperata cylindrica* in the humid lowland of Papua New Guinea." *Plant and Soil*, 230: 115 – 124, 2001.

[21] G.L. Godshalk, "Decomposition of aquatic plants in lakes". PhD. Thesis, Michigan State University. *Department Botany and Plant pathology*: 49-61, 1977.

[22] J. Kortleven, "Quantitative aspects of humus accumulation and decomposition." Landbk, Onderz. 69. 1. *Pudoc, Wageningen.* 109 pp, 1963.

[23] C-H. Wang, "Effects of different organic materials on crop production under a rice–corn cropping sequence." *Communications in Soil Science and Plant Analysis*, 44 : 2987–3005, 2013.

[24] M.K. Abbasi, M. Hina, K.A. Abdul, and S.R. Khan, "Mineralization of three organic manures used as nitrogen source in a soil incubated under laboratory conditions." *Communications in Soil Science and Plant Analysis*, 38: 1691–1711, 2007.

[25] R. Hassan, E. Khadija, D. Belghyti, and M. Hadji, "Physico-chemical waste water unit of sugar SUNABEL Mechraa Belksiri". *ScienceLib* Editions Mersenne : Volume 5, N°130215, 2111-4706, 2013.

[26] E.D. Fiogbé, I. Imorou Toko, J-C. Micha, and C.H.F Van Hove "*Azolla* : une fougère aquatique utilisable comme intrant pour diversifier et intégrer les productions agro-piscicoles des zones humides." *Journée scientifique à l'Université Nationale du Bénin*, 2001.

[27] A.Z.K. Djogbédé, L.C. Hinvi, E.D. Fiogbé, "Effets de substitution des engrais chimiques par *Azolla pinnata* en riziculture au Nord Bénin." *International Journal of Biological and Chemical Sciences,* 6(6) : 3027-3044, 2012.

[28] C.K.H. Smith, L. Gholz and F. De Assis Oliveira, "Fine litter chemistry, early-stage decay, and nitrogen dynamics under plantations and primary forest in lowland Amazonia." *Soil Biology & Biochemistry,* 30(14): 2159-2169. 1998.

[29] T. Tilahun, D. Nigussie, B. Wondimu, and G. Setegn, "Effects of farmyard manure and inorganic fertilizer application on soil physico-chemical properties and nutrient balance in rain-fed lowland rice ecosystem". *American Journal of Plant Sciences*, 4, 309-316, 2013.

[30] G. Villegas-Pangga,"Agri-wastes for soil productivity improvement in a lowland rice ecosystem." Symposium on agricultural and agro-industrial waste management, march 12-14, Sao Pedro, sp, Brazil, 2013.

# Asymmetry effect of intercropping non host crops between cabbage and climatic factor on the population of the diamondback moth (*Plutella xylostella* L.) and yield

**Mohammad Feizal Daud[1], Fauziah I.[1], Mohd Rasdi Z.[1], Fairuz K.[1], Abu Zarim U.[2], Syed Abdul Rahman S. A. R.[3], Ismail R.[1], Mohd Hanysyam M. N.[1], Norazliza R.[1]**

[1]Faculty of Plantation and Agrotechnology, Universiti Teknologi MARA (UiTM) Melaka, Jasin Campus, Melaka, Malaysia
[2]Malaysian Agriculture Research and Development Institute (MARDI) Cameron Highlands Research Station, Pahang, Malaysia
[3]Taman Agro Al Mashoor, Cameron Highlands, Pahang, Malaysia

**Email address:**
mfeizal@salam.uitm.edu.my (Md. Feizal. D.), fauziah@salam.uitm.edu.my (Fauziah I.), dddpim@salam.uitm.edu.my (Mohd Rasdi Z.), fairuzdk@salam.uitm.edu.my (Fairuz K.)

**Abstract:** Diamondback Moth (DBM) is one of the major constraints for the profitable production of cabbage in Cameron Highlands, Malaysia. The loss in yield of cabbage caused by DBM could reach up to more than 80% if a severe infestation occurred in the field. Due to that problem, farmers in Cameron Highlands are preferred to apply the broad spectrum of synthetic insecticides rather than implementing other pest management strategies. The demand for insecticides in Cameron Highlands seems endless until today, but more aggressive. This paper highlighted the possibility of implementing intercropping cabbage with seven treatments in reducing adults and larvae populations of DBM in the polyculture system. The distributions of adults and larvae population of DBM against the factors of temperature and rainfall also have been addressed in this study. A total of seven treatments was implicated, including onion (*Allium cepa* L.), tomato (*Solanum lycopersicon* L.), marigold (*Tagetes erecta* L.), citronella grass (*Cymbopogon nardus* L. rendle), and mosquito plants (*Pelargonium citrosum*) which arranged in Randomized Complete Block Design (RCBD). The results indicated that insignificant variations (P>0.05) were observed among the population of adults and larvae at all seven treatments tested compared with the control. Moreover, adults and larvae of DBM were negatively correlated with temperature and rainfall factors. The results also showed that a critical period of DBM infestations was detected in the pre-heading stage (< 30 DAT). In terms of performance among treatments, our finding also has shown that mosquito plant was the best treatment acted as a repellent, however mosquito plant treatment contributed to the lowest yield due to the competition of plant growth resources and its wider canopy structure.

**Keywords:** Cabbage, Intercropping, *P. xylostella* (DBM), Polyculture System, Cameron Highlands

## 1. Introduction

In the present day, the Department of Agriculture (DOA), Malaysia is recognized Plutella xylostella as one of the limiting factors for a profitable cabbage production in the Cameron Highlands. Many vegetable farmers in this area are facing a serious cabbage yield loss if there is no earlier mitigation measure has been taken by cabbage farmers after detecting the infestations of P. xylostella in their farm [1]. The problems are arising whilst the larvae of P. xylostella feed undersurface of the leaves of cabbage plants and finally created the transparent areas and holes on the leaves. To overcome these problems, most of cabbage farmers are aggressively applying the broad spectrum of synthetic insecticides in their farms. Generally, cabbage farmers were implementing synthetic insecticides for 20-30 applications at every 3-5 days interval of 90 days of cropping periods [2]. Thus, it has resulted in the increment percentage of insecticide resistance to many insect pests, particularly P. xylostella, additional operation costs, heavy reliance on chemical control and insecticide residues hazardous to farmer's health in the long run. Findings from the study done by [3] stated that the implementation of a broad range of insecticides in the field has been contributed to the development of P. xylostella resistance. [4] also highlighted the similar issues which claimed that the excessive use of pesticides in the farm could be related to the pest incidences in one area. Due to over reliance on pesticides in Cameron Highlands, the trend of rapid developing resistance, environmental pollutions in all rivers and hazard to non targets organisms have become more serious today, the demand for insecticides are also seems to be endless but increasing year by year on.

Implementation of other pest management strategies is now looking more crucial in order to avoid these problems turn into more chronic. Intercropping system is one of the possible methods to employ as the management tactics to reduce the infestations created by P. xylostella. [5] and [6] have mentioned in details in the intercropping system where this cropping system is involved with the different crops in one field. This cropping system is also unlikely to share the same insect pests and disease-causing pathogens, and also creates a great opportunity to conserve the soil. Intercropping system is suitable for vegetable production [4] due to its ability to limits outbreak of insect pests [7]. Previous research by [8] has been revealed that intercropping cabbage with tomato reduced the infestation of P. xylostella in cabbage cultivation. Some evidences of previous research work also have shown beneficial plants such as marigold [9], citronella grass [10] onion and garlic could play as pest repellent plants and they were found to have a potential repellent effect on P. xylostella. Pest repellent plant can be used as an alternative option in controlling pests in the field area [11]. This statement has been supported by Kianmatee and Ranamukhaarachchi [12] where they highlighted that some plant species have been recognized as pest repellent plant for the particular pests. It was further agreed by [13] stated that some plants contained organic substances that keeping pests away from crops and avoiding potential damage. For example, marigold plant was reported to contain bioactive compounds [9]. Intercropping system would be offer a great alternative strategy if the combination of pest and repellent plant successfully identified [12]. Furthermore, climatic conditions such as temperature and rainfall are also playing crucial factors in DBM expansion, migration and reproduction. By correlating these two climatic factors with the population of DBM, the population dynamics of DBM could be patterned and measured. Hence, the present study was carried out with two main objectives: (1) To evaluate the effectiveness of intercropping cabbage with non-host crops on the population of P. xylostella, and (2) To determine the relationship between adults and larvae population of P. xylostella with the climatic conditions throughout cropping periods.

## 2. Materials and Methods

### 2.1. Experimental Site and Experimental Design

The study was undertaken in a farmer's field located at Tanah Rata, Cameron Highland, Pahang, Malaysia from January to July 2012. The study area was consistently cultivated with the various types of highland vegetables. The annual rainfall is an average of 2000 mm with the minimum temperatures means ranging from 11 to 15°C and 21 to 25°C for the maximum. Minimum slope area (less than 5%) of experimental site was chosen for homogeneity of plant growth. The soil pH is found 4.0-5.5 and then corrected to 6.0-6.5 with the liming compound for better plant growth. A total of seven treatments was arranged in the Randomized Complete Block Design (RCBD) with four blocks as replicates.

*Treatments:* T1 (Cabbage + Onion), T2 (Cabbage + Tomato), T3 (Cabbage + Marigold), T4 (Cabbage + Citronella grass), T5 (Cabbage + Insecticides and Fungicide spray=positive control), T6 (Cabbage + Control, cabbage only=negative control), T7 (Cabbage + Mosquito plants).

### 2.2. Plot preparation and Plant Materials

The total size of experiment plot was 8m x 23m (184m²). All raised beds were constructed manually. Each block was divided into seven raised beds. The sizing of each unit of raised beds was 1m x 5m (5m²) with spacing between rows was 0.3m. Each constructed raised bed was occupied with two rows of cabbage plants as a main crop from variety K.Y Cross. Each row consisted of ten cabbage plants. Four-weeks-old of K.Y Cross cabbage was intercropped with onion *(Allium cepa* L.*)*, tomato *(Solanum lycopersicon* L.*)*, marigold *(Tagetes erecta* L.), citronella grass (*Cymbopogon nardus* L. *rendle)*, and mosquito plants (*Pelargonium citrosum*). The planting distance between each cabbage plant was 50 cm and intercrops were 30 cm for onion, 50 cm (tomato), 50 cm (marigold), 35 cm (citronella grass) and 50 cm (mosquito plants). The cabbage seedlings were cultivated in the nursery in a month advanced due to its rapid growth period than other intercrops plants. Three seeds were sown consistently in the polyethylene seedling tray and only healthy cabbage plants with three to four leaf stages were chosen to transplant at all raised beds. Organic manure (treated chicken dung) was mixed with soils as a basal dressing after transplanting cabbage plants in the field. The NPK 15:15:15 and NPK 12:12:17:2 were spread at each cabbage plant stand over thirty days of transplanting period (DAT). At 60 DAT, NPK 12:12:17:2 was applied for the second time fertilization.

### 2.3. Insecticide Application

All treatments were sprayed with fungicides and insecticides using the agricultural knapsack sprayer (Model PB-16) with a tank capacity of 16 litres respectively. Both insecticides and fungicides were sprayed starting at seedling stage until late heading stage. Application of insecticides combined with fungicides sprayed only for treatment five (T5) at every seven day interval based on the following sequence; (1) Florbac (*Bacillus thuringiensis* subspecies *aizawai*) and Succes 25SC (spinosad), (2) Prevathon 5SC (chlorantraniliprole) and Antrocol 70 WP (propineb), (3)

Amistar (azoxystrobin) and Equation (famoxadone and cyamoxanil), and Funguran OH (copper hydroxide). Combination of pesticide and bio-pesticide (*Bacillus thuringiensis* subspecies *aizawai*). Application through cocktail application is a preferred practice applied by most farmers in Cameron Highlands. The cocktail application with two or more mixing pesticides applied in one spraying application done by vegetable farmers in the Cameron Highlands with two reasons; to save the labour time and operation cost in the field. Other treatments except T5 were applied with a single application of fungicide involved Antrocol 70 WP (propineb), Blucozeb M-45 (mancozeb), Terrachlor 75 (quintozene), and Equation (famoxadone and cyamoxanil) based on the product label.

### *2.4. Plant Maintenance*

All standard maintenance practice requirements for cabbage plants, including irrigation, fertilization, weeding, insect pests and disease controls were followed the guidelines from MARDI Research Station in the Cameron Highland and also from DOA. Weekly inspections on crops were carried out starting from a vegetative stage until the harvesting period.

### *2.5. Data Collection and Analysis*

Data collections were embarked consistently at 0800 am, where adults' *P. xylostella* were less active and easier to sample. A total of 14 samplings (S1-S14) by five days interval have been conducted during the experimental period. First sampling was undertaken seven days after the cabbage plants were sown. In each plot, adults and larvae of *P. xylostella* population was counted at all leaves of each selected plant. Each leaf of cabbage plants was checked to detect adult and larvae population at every sampling interval. Visual counting of adults and larvae was done *via* systematic zig-zag pattern from ten cabbages out of 20 cabbages per row per plot. Data recorded are included (1) number of *P. xylostella*s' adult and larvae populations per treatment, (2) climatic data (temperature and rainfall) provided by MARDI Cameron Highland and (3) weight of cabbage head (g). At harvesting stage, a similar number of cabbage plants were incised just above soil level and weighted individually using a weighing scale (Brand MK-01, China) with the capacity of 3 kg. Before weighing, all damaged leaves were removed by a cutter. The mean weight of cabbage heads harvested per treatment was recorded and categorized as a marketable yield (>700 g per cabbage head) as determined by the Federal Agricultural Marketing Authority (FAMA), Cameron Highlands. The mean weight of cabbage head harvested per treatment was recorded as a yield. The relationship between climatic factors and *P. xylostella* population using Pearson Correlation was measured. All data collected from the field were coded, entered, and subjected to analysis of variance (ANOVA) at P<0.05 level and means comparison using the Statistical Analysis System software (SPSS ver. 18 Software Packages). Mean separations were conducted by using the Tukey HSD test. All results are expressed as mean with standard error. Each bar represents mean ± standard deviation. Asterisks (*) show a significant difference from the control at P<0.05.

## 3. Results

### *Population Abundance of Plutella Xylostella (L.) at Different Treatments*

Figure 1 is presented the cumulative total mean adult and larvae populations of *P. xylostella* at each treatment from samplings S1 to S14 for the entire cabbage growth period. No significant difference (P>0.05) was found between the number of adult and larvae populations at each treatment. It was observed that the highest densities of adult were recorded at T2, followed by T1 and T4. According to our data, the lowest adult population among the treatments was found at T7. While T3 and T5 were showed a quite similar in the total cumulative mean number of adult. In comparison, the total mean larvae populations at all treatments were not much different during the cropping period.

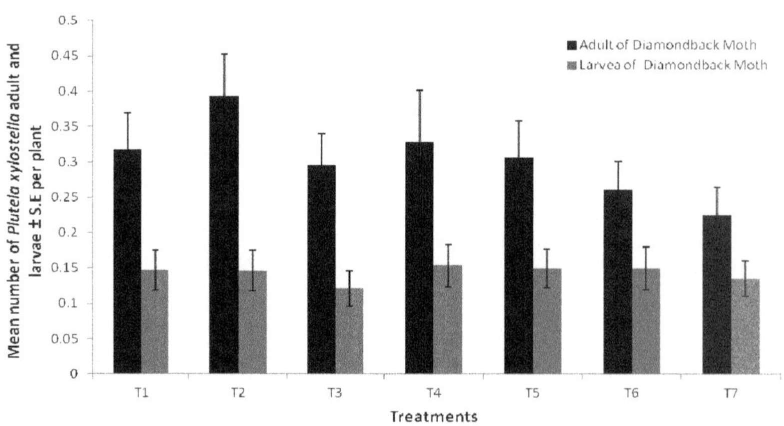


***Figure 1.*** *The overall cumulative total mean adult and larvae populations of P. xylostella on cabbage plants from S1-S14 during cropping period (┬=standard error of the mean). Treatment: T1 (Cabbage + Onion), T2 (Cabbage + Tomato), T3 (Cabbage + Marigold), T4 (Cabbage + Citronella grass), T5 (Cabbage + Insecticides and Fungicide spray=positive control), T6 (Cabbage + Control, cabbage only=negative control), T7 (Cabbage + Mosquito plants).*

### *Adults and Larvae of P. Xylostella at Different Sampling Date*

Figures 2 (A) and (B) showed the trends of average adults and larvae populations on the respective sampling dates at all treatments. In figure 2 (A), results have shown that, the overall mean numbers of adults from S1-S4 was ranging between 0.3 - 2.1 per treatment. Samplings from S1 to S4 were considered as seedling stage of cabbage plants. The adult population was increased at S2-S3. However, the trend mean number of the adult population per treatment was initiated to drastically decrease at heading formation stage to the harvesting stage (S4 - S14) with less varying among treatments. The highest peak of the adult population was recorded at pre-heading stage (S3). At this stage, T4 (2.1) was recorded the highest number of adults, followed by T2 and T5 (1.6), T1 (1.1), T6 (1.0), while T3 and T7 (0.7) respectively, were recorded as the lowest. Stages of S3 and S4 were considered as an early growth stage of cabbage plants where the mean density of adult populations was observed higher (> 1.0) compared to the other treatments. In figure 2 (B), the graph shows the mean values of the larvae population for all treatments at all sampling dates. The numbers of the larvae population in all treatments were started to increase at S1 and S2 and happening decreased from S3 to S10 of lower than 0.8 mean number with a slight fluctuation. The highest peaked of the larvae population was observed at S11 for T1, T4 and T6 and considered slightly higher than the other sampling dates since S11 was the very last of heading stage.

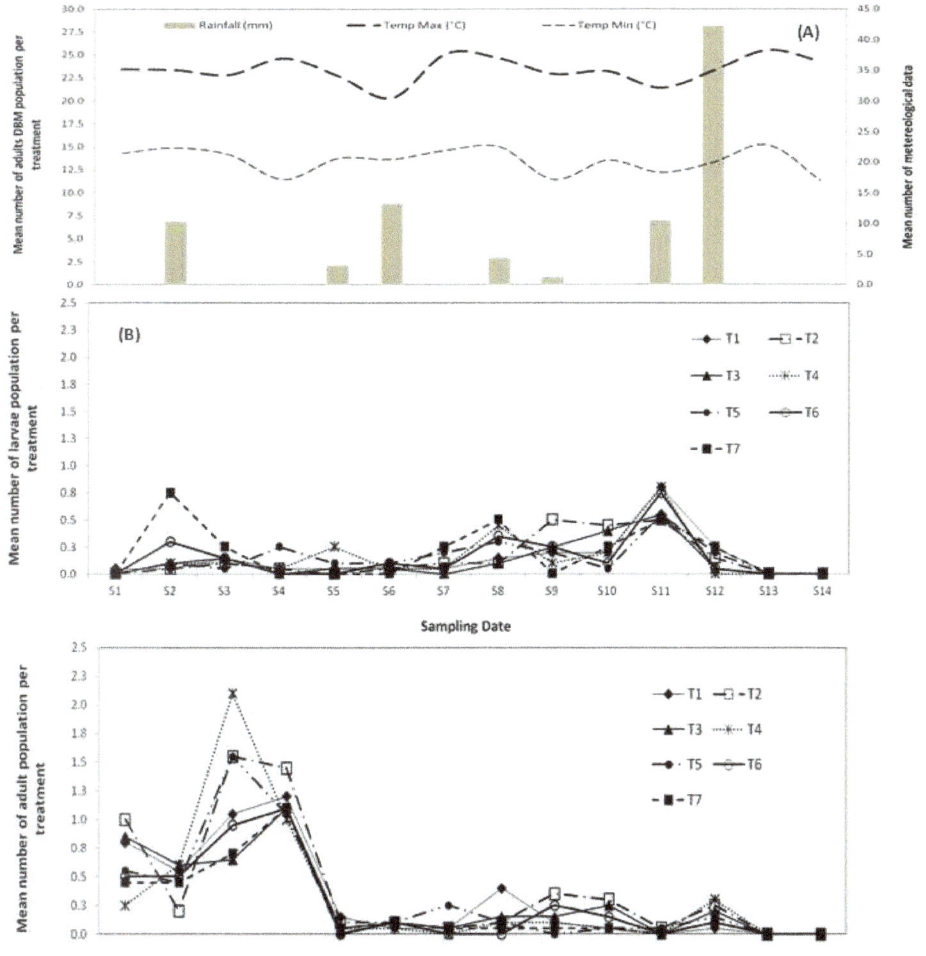


*Figure 2. Relationship between mean population of adults (2A) and larvae (2B) with the climatic conditions (rainfall and temperature) at the respective sampling dates. T1 (Cabbage + Onion), T2 (Cabbage + Tomato), T3 (Cabbage + Marigold), T4 (Cabbage + Citronella grass), T5 (Cabbage + Insecticides and Fungicide spray=positive control), T6 (Cabbage + Control, cabbage only=negative control), T7 (Cabbage + Mosquito plants)*

### *Effects of Climatic Conditions on Population Abundance of P. Xylostella at Different Treatments*

The relationship between rainfall and temperature conditions with the population abundance of adults and larvae *P. xylostella* in the experimental plot also was measured. Pearson's correlation analysis was conducted to evaluate the relationship between adults and larvae population in all treatments with temperature and rainfall and is summarized in Table 1. Through the coefficient of correlation analysis, we found that the mean adult and larvae population in all treatments were negatively correlated with the temperature and rainfall data. The negative relationship between adult and larvae populations with temperature and rainfall suggested that these populations were not affected by temperature and rainfall in the field conditions.

***Effects of Intercropping on the Cumulative Weight (g) of Cabbage Yield***

Results on the effects of intercropped on the cumulative weight of cabbage head produced at each treatment are shown in Figure 3.

***Table 1.*** *Correlation analysis between P. xylostella (adult and larvae) and climatic conditions*

| Stage | Treatment | Rainfall Pearson Correlation | Sig. (2-tailed) | N | Max. Temperature Pearson Correlation | Sig. (2-tailed) | N | Min Temperature Pearson Correlation | Sig. (2-tailed) | N |
|---|---|---|---|---|---|---|---|---|---|---|
| | T1 | -.242 | .405 | 14 | .119 | .685 | 14 | .043 | .884 | 14 |
| | T2 | -.210 | .471 | 14 | .032 | .912 | 14 | -.169 | .564 | 14 |
| | T3 | -.188 | .521 | 14 | .115 | .695 | 14 | -.030 | .920 | 14 |
| Adult | T4 | -.095 | .748 | 14 | -.039 | .895 | 14 | .031 | .916 | 14 |
| | T5 | -.171 | .558 | 14 | .052 | .859 | 14 | .037 | .900 | 14 |
| | T6 | -.216 | .457 | 14 | .031 | .916 | 14 | -.166 | .571 | 14 |
| | T7 | -.200 | .494 | 14 | .122 | .678 | 14 | -.103 | .726 | 14 |
| | T1 | .289 | .316 | 14 | -.454 | .103 | 14 | -.286 | .322 | 14 |
| | T2 | .110 | .707 | 14 | -.375 | .187 | 14 | -.409 | .146 | 14 |
| | T3 | .070 | .812 | 14 | -.551 | .041 | 14 | -.287 | .320 | 14 |
| Larvae | T4 | -.016 | .956 | 14 | -.327 | .254 | 14 | -.060 | .838 | 14 |
| | T5 | .212 | .468 | 14 | -.238 | .412 | 14 | -.310 | .281 | 14 |
| | T6 | .138 | .637 | 14 | -.381 | .179 | 14 | -.086 | .770 | 14 |
| | T7 | .267 | .357 | 14 | -.042 | .887 | 14 | .386 | .173 | 14 |

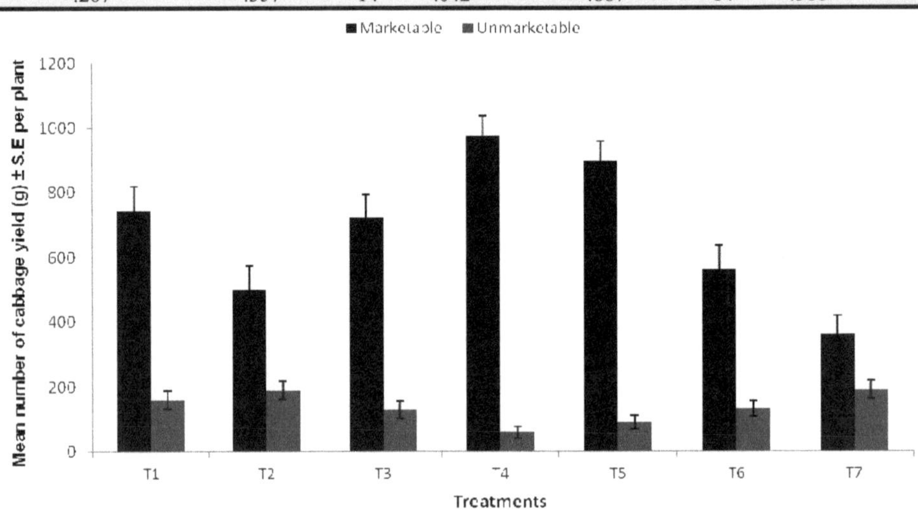


***Figure 3.*** *The overall mean cumulative weight of marketable (>700g) and unmarketable (<700g) cabbage head (⊤ = standard error of the mean). T1 (Cabbage + Onion), T2 (Cabbage + Tomato), T3 (Cabbage + Marigold), T4 (Cabbage + Citronella grass), T5 (Cabbage + Insecticides and Fungicide spray=positive control), T6 (Cabbage + Control, cabbage only=negative control), T7 (Cabbage + Mosquito plants)*

According to Tukey HSD results in Table 2, T4 (973 g) was observed to have a significant difference (<0.05) while compared with sole crop (562 g) and other treatments.

***Table 2.*** *Yield of cabbage head in response to different intercropping treatments*

| Intercrops | | The total mean weight of marketable head cabbage yield at each treatment (>700 g/ cabbage head) | |
|---|---|---|---|
| T1 | (Cabbage+Onion) | 743 | ab |
| T2 | (Cabbage+Tomato) | 500 | c |
| T3 | (Cabbage+Marigold) | 722 | ab |
| T4 | (Cabbage+Citronella grass) | 973 | a |
| T5 | (Cabbage+Insecticides+Fungicide spray) | 895 | b |
| T6 | (Cabbage+Control, cabbage only) | 562 | c |
| T7 | (Cabbage+Mosquito plants) | 361 | bc |

* Numbers with the same letters are not statistically different according to Tukey HSD test (P < 0.05).

# 4. Discussion

The combination of all tested intercrops trials under field conditions were not caused their effective suppression action in controlling adult and larvae populations throughout the experimental period. The suppressing ability performed by all tested intercrops in this study was found reasonably low. Plant distance, densities, growth stage and the layout of testing intercropped in the trial plots could be possibly give some direct or indirect influences to their performance effects. [3] the performance effects of pest repellent plant depend on the crop combinations and the host range of pests. In our experiment, we found that cabbage plants intercropped with the mosquito plant was performed better by giving the lowest population abundance of adult *P. xylostella* followed by other treatments named marigold, insecticides and fungicide sprays, citronella grass, onion and tomato. It can be seen that mosquito plants more repellent to adult *P. xylostella* whilst tomato was the poorest in performance as a pest repellent plant function. Although citronella grass was found to have a moderate performance in repelling adult *P. xylostella*, but we observed that citronella grass still has a high potential to be used as a pest repellent plant. A similar situation as also observed in another study [14].

In response to the treatment insecticides and fungicide spray (T5), it is supposed to have lower total cumulative means of adult and larvae population but the result obtained in this study showed contradiction. From the observation and interviewed authors with workers in the experimental plot, they claimed that the experimental plot has a long story of excessive usage of pesticides. We could anticipate this situation has finally led to the rapid development of resistance to both chemical and biological pesticides (*Bacillus thuringiensis)* by adults and larvae *P. xylostella* as mentioned by [3]; [15]. This phenomenon is similar to the previous studies by [16]. Many researchers consider *P. xylostella* as the most difficult pest to be controlled in crucifers [17] due its capability developed resistance to a wide range of insecticides.

According to our data, the adults were commenced to increase drastically following the day of transplanting cabbage plants in the field. The highest peak of adult attacked was detected at ranged between 12 to 30 days (S3 and S4) and obviously dropped at S5 onwards until the harvesting. The highest peak of the attacker was occurring in a pre-heading stage of cabbage plants. During this period, the numbers of adult population were higher compared to the heading formation and harvesting stages. We considered this stage as a critical stage of adult damage. This finding is also consistent with the study done by [16] who mentioned that the most destructive of adult *P. xylostella* is at a pre-heading stage. It was remarked that this stage is crucial and must be monitored seriously by a cabbage farmer in management of *P. xylostella*. Added to that, a similar trend was also expressed by [18] via their study on the effects of intercropping beans and onions on populations of *Liriomyza* spp. which associated parasitic hymenoptera. They have also observed that the dense populations of insect pest reflect on the changes in leaf area with onion maturation. As we mentioned earlier, the first 30 days after transplanting, cabbage seedlings are considered as a critical attack by adult *P. xylostella*. In order to lessen a critical outbreak on a cabbage farm for the first 30 DAT, we strongly suggest that farmers ought to schedule the transplanting period of cabbage seedlings on the rainfall period as a farm management strategy. We consider this transplanting period as a corrective measure (time) for the cabbage farmer to apply insecticides effectively, particularly on the adult *P. xylostella*. It is essential for cabbage farmers to take immediate actions to shield the pre-heading stage and reduce damage at the growth stage. The results on cumulative mean larvae population,all treatments have shown quite similar in mean larvae population. Findings from this study showed that all treatments were insignificantly effective repelling larvae populations of *P. xylostella*. For marigold (T3), similar trend (graph pattern) on the population of *P. xylostella* was recorded by [19] using marigolds (*Tagetes erecta* L.) to protect sole crop from the onion fly, *Delia antiqua* (Meigen).

Generally from the findings, the rainfall and temperature (as shown in Figure 3) were not the major factors influencing the severity infestation of adult *P. xylostella* to the cabbage plants in the field. As reported by [20] that *P. xylostella* could manage their population to develop and reproduce well in a wide range of temperature between 10 to 30°C since this range is considered as a temperature tolerance for *P. xylostella*. Furthermore, the adult densities (Figure 2A) on a cabbage farm have been slightly reduced after a rainfall. Under field conditions, rainfall intensity either heavy or little may influence changes in the adult population. However, the numbers of larvae (as presented in Figure 2B) did not show any significant difference in population size, even theoretically we can expect that rainfall to possibly lead to decrease the number of larvae population by wash-off them from the cabbage leaves, especially larvae at the lower part of the leaves due to direct or indirect cause of mud-splash [21]. We are notable that the subsequent generation of adult will be reduced if the numbers of larvae decrease due to the rainfall effects.

For the marketable cabbage head, citronella grass has contributed to the highest total cumulative mean of marketable weight cabbage head. On the contrary statement, [14] reported that citronella grass intercropped with Chinese kale had produced the lowest yield. Mosquito plant (T7) was performed better as a pest repellent plant nevertheless has resulted in the lowest production in total cumulative means of the cabbage head. The lowest production in the marketable of cabbage head in the treatment of mosquito plant possibly due to the direct effect of some significant factors include the wide canopy structure, nutrient availability, sunlight and space competition between mosquito plants and cabbage plants.

## References

[1] Corbett, G.H., and Pagden, H.T. (1941) "A review of some recent entomological investigations and observations" *Malay. Agric. J.*, Vol. 29, pp. 347-375.

[2] Susila W, Sumiartha K, Nemoto H, Kawai S (2003) The effect of insecticides on populations of diamondback moth, Plutella xylostella (Lepidoptera: Yponomeutidae) and its parasitoid, Diadegma semiclausum (Hymenoptera: Ichneumonidae) in cabbage. J ISSAAS 9(1):132–138

[3] Adati, T., Talaker, N.S., and Shelton, A.M. (1993). Biology, ecology and management of the diamondback moth, *Annu. Rev. Entomol*, Volume 38, 275-301.

[4] Xu, Q.C., Xu, H. L., Qin, F. F., Tan, J. Y., Liu, G. and Fujiyama, S. (2010). Relay-intercropping into tomato decreases cabbage pest incidence. *Journal of Food, Agriculture & Environment, Vol.* 8 (3&4), pp. 1037-1041.

[5] Gao,Y., Duan, A., Qiu, X., Liu, Z., Sun, J., Zhang, J., and Wang, H. (2010). Distribution of roots and root length density in a maize/soybean strip intercropping system. *Agricultural Water Management*, Volume 98, Issue 1, pp. 199-212.

[6] Nielsen,H., Gooding,M., Ambus,P., Hellou, C.G., Crozat, Y., Dahlmann, C., Dibet, A., Fragstein, V. P., Pristeri, A., Monti, M., and Jensen E.S. (2009). Pea–barley intercropping for efficient symbiotic $N_2$-fixation, soil N acquisition and use of other nutrients in European organic cropping systems. *Field Crops Research*, Volume 113, Issue 1. 64-71.

[7] Obadoni, B.O., Mensah, J.K. and Emua, S.A. (2010). Productivity of Intercropping Systems Using Amaranthus cruentus L and Abelmoschus esculentus (Moench) in Edo State, Nigeria. *World Rural Observations, Vol. 2* (2), 53-60.

[8] Bukovinszky, T., Tréfás, H., van Lenteren, J. C., Vet, L. E. M., & Fremont, Julien. (2004). Plant competition in pest-suppressive intercropping systems complicates evaluation of herbivore responses. *Agriculture, Ecosystems & Environment, Vol. 102* (2), pp. 185-196.

[9] Koon, H.W., Cerruti R. H.,Antoon, P. (2007). Protecting Crops from Nematode Pests: Using Marigold as an Alternative to Chemical Nematicides (Vol. PD-35, pp. 1-6): College of Tropical Agriculture and Human Resources (CTAHR).

[10] Patrícia Fontes Pinheiro, Vagner Tebaldi de Queiroz, Vando Miossi Rondelli, Adilson Vidal Costa, Tiago de Paula Marcelino, and Dirceu Pratissoli (2013). Insecticidal Activity of Citronella Grass Essential Oil on Frankliniella schultzei And Myzus persicae. Ciênc. agrotec., Lavras, vol 37 (2), p. 138-144.

[11] Kianmatee, S. and Ranamukhaarachci, S.L. (2007a) Combining Pest Repellent Plants and Biopesticides for Sustainable Pest Management in Chinese Kale. *Journal of Asia-Pacific Entomology,Vol.10* (1), 69–74.

[12] Zehnder, G. (2004) Insect pest management in the home vegetable garden.

[13] http://www.clemson.edu/scg/ipm/hmeveg.htm#Trap%20 Plants (Accessed June 25, 2012).

[14] Kianmatee, S. and Ranamukhaarachci, S.L. (2007b). Pest Repellent Plants for Management of Insect Pests of Chinese Kale, Brassica oleracea L. *International Journal of Agriculture & Biology, Vol. 9, No. 1*, 64-67.

[15] Fauziah, I. (1990) "Studies on Resistance to Acylurea Compounds in *Plutella xylostella* L. (Lepidoptera:Yponomeutidae)" PhD. Thesis, University of London, United Kingdom. pp. 259.

[16] Ayalew, G. (2006) Comparison of yield loss on cabbage from Diamondback moth, Plutella xylostella L. (Lepidoptera: Plutellidae) using two insecticides. *Crop Protection*, Vol. 25 (9), pp.915–919.

[17] Talaker, N.S., and Shelton, A.M. (1993). Biology, ecology and management of the diamondback moth, *Annu. Rev. Entomol*, Volume 38, 275-301.

[18] Johnson, M.W and Mau, R.F.L. (1986) Effects of intercropping beans and onions on populations of liriomyza spp. and associated parasitic hymenopteral, *Proceedings of the Hawaiian Entomological Society,* Department of Entomology, University of Hawaii, Honolulu, Vol. 27, pp. 95-103.

[19] Finch, S., Billiald, H., and Collier RH., 2003. Companion planting-do aromatic plants disrupt host-plant finding by the cabbage root fly and the onion fly more effectively than non-aromatic plants? *Entomologia Experimentalis et Applicata, Vol.* 109, pp. 183–195.

[20] Ali, G., Karim,K., Yaghoub, F. and Habib, A. (2009). Effect of temperature on life table parameters of Plutella xylostella (Lepidoptera: Plutellidae) on two brassicaceous host plants. *Journal of Asia-Pacific Entomology,, 12*, 207-212.

[21] Sivapragasam. A., Yosiaki. I. and Tetsuo. S. (1988). Population fluctuations of the diamondback moth, plutella xylostella (L.) on cabbages in Bacillus Thuringiensis sprayed and non sprayed plots and factors affecting within-generation survival of immatures. *Res. Popul. Ecol., 30*, 329-342.

# The most important factors affecting yield of the *Eucalyptus* species

**Siti Latifah[1], Teodoro Reyes Villanueva[2], Myrna Gregorio Carandang[2], Nathaniel Cena Bantayan[2], Leonardo M. Florece[3]**

[1]Forestry Program Study, Sumatera Utara University, Medan, Indonesia
[2]Forest Resources Management, University of the Philippines Los Banos, Los Banos, Philippines
[3]Environmental Science, University of the Philippines Los Banos, Los Banos, Philippines

**Email address:**
sitilatifah 164@yahoo.co.id (S. Latifah), teodoro_villanueva@yahoo.com (T. R. Villanueva), mgc405@yahoo.com (M. G. Carandang), ncbantayan@gmail.com (N. C. Bantayan)

**Abstract:** This study was conducted primarily to obtain factors affecting yield for forest plantation *Eucalyptus* spp in Aek Nauli, North Sumatera, Indonesia. The data in this study used 650 rhombic plots consisting of 106 PSPs and 544 inventory plots with several variations of plot size. Stands features referred to diameter, height, merchantable volume, age, species, spacing, site index, basal area, and density of *Eucalyptus* species. Geographical features referred to slope, elevation, rain fall and soil of the study area .The most important factors affecting yield of the E. hybrid, *E. urophylla, and all Eucalyptus* species are rainfall followed by soil group A while for *E. grandis, E. pellita* and mixed *Eucalyptus* are stand density, elevation, interaction site index and age, respectively.

**Keywords:** Stand, Geografhic, Yield, *Eucalyptus*

## 1. Introduction

The global forest plantations have increased by nearly 12 million ha/year in 1990s, less than 4 million ha/year compared in 1980s (FAO, 2001; ITTO, 2001). This increase in forest plantations establishment indicates that there is steadily increasing investment in forestry. Therefore, the current problem is not only the availability of the investment but also the ability to plan, manage, control and sell the forest plantations products successfully [2].

Growth modeling of plantation timber species is a vital tool for the prediction of yield from future harvesting and estimating financial returns. One criterion of sustainable forest management is sustainability of yield [3]. These requirements have implications for model design, implementation and use. Reliable predictions depend on reliable models and reliable input data. The yield is dependent on several factors, the most important being the species composition, stand age, site quality, genetic variation, stand density, management regime and environmental conditions [ 4].

Forest management decisions are predicated on information and projection about current and future resource conditions. Inventories taken at one point in time provide information on current volumes and related statistics. Managers of these plantations require precise and reliable predictions of future stand yield. [3].

There are many complex and interrelated factors that influence the quantity, quality, and distribution of growth in a forest stand. Some of the more important factors are stand age, site quality, genetic variation, stand density, management regime and environmental conditions. Thus the purpose of this study is to obtain factors affecting yield of the *eucalyptus.*

## 2. Material and Method

### *2.1. Study Site*

The study was conducted in Aek Nauli sector of the Industrial Timber Estate (*Hutan Tanaman Industri* - HTI) namely, PT. Toba Pulp Lestari (PT. TPL), from December 2008 – March 2009. It is in the territorial jurisdiction of

Porsea subdistric, Simalungun regency, North Sumatera Province, Indonesia. The study area has the highest annual rainfalll in October and the lowest in August. Based on its topography, the study area has a rainfall range of 85 – 434 mm/month, it has an elevation range from 350 to 1400 msal and it has top soil depth ranges from 10 to 44 cm. The study area covered an area 189,975 ha with Aek Nauli sector has an area 21341.8 ha.. Geographically, it is located between 02° 40'00" to 02° 50'00" north latitude and 98° 50'00" to 99° 10'00" east longitude .[ 7 ]

### *2.2. Materials*

The concerned institutions and organizations were requested to facilitate data gathering for the research. These institutions included agencies in Indonesia such as PT. TPL, National Coordination Agency for Survey and Maping (BAKOSURTANAL , *Badan Kordinasi Survei dan Pemetaan Nasional* ), Agency for Meteorology and Geophysics-regional I Sampali , Medan, Agency for Watershed Management in Asahan Barumun, North Sumatera. Secondary data were collected from the libraries and internet-base sources of the institutions/ agencies in the Philippines and Indonesia. Libraries used included SEARCA, College of Forestry and Natural Resources (CFNR), UPLB and University of North Sumatera

The data in this study used 650 rhombic plots consistin of 106 permanent sample plots (PSPs) and 544 inventory plots with several variations of plot size, from 0.02 to 0.04 ha.. PSPs have an area of 3.44 ha. It consisted of 106 plot or 273 observations. Inventory plot has an area of 21.76 ha. It consisted of 544. Thus, the total area of the plots in this study site is 25.2 ha.

The primary data and secondary data were digitally encoded. Field observation was conducted to gather stands and geographic data. Stands features referred to diameter, height, merchantable volume, age, species, spacing, site index, basal area, and density of *Eucalyptus* plantations. Geographical features referred to slope, elevation, rain fall and soil of the study area.

Species were grouped into *E. hybrid* (*E.*hyb), *E. grandis* (*E.gra*), *E. pellita, E. urophylla* (*E. uro*), mixed *Eucalyptus,* and all *Eucalyptus species.* The data sets derived from 470 *E.* hybrid, *94 E.grandis*, 23 *E. pellita*, 46 *E. urophylla* and 119 mixed *E.* observations. *E.* hybrid is breeding from E. *urophylla* vs *E. grandis* and *E. pellita* vs *E. grandis*. Mixed *Eucalyptus* observations consisted of *E.*hybrid, *E.grandis*, *E. pellita*, and *E. urophylla species.*

### *2.3. Data Analysis*

To determine the factors that most influence the yield of eucalyptus, partial coefficient determination ($R^2$) of the final models for Each *Eucalyptus* species was used [8]. The coefficient of partial determination can be defined as the percent of variation. This coefficient is used to provide insight into whether or not one or more additional predictors may be useful in a more fully specified regression model. It helps to provide the proportion of the variance of one variable which is predictable from the other variable. It is a way of measurement which allows determining how clear it can be in making predictions from a certain data provided. [1].

Coefficients of determination ($R^2$) represent the proportion of SST (Total Sum of Square) that is explained by use of the regression model. It is a measure of how well the regression line represents the data [1]. The computational formula for $R^2$ as follow:

$$R^2 = \frac{SS\ reg}{SS\ tot} = \frac{b\ SPxy}{SS\ y}$$

Where:
SS reg = the regression sum of squares
SS tot = the total sum of square

Usually, the higher the value of $R^2$ is, the better the regression model will be. This is because if $R^2$ is larger, a greater portion of the total error is explained by the included independent variable and a smaller portion of errors is attributed to other variables and randomness. The range of r-square is from 0 to 1. An $R^2$ = 90%, implies that 90% variation in Y is attributable to the variation in X.

## 3. Results and Discussion

### *3.1. Descriptive Statistic for Stand Variables of the Study Area*

Stand variables used in this study consisted of merchantable volume in cubic meter per hectare, basal area in square meters per hectare, age in year, diameter at breast height in centimeter, height in meter, site index in meter, stand density in number of tree per hectare, and spacing in square meters.

The summary descriptive statistics of the actual stand variables to develop the models are shown in Table 1. The merchantable volume ranges from 0.0003 to 289.5 $m^3$/ha. Merchantable volume averages for *E.* hybrid, *E.grandis*, *E. pellita*, *E. urophylla*, mixed *Eucaylptus* and all species observations are 37.532 $m^3$/ha, 60.965 $m^3$/ha, 45.408 $m^3$/ha, 29.557 $m^3$/ha, 11.726 $m^3$/ha and 35.796 $m^3$/ha, respectively. *E. grandis* has the highest average volume (60.965 $m^3$/ha) and m*ixed Eucalyptus* has the lowest average volume (11.726 $m^3$/ha).

The study site has a basal area ranges of 0.0128 to 28.064 $m^2$/ha. The basal area averages for *E. hybrid, E.grandis*, *E. pellita*, *E. urophylla*, mixed *Eucaylptus* and all species are 6.59731$m^2$/ha, 10.076 $m^2$/ha, 10.831 $m^2$/ha, 6.369 $m^2$/ha, 2.372 $m^2$/ha and 6.479 $m^2$/ha, respectively.

The stand ranges in age from 0.17 to 6.7 years. The stand age averages for *E. hybrid, E.grandis*, *E. pellita*, *E. urophylla*, mixed *Eucaylptus* and all species are 2.0 year, 3.1 year, 3.03 year, 2.048 year, 1.34 year, and 2.08 year, respectively. In terms of stand age, *E. hybrid* has both the highest age (6.7year) and the lowest age (0.17 year).

Table 1 indicates that diameter ranges from 0.3438 to

24.80 cm. The average diameter for *E.* hybrid, *E.grandis*, *E. pellita*, *E. urophylla*, mixed *Eucalyptus* and all species observations are 7.274 cm, 11.213 cm, 10.901cm, 7.586 cm, 4.294 cm and 7.424 cm, respectively. In terms of stand diameter, *E.* hybrid has the highest diameter (24.80 cm) and mixed *Eucalyptus* has the lowest stand diameter (0.34 cm).

Regarding stand height in this study area, the stand height ranges from 31.5 to 0.14 m. The stand height averages for *E.* hybrid, *E.grandis*, *E. pellita*, *E. urophylla*, mixed *Eucalyptus* and all species are 10.234 m, 15.514 m, 12.151 m, 11.227 m, 6.667 m, and 10.449 m, respectively. In terms of stand height, *E. grandis* has the highest average stand height (5.514 m) and mixed *Eucalyptus* has the lowest average stand height (6.667 m).

The study site has a site index ranges from 25.936 to 0.154 m. The site index averages, respectively, for *E.* hybrid, *E.grandis*, *E. pellita*, *E. urophylla*, *Mixed Eucalyptus* and all species are10.323 m, 15.747 m, 16.359 m, 12.255 m, 6.464 and 10.693 m. *E. pellita* have the highest average site index (16.359 m ) and mixed *Eucalyptus* have the lowest average site index (6.464 m ).

***Table 1.*** *Descriptive statistics for actual stand variables of models E. hybrid, E. grandis, E.pellita, mixed Eucalyptus and all species in Aek nauli sector*

| Species | Variables | Max | Min | Average | Variance | Standard Deviation |
|---|---|---|---|---|---|---|
| *E. hybrid* | Volume($m^3$/ha) | 289.500 | 0.0003 | 37.532 | 0.710 | 0.843 |
| | Basal area ($m^2$/ha) | 28.064 | 0.0128 | 6.597 | 0.196 | 0.443 |
| | Age (year) | 6.700 | 0.1700 | 2.007 | 2.225 | 1.492 |
| | Diameter (cm) | 24.800 | 0.3438 | 7.274 | 24.504 | 4.950 |
| | Height (m) | 31.500 | 0.1385 | 10.234 | 0.102 | 0.319 |
| | Site index (m) | 25.936 | 0.1542 | 10.323 | 0.106 | 0.325 |
| | Density (tree/ha) | 1751.000 | 290.0000 | 1156.291 | 0.011 | 0.107 |
| | Spacing ($m^2$) | 9.000 | 3.0000 | 7.293 | 1.269 | 1.126 |
| *E.grandis* | Volume($m^3$/ha) | 152.307 | 0.3149 | 60.965 | 0.282 | 0.531 |
| | Basal area ($m^2$/ha) | 20.277 | 0.4829 | 10.076 | 0.067 | 0.259 |
| | Age (year) | 6.120 | 1.0800 | 3.119 | 1.585 | 1.259 |
| | Diameter (cm) | 16.000 | 2.3938 | 11.213 | 12.470 | 3.531 |
| | Height (m) | 25.400 | 2.6188 | 15.514 | 0.047 | 0.216 |
| | Site index (m) | 25.004 | 4.6322 | 15.747 | 0.040 | 0.200 |
| | Density (tree/ha) | 1280.000 | 575.0000 | 925.532 | 0.006 | 0.079 |
| | Spacing ($m^2$) | 9.000 | 6.7500 | 8.928 | 0.158 | 0.398 |
| *E. pellita* | Volume($m^3$/ha) | 116.005 | 5.3705 | 45.408 | 0.172 | 0.415 |
| | Basal area ($m^2$/ha) | 20.245 | 3.6374 | 10.831 | 0.040 | 0.201 |
| | Age (year) | 5.010 | 1.5000 | 3.029 | 1.402 | 1.184 |
| | Diameter (cm) | 15.400 | 5.8217 | 10.901 | 8.529 | 2.921 |
| | Height (m) | 19.300 | 5.5375 | 12.151 | 0.040 | 0.201 |
| | Site index (m) | 23.724 | 8.3147 | 16.359 | 0.023 | 0.152 |
| | Density (tree/ha) | 1200.000 | 675.0000 | 1047.000 | 0.003 | 0.055 |
| | Spacing ($m^2$) | 9.000 | 9.0000 | 9.000 | 0.000 | 0.000 |
| *E.urophylla* | Volume($m^3$/ha) | 135.520 | 0.0120 | 29.557 | 0.584 | 0.764 |
| | Basal area ($m^2$/ha) | 17.652 | 0.3000 | 6.369 | 0.127 | 0.356 |
| | Age (year) | 5.000 | 0.6700 | 2.048 | 1.346 | 1.160 |
| | Diameter (cm) | 15.278 | 1.7000 | 7.586 | 15.735 | 3.967 |
| | Height (m) | 22.004 | 2.2693 | 11.227 | 0.061 | 0.246 |
| | Site index (m) | 23.539 | 2.5337 | 12.255 | 0.067 | 0.259 |
| | Density (tree/ha) | 1625.000 | 725.0000 | 1201.152 | 0.008 | 0.091 |
| | Spacing ($m^2$) | 9.000 | 6.0000 | 6.799 | 1.160 | 1.077 |
| mixed *Eucalyptus* | Volume($m^3$/ha) | 170.000 | 0.0600 | 11.726 | 0.366 | 0.605 |
| | Basal area ($m^2$/ha) | 19.000 | 0.0320 | 2.372 | 0.103 | 0.321 |
| | Age (year) | 6.100 | 0.4900 | 1.374 | 1.678 | 1.295 |
| | Diameter (cm) | 15.900 | 0.5000 | 4.294 | 14.508 | 3.809 |
| | Height (m) | 27.900 | 0.4000 | 6.667 | 0.080 | 0.283 |
| | Site index (m) | 23.350 | 1.5896 | 6.464 | 0.061 | 0.247 |
| | Density (tree/ha) | 1476.000 | 625.0000 | 1092.261 | 0.006 | 0.078 |
| | Spacing ($m^2$) | 9.000 | 6.0000 | 7.412 | 0.240 | 0.490 |
| All species | Volume($m^3$/ha) | 289.500 | 0.0003 | 35.795 | 0.691 | 0.831 |
| | Basal area ($m^2$/ha) | 28.064 | 0.0128 | 6.479 | 0.187 | 0.433 |
| | Age (year) | 6.700 | 0.1700 | 2.080 | 2.216 | 1.489 |
| | Diameter (cm) | 24.800 | 0.3438 | 7.424 | 24.055 | 4.905 |
| | Height (m) | 31.500 | 0.1385 | 10.449 | 0.097 | 0.312 |
| | Site index (m) | 25.936 | 0.1542 | 10.693 | 0.099 | 0.315 |
| | Density (tree/ha) | 1751.000 | 290.0000 | 1116.715 | 0.010 | 0.102 |
| | Spacing ($m^2$) | 9.000 | 3.0000 | 7.538 | 1.300 | 1.140 |

Table 1 indicates that stand density ranges from 290 to 1751 tree/ha. Results show that *E. hybrid* has the highest average stand density (1156.3 tree/ha) and *E. grandis* has the lowest average stand density (925.5 tree/ha).

Spacing in this study site ranges from 3 to 9 $m^2$. Results show that average spacing for *E.* hybrid, *E.grandis, E. pellita, E. urophylla*, mixed *Eucalyptus* and all species is 7.29 m, 8.923 m, 9.00 m, 6.799 m, 7.41 m and 7.538 m, respectively. In terms of spacing, *E. pellita* observations have the highest average spacing (9 m) and *E. urophylla* observations has the lowest average spacing (6.799 m).

### *3.2. Descriptive Statistic for Geographic Variables of the Study Area*

Geographic variables were obtained and these were top soil depth in centimeter, monthly rainfall in millimeter, slope in percent, elevation in masl, and groups of soil. In regression analysis, the dependent variables are frequently influenced not only by quantitative variables but also by variables of spatial, temporal or qualitative nature. Descriptive statistic about max, min, mean, variance and standard deviation for soil groups were not needed because of a nominal scale. It is no more than giving each feature a name and is thus purely subjective and descriptive in nature. It divides the data based on qualitative considerations. It is usually coded numerically by ones or zeroes to indicates presence or absence of an attribute. Descriptive statistic geographic variables to develop the models are shown in Table 2. Dummy variables for group of soil are shown in Table 3.

Table 2 indicates that top soil depth is from 10 to 44 cm. Results show that average top soil depth for *E.* hybrid, *E.grandis, E. pellita, E. urophylla*, mixed *Eucalyptus* and all species observations are 23.455 cm, 22.287 cm, 21.957 cm, 21.826 cm, 22.143 cm and 22.956 cm, respectively. In terms of top soil depth, *E.* hybrid observations have the highest average top soil depth (23.455 cm) and mixed *Eucalyptus* observations have the lowest average top soil depth (22.143 cm).

The study area has a rainfall from 85 – 434 mm/month. The average monthly rainfall for *E.* hybrid, *E.grandis, E. pellita, E. urophylla*, mixed *Eucalyptus* and all species observations are 222.30 mm, 201.46 mm, 154.00 mm, 209.24 mm, 221.05 mm, and 216.61 mm, respectively. Regarding monthly rainfall, Table 2 indicates that *E.* hybrid observations has the highest average monthly rain fall ( 222.3 mm) and *E. pellita* observations has the lowest monthly rainfall (154 mm).

Table 2 shows that Aek Nauli sector has a field slope of 4 – 45 %. The average field slope for *E.* hybrid, *E.grandis, E. pellita, E. urophylla, Mixed Eucalyptus* and all species observations are 15.51% cm, 15.29%, 28.47%, 14.47%, 11.78% cm and 15.23%, respectively. In terms of field slope, *E. pellita* observations have the highest average field slope (28.47%) and *Mixed Eucalyptus* observations have the lowest average field slope (11.78 %).

Aek Nauli sector has an elevation ranges from 350 to 1400 masl. Elevation in the research area averages for *E. hybrid, E.grandis, E. pellita, E. urophylla*, mixed *Eucalyptus* and all species observations are 1193.511 masl, 1225.532 masl, 1256.522 masl, 1175.000 masl, 1179.832 masl, and 1196.144 masl, respectively. *Eucalyptus grandis* observations have the highest average elevation.

The type of soil in Aek nauli sector is shown in Table 3, majority of the areas has type soil C(55.85 %) followed by and soil type A (26.33%), soil type E (10.77%), soil type D (5.32%) and soil type B (1.73%)

***Table 2.*** *Descriptive statistics for geographic variables of the models E. hybrid, E. grandis, E. pellita, mixed Eucalyptus and all species*

| SPECIES | VARIABLES | MAX | MIN | AVERAGE | VARIANCE | STANDARD DEVIATION |
|---|---|---|---|---|---|---|
| *E.* hybrid | Top soil (cm) | 44.000 | 10.000 | 23.455 | 0.010 | 0.098 |
| | Rain fall (mm/month) | 434.000 | 85.000 | 222.300 | 0.028 | 0.168 |
| | Slope (%) | 45.000 | 4.000 | 15.512 | 0.094 | 0.307 |
| | Elevation (masl) | 1400.000 | 350.000 | 1193.511 | 0.025 | 0.159 |
| *E.grandis* | Top soil (cm) | 35.000 | 15.000 | 22.287 | 17.669 | 4.203 |
| | Rain fall (mm/month) | 361.000 | 105.000 | 201.457 | 0.029 | 0.170 |
| | Slope (%) | 45.000 | 4.000 | 15.298 | 0.098 | 0.313 |
| | Elevation (masl) | 1400.000 | 350.000 | 1225.532 | 0.016 | 0.125 |
| *E. pellita* | Top soil (cm) | 30.000 | 12.000 | 21.957 | 0.008 | 0.090 |
| | Rain fall (mm/month) | 275.000 | 105.000 | 154.000 | 0.039 | 0.196 |
| | Slope (%) | 45.000 | 4.000 | 28.478 | 0.088 | 0.297 |
| | Elevation (masl) | 1400.000 | 850.000 | 1256.522 | 0.022 | 0.149 |
| *E.urophylla* | Top soil (cm) | 35.000 | 15.000 | 21.826 | 0.008 | 0.090 |
| | Rain fall (mm/month) | 434.000 | 89.000 | 209.239 | 0.039 | 0.196 |
| | Slope (%) | 45.000 | 4.000 | 14.467 | 0.088 | 0.297 |
| | Elevation (masl) | 1400.000 | 350.000 | 1175.000 | 0.022 | 0.149 |
| Mixed | Top soil (cm) | 35.000 | 10.000 | 22.143 | 0.009 | 0.094 |
| *Eucalyptus* | Rain fall (mm/month) | 356.000 | 89.000 | 221.050 | 0.031 | 0.176 |
| | Slope (%) | 45.000 | 4.000 | 11.782 | 0.070 | 0.264 |
| | Elevation (masl) | 1400.000 | 350.000 | 1179.832 | 0.019 | 0.139 |
| All species | Top soil (cm) | 44.000 | 10.000 | 22.956 | 0.009 | 0.096 |
| | Rain fall (mm/month) | 434.000 | 85.000 | 216.609 | 0.029 | 0.172 |
| | Slope (%) | 45.000 | 4.000 | 15.227 | 0.095 | 0.308 |
| | Elevation (masl) | 1400.000 | 350.000 | 1196.144 | 0.022 | 0.150 |

***Table 3.*** *Number and proportion of geographic variables on ranking type of soil for all species in Aek Nauli sector*

| Dummy –Soil Group | | | Frequency | Percent |
|---|---|---|---|---|
| Soil A | 1 | Rating = 1 | 198 | 26.330 |
| | 0 | otherwise | 554 | 73.670 |
| Soil B | 1 | Rating = 2 | 13 | 1.729 |
| | 0 | otherwise | 739 | 98.271 |
| Soil C | 1 | Rating = 3 | 420 | 55.851 |
| | 0 | otherwise | 332 | 44.149 |
| Soil D | 1 | Rating = 4 | 40 | 5.319 |
| | 0 | otherwise | 712 | 94.681 |

Where:
Soil A : for group of soil dystrandepts, hydrandepts
Soil B : for group of soil dystropepts, dystrandepts
Soil C : for group of soil dystropepts, hapludults
Soil D: for group of soil dystropepts, humitropepts

### *3.3. Factors Affecting Yield*

Large areas of *Eucalyptus* plantations have the potential to alter the diversity of plant across landscapes. The effects of plantations differ strongly based on the characteristics of the surrounding landscapes and management regime (especially understory vegetation control) [11]

The main target of sustainable management of *Eucalyptus* plantations is maintaining long-term productivity of the land [4]. As the *Eucalyptus* plantations area expands, rotation lengths shorten, and inappropriate plantation establishment and management techniques are applied. Based on the final models for each eucalyptus species, there are several factors affecting the yield of *Eucalyptus* plantations which can be shown in Table 4.

The most important factor affecting yield of the *E.* hybrid is rainfall ($R^2$ = 89.94%), followed by dummy soil group C ($R^2$ = 89.86%) and top soil ($R^2$ = 89.40%). Partial coefficient Determination ($R^2$) for rainfall is equal to 89.94 %, it means that error sum of square (ESS) can be reduced by 89 % if rainfall is added to the regression model already containing age, top soil and dummy soil group C .

Stand density is the most important factor affecting yield of the *E. grandis* followed by site index, because it had the highest partial $R^2$ = 0.9808. It means that error sum of square can be reduced by 98.08 % if stand density is added to the regression model already containing basal area, top height, and age. The final model of *E. grandis* had the regression coefficient for stand density - 0.00054. It means that there is a decrease in volume by0.00054 per unit increase in stand density holding other independent variables constant. Stand density is a measure of how many trees are growing per unit area. Together, site and density tell us how much timber we can produce, as well as what kind of wood quality we can expect at harvest time.

Elevation is the most important factor affecting yield of the *E. pellita* followed by age, because it had the highest partial $R^2$ = 0.9952. It means that error sum of square can be reduced by 99.52 % if elevation is added to the regression model already containing basal area, top height, and age. The final model of *E. pellita* had the regression coefficient for elevation - 0.00023. It means that there is an increase in yield by 0.00023 per unit decrease in elevation holding other independent variables constant. Elevation affects on plants are mostly indirect, because they modify other site factors especially climate and soil. A rise in elevation is accompanied by a fall in temperature. Since the distribution of plants is limited by temperature, elevation likewise sets boundaries on species distribution [9]

Rainfall is the most important factor affecting yield of the *E. urophylla, followed by* elevation, because it had the highest partial $R^2$ = 0.9224. It means error sum of square can be reduced by 92.24 % if rainfall is added to the regression model already containing basal area, top height, and elevation. The final model of *E. urophylla* had the regression coefficient for rainfall 0.00351. It means that there is an increase in yield by 0.00351 per unit increase rainfall.

Interaction site index with age is the most important factor affecting yield of the mixed *Eucalyptus,* followed by stand density because it had the highest partial $R^2$ = 0.9153. It means that error sum of square can be reduced by 91.53 % if site index and age are added to the regression model already containing basal area, age, stand density and top soil. The final model of mixed *Eucalyptus* had the regression coefficient for interaction site index with age 0.81678. It means that there is an increase in yield by 0.81678 per unit increase in interaction site index with age holding another independent variables constant.

Likewise, result of analysis for all *Eucalyptus* species show that rainfall is the most important factor affecting the yield followed by soil group A, because it had the highest partial $R^2$ = 0.9032. It means that error sum of square can be reduced by 90.32 % if rainfall are added to the regression model already containing basal area, top height, age, spacing, top soil and dummy soil group A. The final model of the *E.* all species had the regression coefficient for rainfall 0.00103. It means that there is an increase in yield by 0.00103 per unit increase rainfall. So, in general, rainfall is most important factor affecting yield of the *Eucalyptus* plantation in study site.

***Table 4.*** *Partial coefficient determination ($R^2$) of the final models for Each Eucalyptus species.*

| Species | Model | Independent Variables | $R^2$ |
|---|---|---|---|
| *E.* hybrid | 2 | Interaction basal area and top height | 0.3892 |
| | | Age | 0.6191 |
| | | Top soil | 0.8940 |
| | | Rainfall | 0.8994 |
| | | Dummy soil group C | 0.8986 |
| *E grandis* | 1 | Basal area | 0.6611 |
| | | Age | 0.8927 |
| | | SI | 0.9224 |
| | | Stand density | 0.9808 |
| *E.pellita* | 2 | Interaction basal area and top height | 0.4486 |
| | | Age | 0.9667 |
| | | Elevation | 0.9952 |
| *E. urophylla* | 2 | Interaction basal area and top height | 0.3150 |
| | | Rainfall | 0.9224 |
| | | Elevation | 0.9218 |
| Mixed *Eucalyptus* | 3 | Interaction basal area and age | 0.7975 |
| | | Age | 0.8515 |
| | | Interaction site index and age | 0.9153 |
| | | Stand density | 0.8869 |
| | | Top soil | 0.8789 |
| All species | 2 | Interaction basal area and top height | 0.3598 |
| | | Age | 0.6504 |
| | | Site index | 0.8659 |
| | | Spacing | 0.8821 |
| | | Top soil | 0.8982 |
| | | Rainfall | 0.9032 |
| | | Dummy soil group A | 0.9027 |

Rainfall is the primary of water in the forest plantations. The amount of rainfall actually reaching the forest floor depends on the characteristics of the canopy, rain, and other climatic factors. Rainfall in *Eucalyptus* plantations plays two major roles, namely: as a factor causing soil detachment, dispersal and transport and a supplier of moisture to the soil. Water is considered to be a vital factor in the life of *Eucalyptus* plantations affecting growth and reproduction of the various organisms found therein

## 4. Conclusions

The most important factors affecting yield of the E. hybrid, *E. urophylla, and all Eucalyptus* species are rainfall followed by soil group A while for *E. grandis, E. pellita* and mixed *Eucalyptus* are stand density, elevation, interaction site index and age, respectively. The main target of sustainable management of *Eucalyptus* plantations is maintaining long-term productivity of the land. Resource managers and decision makers should consider the most important factors affecting the yield namely; for, *E. urophylla, and E.* all species are rainfall followed by soil group A while for *E. grandis, E. pellita* and mixed *Eucalyptus* is stand density, elevation, interaction site index and age, respectively.

## Acknowledgement

This research is part of PhD thesis of the first author funded by SEAMEO-BIOTROP DIPA 2008. The authors extend deeply acknowledgment to the Ministry of Education and Culture – Republics Indonesia for scholarship and support to accomplish this paper. Sincerely appreciation is also extended to anonymous reviewer for correction and comments. Gratitude is also extended to PT. Toba Pulp Lestari which gave permission to conduct this research in its area.

## References

[1] ANALYST SOFT. 2008. Stat Plus. http://www.analystsoft.com/en/products/statplus

[2] Aruan a. l. P. 2000. The Future Role of Plantation Forests and Forest-Based Industry. Http://Www.Fao.Org/Docrep/Article/Wfc/Xii/0142-A1.Htm

[3] AVERY, T.E and H.E BURKHART. 1994. Forest Measurement. Fourth Edition. New York: McGraw-Hill, Inc. 408 pp.

[4] DAVIS, L.S., K.N. JOHNSON, P.S. BETTINGER and T.E. HOWARD. 2001. Forest Management to Sustain Ecological, Economic and Social Values. McGraw-Hill Companies, Inc. 1221 Avenue of the Americas, New York. NY 10020. 804 p.

[5] FAO. 2001. Global Forest Resources Assessment.2000. Main report.FAO, Rome, Italy. 479 p

[6] ITTO. 2001. Annual Review and Assessment of the World Timber Situation.

[7] PERSEROAN TERBATAS TOBA PULP LESTARI (PT. TPL). 2008. Summary of the Annual Work Plan Efforts Timber Forest Products Utilization in the Industrial Timber Plantation Forests.

[8] LATIFAH, S. , VILLANUEVA TR, CARANDANG, MG, BANTAYAN, N.C, FLORECE, L.M. 2014. . Predicting Growth and Yield Models for *Eucalyptus* Species in Aek Nauli, North Sumatera, Indonesia. Journal of *Agriculture, Forestry and Fisheries.* Vol. 3, No. 4, 2014, pp. 209-216. doi: 10.11648/j.aff.20140304.11

[9] VILLANUEVA, T. R. 2005. Upland Ecosystem Management. Second Edition. University of the Philippines. Open University, Laguna, Philippines.

[10] WEI, X, and NUNO M.G.B. 2000. Genetic Gains and Levels of Relatedness from Best Linier Unbiased Prediction of *Eucalyptus Europhylla* for Pulp Production in Southeastern China. Canadian Journal of Forestry Research. 30 (10): 1601-1607

[11] YUE, W., VALERIE, M., BAKER, and THOMAS, G. 2007. Modeling and Prediction of Dominant Height and Site Index of *Eucalyptus globulus* Plantations Using a Nonlinear Mixed-Effects Model Approach. Canadian Journal of Forest Research, Volume 37, Number 8, 1 August 2007 , pp. 1390-1403(14)

# 18

# An assessment of the socio-economic and ecological impacts of environmental changes on rural livelihood: A study across Addado, Buhodle and Northern Galkaayo of central and northern Somalia

**Badal Ahmed Hassan[1], Edinam K. Glover[2, *], Olavi Luukkanen[1], Ramni Jamnadass[3], Ben Chikamai[4]**

[1]Viikki Tropical Resources Institute, Department of Forest Sciences, University of Helsinki, Helsinki, Finland
[2]Faculty of Law, University of Helsinki, Helsinki, Finland
[3]World Agroforestry Centre (ICRAF), United Nations Avenue, Gigiri, Nairobi, Kenya
[4]Kenya Forestry Research Institute, Nairobi, Kenya

**Email address:**
badal.hassan@helsinki.fi (B. A. Hassan), eddie.glover@helsinki.fi ( E. K. Glover), Olavi.Luukkanen@helsinki.fi (O. Luukkanen), r.jamnadass@cgiar.org (R. Jamnadass), b.chikamai@kefri.org (B. Chikamai)

**Abstract:** The international community has long recognized desertification as a significant global social, economic and environmental problem of concern with detrimental effects on the livelihoods of people in many countries world-wide. To explore this critique, this study examined local people's views on the causes, effect and socio-economic impact of desertification and degradation to the community. The study involved a cross-sectional survey conducted with four community categories, namely nomadic pastoralist, agro-pastoralists, villagers and Internally Displaced Persons (IDPs), from Addado, Buhodle and northern Galkaayo of central and northern part of Somalia; through questionnaire-based interviews. Descriptive statistics attribute environmental changes to negative implications of human activities on the environment: the menace of which led to a decline in forage, current loss of biodiversity and the related changes in the environment, permanent migration, increased poverty and health problems. The startling results of this study conclusively demonstrate that the challenges involve proper interventions for tree planting against desertification as well as for community empowerment through public education and formation of community based environmental associations to coordinate the overall environmental management activities and to raise public awareness.

**Keywords:** Addado, Buhodle, Central Somalia, Desertification, Northern Galkaayo, Nordern Somalia, Rural Livelihoods, Socio-Economic Impacts

## 1. Introduction

### *1.1. Drylands*

As defined by United Nations Convention to Combat Desertification (UNCCD) drylands comprise land within the arid, semi-arid and dry sub-humid areas. Aridity is defined based on an Aridy Index, which is determined by the ratio of the precipitation (P) to the potential evapo-transpiration (PE).

Most arid, semi-arid and dry sub-humid zones simultaneously experience rises in temperature and declines in rainfall pattern. Drylands may be defined in several ways but the simplest way, according to [28], is to use a measure of aridity which represents a lack of moisture in the average climatic conditions. Dryness can be classified using the "Aridity Index (AI)", or "dryness ratio (P/PET)", which is the ratio of the mean annual precipitation (P) and the mean annual potential evapotranspiration (PET). PET is the maximum amount of moisture which can evaporate from a surface and transpire through vegetation under given atmospheric conditions, assuming there is no limit to the amount of moisture available. According to [8], the arid, semi-arid and dry sub-humid climatic zones, which fall within UNCCD focus, are areas, other than polar and sub-polar regions, in which the ratio of annual precipitation to potential evapotranspiration falls within the range from 0.05 to 0.65 [13]. Aridity zones are classified according to aridity index: <0.05 hyper-arid (7.5 % of the world covered); 0.05-0.20 arid (12.1 % of the world covered); 0.2-0.50 semi-arid (17.7 % of the world covered); 0.50-0.65 dry sub-humid (9.9 % of the world covered) [8].

Drylands cover 40 percent of the world's land area and host one third of world's population [24]. Two-thirds of rural population in sub-Saharan Africa (SSA) live in arid and semi-arid areas or area with poor infrastructure [15; 40].

### *1.2. Sahel, Drought and Desertification*

Climatologically, the Sahel describes a period of abnormally dry weather sufficiently prolonged for the lack of water to cause serious hydrologic imbalance in the affected area [14]. It is characterized by frequent droughts, low precipitation, poorly distributed and highly variable monthly and seasonal unpredictable rainfall. During drought periods where there is such long spell of dry weather, desertification[1] becomes harsher, as well as more noticeable, and its rate of incidence increases abruptly. As the likelihood of droughts rises as one moves from the humid to the more arid regions, so too, does the susceptibility to desertification. It is known that natural resources such as land forms, soils and vegetation are often transformed during such long spell of dry weather. In this way drought epitomizes an "indirect" cause of desertification owing to its possible exacerbation of poor land use, a direct cause of desertification as it induces degradation[2] of soil and vegetation [26;8]

Dryland ecosystems tend to be naturally dynamic instability systems. According to [36], several dryland communities experience a high level of responsiveness to precipitation when it falls but in turn exhibit very insignificant productivity. It is sometimes difficult to determine if a particular vegetation condition is a direct reaction to unfavorable short-term environmental conditions (mostly drought) or is actually a reflection of long duration or irreversible degradation. According to [10] and [26], desertification is influenced by factors such as prevalent climate of rural poverty, ignorance, greed, social and economic changes or misguided government policies while [32] added that protracted drought may spur the rate of desertification.

According to [16] the United Nations Environment Programme (UNEP) has estimated that a total of 35 million km² of the world's range, rainfed crop land and irrigated land - an area about the land area of North and South America combined - undergoes the desertification and degradation processes. According to a 1991 UNEP assessment, about 850 million people (one-sixth of the world's population) suffer from desertification and land degradation processes [17]. [10] reported that desertification and land degradation affect drylands worldwide but tends to be more pronounced in Asia and Africa, each of which accounts for 37 % of all desertified land [8].

### *1.3. Somalia's Environmental Challenges*

Agro-ecologically, Somalia is arid and semi-arid land with the rainfall ranging from less than 50 mm/year along the northern coast to about 600 mm/year in the southernmost regions of the country and the potential evapo-transpiration between 1900 and 2500 mm [5].

With abundant grazing land, over 70 % of the Somali population are nomadic pastoralists; thereby pastoralism[3] and agro-pastoralism [4] are common systems of land production and the primary source of livelihoods; therefore, these natural resources form the basis of rural livelihoods in Somalia.

The terrestrial ecosystems of the country are threatened by continuous deforestation, degradation and diminution, with far-reaching ecological, economic and social consequences. The most common vegetation types in Somalia are *Acacia-commiphora* woodland which is defined as open stands of trees of at least 8 m tall with a canopy cover of 40 percent or more dominated by *Acacia* and *Commiphora* genus, bush-land[5] and shrub-land[6] [21].

Somalia has witnessed the great damage caused by the long civil war, especially in the fields of agriculture, forestry and education. There is a great deal of challenges of environmental concerns facing Somalia, which is far less studied. The country suffers from almost all types of environmental issues such as famine; use of contaminated water contributes to human health problems; deforestation; overgrazing; soil erosion; and, desertification. The country is prone to recurring droughts; frequent dust storms over eastern plains in summer; and floods during rainy season. In most areas of the country particularly Addado and Galkaayo, woodlands have turned into scrub and the ground vegetation has disappeared from extensive areas. The strong indicators emphasizing the seriousness of the problems are: reduced agricultural productivity and food insecurity, combined with dominance of less vegetative cover, as well

---

[1] "Desertification is land degradation in the arid, semi-arid and dry sub-humid regions resulting from various factors, including climatic variations and human activities" (UNEP 1994).

[2] "Degradation implies reduction of resource potential by one or combination of processes acting on land. These processes include water erosion, wind erosion and sedimentation by those agents, long-term reduction in the amount of diversity of natural vegetation, where relevant, and salinization and sodication" [17].

[3] Pastoral production systems are those in which 50 % of gross household revenue (i.e. the total value of marketed production plus the estimated value of subsistence production consumed within the household) comes from livestock-related activities (Marton and Meadows 2000).

[4] Agro-pastoralists are those in which more than 50 % of household gross revenue comes from farming, and 10-50% from livestock (Morton and Meadows 2000).

[5] Bushlands are open stands of bushes (usually between 3 and 7 m tall) with a canopy cover of 40 percent or more

[6] Shrublands are open or closed stands of shrubs up to 2 m tall

as extinction of wildlife species. In short, the area could best be described as "an overgrazing endpoint where most of the grass and topsoil has already gone" [9].

There are hardly any studies that seek to understand the implications of such causes and impacts of desertification on the environment, and on socio-economic conditions of local people and how to tackle the major environmental problems faced in Somalia. The purpose of this study therefore is to fill in this gap in our knowledge and to draw out recommendations for social development. In view of these developments, this study focuses on describing and analyzing the subject in relation to Addado, Buhodle and Galkaayo in Somalia. The study adds one small case study to our understanding of environmental change, in this case, Somalia. In an attempt to continue the transformation toward stability and sustainability in Somalia, it is important that the connection between social and environmental issues relating to resource extraction practices are more clearly understood.

## 2. Material and Methods

### *2.1. Bio-Physical and Socio-Economic Conditions*

#### *2.1.1. Study Area*

This research was conducted in three SAHANSAHO project[7] areas, namely, Addado, Buhodle and Galka,ayo of central and northern Somalia (Fig. 5) between April and May 2012.

Addado lies between latitude 05° 53′ 64″ N and longitude 46° 37′65″ E; Buhodle lies between latitude 08° 13′ 59″ N and longitude 46° 21′80″ E and Galkaayo lies between latitude 06° 50′ 91″ N and longitude 47° 26′49″ E. The three study locations have in general similarities in terms of climate, topography and socio-economic conditions.

#### *2.1.2. Bio-Physical Conditions*

##### *2.1.2.1. Climate*

Climatologically the project areas lie in the arid zones prone to drought. The rainfall is scarce and erratic in the region with two rainy seasons, the spring season locally known as "*gu*" from March to May and the autumn season locally known as "*dayr*" lasting from October to December.

The annual average rainfall lies between 150 mm to 300 mm [23], (Fig. 4). Due to relatively high altitude, Buhodle receives a higher rainfall ranging between 250 and 400 mm per annum. The higher precipitation in these areas enables more vegetation cover compared to the other two areas. The major rainfall season lasts from April-May and minor rainfall season lasts from October-November. There are no perennial rivers, and there is heavy reliance on boreholes and rainwater harvesting as permanent sources of water for human beings and livestock. The area has a mean annual temperature of 28 °C. The hottest months are January to March, during these months temperatures exceed 30 °C. June to September is the coolest time of the year.

##### *2.1.2.2. Vegetation*

Generally, the vegetation of the zone is bushland and shrubland (see Fig.1, Fig.2 and Fig. 3). To some extent, *Acacia-Commiphora* woodland can be found in Buhodle (Fig. 3). The three most common tree genera are *Acacia*, *Boswellia* and *Commiphora.*

***Figure 1.*** *Addado (Photo: Abdikadir M. Abdi)*

***Figure 2.*** *Galkaayo (Photo:Mustafa A/asis)*

***Figure 3.*** *Buhodle (Photo: Badal A. Hassan).*

[7] SAHANSAHO project is a Finnish Somalia Network initiative in response to the environmental changes in Somalia; financed by the Ministry for Foreign Affairs of Finland.

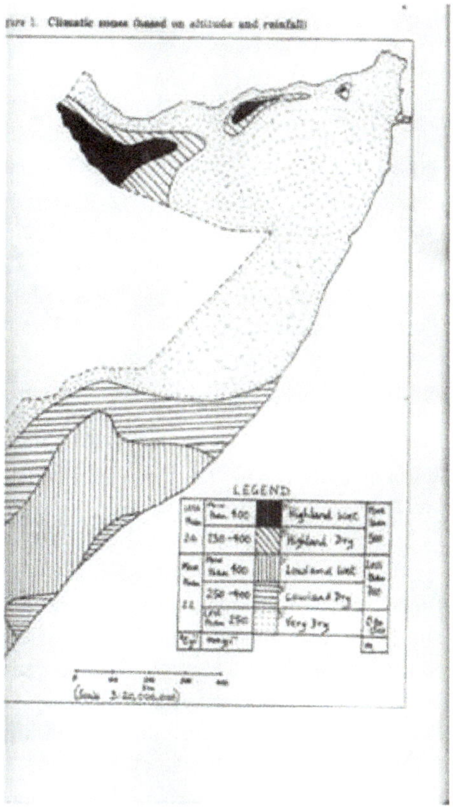


*Figure 4.* Climate characterization map of Somalia

*Figure 5.* Map of Somalia showing the study area (modified from [23].

### *2.1.2.3. Topography and Soils*

The area consists largely of featureless plain, prone to flooding during rainy season often making roads impassable. However, there are some scattered hills in Buhodle and Galkaayo.

The plain rises from an altitude of 309 m above sea level in Addado and Galkaayo and 366 m above sea level in Buhodle. The area is flat with three different types of soils: sandy soil, clay soil and lime soil.

### *2.1.2.4. Social and Land-Use Characteristics*

*Population and Main economic activity*

The estimated population of the three study areas is 315,000, with annual population growth rate of 3. 7 %; with more than 50 % of the population is younger than 26 years old. The population has an average family size of six members. Over 70 % of the population of the area are nomadic pastoralists of the Somali ethnic group. They rear camels, cattle, sheep and goats and their livelihood activities revolve around sale of livestock and livestock products. Other sources of income are purchase of grain and other commodities, formal employment, and non-timber forest products (NTFPs) collection. Their diet consists mainly of meat, milk and grain.

The pastoral people in the area practice seasonal migration to access pasture, settling near to water sources and good pasture for a few weeks before moving on. Those who cannot travel long distances, mostly women, children and the elderly are left behind and are settled around boreholes/reservoirs or settlements. Due to prolonged civil war and frequent droughts, large number of nomadic pastoral communities lost their livestock, thereby falling back to sedentary life and settling in the outskirts of the major towns as Internally Displaced Persons (IDPs) and depending on food aid.

## *2.2. Methodology*

### *2.2.1. Sampling and Data Collection*

Data were collected based on a questionnaire as well as group discussions to gather information from the households and from village leaders, special groups and key informants. Data collection involved standardised as well as open-ended questions. The informants were asked to list the declining tree species in their own language. The information given by the respondents was clarified and completed by observation in and near the project site.

Data were gathered from both primary and secondary sources. Primary data were obtained by using a pre-tested semi-structured questionnaire. Interviews were conducted with key informants,[8] village leaders and special groups among study village residents in Addado, Buhodle and northern Galkaayo. Data was collected on household and land-use characteristics: Main source of household income, sources of water, sources of energy, land tenure security, environmental degradation and the causes of the degradation in the project area, socio-economic impacts of degradation, management of forest resources and participation of farmers in tree planting activities.

[8] Key informants are experts, elderly people and most knowledgeable people on particular topics. Key informants' impressions were used to cross-reference the validity of the different variables employed in the questionnaire [8].

To every interviewee, the purpose of the research and its importance was explained by the interviewer in order to build confidence in the participants to respond to all questions. The importance of the key informants' interview lay in obtaining reliable, thorough and in-depth results on aggregate data at the community level for the study area. Key informants' impressions were used to cross-reference the validity of the different variables employed in the questionnaire in the study area [8].

The household questionnaire was collected from the households dependent or not dependent on plant resources. During the field survey, four distinctive groups were selected to carry out the socio-economic questionnaire survey using structured questionnaire: A survey of 90 households consisting of 30 households in each study area was carried out at Buhodle, Galkaayo and Addado between 20 April and 14 May, 2012. A representative sample of 33 nomadic pastoralists, 9 agro-pastoralists, 19 Villagers and 29 Internally Displaced Persons (IDPs) were randomly selected for the study. Respondents were selected by a simple random sampling procedure, which was representative of the farming system categories in the three project areas.

Brief description is given of the categories in the following paragraphs.

#### 2.2.1.1. Pastoralists

According to [1], nomadic pastoralists; depend on domesticated livestock and migrate in an established territory to find pasture for their livestock. Most groups have focal sites that they occupy for considerable periods of the year. Pastoralists may depend entirely on their herds or may also hunt or gather, practice some agriculture or trade with agricultural peoples for grain crops between seasonal migrations. The patterns of pastoral migration are many, often depending on the type of livestock, the topography, and the climate. The pastoral communities in the Sahelian zone (from Niger to Somalia) are the best example of this type of nomadism.

In this study, nomadic pastoralists were defined as those who get more than 50 % of their livelihood from livestock and livestock-related products. Pastoralism has long been an integral part of dryland inhabitants' history, land use and economy.

#### 2.2.1.2. Agro-Pastoralists

As mentioned earlier, the three project areas are arid and semi-arid lands, which give little opportunity for crop farming, however, there are some few places where some locals practice agriculture. Agro-pastoralists are farmers who practice mixed farming with livestock and rainfed agriculture. They are also considered in this study as those who earn more than 50 % of their livelihood from crop production and the rest from livestock products. They represent the second most important category and encompass both the original inhabitants and immigrants from high- potential areas. Rainfed agriculture refers to the type of agriculture practiced during the rainy season, using early maturing crops and/or along riverine flood plains and adjoined interfluves.

#### 2.2.1.3. Villagers

Apart from pastoralism and agro-pastoralism, most of the urban dwellers, in either bigger towns like Galkaayo or other towns like Addado and Buhodle, earn their livelihood from other sources, other than livestock or crop production. They are either formal employees earning salary or informal employees e.g. shop owners, firewood collectors, charcoal burners, resin collectors etc. On the basis of foregoing, villagers are classified as those who derive more than 50 % of their daily livelihood from either formal or informal employment, other than livestock rearing or crop production.

#### 2.2.1.4. Internally Displaced Persons (IDPs)

Internally Displaced persons (IDPs), are persons or group of persons who have being forced or obliged to flee or to leave their homes or places of habitual residence in particular, as result of or in order to avoid the effects of armed conflicts, situation of generalised violence, violation of human rights, or natural or human made disasters, and who have no crossed an internationally recognised state border. Due to prolonged droughts and recent conflicts between Khatumo State and Somaliland near Buhodle many nomadic pastoralists lost their livestock or were displaced from their customary habitation area, and migrated to larger towns like Addado, Galkayo and Buhodle. The IDPs depend mainly on food aid or NTFPs. Therefore, this investigator considered them as a separate livelihood group.

#### 2.2.1.5. Group Discussions

Group discussion was applied in order to give interviewees the chance to express themselves freely on issues under investigation and to discuss amongst themselves the problems or opportunities they might have. The group discussion served to elucidate important points and helped to verify and check the accuracy and validity of information gathered from the household questionnaire, key informants interview, and the literature review.

Sources of secondary data used in the study included grey literature and peer-reviewed documents as well as institutions involved in natural resource management in the Horn of Africa. Some of these institutions included the Viikki Tropical Resources Institute (VITRI), the Kenya Forestry Research Institute (KEFRI) and the World Agroforestry Centre (ICRAF); literature was also retrieved from libraries in Finland and Sweden and from internet sources, and the universities in Kenya. Information was also obtained from reports and files maintained by the Addado, Buhodle and northern Galkaayo.Local Councils.

### 2.2.2. Data Analysis

Data gathered from questionnaire were coded, computerized and analyzed using the Statistical Package for Social Sciences (SPSS) software. Statistical procedures allow the researcher to "reduce, summarize, organize, evaluate, interpret, and communicate numeric information". In this study, descriptive statistics were used to describe the demographic characteristics of the respondents and to find out the distribution of respondents over the different categories. Percentages were used as a tool of analysis for interpreting the qualitative information collected from the respondents. Descriptive statistics is a useful analytical tool that allows the researcher to examine the characteristics, behaviours, and experiences of study participants [8].

# 3. Results

## *3.1. Socio-Economic Household Characteristics*

### *3.1.1. Livelihood Characterization of the Respondents*

A survey of 90 households was carried out at Buhodle, Galkaayo and Addado between April and May 2012. From the target population, a representation sample of 36.3 % pastoralists, 31.9 % IDPs, 20.9 % Villagers and 9.9 % agro-pastoralist, were randomly selected for the study (Table 1).

***Table 1.*** *Characterization of the livelihood sources of the respondents.*

| | | Frequency | Percent | Valid Percent | Cumulative Percent |
|---|---|---|---|---|---|
| Valid | Pastoralist | 33 | 36.7 | 36.7 | 36.7 |
| | Agro-pastoralist | 10 | 11.1 | 11.1 | 47.8 |
| | Villager | 19 | 21.1 | 21.1 | 68.9 |
| | Internally Displaced Persons (IDPs) | 28 | 31.1 | 31.1 | 100.0 |
| | Total | 90 | 100.0 | 100.0 | |

### *3.1.2. Level of Education*

The result shows that, 75.5 % of the women who were interviewed were illiterate; 10.8% attained primary school level education, 5.4% had attended secondary school and 2.7% attained diploma level education. None of the women respondents were semi-literate or had attained university level education. The study also shows that 54.7 % of the men respondents were illiterate, 17% attained primary school level education, 13.2 % were semi-literate, 11.3 % had attended secondary school and 1.9 % had diploma level education; while 1.9 % attended university level (Table 2).

***Table 2.*** *Level of Education in Relation to Gender.*

| | | | Level of Education | | | | | | Total |
|---|---|---|---|---|---|---|---|---|---|
| | | | Illiterate | Semi-literate | Primary school | Secondary school | Diploma | University degree | |
| Gender | Male | Count | 29 | 7 | 9 | 6 | 1 | 1 | 53 |
| | | % within Gender | 54.7 | 13.2 | 17.0 | 11.3 | 1.9 | 1.9 | 100.0 |
| | | % within Level of Education | 50.0 | 100.0 | 64.3 | 75.0 | 50.0 | 100.0 | 58.9 |
| | | % of Total | 32.2 | 7.8 | 10.0 | 6.7 | 1.1 | 1.1 | 58.9 |
| | Female | Count | 29 | 0 | 5 | 2 | 1 | 0 | 37 |
| | | % within Gender | 78.4 | 0.0 | 13.5 | 5.4 | 2.7 | 0.0 | 100.0 |
| | | % within Level of Education | 50.0 | 0.0 | 35.7 | 25.0 | 50.0 | 0.0 | 41.1 |
| | | % of Total | 32.2 | 0.0 | 5.6 | 2.2 | 1.1 | 0.0 | 41.1 |
| Total | | Count | 58 | 7 | 14 | 8 | 2 | 1 | 90 |
| | | % within Gender | 64.4 | 7.8 | 15.6 | 8.9 | 2.2 | 1.1 | 100.0 |
| | | % within Level of Education | 100.0 | 100.0 | 100.0 | 100.0 | 100.0 | 100.0 | 100.0 |
| | | % of Total | 64.4 | 7.8 | 15.6 | 8.9 | 2.2 | 1.1 | 100.0 |

### *3.1.3. Main Source of Household Income*

From the economic related information of the respondents, there were various categories of alternative income sources. The result shows that 38.9 % of the respondents got their livelihood support from pastoralism, 24 % from informal employment, and 15.6 % from formal employment, 13.3 % depended on food aid and 7.8 % on agro-pastoralism (Fig. 6).

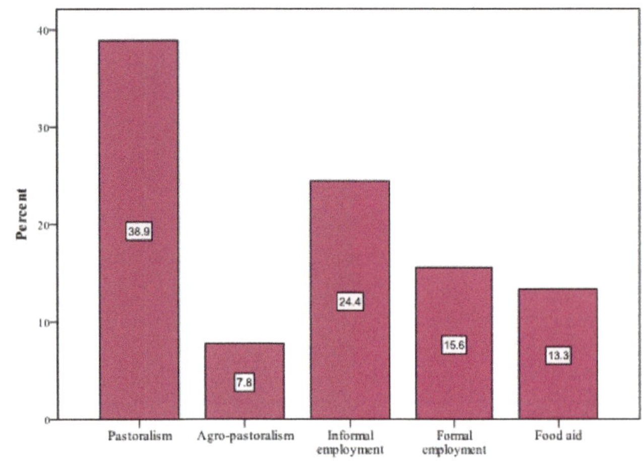


***Figure 6.*** *The Sources of Subsistence of the Respondents in the Study Area*

### *3.1.4. Other Sources of Household Income*

When the interviewees were asked to indicate other sources of household incomes, 38.9 % mentioned remittances, 44.4 % Non-Timber Forest Products (NTFPs) and 13.3 casual work(Fig.7).

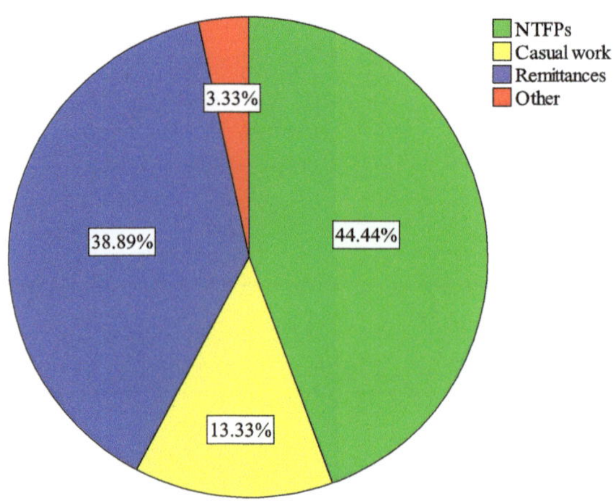


***Figure 7.*** *Other Sources of Household Income as indicated by the Respondents.*

### *3.1.5. Sources of Water*

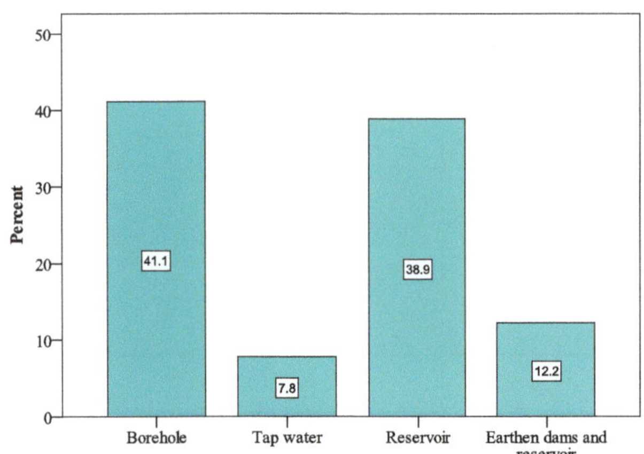


***Figure 8.*** *Respondent's Main Sources of Water.*

The results reveal that 41.1 % of households interviewed use borehole as main sources of water, 38.9 % rely on reservoir as primary sources of water, 12.2 % use earthen dams and reservoirs and only 7.8 % receive water from tap (Fig. 8).

### *3.1.6. Sources of Energy*

The results in Figure (9) show that, 77.8 % of the respondents use firewood as source of cooking energy, 18.9 % use charcoal, 2.2 % kerosene and only 1.1 % use electricity for cooking purpose.

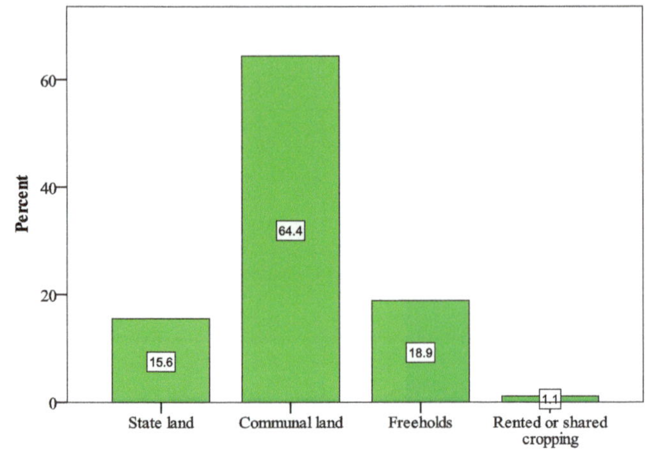


***Figure 9.*** *Sources of Energy as indicated by the Respondents.*

### *3.1.7. Land Tenure*

The study result shows that, four major land tenureships exist in the study areas, namely, communal land[9] (64.4 %), freehold[10] (18.9 %), state land (15.6 %) and rented land (1.1 %) (Fig.10)

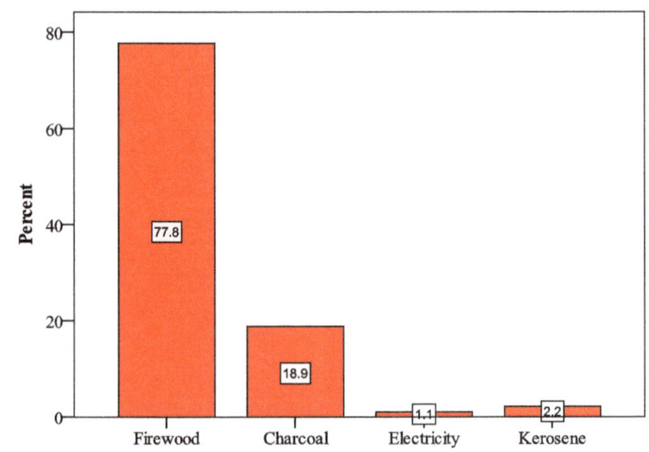


***Fig 10.*** *Land tenure-ship in the study area*

### *3.1.8. Causes of Environmental Degradation in the Project Area*

Ninety percent (99 %) of the respondents agreed that there is high resources and environmental degradation in each project area. There were also various causes of the resources

[9] Communal land: land collectively owned by particular tribes or clans.

[10] Freehold: Land held freely. Simply, it represents absolute ownership of real estate.

reduction and the environmental changes as perceived by local people in the study areas. The number of people who referred to certain cause was interpreted as indicating the strength of that particular factor as shown in (Fig.11). The results indicate that respondents ranked the causes of resources degradation in the following order: Drought (31.11 %), charcoal burning (28.89 %), reservoir expansion (21.11 %) and Overgrazing (18.89 %).

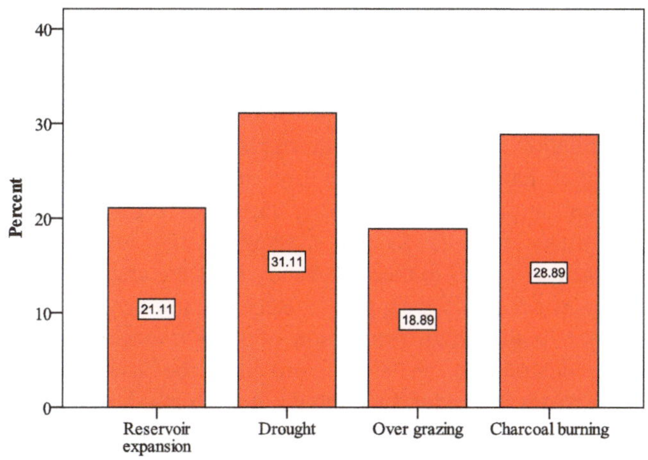


*Figure 11. Major Causes of Resource Environmental Degradation.*

### *3.1.9. Existing Environmental*

#### *3.1.9.1. Management Practices*

The respondents were asked if there is any ongoing environmental management programs in response to the existing environmental changes, 71.1 % said nothing is going on, 8.9 % said tree planting activities is going one, 8.9 % said community mobilization, 6.7 % mentioned new environmental legislation and 4.4 % said waste management (Fig. 12).

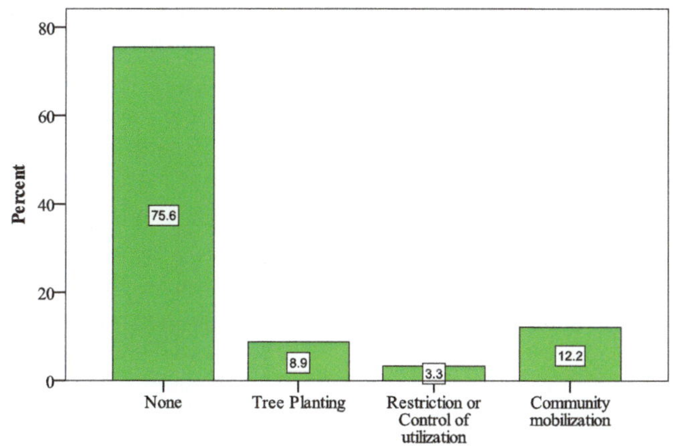


*Figure 12. Natural Resources Management Practices in the Study Areas*

### *3.1.10. Willingness to Participate in NTR and Tree Planting for Agroforestry Purpose*

When the interviewees were asked about their willingness to participate in natural regeneration and tree planting activities for environmental rehabilitation and for agroforestry purposes 80% of men respondents and 90 % of women respondents answered in affirmative that they were willing to participate. When asked which land-use zone (s) they would like to plant, however, they gave different views, 56.7 % said in communal land, 26.7 % said in homestead and 16.7 % said in farm (Table 3).

*Table 3. Desired Land-use Zones to Plant Trees*

| | | Frequency | Percent | Valid Percent | Cumulative Percent |
|---|---|---|---|---|---|
| Valid | Farm | 15 | 16.7 | 16.7 | 16.7 |
| | Homestead | 24 | 26.7 | 26.7 | 43.3 |
| | Communal land | 51 | 56.7 | 56.7 | 100.0 |
| | Total | 90 | 100.0 | 100.0 | |

# 4. Discussion

## *4.1. Socio-Economic Household Characteristics*

### *4.1.1. Level of Education*

The survey found that, majority of the respondents had never attended school of any kind. A possible explanation is the fact that the three study areas are within the arid land with large nomadic pastoral community. The nomadic pastoralist, have no permanent settlements, therefore, the majority of these children have no possibility to go to school. Those who have formal education e.g. primary, secondary school or tertiary are either villagers or those having relatives in the towns or being in boarding school. Somalia's nomadic people do not possess basic literacy skills. Their children do not have access to basic education. Somalia's nomads, therefore, arguably need a better understanding of their socio-cultural predicament, which many consider as less developed.

According to [27], it is estimated that there are between 25 million and 40 million children of school age living in nomadic or pastoralist households of whom only between 10% and 50% attend school. Between 15 million and 25 million of the estimated 100 million of outof-school children are probably nomads and pastoralists. While rates of participation and completion of basic education for pastoralist boys are very low, the rates for girls are far lower.

[37] noted that since the colonial era the education system was never conducive to Somalia's pastoral communities as it has been generally characterized by low enrolment, male-biased and urban (or settlement) orientation features. They added that 5-10% of all school children were enrolled, of which girls constituted less that 20% of the total enrolment.

According to [28], there is little evidence, however, that pastoralist education has been addressed through major national initiatives in any country, with the exceptions of Uganda and Mongolia, since the World Education Forum in 1990.

Women's lack of education and resulting high rates of illiteracy were revealed in this survey. The possible explanation is that girls in pastoralist households have little chance to get education opportunity due to, early marriage, housework or cultural barriers. The main reason for early marriage is to get wealth return called "*Yarad*" in Somali from the bridegroom; mainly camel or cattle, since they are the most important wealth among the pastoralists. Besides the economic and the early marriage factors, most of the pastoral communities have strong cultural belief and they are afraid that girls would behave immorally if sent to school.

Results of the present study are consistent with the findings of previous Participatory Rural Appraisal studies carried out by the Alternative Basic Education for Karamoja programme working with pastoralist groups in Uganda [29]. Their findings show that early marriages have denied many girls the chance to realise their potential, and men dominate decision making at household and community levels.

Similar research carried out in among pastoralist communities in north-eastern Mali and western Niger, Oxfam GB indicate that, attendance and participation by pastoralist girls in formal schooling is hindered by a range of issues, including early marriage, their excessive workloads, popular beliefs that women are inferior to men and less intellectually able, and widespread poverty (ibid).

Education plays an indispensable role in addressing critical challenges of development. A sound and effective system of higher education provides solutions to a whole range of challenges plaguing society, globalization, poverty alleviation and democracy.

#### *4.1.2. Main Source of Household Income*

Occupations and associated total annual household income level of respondents varied greatly between different categories in the study areas. Result indicates that pastoralism was the main source of household income in the study areas, as over 70% of the population are nomadic pastoralist. In spite of fragile ecosystems of the area, results revealed that woody plants and woodlands constitute other important sources of household income including essential goods and services, income, employment and capital. Although the communities continued to rely on livestock production, it was realized that communities lacked diversity in utilization patterns, and technologies for promoting multi-production systems.

Informal employment (e.g. petty trading) was also noted as a major means of household income. A possible explanation is that the study areas are close to the borders of Somalia and Ogaden (eastern Ethiopia), therefore, there are potential markets and people are trading between the borders.

Fuelwood supports lucrative local trade. Trade in charcoal is a major source of income for many households. Other important source of household income as indicated by the respondents is remittance. Since the Somali civil war began in 1988, there have been large movements of people from Somalia to neighbouring countries, and unprecedented levels of migration to Europe and North America. Remittances have assumed great importance in this context [3]. At the household level, recent studies show that remittances constitute 40% of the income of urban households [31].

#### *4.1.3. Sources of Water*

Water scarcity is the major problem in the area. Since the area is dry by nature; low precipitation and high evapo-transpiration with no perennial rivers, the people in the study area primarily depend on man-made water sources like boreholes and rainwater harvesting namely constructed ground reservoirs which is expensive for the average households and earthen dams occasionally. However, these sources do not provide water throughout the year; the reservoirs dry up during the dry season due to decrease in precipitation or prolonged droughts. Other methods of rainwater harvesting like the use of earth dams is not efficient and not applicable in most parts of the area, particularly in Addado and Galkaayo, due to a high proportion of sand in the soil. Sandy soils are characterized by low water holding capacity. Recent studies by [7] show that soil with a high proportion of sand is unsuitable for dam construction. A sandy soil may be used in the downstream shoulder but should not be used elsewhere unless there is no alternative. In case a sandy soil is used in the rest of the dam, careful attention must be paid to compaction, the best soil reserved for the core, and some consideration given to obtaining embankment water tightness by other means. Sands play important role in larger dams as a filter material [7].

#### *4.1.4. Sources of Energy and Alternative Energy Preferences*

As common in the vast majority of the rural communities in sub-Saharan Africa, wood is the main sources of energy in the study area. The study area residents use firewood or charcoal as main sources of energy for cooking (Fig. 9). Results of the present study are consistent with the findings of [18] and [43] who found that approximately 90 % of the population in sub-Sahara Africa relies on fuelwood and charcoal as sources of energy. Fuelwood and charcoal do not provide only energy but also employment and income generation in the Horn of Africa e.g. one-third of the wood-based energy consumption of Addis Ababa come from fuelwood [18; 11]. Similar researches conducted in Zambia by [18] and [12] also indicate that charcoal production provides 41,000 jobs in Zambia. The same authors reported that Zambian forest resources contribute nearly three-quarters of the total poor households income and one-third of the richer households' income.

Firewood problem, formerly of women has become a common issue in the study area affecting every member of the family. Women spend an average of one to two hours per day collecting firewood. In order to solve the problem of wood based energy, in terms of quality and availability, the group proposed to plant high energy efficient local species for firewood in community forest to overcome wood scarcity.

Majority of the study area residents indicated solar cooking as their most preferred alternative energy for cooking. They found it the most affordable alternative energy as compared to other alternatives.

#### *4.1.5. Land Tenure*

In the study area, four different types of land tenure systems were discovered, namely; communal land, freehold, state land and rented land. Communal land ownership and freehold land right were the two most common land tenure systems in the area. Some earlier studies confirm that customary systems are the de facto systems of land tenure in operation in many dryland zones, rather than statutory laws. In Africa, for example, most people hold their land under indigenous customary systems irrespective of formal legal position [34]. By clearing a piece of land or fencing, gives an individual the right to own that land under freehold. However, the study areas residents confessed that the land tenure systems were sometimes complicated and necessitated mutual understanding and relationship between pastoralists in the area and neighbouring pastoral communities.

#### *4.1.6. Major Causes of Environmental Changes*

The natural resource base in the study areas continue to be seriously threatened, disappearing or degrading at alarming rate due to deforestation, long spells of droughts, expansion of reservoir constructions and over-grazing. Facts distilled from the analysis undertaken at the household level revealed the following:

##### *4.1.6.1. Drought*

Information obtained from study areas residents revealed that droughts endangered and exacerbated poverty in the area; where an average household was said to lose 40% of its livestock in each drought event. The effects of the drought are not only limited to the loss of livestock, but comprise also food deficiencies, food quality, high rate of land degradation, loss of life and a drastic reduction of the water resources particularly of pastoral communities. By their very nature, agriculture in the drylands is exposed to the vicissitudes of an irregular rainfall pattern or frequent drought. Uncertainty in production due to fluctuations in total rainfall and changes in its distribution, decrease in relative productivity in rainfed lands etc. affect the livelihoods of many poor and marginalised pastoralists in the area. [31], studied the causes of desertification in the Horn of Africa and found that, recurring droughts leading to desertification in the Horn of Africa have created grave ecological catastrophes prompting massive food shortages.

##### *4.1.6.2. Deforestation Caused by Charcoal Burning*

Deforestation occurs in two forms: land conversion for other uses, mainly crop production/building up (urbanization), and cutting of woody plants in search of extra incomes or household consumption, i.e. for energy, construction, fences etc. In the area charcoal burning is graded as the second main cause of deforestation and desertification. For the last fifteen years charcoal trading was among the most important export commodity in the country, generating millions of dollars per year. Forests and woodlands cover approximately 23% of the country and are dominated by Acacia species. There has been significant, but localized, deforestation for agriculture, firewood and charcoal. Trees are the most important terrestrial resource because of the goods they provide, particularly in dry times [41].

As a result of prolonged civil war and frequent droughts many pastoralists and agro-pastoralist ended up as IDPs. Therefore, due to lack of other opportunities, many poor men (IDPs) found charcoal burning as an option of income generating (tragedy of the commons). The price of the charcoal depend o the quality of the charcoal; the best quality is the one with high energy efficiency and of course every tree cannot produce good quality charcoal, thereby the charcoal burners target specific tree species. Among the most extensively logged trees for charcoal production are the centenarian trees of *Acacia tortilis* and *Acacia busei.* These trees are locally important multi-purpose trees with high socio-economic and ecological benefits. Thereby, such exploitation activities of the aforementioned species caused numerous social and environmental problems. Hence, the civil societies, scholars and international NGOs were writing about the charcoal issue for the last ten years or so. In response, the local administrations in these areas issued a law banning cutting of *Acacia tortilis and Acacia busei* species. However, due to lack of capacity for law enforcements the problem still continues. On the other hand, the law permits utilization of dead tree (s) of these species for charcoal production; therefore, the charcoal burners developed new techniques. They started contaminating the tree's root system in order to kill the tree (s) by putting salt and used engine lubricants in the root level of the trees and eventually the trees die and they use it for charcoal production. Fortunately the issue of charcoal industry in Somalia is now highly politicized in the international level. Earlier this year the United Nations Security Council passed a resolution banning Somali exports of charcoal [42].

##### *4.1.6.3. Expansion of Reservoir Construction*

As means of coping strategies the local people adapted rainwater harvesting through constructed underground reservoirs as a source of water. However, during the field assessment it was observed that reservoirs cause tremendous environmental impact. Problems around these areas consist of overstocking, large-scale vegetation depletion and changes, especially around water points, and accelerated soil erosion by wind, sheet and gully. Bare soil areas are conspicuous around water points in Addado, Buhodle and Northern Galkaayo of Central and Northern Somalia as a result of heavy grazing. Similar studies conducted in other parts of the Horn of Africa revealed that water points are starting points of desertification. Soil and vegetation degradation around watering points has been observed in many drylands around the world. It can be recognized in space-borne imagery as radial brightness belts fading as a function of distance from the water wells [20].

It is known that reservoir expansion in the area is the third major problem causing environmental degradation for the following reasons;

1) Since the land is communally owned, there is big resources competition among the people in the area, each one trying to get the most appropriate place to construct his/her reservoir; therefore, the reservoirs are neither collectively owned nor concentrated in one place, but individually owned and scatted in the grazing land. That means there are tens or even hundreds of water points in each project area.
2) Every reservoir owner fences approximately one hectare of land around the reservoir as watershed area using dead fencing which causes huge deforestation.
3) Branches, shrubs and *Andropogen kelleri* locally known as (duur), are used to cover reservoir to avoid water loss through evaporation. These cause fodder shortage and soil exposure to wind and water erosion, resulting in desertification.

#### *4.1.6.4. Overgrazing*

***Table 4.*** *List of over exploited species as indicated by the respondents*

| Species (Scientific Name) | Somali Name | Species Type | Specific Area | Reason (S) of Exploitation |
|---|---|---|---|---|
| Andropogen kelleri | Duur | G | B | Thatching and covering reservoirs |
| Chrysopogon plumulosus | Dareemo | G | A,B,GK | Overgrazing, drought and thatching |
| Sporobolus marginatus | Dixi | G | B | Overgrazing, making mats and drought |
| Ziziphus mauritaniana | Gob | T | A,B,GK | Construction, fencing, drought |
| Riurosoefigyy dkikdjtiujAca cia tortilis | Qudhac/Qurac | T | A,GK | Charcoal production |
| Acacia busei | Galool | T | A,GK | Charcoal production |
| Commiphora myrrha | Dhidin/geed malmal | T/Sh | A, GK | Fencing, drought, tapping |
| Commiphora stellatoubescens | Xoday | T/Sh | A, GK | Carving, drought, fencing |
| Cordia sinesis | Madheedh/Mareer | Sh | A, GK | Construction, overgrazing, drought |
| Dobera glabra | Garas | T | A, B, GK | Poles, carvings and seed over harvesting |
| Balanites aegyptiaca | Quud/Iriiri | T | A, GK | Charcoal production, construction poles |
| Grewia villosa | Gomoshaa | Sh | A,B,GK | Overgrazing, fruit overharvesting, drought |
| Grewia tenax | Dhafaruur | Sh | A,B,GK | Fruit overharvesting, drought, overgrazing |
| Grewia bencilities | Hohob | Sh | A,B,GK | Fruit overharvesting, drought, overgrazing |
| Acacia milifera | Bilcil | T/Sh | A, GK | Fencing |
| Terminalia spinosa | Lebi | T | B | Curving and low germination |

Note: A = Addado, B = Buhodle, GK = Galkaayo, G = Grass, Sh= Shrub and T = Tree

Overgrazing is another important factor causing desertification. The possible explanations of overgrazing is that both livestock from the area and from the neighbouring communities make excessive use of the limited fodder resources, in the silvopastoral areas, which is a communal resource, freely available for everyone. Livestock and other wildlife are mostly browsers; damage the barks of woody perennials leading to outbreak. Land tenure arrangements was a disincentive to conservation i.e. fencing off some areas was not acceptable at the individual level.

Pastoralism in the three study areas were rapidly worsening at an alarming rate due to frequent droughts, charcoal production, reservoir construction and overgrazing which continued to adversely affect the quality of the pastures in the area. The situation was also aggravated by influx of livestock from neighboring districts as well as from neighboring country of Ethiopia. This influx continued to exert increased pressure on an already fragile eco-system. It is estimated that over 60% of pastoral communities within the study areas live below the poverty line. Poverty in the study areas may be attributed to declined soil fertility, soil erosion, shortage of fodder for livestock, shortage of fuel-wood and other wood products, limited income and unsuccessful land use practices. The forests in Somalia are suffering heavy exploitation and degradation reflected in negative effects on species composition and ecosystem services [2]. The following species are most exploited species as indicated by study area residents respondents (Table 4).

There are also many disappeared edible plant species which were locally important sources of nutrition. These species naturally which grow in association with grasses like *Andropogen kelleri and Chrysopogon plumulosus,* but as a result of habitat destruction these species disappeared. Such important vegetables including several of *Cucumis* species among them is *Cucumis metuliferus* syn. Horned cucumber and other herbs with edible roots.

### *4.1.7. Willingness to Participate in NTR and Planting Trees for Agroforestry Purpose*

Generally, a considerable number of local people already had considerable amount of experiences of benefits acquired from trees and other vegetation resources, and expressed their willingness to adopt agro-silvo-pastoral system. The most important result found on this issue was the willingness among the majority of the local people, especially women, to participate in tree planting. The emphasis was stronger among women as compared to men. Clearly, over 90 % of women wanted to have more trees, for income generation and satisfaction of their basic needs from the sale of products.

Majority of men also found tree planting acceptable; although some had different perceptions about its practicality and how it could be managed. Likewise, responses received from a cross-section of respondents showed that the most important restraint were livestock grazing and unclear ownership. Results also indicate that men who showed interest in planting trees, found integration of trees with their own farmlands acceptable. One reason is that the Somali community respects ownership. Similar studies by [8] in the Sudan showed that land tenure and land security are crucial issues for the success of any tree planting projects. However, the result of the current study shows that, land availability was not the limiting factor; because over sixty percent of the land was considered as communal, where every member of the community had right to plant trees.

# 5. Conclusion

The present study examined the state of the environment as well as the socio-economic and ecological impacts of environmental degradation in Addado, Buhodle and Galkaayo in central and northern Somalia. The mounting pressure on the area's natural resources due to increasing drought events, deforestation, expansion of reservoirs and overgrazing has resulted in a high level of resource exploitation and environmental degradation which adversely affect the livelihood of the people in the areas.

Against this backdrop, there is an immediate need of action for seeking newer approaches in the farming systems to meet the basic needs of the local people. Adoption of tree planting/agroforestry on a wide range of land in the area and incorporation of multipurpose trees in the existing farming systems seem to be an apt option for reducing the pressure on the natural vegetation and for promoting self-sustaining. Agroforestry is the most preferred resource management system, because it has the potential to stabilize the production of food, forage, firewood and timber and protecting the environment. In agroforestry, crop productivity is enhanced under tree canopy due to improved soil fertility and the ameliorative influence of shade in a hot dry environment by reducing the understory temperature and mitigating evapo-transpiration. However, this requires development of appropriate skills, implementation capacity and the presence of institutions at the local, regional and national levels so as to ensure the effective use of existing resources and to foster the mobilization of additional financial and human resources for continued development of the subsistence sector.

# References


[1] Benton, H. H. 1973-1974. Encyclopaedia Britannica, Inc. Ready Reference and Index volume VII, Publisher University of Chicago.

[2] Bongers F. and Tennigkeit T. (eds), 2010. Degraded Forests in Eastern Africa: Management and Restoration. Earthscan, UK.

[3] DFID, 2005. Somalia Country Study. European Community's Poverty Reduction Effectiveness Programme (EC-PREP). University of Oxford, Oxford, UK.

[4] Dixon, R.K., Winjum, J.K., Andrasko, K.J., Lee, J.J. and Schroeder, P.E., 1994. Integrated land-use systems: assessment of promising Agroforestry and alternative land-use practices to enhance carbon conservation and sequestration. *Climate Change* 27:71-92.

[5] Faillace, C. And Faillace, E. R. 1987. Water quality data book of Somalia. General report. Deutsche Gesellschaft für Technische Zuzammenarbeit (GTZ) GMBH. Project No. 80.2193.3-09.112.

[6] FAO, 1993. Key aspects of strategies for the sustainable development of Drylands. Rome.

[7] FAO, 2010. Manual on Small Earth Dams: A Guide to Siting, Design and Construction. FAO, Rome. Irrigation and Drainage Paper 64.

[8] Glover, E. K. 2005. Tropical dryland rehabilitation: Case study on participatory forest management in Gedaref, Sudan. Doctoral Dissertation, University of Helsinki Tropical Forestry Report No. 27. 183 p.

[9] Glover, E.K. 2010. Approaches to Halt and Reverse Land Degradation in Kenya: Agroforestry Development and Environmental Sustainability. , 2010 VDM Verlag, Germany

[10] Grainger, A. 1990. The threatening desert: Controlling desertification. Earthscan Publications Ltd., London. 359 p.

[11] Haile, F. 1991. Women Fuelwood Carriers in Addis Ababa and the Peri-Urban Forest. Report to the Government of Norway, the International Development Research Centre and the National Urban Planning Institute, Geneva, International Labour Office.

[12] Hibajene, S. J. and A. Ellegard. 1994. Charcoal Transportation and Distribution: A Study of the Lusaka Market. Energy, Environment and Development Series 33, Stockholm Environment Institute, Stockholm.

[13] Hoffman, M. T. & Todd, S. W. 2000. A national review of land degradation in South Africa: The influence of biophysical and socio-economic factors. *Journal of Southern African Studies*. 26 (4): 743-758.

[14] Huschke, R.E. (ed.) 1959. Glossary of meteorology: Boston, American Meteorological Society, 638 p.

[15] Iiyama, M., Kristjanson, P., Ogutu, J., Maitina, J., Kariuki, P., Morimoto, Y. And Baur, H., 2008. Conservation, natural resource management and development challenges in rural Africa: Evidence from east Africa. In: Natural Resources. Nover Science Publisher, Inc.

[16] IUCN 2001. Regional workshop on community involvement in forest management in Eastern and Southern Africa. IUCN Eastern Africa Programme, Kampala, Uganda. Forest and Social Perspectives in Conservation No. 6. 25 p.

[17] Kadomura, H. (ed.) 1997. Data book of desertification/land degradation. Centre for Global Research, National Institute for Environmental Studies (NIES), Japan. 68 p.

[18] Kaimowitz, D., 2002. Forests and Rural Livelihoods in Sub-Sahara Africa. Forests in Poverty Reduction Strategies: Capturing the potential. Tuusula, Finland EFI Proceedings 47 (2003),

[19] Karnieli A., U. Gilad, M. Ponzet, T. Svoray, R. irzadinov, O. Fedorina, 2008. Assessing land-cover change and degradation in the Central Asian deserts using satellite image processing and geostatistical methods. *Journal of Arid Environments*, 72: 2093–2105.

[20] Kindt, R., van Breugel, P., Lillesø, J.-P.B., Bingham, M., Sebsebe Demissew, Dudley, C., Friis, I., Gachathi, F., Kalema, J., Mbago, F., Minani, V., Moshi, H.N., Mulumba, J., Namaganda, M., Ndangalasi, H.J., Ruffo, C.K., Jamnadass, R. and Graudal, L., 2011. Potential natural vegetation of Eastern Africa. Volume 3: Description and tree species composition for woodland and wooded grassland types. Forest & Landscape Working Paper 63-2011

[21] Kürsten, E., 1999. Fuelwood production in Agroforestry systems for sustainable land use and $CO_2$ mitigation. Ecological Engineering 6: 69-72.

[22] Mahony, D., 1990. Trees of Somalia. Field guide for development workers. Oxfam.

[23] Millennium Ecosystem Assessment, 2005. Ecosystems and Human Well-being. Dryland Systems. World Resources Institute, Washington, DC

[24] Mogaka, H., Gacheke, S., Turpie, J., Emerton, L. and Karanja, F., 2001. Economic Aspects of Community Involvement in Sustainable Forest Management in Eastern and Southern Africa.

[25] IUCN, 2001. Eastern Africa Regional Office Forest and Social Perspectives in Conservation, No. 8, Nairobi.

[26] Olsson, K. 1985. Desertification and land degradation on perspective. In: Poulsen, E. and Lawesson, J. (eds.). Dryland degradation: causes and consequences. Aarhus University Press. pp. 5-13.

[27] Oxfam GB, 2003. Achieving EFA through Responsive Education Policy and Practice for Nomadic and Pastoralist Children: What can Agencies do? Unpublished paper, Oxford: Oxfam GB .

[28] Oxfam GB 2005a. Pastoralist Education Learning Resource: A compilation of information on pastoralist education programmes in Horn and East Africa', Nairobi.

[29] Oxfam GB, 2005b. Beyond the Mainstream, Education and Gender Equality Series, Programme Insights. Oxford: Oxfam GB.

[30] Pantuliano S. and Pavanello S. 2009. Taking drought into account: Addressing chronic vulnerability among pastoralists in the Horn of Africa, Overseas Development Institute, UK .

[31] Social Development Papers (SDP), 2006. Remittances and Economic Development in Somalia: An Overview. Conflict Prevention and Reconstruction, Paper No. 38/ November.

[32] Tyson, W., 1995. Ecosystem restoration: theory, practice, and evidence. In: Roundy, B.A., McArthur, E.D., Haley, J.S., Mann, D.K. (Compilers), Proceedings of the Wildland Shrub and Arid Land Restoration Symposium, Las Vegas, NV, October 19-21, 1993. General Technical Report INT-GTR-315. US Department of Agriculture, Forest Service, Intermountain Research Station, Ogden, UT, pp. 116-118.

[33] UNCED 1992. Report of the United Nations Conference on Environment and Development; Earth Summit Agenda 21. Rio de Janeiro, 3-14 June 1992. New York: United Nations. 300 p.

[34] UNDP, 2003. The Global Drylands Imperative: Land tenure reform and the Drylands. Drylands Development Centre. Nairobi, Kenya.

[35] United Nations Development Programme (UNDP) 2010. Assessment of Development Results: Evaluation of UNDP Contribution. Somalia Evaluation Office.

[36] UNEP, 1997. World atlas of desertification. Edward Arnold, London, UK.

[37] UNICEF-Somalia 1993. Strategies on Nomadic Education Delivery. Education Unit, United Nations Children's Fund (UNICEF) Somalia, UNICEF Somalia, Document 1103.

[38] World Bank, 2007. World Development Report 2008: Agriculture for Development. Washington DC: The World Bank.

[39] World Conservation Union (IUCN), 2006. Country Environmental Profile for Somalia. The European Commission Somalia Operations Office, Nairobi, Kenya.

[40] The National 2012. Charcoal exports from Somalia banned. Available at: http://www.thenational.ae/news/uae-news/charcoal-exports-from-somalia-banned Accessed on 30.3.2013

# 19

# Strategies of eco-friendly stimulant use policy in Perum Perhutani, Indonesia

**Sukadaryati**

Ph D Student of Forestry Faculty of Bogor Agricultural University, Jalan Raya Dramaga, Kampus IPB Dramaga, Bogor, Indonesia

**Email address:**
daryatielin@yahoo.co.id

**Abstract:** The use of strong acid H2SO4 stimulant (AN-ORGANIC) in effort to improve resin results in Perum Perhutani area is proved not safe for workers and the environment. On the other hand, eco-friendly stimulants such as ETRAT has been used in most areas of Perhutani and wood vinegar were developed. This research is aimed to identify strategies of eco-friendly stimulant use policies in tapping pine. Data collection was conducted through in-depth interviews to several informants consisting of tapper community, foreman, Ranger, Head of RPh, Head of KPH, Deputy head of KPH, NTFPs section chief, Head and deputy chief Production Unit. Selection of informants is carried out purposively that is derived from pine resin producing areas 3 largest in Perum Perhutani (KPH Banyumas West, KPH Kediri and KPH Sukabumi). The results showed that the strategy of eco-friendly stimulant use policies could be adopted by Perhutani with four strategies: 1) Increase the use of eco-friendly stimulant by utilizing eco-friendly wood vinegar-based stimulant produced by Perhutani; 2) Implement eco-friendly wood vinegar-based stimulant produced by Perhutani to press the price of an expensive eco-friendly stimulant; 3) Increase the use of eco-friendly and safer stimulant innovation to convince Perhutani so that eco-friendly stimulant can be accepted and used extensively; and 4) Press eco-friendly stimulant prices through innovations that could be accepted usage in Perhutani thoroughly.

**Keywords:** Stimulant, Eco-Friendly, Tapping, Pine, Policy Strategy

## 1. Introduction

Pine resin trade Indonesia in the International ranks third after China [5]. Perhutani as one of the producers of pine resin in Indonesia is capable of exporting 80% of its production to foreign countries, namely Europe, India, South Korea, Japan and United States [13]. Increased effort resin results continue to be conducted by Perhutani in line with consumer demand for pine resin. One method to increase the production of pine resin by administering stimulant or aphrodisiac in pine trees tapping, ie a particular liquids are given in the wound of tapping pine useful to further smoothen out the resin from the stem of pine. The liquid used as a stimulant usually made from acid. Almost the entire area of Perhutani uses AN-ORGANIC (made from strong acids H2SO4) as a stimulant in tapping pine with composition grant of stimulant which varies depending on the altitude of the place.

The use of AN-ORGANIC is considered unsafe for both the health of the pines tapped as well as the tappers as well as the environment. Therefore, it is developed organic-based stimulant products namely ETRAT safer and eco-friendly. In addition, it gives a positive effect on the resin production, which can increase the production of resin. However, the community does not use immediately the stimulant in the field. This is related to Perhutani policy that does not provide instructions on the use of ETRAT in tapping pine for the whole area of pine forest.

Other organic stimulant made from wood vinegar has good prospects in increased production of pine resin. Wood vinegar is a liquid smoke formed through the combustion process that condensed on the cold temperatures. Wood vinegar stimulant can be made from the trunks of wood waste or other materials that have lignoselulosa and can actually be self-produced by the Perhutani because it is easy and cheap. Based on the research results of field trials showed that stimulant made from wood vinegar is able to increase the production of pine resin on average by 17.8%

comparison with does not use stimulant.

Perhutani as pine forest managers have great power in the procurement of stimulant for resin tappers. For Perhutani, the increase in resin production through the introduction of stimulant required to meet the production targets that have been set. Therefore procurement stimulant used and its responsibility to its use in the field. Tapper communities have already using the stimulant that has been provided. Pine tappers have been aware of the negative impacts of the use of AN-ORGANIC stimulant prefer leaving it and replace it with a stimulant that is safe and eco-friendly.

This study is aimed to identify strategies eco-friendly stimulant use policies in tapping pine. Policy strategy is needed to accommodate the interests acquired in tapping pine Perhutani without compromising people's desire tapper.

## 2. Research Method

The data collection was conducted in September-November 2014 in the area of Perum Perhutani (Unit I Central Java, East Java Unit II and Unit III West Java-Banten). This is a qualitative research and data collection was conducted through in-depth interviews of some informants consisting of people who work in the area of tappers Perum Perhutani (89 people), foreman (5 persons), Ranger (1 person), Head of RPh (4 people), Head of KPH (2), Deputy Chief of KPH (2), Section Head of NTFPs (3 people), Head of Production (3) and Unit Deputy Chief (1 person). Selection of informants is carried out purposively that is derived from pine resin producing areas 3 largest in Perum Perhutani (KPH Banyumas West, KPH Kediri and KPH Sukabumi

SWOT analysis is used to achieve the research objectives. According to [3, 13, 14] SWOT analysis is identification of the various factors systematically to formulate corporate strategy by maximizing strengths and opportunities, but at the same time can minimize the weaknesses and threats. SWOT Analysis carried out by:

A. Identify the internal factors of strengths and weaknesses as well as opportunities and threats that can influence the adoption of eco-friendly use of stimulant in tapping pine in Perum Perhutani.

B. Determining Key Success Factors (KSFs) and Map of Strength Position, namely:

1 Determining Key Success Factors

2 From 6 (six) and the main internal factors and 6 (six) main external factors, then selected key success factors (KSFs). To select and set KSF, evaluation of internal and external factors that have been identified previously. Factors that give urgency, support and high exposure to the success for the present and the future, defined as KSF, with each aspects factor assessed are:

3 Urgency factor to the achievement of the targets, including the value of urgency (NF) and weighting factor (BF). NF assessment in achievement is done by changing the value of qualitative into quantitative values with a scale of 1 to 5, with details as following Table 1.

***Table 1.*** *Urgency Assessment Scale (NF)*

| No | Value Scale | Description |
|---|---|---|
| 1 | 5 | Very large/high |
| 2 | 4 | Large/High |
| 3 | 3 | Medium/Enough |
| 4 | 2 | Low/Small |
| 5 | 1 | Very low/Small |

BF value is the ratio between NF and the number/total of NF (for each internal and external factors), which are expressed in percent. The formula for calculating BF [12, 4] is:

$$BF = (NF \times 100)/\Sigma NF$$

a Support to the achievement of objectives factors, including support values (ND) and weight support values (NBD). Support values (ND) each factor is determined on the basis of the amount of support the achievement of the target with a scale of 1 to 5. Scale details ND value equal to the value of NF-scale details. NBD value of each factor is calculated by the formula:

$$NBD = ND \times BF$$

b Linkages between factors to the achievement of targets, including the value of the relationship (NK), the average of relationship value (NRK) and weight relevance value (NBK). Internal factors and external inter-related or related in the achievement of targets. With the linkage will create synergy in supporting the objectives achievement. Linkage values of these factors are as follows:

- The linkages value (NK) each factor to another factor determined on a scale of 0 to 5. Scale of 5 shows a very close relationship between the factors in support of the target achievement; scale value of 1 is very weak, while the scale value of 0 indicates absence of linkage relationships.
- The average value of the relationship (NRK) each factor is divided by the number of NK-related factors (n-1) or the formula $NRK = \Sigma NK / (\Sigma n-1)$
- The NBK value is calculated by the formula $NBK = NRK \times BF$

c The total value of the weight (TNB) is the sum value of NBD and NBK or the formula $TNB = NBD + NBK$

Key success factors (KSFs) is determined based on factors that have the highest value of TNB [9, 10]. Then selected four (4) KSF consisting of one factor with the highest value of TNB from each factor, strengths, weaknesses, opportunities and threats.

# 3. Results and Discussion

Analysis of internal factors (strengths and weaknesses) and external factors (opportunities and threats), which can influence the adoption of eco-friendly use of stimulants in tapping pine in Perhutani can be seen in Table 2.

***Table 2.*** *Identification of Internal and External Factors*

| **Internal Factors** | |
|---|---|
| Strengths | Weaknesses |
| 1. Use of eco-friendly safer stimulant | 1. Productivity of resin is lower |
| 2. Pine forest sustainability is guaranteed | 2. There has been no price difference resin resulting from the use of eco-friendly stimulant and AN-ORGANIC |
| 3. Availability of eco-friendly stimulant | 3. The Price of eco-friendly stimulant is expensive |
| **External Factors** | |
| Opportunities | Threats |
| 1. Stimulant made from eco-friendly wood vinegar may be produced by Perhutani | 1. Use of stimulant made from a strong acid (AN-ORGANIC) is a fixed price for pine resin activities in Perhutani |
| 2. Pine resin tapper desperately need a safe and eco-friendly stimulant | 2. Innovations stimulant that is safe and eco-friendly cannot be accepted for use in Perhutani as a whole |
| 3. Innovation of eco-friendly stimulant used to increase the production of pine resin has been developed | 3. Difficulty of changing the usage habits of AN-ORGANIC-based stimulant that has been done for generations |

### *3.1. Strength Factors*

1. Use of eco-friendly safer stimulant

Until now this stimulant used in tapping pine in Perhutani in the area mostly using stimulant-based inorganic strong acid is H2SO4 or also known as AN-ORGANIC or trademarks SOCEPAS (consisting of sulfuric acid and Chloroastyl Phosponil Acid). The use of stimulant made from a strong acid such as complained by tapper has negative health effects such as hands become itchy and rough, stinging when in contact with the eyes and cause coughing. Further in [11, 15] states that when the short-term effects of inhaling vapors of sulfuric acid can cause irritation of the nose and throat and irritate the lungs. The acid can damage the skin and cause injury very hurt. It can cause blindness if splashed in eyes. Inhalation of vapors acid levels in the long run result in minor irritation of the nose, throat and lungs. For the pine tree itself, the use of strong acids may cause discoloration of the former leads rods, which appear red to black (scorched) and in continuous use in the long time period to cause decay tracheas walls in large quantities so that the trees become fragile. This condition will contribute negatively to the fall of the trees more so because of injury leads that are too deep and the wind reaches it. Even suspected chemical ingredients in the resin of the results leads derived from an organic stimulant that is used so as to interfere with the use or further processing of the resin. Based on some literature also mentions that the use of strong acid stimulant such as H2SO4 also cause soil and water pollution. The use of eco-friendly stimulant here is intended as a natural stimulant (biological, from plants). In Perhutani itself has been taking stimulant ETRAT in tapping pine although still used partially (not thorough) in certain circumstances (leads open and on location with altitude below 500 msl / low). ETRAT stimulant derived from ethylene substances that are useful in the resin formation in the plant. Tappers who use ETRAT stimulant do not feel the complaints as if they were using AN-ORGANIC. Perhutani itself already know that the use of ETRAT safer for the health of the tree tappers and tapped. However, due consideration of the ETRAT purchase price is more expensive than the AN-ORGANIC so that the use of ETRAT is limited. Other organic stimulant made from wood vinegar also had been tested in this study and may increase the production of pine resin. Excess wood vinegar is a stimulant can be produced easily and inexpensively as well as raw materials derived from waste wood or tree or any material containing lignocellulose. The use of wood vinegar stimulant in the field also is not complained by tappers either their health or tree tapped. Actually for Perhutani has realized that the use of safer and eco-friendly stimulant both to the tappers' health and tree tapped.

2. Pine forest sustainability is guaranteed

The use of eco-friendly stimulant, as already described in point 1 gave a positive impact on the sustainability of the pine forest management. Sustainability here includes sustainability of pine stands are supported by adequate tree health and also the tapper as workers. The health of trees here relate to the excessive exploitation of harvesting resin because only about economic oriented so that the trees are "forced" to produce more resin to achieve the target, one of them by using strong acid-based stimulant. For the tappers who work as freelancer in Perum Perhutani, actually also have a choice to work other than as farmer as factory workers or worked in plantation sector or mining. When the income of tappers pine uncompetitive plus the health hazard that continues to overshadow, cannot be denied if one day the tappers will turn to other work other than as a tapper as mentioned above. These conditions will make it difficult for Perhutani to obtain employment forest communities, so that if a shortage of tappers will lower latex production results. Thus the sustainability of forest management as a producer of pine resin would be threatened halfway [1, 6].

3. Availability of eco-friendly stimulant

Eco-friendly stimulant like ETRAT has already available in the market and produced by CV Permata Hijau in Bogor (West Java). Meanwhile considering the ease in manufacturing wood vinegar-based stimulant, it can be produced by Perhutani even by the people of tappers. Just stay committed to apply it in the field.

### 3.2. Weaknesses Factors

1. Productivity resin lower

The use of eco-friendly organic-based stimulant and used in tapping pine resin can improve and extend the time flow of the resin so that resin obtained per harvest. However, the use of organic stimulant produces resin that is lower in comparison with the AN-ORGANIC. At low altitude (<500 msl), pine resin yield improvement using AN-ORGANIC stimulant, ETRAT and wood vinegar respectively of 49.15%; 33.57% and 15.5%. Similarly, at altitude > 500 msl, the increase pine resin results due to the use of AN-ORGANIC, wood vinegar and ETRAT respectively 51.7%; 28.1% and 17.8%.

2. There has been no price difference resin resulting from the use of eco-friendly stimulant and AN-ORGANIC

Wage system is applied in Perhutani based on achievement, that tappers receive wages depend on the achievements that have been made within the time provided. Thus wages are received each month will vary depending tappers' willingness and ability in intercepting pine. Wage rates (known as the carry rate) established Perhutani based on the distance from the location of the leads (subplot) to place Resin Gathering (TPG). The wage rates are also distinguished by the quality of the resin, the quality of A and B where quality A higher rates of wage than quality B at the same distance. The quality here is based on the cleanliness of resin from impurities that goes in like the rest of the resin mixed leaves, twigs, dirt, and other impurities. On the other, as the development time and human awareness of the importance of health, all products which are directly related to humans should be safe for human health. No exception for pine resin products are processed into various products such as Gum Rosin and its derivatives, turpentine, alphapinene, bethapinene, and so forth. For the market share of these products is divided into 2 foodgrade products (which means the product is safe for human health) and non-foodgrade (these products contain substances that are not safe / harmful to human health). Gum rosin derivative products which include foodgrade categories used as tooth clamp bandages, food preservatives, a mixture of soap, cosmetics, and others. Non-foodgrade products found as paper sizing material, a mixture of turpentine, ink, and so forth. Regardless of sorting the two products should indeed all the products meet safety standards for human health. Therefore, one of the requirements is a safe product that is free from B3 (Hazardous Material), including strong acidic chemicals used in tapping pine. Considering the corrosive nature of the danger and possessed of a strong acid, then pine resin resulted is thought to contain harmful substances to human health. While the price of pine resin products in the world market does not see it and do not differentiate pricing based stimulant used. Yet look at the result of negative effects should no price difference between the resin produced by using inorganic and organic stimulant. There's no denying that the Indonesian position though as the third of the largest manufacturers of pine resin but does not act as a "price taker" or determinants of price. This position is difficult and adverse pine resin manufacturers. However along the changing times, everything can change all the time and the market share will definitely look into that direction and will price discriminate based on the type of latex used stimulant.

3. The prices of eco-friendly stimulant is expensive

During this stimulant used in tapping pine in Perum Perhutani area is held by way of buying into the company providing stimulant. AN-ORGANIC stimulant purchased from CV pine in Jember (East Java) was purchased from ETRAT stimulant CV Permata Hijau sustainable in Bogor (West Java). Based on interviews with Perhutani, procurement stimulant given to each unit (I, II and III), and each unit has the authority under the provision that the policies adopted by each unit may vary. AN-ORGANIC stimulant price range between IDR 6000-11.000/litre depends on the distance from the place of production whereas ETRAT stimulant ranged between IDR 10.000-12.000/liter (based on interviews with Perhutani in 2013). Based on price per liter, the price of ETRAT stimulant or organic stimulant more expensive than AN-ORGANIC.

### 3.3. Opportunities Factor

1. Eco-friendly wood vinegar stimulant can be produced by Perhutani

Wood vinegar-based stimulant can be produced by Perhutani in an easy way and inexpensive. How to make a wood vinegar can be done with furnace ground methods and drum method [8], in which the manufacture of wood vinegar can be done together with the manufacture of charcoal. The smoke of burning the charcoal-making process can be used as raw material for the manufacture of wood vinegar (wood vinegar). By using fresh bamboo as a condenser in the cooling process, it hollowed and cut attached to the top of the chimney in the process of making charcoal and then arranged so that most of the smoke coming through the bamboo in order to obtain the final results of the process of cooling the smoke

Burning wood in the form of wood vinegar. The materials used in the manufacture of wood vinegar is easy to find and cheap price. The raw materials consist of sewage trunk, branches twigs or other parts of the tree that are not used are prepared by means of shredding up into 5-10 cm diameter and 15-20 cm long. Furnace ground method made by digging up the soil and incorporates raw materials as mentioned above, burned in such a way so as to emit smoke. Meanwhile the drum method, raw materials are arranged inside the drum and burning is done that removing the acid. This acid is captured and condensed with bamboo rod-shape liquid so called liquid acid. This liquid contains many acidic vinegar / acetic acid ($CH_3COOH$) that is known as wood vinegar. In R & D Perhutani in Cepu, Central Java has also been building a vinegar tool maker with wood dome method and has even been tested for produce wood vinegar, only now being stopped producing.

2. Pine resin tapper desperately need a safe and eco-friendly stimulant

Based on the facts on the ground, the use of strong acid (H2SO4) stimulant or AN-ORGANIC complained of by the tappers because negative health effects such as hands become itchy and rough, stinging when in contact with the eyes and cause coughing. For the tappers that actually do not want the stimulant that has bad effect. It's just because of the resulting latex production is not as high as when using stimulant AN-ORGANIC; they survive using AN-ORGANIC with such risks. But if there is another stimulant that are safer and can improve the production of latex, they prefer to use it. In addition the use of AN-ORGANIC also stimulant also less good effect against pine stands. If the situation is allowed to continue, it is possible for the tappers will turn to work in another place that is seemed safe for them.

Innovation of eco-friendly stimulant used to increase the production of pine resin has been developed

3. Innovation of eco-friendly stimulant used to increase the production of pine resin has been developed

Stimulant that is safe and does not create a significant environmental disturbance or eco-friendly has been developed and tested in Perum Perhutani area. Based on the interviews, in addition to ETRAT, stimulant that is introduced and tested in Perum Perhutani is SUPERFARM, GREEN ONE (GO) 1, 3 GO, GO 5 and BIOCAS. BIOCAS is a new product was introduced in 2014 in the area of East Java Perhutani Unit III and produced by the same factory as the AN-ORGANIC. Stimulants were tested in the Perhutani, claimed by the producers as organic and eco-friendly stimulant. However, with certain considerations, eco-friendly stimulant used in Perhutani area although not all areas of pine stands, chosen ETRAT stimulant. By looking at the number of stimulants offered in Perhutani office shows that up to now, the use of eco-friendly stimulant has become a necessity in tapping pine. All society life have been more aware of the importance of using eco-friendly stimulant.

### *3.4. Threats Factors*

1. The use of strong acid stimulant (AN-ORGANIC) is a fixed price for pine resin activities in Perhutani

So far, most of the widened Perhutani area uses AN-ORGANIC in tapping pine. Based on the interviews, comparison of the use of ETRAT stimulant and AN-ORGANIC respectively 80% and 20%. It is the consideration of the stimulant purchase price where AN-ORGANIC is cheaper than ETRAT and also because of the habit of tapper people who already "addicted" must use a stimulant in any tapping activity especially stimulant that can produce more resin (AN-ORGANIC). Based on the use of AN-ORGANIC stimulants Director's Decree No. 527/045.1/PROD/Dir September 5$^{th}$ 2007 on the recommended use of AN-ORGANIC, while the basic use of ETRAT contained on Directors Decree 220/045.9/PROSAR/DIR April 25$^{th}$ 2011 regarding the use of ETRAT stimulant in tapping pine.

2. Innovations stimulant that is safe and eco-friendly cannot be accepted for use in Perhutani extensively

Innovation stimulants continue to be developed adapted to present demands that are safe and eco-friendly. For Perhutani, it has an open attitude towards all innovations related to tapping pine. Only certain requirements Perhutani filed against the adoption of stimulant innovation, the conditions of production and price. Production terms associated with stimulant' ability to produce pine resin, which should be able to increase the yield of pine resin at least equal to if using AN-ORGANIC. Rate terms associated with stimulant price cheaper or at least equal to the price per liter of AN-ORGANIC [7]. On the other hand Perhutani also aware of the importance of the use of safe and eco-friendly stimulants. Moreover, later all the products produced from the resin of pine required to be safe for human health. With these considerations, Perhutani take steps using eco-friendly stimulant innovation in tapping pine although not yet used throughout Perhutani area.

3. Difficulty of changing the usage habits of AN-ORGANIC-based stimulant that has been done for generations

The use of stimulants in pine resin has been conducted since 1933 in Germany and Russia and America followed in 1936. The use of stimulants and is being adopted in Indonesia, although it has not yet simultaneous use and the legal basis for its use until the adoption decree on the Board of Directors recommended the use of AN-ORGANIC stimulant (No. 527/045.1/PROD/Dir) in 2007. Use of AN-ORGANIC was originally just a suggestion because tapping pine conditions at that time are still many who do not want to or use stimulant. This is related to the main orientation of the pine forest at the time taken shall not resin. Especially if it is connected with the main focus of Perhutani, which is focused more wooden results mainly teak. But with over time, non-timber forest products that are pine resin began to recognize because the outlook is promising. Moreover, the more years the number of request of the resin increased, thus encourage the use of stimulant to improve outcomes resin. For tapping, the more resin obtained will be more and more revenue as well. From the beginning as far as they are only suitable AN-ORGANIC stimulant used to increase production. Knowledge of AN-ORGANIC is what they are bringing to the posterity and siblings. Therefore if there is a new stimulant innovation, they feel free to try it out. They worry that the results of resin they obtain not as much when using AN-ORGANIC. Comparison of the factors internal and external to determine the value of NF and BF is presented in Table 3 below.

FKK evaluation results are presented in Table 4. Based on Table 4 it can be seen that internal factors such as strengths and weaknesses that have the highest relevance value of each stimulant is the use of safer and eco-friendly stimulant and the Price of eco-friendly stimulant is expensive, the weight value respectively of 2.67 and 1.58. External factors such as opportunities and threats that have the highest relevance value of each is the eco-friendly wood vinegar-based may be produced by Perhutani and the innovation of safe and eco-friendly stimulant cannot be accepted for use in a whole Perhutani, with respective weights of 2.64 and 2.01. Based on Table 4 were then created Success Key Factors Table shown in Table 5.

***Table 3.*** *Urgency Comparison Matrix between Internal and External Factors*

| No | Internal And External Factors | More Urgent Factors | | | | | | Total | Bf |
|---|---|---|---|---|---|---|---|---|---|
| | | a | b | c | d | e | f | NF | |
| I | INTERNAL | | | | | | | | |
| A | Strengths(S) | | | | | | | | |
| a | The use of safer and eco-friendly stimulant | | a | a | a | a | a | 5 | 33 |
| b | Pine forest sustainability is guaranteed | a | | c | b | b | f | 2 | 13 |
| c | Availability of eco-friendly stimulant | a | c | | c | c | f | 3 | 20 |
| B | Weaknesses (W) | | | | | | | | |
| d | Resin productivity is lower | a | b | c | | d | f | 1 | 7 |
| e | There has been no price difference resin resulting from the use of eco-friendly stimulant and AN-ORGANIC | a | b | c | d | | e | 1 | 7 |
| f | The price of eco-friendly stimulant is expensive | a | f | f | f | e | | 3 | 20 |
| | Total | | | | | | | 15 | 100 |
| II | EKSTERNAL | | | | | | | | |
| A | Opportunities (O) | | | | | | | | |
| a | The eco-friendly wood vinegar-based may be produced by Perhutani | | a | a | a | a | a | 5 | 33 |
| b | The pine resin tapper desperately need safe and eco-friendly stimulant | a | | c | b | e | f | 1 | 7 |
| c | The innovation of eco-friendly stimulant used to increase pine resin production has been developed | a | c | | c | e | f | 2 | 13 |
| B | Threats (T) | | | | | | | | |
| d | The use of strong acid stimulant (AN-ORGANIC) is fixed price for pine resin activities in Perhutani | a | b | c | | e | d | 1 | 7 |
| e | The innovation of safe and eco-friendly stimulant cannot be accepted for use in Perhutani extensively | a | e | e | e | | e | 4 | 27 |
| f | Difficulty of changing the usage habit of AN-ORGANIC-based stimulant that has been done for generations | a | f | f | d | e | | 2 | 13 |
| | Total | | | | | | | 15 | 100 |

***Table 4.*** *Evaluation of Internal and External factors*

| No. | Internal & Eksternal Factors | BF (%) | ND | NBD | Relevance Value | | | | | | | | | | | | NRK | NBK | TNB | RANK |
|---|---|---|---|---|---|---|---|---|---|---|---|---|---|---|---|---|---|---|---|---|
| | | | | | 1 | 2 | 3 | 4 | 5 | 6 | 7 | 8 | 9 | 10 | 11 | 12 | | | | |
| A. | Internal | | | | | | | | | | | | | | | | | | | |
| I | Strengths (S) | | | | | | | | | | | | | | | | | | 4,63 | |
| 1 | The use of safer and eco-friendly stimulant | 33 | 5 | 1,65 | | 2 | 4 | 4 | 3 | 3 | 4 | 2 | 2 | 3 | 4 | 3 | 3,09 | 1,02 | 2,67 | I |
| 2 | Pine forest sustainability is guaranteed | 13 | 3 | 0,39 | 2 | | 3 | 2 | 1 | 1 | 4 | 2 | 2 | 3 | 2 | 3 | 2,27 | 0,30 | 0,69 | III |
| 3 | Availability of eco-friendly stimulant | 20 | 4 | 0,80 | 4 | 3 | | 3 | 4 | 1 | 4 | 1 | 1 | 1 | 2 | 2 | 2,36 | 0,47 | 1,27 | II |
| II | Weaknesses (W) | | | | | | | | | | | | | | | | | | 2,43 | |
| 4 | Resin productivity is lower | 7 | 4 | 0,28 | 4 | 2 | 3 | | 2 | 3 | 3 | 2 | 2 | 2 | 2 | 4 | 2,64 | 0,18 | 0,46 | II |
| 5 | There has been no price difference resin resulting from the use of eco-friendly stimulant and AN-ORGANIC | 7 | 3 | 0,21 | 3 | 1 | 4 | 2 | | 4 | 3 | 2 | 2 | 2 | 3 | 2 | 2,55 | 0,18 | 0,39 | III |
| 6 | The price of eco-friendly stimulant is expensive | 20 | 5 | 1,00 | 3 | 1 | 1 | 3 | 4 | | 4 | 4 | 4 | 2 | 4 | 2 | 2,91 | 0,58 | 1,58 | I |
| B. | Eksternal | | | | | | | | | | | | | | | | | | | |
| III | Opprortunities (O) | | | | | | | | | | | | | | | | | | 3,74 | |
| 7 | The eco-friendly wood vinegar-based may be produced by Perhutani | 33 | 5 | 1,65 | 4 | 4 | 4 | 3 | 3 | 4 | | 2 | 2 | 2 | 3 | 2 | 3,00 | 0,99 | 2,64 | I |
| 8 | The pine resin tapper desperately need safe and eco-friendly stimulant | 7 | 3 | 0,21 | 2 | 2 | 1 | 2 | 2 | 4 | 2 | | 2 | 2 | 1 | 1 | 1,91 | 0,13 | 0,34 | III |

| No. | Internal & Eksternal Factors | BF (%) | ND | NBD | Relevance Value | | | | | | | | | | | | NRK | NBK | TNB | RANK |
|---|---|---|---|---|---|---|---|---|---|---|---|---|---|---|---|---|---|---|---|---|
| | | | | | 1 | 2 | 3 | 4 | 5 | 6 | 7 | 8 | 9 | 10 | 11 | 12 | | | | |
| 9 | The innovation of eco-friendly stimulant used to increase pine resin production has been developed | 13 | 4 | 0,52 | 2 | 2 | 1 | 2 | 2 | 4 | 2 | 2 | | 1 | 1 | 1 | 1,82 | 0,24 | 0,76 | II |
| IV | Threats (T) | | | | | | | | | | | | | | | | | | 3,17 | |
| 10 | The use of strong acid stimulant (AN-ORGANIC) is fixed price for pine resin activities in Perhutani | 7 | 3 | 0,21 | 3 | 3 | 1 | 2 | 2 | 2 | 2 | 2 | 1 | | 1 | 1 | 1,82 | 0,13 | 0,34 | III |
| 11 | The innovation of safe and eco-friendly stimulant cannot be accepted for use in a whole Perhutani | 27 | 5 | 1,35 | 4 | 2 | 2 | 2 | 3 | 4 | 3 | 1 | 1 | 1 | | 4 | 2,45 | 0,66 | 2,01 | I |
| 12 | Difficulty of changing the usage habit of AN-ORGANIC-based stimulant that has been done for generations | 13 | 4 | 0,52 | 3 | 3 | 2 | 4 | 2 | 2 | 2 | 1 | 1 | 1 | 4 | | 2,27 | 0,30 | 0,82 | II |

Before determining the strategy must first know the strength position map. Strength position map is determined by the total weight value (TNB) all strengths, weaknesses, opportunities and threats as in Fig 1.

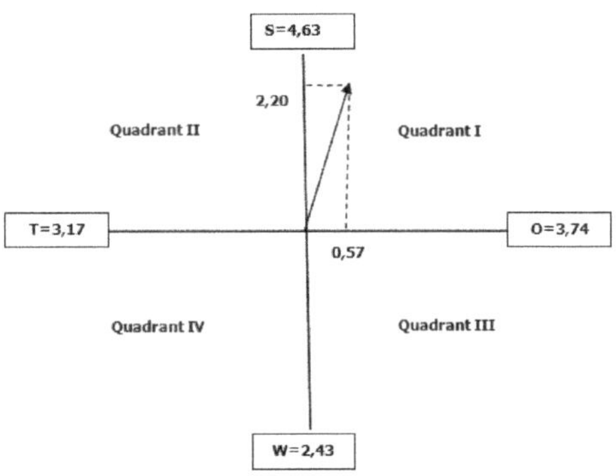


***Figure 1.*** *Strength Position Map*

Fig. 1 above shows the power position is in quadrant I. This shows that the use of eco-friendly stimulant have a competitive advantage, which means that the strength and opportunity factors play a major role in decision strategies. General strategies that can be taken are to use the strengths of (S) to take advantage of opportunities (O) there.

***Table 5.*** *Success Key Factors*

| Internal Factors | |
|---|---|
| ***Strengths* (S)** | ***Weaknesses* (W)** |
| The use of safer and eco-friendly stimulant (S1) | Price of eco-friendly stimulant is expensive (W3) |
| Faktor Eksternal | |
| ***Opportunities* (O)** | ***Threats* (T)** |
| The eco-friendly wood vinegar-based may be produced by Perhutani (O1) | The innovation of safe and eco-friendly stimulant cannot be accepted for use in a whole Perhutani (T2) |

The strategy is a combination or interaction of the success key factors for achieving synergy achievement of goals and objectives. The strategy is a general approach that is done in solving a problem. In preparing the strategy should be considered the driving force and the power resistor. Appropriate strategy is a strategy that can increase strengths and minimize weaknesses.

By combining the four (4) key to success priorities selected, arranged formulation of strategies that can be used to solve the problem. Formulation of the strategy drawn up in order to achieve the goals that have been set by using the SWOT analysis matrix as Table 6 below.

***Table 6.*** *Formulation Strategies Using SWOT Analysis*

| Internal Fkk | Strength (S) | Weaknesses(W) |
|---|---|---|
| External Fkk | The use of eco-friendly safer stimulant | The price of eco-friendly stimulant is expensive |
| Opportunities(O) | | |
| Eco-friendly wood vinegar-based stimulant may be produced by Perhutani | Increase the use of eco-friendly stimulant by using eco-friendly wood vinegar-based produced by Perhutani | Implement eco-friendly wood vinegar-based stimulant produced by Perhutani to press the price of an expensive eco-friendly stimulant |
| Threats (T) | | |
| The innovation of safe and eco-friendly stimulant cannot be accepted for use in a whole Perhutani | Increase the use of eco-friendly and safer innovations stimulant to assure Perhutani so that eco-friendly stimulant can be accepted and used as a whole | Press the stimulant prices through eco-friendly innovations that can be accepted for use in Perhutani as a whole |

Use eco-friendly stimulant policies can be achieved by using 4 (four) strategy, namely:

1. Increase the use of eco-friendly stimulant by using eco-friendly wood vinegar-based stimulant produced by Perhutani. This is an ideal strategy that takes advantage of opportunities that exist with the strength that held (Strategy S - O). Wood waste in the form of the branches and twigs of pine trees or others lignocelluloses material is relatively abundant and can be used by Perhutani to produce wood vinegar in the manufacture of eco-friendly stimulant. This is a great opportunity and potential for Perhutani to develop eco-friendly stimulant due to the production of wood vinegar as a base for eco-friendly stimulant is very easy to manufacture and can be done by Perhutani itself. With abundant production of wood vinegar produced by Perhutani, the use of stimulant is possible applied over the entire area with several advantages such as not to pollute the soil and surrounding environment, safe for farmers/tapper and tree are tapped and could depress the price of an expensive stimulant on the market. These opportunities should really be used by Perhutani to make pine resin policy all over the region in promoting eco-friendly era in the management of natural resources in the Perhutani sector.
2. Implement from eco-friendly wood vinegar-based stimulant produced by Perhutani to press the price of an expensive eco-friendly stimulant. This is a strategy of utilizing the existing opportunities to overcome weaknesses (Strategy OW). This is related to the first strategy above, that by producing wood vinegar itself from existing materials in its own territory, and implement in tapping activities, Perhutani will greatly benefit because it can press the price of an expensive stimulant on the market that had been purchased by Perhutani. By pressing this eco-friendly stimulant price the selling price of pine resin will be more competitive because production costs can be reduced and the results resin obtained from the use of stimulants leads that are eco-friendly. This strategy needs to be taken into consideration for Perhutani to adopt and implement policies eco-friendly stimulant use throughout their working area.
3. Increase the use of eco-friendly and safer stimulant innovations to assure Perhutani that eco-friendly stimulant can be accepted and used as a whole. This is a strategy of using or optimizing the strength to overcome the potential threats facing (Strategy S - T). Innovations for eco-friendly stimulant development need to be done to obtain an eco-friendly stimulant that could further improve the results of pine resin. To further assure Perhutani in any development activity of this innovation, Perhutani should be actively involved so that eco-friendly stimulant innovation totally acceptable and thoroughly implemented throughout the region of Perhutani. It is very possible to do because Perhutani has research institution that is the Research and Development Center in Cepu that could be involved in the research development innovation of eco-friendly stimulant.
4. Press stimulant prices through eco-friendly innovations that can be accepted for use in Perhutani as a whole. This is a strategy of pressing weaknesses to overcome the potential threats facing (Strategy W - T). The price of eco-friendly stimulant is still expensive in the market so that Perhutani have not implemented the use of it's for overall because of price considerations. The expensive price of eco-friendly stimulant can actually be pressed because Perhutani has the potential and opportunity to be used as described in the strategies mentioned above. Prices will be more pressed again when the innovations of eco-friendly stimulant being developed that will be obtained stimulant that can increase resin productivity, produced by Perhutani, safe for farmers/tappers and eco-friendly. Thus Perhutani will be more confident and receiving the use of eco-friendly stimulant innovation in the entire region.

# 4. Conclusions and Recommendations

### *4.1. Conclusions*

Strategies stimulant use eco-friendly policies could be adopted by Perhutani with 4 strategies as follows:

1. Increase the use of eco-friendly stimulant by using eco-friendly wood vinegar-based produced by Perhutani;
2. Implement eco-friendly wood vinegar stimulant produced by Perhutani to press the price of an expensive eco-friendly stimulant;
3. Increase the use of eco-friendly and safer stimulant innovations to assure Perhutani so that eco-friendly stimulant can be accepted and used as a whole;
4. Press eco-friendly stimulant prices through innovations that can be accepted for use in Perhutani as a whole.

### *4.2. Recommendations*

Based on the above conclusions and the map position of power there, the Perhutani advised as follows:

1. For short term, Perhutani could adopt the use of eco-friendly stimulant innovation with strategies utilizing the existing opportunities with the strength that held(Strategy SO) is "Increase the use of eco-friendly stimulant by using eco-friendly wood vinegar-based that produced by Perhutani ";
2. For long term, Perhutani need to develop innovations to obtain eco-friendly stimulant that can increase pine resin productivity, and press expensive stimulant prices in the market.

## References


[1] Cahyo, Andy S. "Faktor-Faktor yang Mempengaruhi Petani Penyadap Pinus di Kawasan Hutan Dengan Tujuan Khusus (KHDTK) Gombong. Tekno Hutan Tanaman". Pusat Penelitian dan Pengembangan Peningkatan Produktivitas Hutan, 4(2), pp, 49-56, 2011.

[2] Darmadji, P. "Antibakteri Asap Cair dari Limbah Pertanian". Jurnal Agritech, 6(4), pp, 19 – 22, 1996.

[3] David FR. Manajemen Strategis. Terjemahan dari *Strategic Management: Concepts and An-organices*. Sunardi D (penerjemah). Jakarta (ID), Salemba Empat, 2009.

[4] Dubois O. Capacity to manage role changes in forestry. International Institute for Environment and Development (IIED). London (GB). 1998.

[5] FAO. Gum Naval Stores : Turpentine and Rosin from Pine Resin. Roma (IT), 2010.

[6] Handayana, R. Percepatan Alih Teknologi Sistem Usahatani Berbasis Padi Berwawasan Agribisnis (SUTPA) dan Kendala Pengembangannya. Prosiding Dinamika Ekonomi Pedesaan dan Peningkatan Daya Saing Sektor Pertanian. Pusat Penelitian sosial Ekonomi, Badan Litbang Pertanian, Departemen Pertanian. Jakarta, 1998.

[7] Irawan, B dan B. Rachman. Efisiensi Usahatani Padi Sawah Di Jawa Barat. Prosiding Patanas Perubahan Ekonomi Pedesaan Menuju Struktur Ekonomi Berimbang. Pusat Penelitian Sosial Ekonomi Pertanian. Badan Penelitian dan Pengembangan Pertanian. Bogor, 1988.

[8] Iskandar, H., D Kresno., Santoso. Panduan Singkat Cara Pembuatan Arang Kayu Alternatif pemanfaatan limbah kayu oleh masyarakat. Proyek ITTO di Malinau. CIFOR, 2004.

[9] LAN. Isu Aktual Sesuai Tema. Lembaga Administrasi Negara. Jakarta, 2008.

[10] LAN. Teknik Analisis Manajemen. Lembaga Administrasi Negara. Jakarta, 2008.

[11] LIPI. Lembar Data Keselamatan Bahan. Sumber: http://www.kimianet.lipi.go.id. Diakses tanggal 2 Jul 2013, 2004.

[12] Nurrochmat, D.R. Dampak Krisis Ekonomi dan Moneter Terhadap Usaha Kehutanan Masyarakat: Kemenyan di Tapanuli Utara dalam Resiliensi Kehutanan Masyarakat Indonesia.. Yogyakarta (ID), Debut Press, 2001.

[13] Perum Perhutani. "Current Situation of Indonesian Gum Rosin In The World Market". Praque: *Pine Chemicals Association International Conference*, 2010.

[14] Rangkuti, F. Analisis SWOT Teknik Membedah Kasus Bisnis, Reorientasi Konsep Perencanaan Strategis untuk Menghadapi Abad 21. Jakarta (ID), PT. Gramedia Pustaka Umum, 2008.

[15] Santosa, G. Pengaruh Pemberian Etrat terhadap Peningkatan Produktivitas Penyadapan Getah Pinus (Studi Kasus di KPH Sukabumi Perum Perhutani Unit III Jawa Barat dan Banten), Researh report. Fakultas kehutanan IPB, Bogor. unpublished, 2011.

# Connectivity and ecological indicators analysis of tropical forest landscape in Batang Toru watershed - Indonesia

**Samsuri[1], I. Nengah Surati Jaya[2], Cecep Kusmana[3], Kukuh Murtilaksono[4]**

[1]Forestry Department, Sumatra Utara University, Medan, Indonesia
[2]Forest Management Department, Bogor Agricultural University, Bogor, Indonesia
[3]Silviculture Department, Bogor Agricultural University, Bogor, Indonesia
[4]Soil and Land Resources Department, Bogor Agricultural University, Bogor, Indonesia

**Email address:**
gsamsuri@gmail.com (Samsuri), insjaya@cbn.net.id (I. N. S. Jaya), ckusmana@ipb.ac.id (C. Kusmana), kmurtilaksono@ipb.ac.id (K. Murtilaksono)

**Abstract:** Connectivity is one of the important issues in the context of natural resources due to its potential in preventing the impact of habitat fragmentation. Landscape forest connectivity facilitates organism movement, genetic exchange, and other ecological material flows. Loss of connectivity may result declining of ecosystem production and cut the material flows within the forest ecosystems. Connectivity degree is needed to determine the management strategy of forest landscape as a wildlife habitat. This paper defines connectivity index of forest landscape in Batang Toru watershed, and describes correlation between connectivity with ecological indicators, biophysical and anthropogenic factors. Landsat satellite imageries acquired in 1989, 2001 and 2013 were used to detect land cover in several different years. Fragstat was used to generate landscape metrics. Landscape metrics were analyzed using a scoring method to determine the connectivity index of forest landscape. A Pearson correlation analysis was performed to obtain a correlation between connectivity index and the distance from roads, the distance from rivers, elevation and slope. The study found that the landscape connectivity tend to decline over the period from 1989 to 2013. The lowest connectivity index was found in the downstream area of Batang Toru watershed. Areas with low connectivity index were identified as having a relatively low diversity index of tree species. The connectivity index of forest landscape has a positive correlation with the distance from roads and the distance from rivers.

**Keywords:** Fragmentation, Species Diversity Index, Fragstat, Ecosystem

## 1. Introduction

Connectivity has become a primary issue in various studies due to its potential in mitigating the impacts of habitat fragmentation [1, 3]. Improving ecosystem connectivity is one of the main objectives of forest landscape management [36] besides maintaining the stability and integrity of natural ecosystems [10]. As was shown in the study [41], the establishment of certain bird community was supported by the change in the surrounding landscape, not in the site where the animals live. Forest landscape connectivity can be evaluated and improved through reforestation around the forest [18, 19].

The measures of landscape connectivity were influenced by different aspects of landscape structure [20] and landscape connectivity is a poorly defined concept and the same landscape may have different connectivity values when different measures of landscape connectivity were used. Nevertheless there are two general predictions that are able to explain landscape connectivity measures, i.e. (1) a significant increase in inter-patches distance decreased landscape connectivity, and (2) the effect of constituent elements of landscape connectivity was smaller than the effect of habitat elements. Landscape connectivity is as a degree of spatial connectivity among landscape elements such as patches, corridors, and matrix [16, 17]. Patch connectivity emphasized on a number as well as a series of habitat patches and the Euclidean distance or effective distance between the patches [4]. Corridor connectivity indicated a linear connection and its distribution can be improved through connectivity restoration [9, 21]. Connectivity matrix can be used to evaluate overall landscape mosaic, including landscape matrix to maintain maximum landscape continuity of non-built areas [27]. Thus, landscape mosaic is important as a whole, not only as landscape counterparts [1].

Connectivity is related to the functional connectivity within the landscape and not merely a physical connection among landscape elements. Connectivity is actually more than just a physical connection but also include a resistance to movement due to barrier or land use type. In general, landscape connectivity emphasized not only the spatial characteristics but also the ecological processes and the organism movement (functional connectivity). There is a tendency to focus how landscape structure was spatially and simply managed through its mapping, than to see what the landscape is and how will it be. However the challenge is the lack of understanding of ecology during planning, while the ecological effect within the network is quite real. This can be resolved by linking the scientific characteristics of functional connectivity when designing sustainable landscape planning.

Some studies and literatures only emphasized natural landscape study at local scale. However if we take a look at its relation as well as its effects on existing lives inside it, there is connection that geographically cover a much larger area, for instance at regional scale or a watershed. The restoration of habitat connectivity is the application of landscape ecology concept and metrics. Connectivity is extremely important and is a tangible characteristic of landscape. This is a parameter of landscape functions and is a major issue in assessing as well as planning biodiversity conservation. A well-understood fact is that connectivity is important for the disturbance on plants and animals in a fragmented landscape [27, 28].

Connectivity is fundamental to spatial concept that supports some land-use planning and conservation strategy [42]. Connectivity metrics can be applied to model ecological processes, e.g. to obtain average isolation and predict relative connectivity of habitat islands [22, 37]. Connectivity metrics are based on network theory [27, 28]. Connectivity can be improved through landscape restoration. Restoration can be considered to speed up the succession. A restoration decision making can be made through landscape modeling [44] as well as landscape connectivity approach. In some cases, most of the analysis methods of connectivity metrics were supported by spatial data [38]. Sites in areas of high landscape connectivity level will be given a priority in conducting restoration activity. These can be identified through the application of landscape ecology principles focusing on population dynamics, to provide information on each stage of restoration decision-making process [32]. Nevertheless, it is practically difficult to assess which landscape ecosystem that deserves to be main priority for restoration. However this can be solved if the potential ecological characteristics in the context of its ecosystem structure and function can be recognized during ecological restoration [7, 21]. Thus the purpose of this study is to obtain landscape connectivity indices and its correlation with forest landscape ecology indicators, biophysical and anthropogenics factor.

## 2. Material and Methods

### *2.1. Study Area*

The study was conducted in Batang Toru watershed which consists of Puli, Sarula and Batang Toru Hilir sub-watersheds, Sumatera Utara Province-Indonesia. The study area was located between 1° 10' 36.6" – 1° 10' 36.47" N and 98° 23' 48.22" - 98° 49' 15" E (Fig. 1). The study was conducted from September 2013 to January 2014. The study area covered an area of 202,277 ha at elevation of around 0 to 2,000 m above the sea. Based on its topography, the study area consists of flat area (0% - 1 %) of around 75.66% of the total area and a quite steep to steep area (> 15%) of around 24.34% of the total area. Oldeman's climate classification type divides the study area into 3 climate types, namely A (Tapanuli Utara), D1 (Tapanuli Selatan) and A (Tapanuli Tengah). Based on the Decree of the Indonesia Minister of Forestry No. 44 Year 2005, the study area consists of Conservation Forest (0.06%), Nature Reserve Forest (6.27%), Production Forest (39.10%), Protection Forest (2.93%), Limited Production Forest (12.55%) and other uses (39.08%). Nature reserve forest, protection forest and conservation forest were managed by the Ministry of Forestry and were important sites for the diversity in the Sumatra Island. It is also a habitat of Sumatera orang utan (*Pongo abelii*). Batang Toru forest has biodiversity richness as a home of thousand species of flora and fauna such as 67 mammals species, 287 birds species, 110 types of herpetofauna, and 688 plant species [33, 34]

### *2.2. Materials*

The study used satellite imagery, i.e. Landsat TM 1989, Landsat ETM 2001 and Landsat 8 OLI 2013 path/row 128/059, contour map, road network map, and river map. In addition, vegetation data from field measurements was used as supporting data. Data on vegetation was measured on a plot that has a size of 50 m x 50 m which was divided into 4 quadrants that has a size of 25 m x 25 m respectively. Field survey activity used GPS, haga hypsometre, phi band, compass, and talley sheet. While data analysis tools consisted of Erdas Imagine 9.1 that was used to interpret satellite imagery, Arc GIS to perform spatial data analysis, Fragstat 3.3 to create landscape metrics, Excel and SPSS vers 16 to perform statistical data analysis.

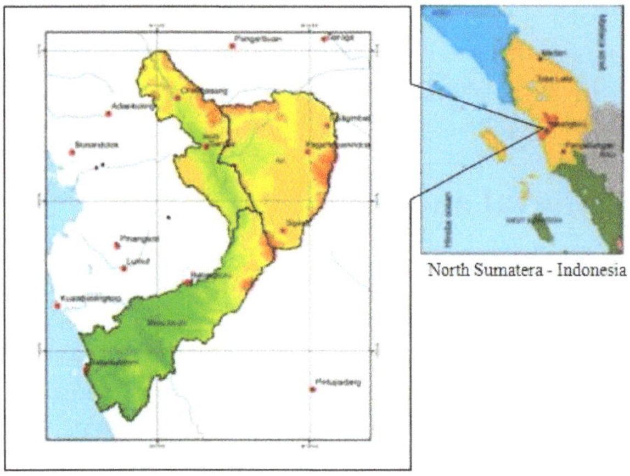


***Figure 1.*** *Study area*

### *2.3. Data Analysis*

For the interpretation of satellite imagery, a supervised classification method was used. The result of satellite imagery interpretation was examined for its accuracy using *Overall accuracy* and *Kappa accuracy* [23]. Accuracy test was performed using 140 field check-points to test the accuracy of 2013 landsat image classification. Land cover map data of 1990 and 2001 published by the Ministry of Forestry of Republic of Indonesia were used to validate accuracy of image classification of 1989 and 2001.

Forest landcover type was analyzed using Fragstat 3.3, to obtain forest landscape metrics [29, 30]. Then each forest landscape metrics was classified into 5 (five) classes and was scored using Likert scale. Landscape metrics for determining the connectivity of forest landscape was the interconnectedness between forest patches (*connectance*) as well as the extent and compactness of forest patches (*radius of gyration*) (Table 1). The scores were summed algebraically. Equation 1 is used to convert the total score into 0 – 1 value [24]. Meanwhile the vegetation measurement data was analyzed to obtain species diversity index value, basal area factor and stand density [25, 26].

$$Ind\_FLC = \frac{(score_{total} - score_{min})}{(score_{tot-max} - score_{tot-min})} (ind\_FLC_{max} - ind\_FLC_{min}) \quad (1)$$

Notes :

| | | |
|---|---|---|
| Ind_FLC | = | index value of forest landscape connectivity |
| $score_{total}$ | = | total score as input |
| $score_{tot-min}$ | = | minimum value of total score |
| $score_{tot-max}$ | = | maximum value of total score |
| $ind\_FLC_{max}$ | = | maximum index of forest landscape connectivity (converted value) |
| $ind\_FLC_{min}$ | = | minimum index of forest landscape connectivity (converted value) |

The result of connectivity index calculation was further classified into 5 (five) forest landscape connectivity degree (FLC). Each class used the same value range of 0.2 so that the five classes are as follows: very low FLC (0 – 0.2), low FLC (0.21 – 0.40), moderate FLC (0.41 – 0.60), high FLC (0.61 – 0.80) and very high FLC (0.81 – 1.00). In order to obtain a correlation pattern between FLC and the ecological indicators, trend analysis correlation between diversity indices and forest patch size with FLC value was conducted.

***Table 1.*** *Landscape metrics used in the connectivity analysis of forest landscape.*

| Metric | Code | Description | Value | Score |
|---|---|---|---|---|
| Connectance | CONN | Connectance is defined on the number of functional joining between patches of the same type, where each pair of patches is either connected or not based on a user-specified distance criterion. Connectance is reported as a percentage of the maximum possible connectance given the number of patches. | <20 | 1 |
| | | | 20 – 40 | 2 |
| | | | 40 – 60 | 3 |
| | | | 60 – 80 | 4 |
| | | | >80 | 5 |
| Radius of gyration | GYRATE | Radius of gyration is a measure of patch extent; thus it is effected by both patch size and patch compaction. Note that the choice of the 4-neighbor or 8-neighbor rule for delineating patches will have an impact on this metric. | <200 | 1 |
| | | | 200-400 | 2 |
| | | | 400-600 | 3 |
| | | | 600 – 800 | 4 |
| | | | >800 | 5 |

## 3. Results and Discussions

### *3.1. Landscape Metrics of Batang Toru Watershed*

Fragstat analysis generated value of connectivity parameter, i.e. connectan index of landscape, and radius of gyration. It can explain connectivity of forests landscape [11]. Forest connectivity decrease during period of 1989 - 2013. The decrease was depicted by the landscape metric value of *radius of gyration* that decrease from the range of 400-700 m in 1989 to around 50 – 200 in 2013 (Fig. 2-a). The *connectance* value also tend to decrease from around 3 – 9 % in 1989 to around 0-2 % in 2013 (Fig. 2-b).

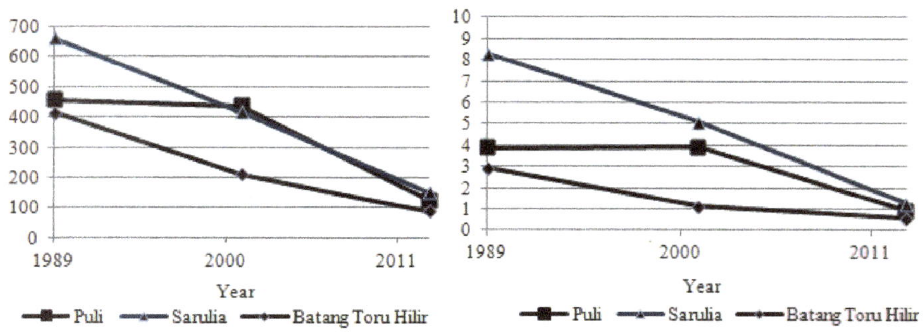


(a) (b)

***Figure 2.*** *Landscape metric values of radius of gyration (a) and connectan (b) during 1989 – 2013*

### *3.2. Connectivity Index of Forest Landscape and Ecological Indicators*

Connectivity index was developed based on landscape metrics of c*onnectan* and *radius of gyration index.* Forest landscape connectivity map show that most of the forests in sub-watershed of Puli have high (index value of around 0.6-0.8) and very high connectivity (connectivity index > 0.80). High forest landscape connectivity (0.6-0.8) was found distributed along Sumatra road that through Sipirok and Sarula city. If the forest landscapes have been restored, it will connect forests around them. The connectivity will enhance the functional relationship of forest ecosystem. High connectivity degree was usually found in more compact forest, while low connectivity was found in the fragmented forests.

Connectivity forest landscape index was mapped as illustrated in Fig. 3. Connectivity index map of 1989 show distribution of connectivity degree of forest patch of Batang Toru watershed. High connectivity index was found mostly in sub-watershed of Sarula while very low connectivity index was found in sub-watershed of Batang Toru Hilir. Low connectivity index was found in small size forest patches. However, connectivity index map of 2013 show that most of Puli sub watershed area have medium connectivity degree, while high connectivity was found in forest patches in sub-watershed of Sarula.

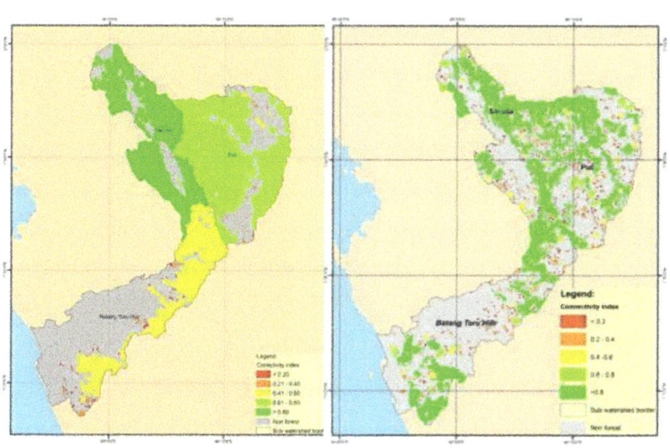


***Figure 3.*** *Forest connectivity index in 1989 (a) and 2013 (b) in Batang Toru watershed*

Table 2 show that sub-watershed of Sarula has a relatively large area of very high connectiviy degree compared to sub-watersheds Puli and Batang Toru Hilir. While sub-watersheds of Batang Toru Hilir and Puli have connectivity degree about 2,490 ha and 2,533 ha respectively. Due the location of sub-watershed of Batang Toru Hilir which is in the downstream part of watershed, the distance between the remaining forests are getting farther. It is similar to sub-watershed Puli which is located in the upstream and middle parts of Batang Toru watershed.

***Table 2.*** *Total area distribution of forest connectivity degree in each sub-watershed in 2013*

| Degree of connectivity | Total area (ha) | | | Total |
|---|---|---|---|---|
| | Batang Toru Hilir | Puli | Sarulla | |
| Highest connectivity | 2 | 4 | 16,646 | 16,655 |
| High connectivity | 17,306 | 2 | 1,560 | 18,878 |
| Moderate connectivity | 857 | 21,767 | 818 | 23,445 |
| Low connectivity | 2,490 | 2,533 | 0 | 5,024 |
| Lowest connectivity | 62,096 | 50,438 | 26,699 | 139,276 |
| **Total** | 82,751 | 74,743 | 45,724 | 203,277 |

Correlation analysis between ecological indicators and connectivity index shows correlation between connectivity index and ecological responses, that is species diversity index of Shannon-Whinner (H'). The relationship follow polynomial model. The equation was $y = 5.437x^2 + 6.967x - 0.217$ with a determination coefficient value of 64.4 %. The $y$ was species diversity index value and $x$ was forest connectivity index of each patch. Fig. 4 ilustrated that the species diversity index value increase while the connectivity index increase close to 0.7. High connectivity will enhance the material flow and change the biological processes [12], reduce the threat of extinction and provide protection from interferences [35]. The connectivity index describes the existing condition of forest landscape. It can't ilustrate clearly the effect of connectivity index on species diversity in certain forest landscape. Landscape history is needed to more explain the correlation between species diversity and forest landscape connectivity [28].

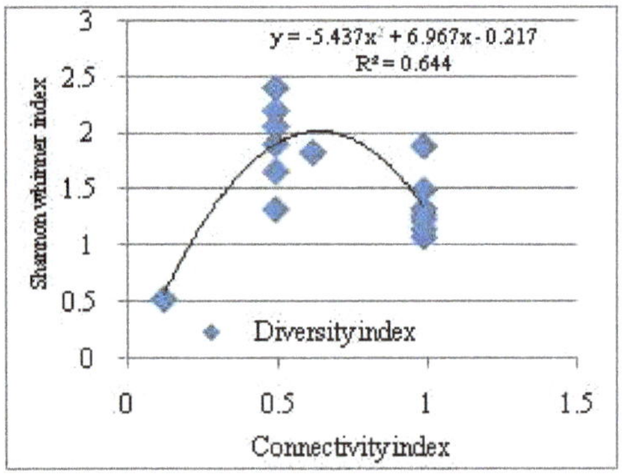


*Figure 4. Correlation between forest connectivity index and diversity index*

The loss of natural connectivity of an ecosystem is the biggest threat in the distribution of wildlife and the living capability, and biodiversity conservation in general. This required more serious attention especially on improving of connectivity in habitat conservation and landscape. The maintenance and restoration of landscape connectivity has become a central issue in ecology and biodiversity conservation because landscape connectivity facilitates organism movement, genetic exchange, and the flow of other ecological materials [8]. Organism movement that is considered in determining forest landscape connectivity of Batang Toru watershed is orangutan as its endemic species. The need for habitat among endemic flora and fauna is a central key in biodiversity conservation as well as the stability and integrity of natural ecosystem [6, 39, 40]. Thus it is highly important to consider connectivity as a basis in conservation planning and landscape change analysis.

### 3.3. Forest Landscape Connectivity Index and Accessibility

Human factor and biophysical condition affect the forest landscape connectivity index. Human activities are generally supported by the presence of infrastructure to access the forest, i.e. road and river networks, as well as field biophysical condition that is slope and elevation. Pearson correlation can explain the correlation level forest landscape connectivity with the distance from road, distance from river, elevation and slope (Table 3).

*Table 3. Pearson correlation between connectivity index and species diversity index, distance from road, distance from river, elevation, and slope*

| Variable | Sub watershed | Pearson correlation | Significant level |
|---|---|---|---|
| Shannon-winner index | Batang Toru Hilir | 0.240 | ns |
| | Puli | 0.240 | ns |
| | Sarula | 0.240 | ns |
| Distance from road | Batang Toru Hilir | 0.863** | 0.010 |
| | Puli | 0.868** | 0.010 |
| | Sarula | 0.690* | 0.050 |
| Distance from river | Batang Toru Hilir | 0.934** | 0.010 |
| | Puli | 0.957* | 0.050 |
| | Sarula | 0.944* | 0.050 |
| Elevation | Batang Toru Hilir | -0.563* | 0.050 |
| | Puli | 0.539 | ns |
| | Sarula | 0.749** | 0.010 |
| Slope | Batang Toru Hilir | -0.551* | 0.050 |
| | Puli | 0.540 | ns |
| | Sarula | 0.749** | 0.010 |

The high correlation between connectivity index and the distance from road was found in Puli sub-watershed while the low correlation was found in Sarula sub watershed. However, the high correlation with the distance from river was found in Puli sub-watershed, the lowest correlation is in Batang Toru Hilir sub-watershed. Forest near the road and river has high distubance to the forest. The farther away from the road and river, the connectivity tends to be higher (Fig. 5 and Fig. 6). This is triggered by the river and road that has become an indirect cause of forest damage [39]. The road will attract human to change land use and land cover. Human will change the forest into cultivation land and gather forest products [43], thus trigger land use change. The high of deforestation degree is trigger by closeness of connection between forest and human settlement [31]. Its distribution and pattern follow the road, since the road trigger settlement establishment. Land forest clearing for agriculture was conducted near road and has left only a small part of forest among cleared land [2, 13, 15]. Further, the road increase connectivity among settlement centre that could threat forests sustainability [14, 31].

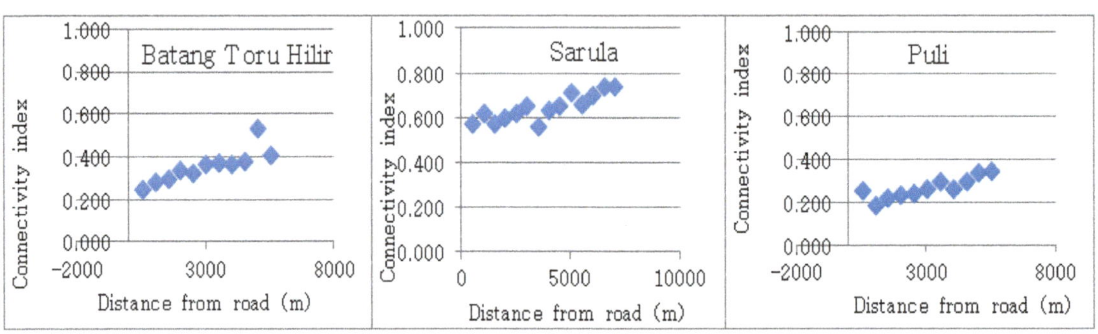


*Figure 5. Connectivity index and distance from main road*

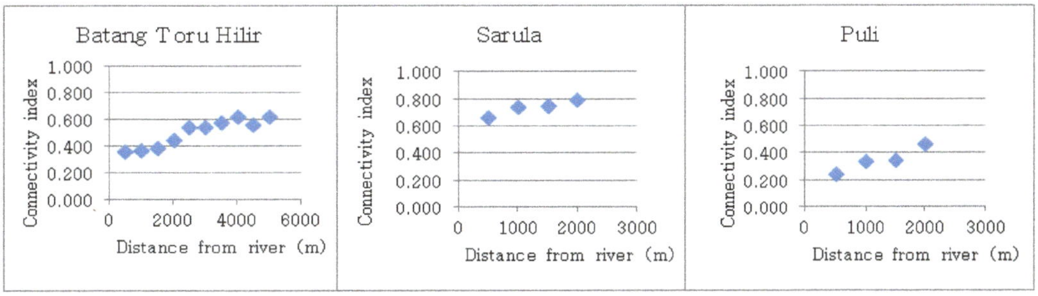


***Figure 6.*** *Connectivity index and distance from river*

In Sarula sub watershed, connectivity index of forest landscape raise with the increasing of elevation (Fig. 7). Forest in steep and difficult accessibility area have less human disturbance [5]. A contrast situation was found out in Batang Toru Hilir sub-watershed that was the higher its elevation the lower its connectivity index. The remaining forests in Batang Toru Hilir sub-watershed have high elevation and were disturbed, while forests in low elevation were more maintenaced and monitored. Slope has no correlation with the connectivity index of forest landscape in Puli sub-watershed. A significant positive correlation between connectivity index of forest landscape and slope was found in Sarula sub-watershed. The higher of slope, the higher of connectivity index of forest landscape. Meanwhile, a negative correlation was found in Batang Toru Hilir sub-watershed that is the higher of slope, the lower of the connectivity index of forest landscape (Fig. 8). Forest landscape as a wildlife habitat should have high connectivity. Connectivity will be obtained from a relatively large and compact forest ecosystem. Effective conservation and restoration strategy would reinstate the forest ecosystem function [39].

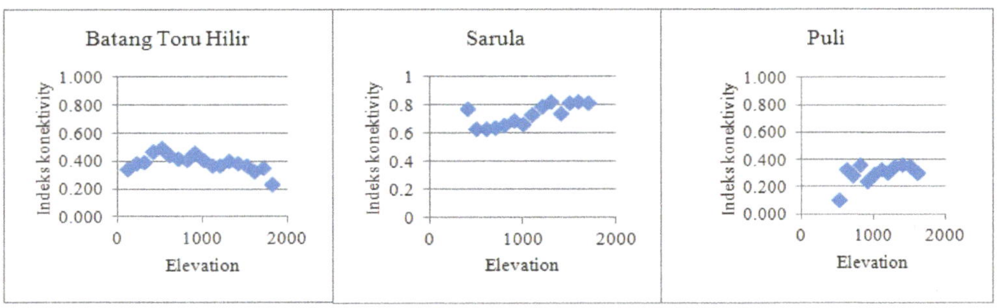


***Figure 7.*** *Connectivity index and elevation*

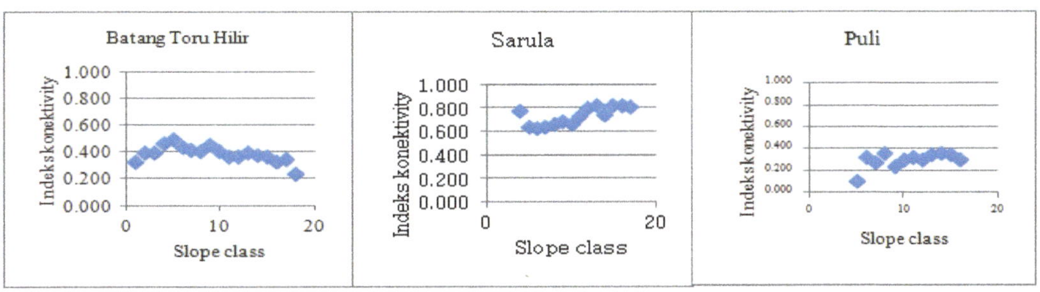


***Figure 8.*** *Connectivity index and slope class*

# 4. Conclusion

The degree of forest landscape connectivity at Batang Toru watershed tend to decrease during 1989-2013. Sub watershed of Batang Toru Hilir has the low degree of forest connectivity. Ecological indicator i.e diversity index could be indicated by the degree of forest landscape connectivity. The accessibility to the forest increases the disturbance to the forest so that it will reduce the connectivity degree of forest landscape. However, as the wildlife habitat, Batang Toru forest landscape connectivity could be maintenance and enhanced by restoring or rehabiliting of fragmented forest.

# Acknowledgement

This research is part of PhD thesis of the first author funded by SEAMEO-BIOTROP DIPA 2013. The authors extend deeply acknowledgment to the Ministry of Education and Culture – Republics Indonesia for scholarship and support to accomplish this paper. Sincerely appreciation is also extended to anonymous reviewer for correction and comments.

## References


[1] Anderson B, and C.N. Jenkins, Applying nature's design: corridors as a strategy for biodiversity conservation. Columbia University Press, New York USA, 2006.

[2] Arima, E.Y., R.T. Walker, S.G. Perz and M. Caldas, "Looger and forest fragmentation behaviour model of road building in the Amazon Basin". Annal of the Association of American Geography, 95, pp, 525 – 541, 2005.

[3] Batistella, M., S. Eduardo, Brondizio, F. Emilio and Moran, "Comparative analysis of landscape fragmentation in Rondônia, Brazilian Amazon". International Archives of Photogrammetry and Remote Sensing, 33, pp, 148-155, 2000.

[4] Broquet, T., N. Ray, E. Petit, H.M. Fryxell and F. Burel, "Genetic isolation by distance and landscape connectivity in the American marten (Martes americana)". Landscape Ecology, 21, pp: 877-889, 2006.

[5] Cabral, D. C., S.R. Freitas and J.T. Fiszon, "Combining sensors in landscape ecology: imagery based and farm level analysis in study human driven forest fragmentation". Sociedade & Natureza, 19, pp, 69-87, 2007.

[6] Collinge, S.K. "Effects of grassland fragmentation on insect species loss, colonization, and movement patterns". Ecology, 81, pp, 2211–2226, 2000.

[7] Cortina J. "Ecosystem structure, function, and restoration success: are they related". Journal for Nature Conservation, Elsevier, 14, pp, 152 – 160, 2006.

[8] Crooks, K.R. and M. Sanjayan (Eds.), Connectivity conservation. Cambridge University Press, New York. Clergeau and Burel 1997, 2006.

[9] Davies, Z. and A. Pullin, "Are hedgerows effective corridors between fragments of woodland habitat? An evidence-based approach". Landscape Ecology, 22, pp: 333-351, 2007.

[10] Decout, S., S. Manel, C. Miaud, and S. Luque, "Connectivity and landscape patterns in human dominated landscape: a case study with the common frog Rana temporaria", 2010. www.symposcience.org [2 December 2011]

[11] Fahrig, L. "Effect of habitat fragmentation on biodiversity". Annual review of Ecology, Evolution, and Systematics, 34(1), pp, 487-515, 2003

[12] Fahrig, L. and G. Merriam, "Habitat patch connectivity and population survival" Ecology, 66, pp: 1762-1768, 1985.

[13] Fearnside, P.M. "Brazil's Cuiaba – Santarem (BR-163) Higway". The Environmental Management, 39, pp, 601-614, 2007

[14] Fearnside, P.M. "Will urbanization cause deforested area to be abandoned in Brazilian Amazon?" Environmental Conservation, 35, pp: 197 – 199, 2008.

[15] Feraz,, S.F.B., C.A. Vettorazzi, D.M. Theobald, and M.V.R. Ballester, "Landscape dynamic of Amazonia deforestation between 1984 and 2002 in Central Rendonia Brazil: Assessment and Future Scenario". Forest Ecology and Management, 204, pp, 67-83, 2005.

[16] Forman, R.T.T. and M. Godron, Landscape Ecology. New York, Wiley, 1986

[17] Forman, R.T.T. Land Mosaics: The ecology of landscapes and regions (2nd Ed.), Cambridge, Cambridge University Press, 1995.

[18] Garcia, S. "Estimating landscape fragmentation form satellite images; the high of sensor spatial resolution. Remote Sensing for Agriculture". Ecosystem Hydrology, Proc. SPIE 5232, pp 668-675, 2004.

[19] García-Feceda, C., S. Saura and R. Elena-Rosselló́a, "Improving landscape connectivity in forest districts: A two-stage process for prioritizing agricultural patches for reforestation". Forest Ecology and Management, 261, pp. 154–161, 2011.

[20] Goodwin, B.J. and L. Fahrig, "How does landscape structure influence landscape connectivity?" OIKOS 99, pp, 552–570, 2002.

[21] Graves, T., S. Farley, M. Goldstein and C. Serheen, "Identification of functional corridors with movement characteristics of brown bears on the Kenai Peninsula, Alaska". Landscape Ecology, 22, pp, 765-772, 2007.

[22] Gustafson, E.J. "Quantifying landscape spatial pattern: what is the state of the art". Ecosystems, 1, pp, 143-156, 1998.

[23] Jaya, I.N.S. Analysis citra dijital: perspektif penginderaan jauh untuk pengelolaan sumberdaya alam. Bogor (ID), Institut Pertanian Bogor, 2009.

[24] Jaya, I.N.S., R. Boer and Samsuri. Developing fire risk index in Central Kalimantan. International Research Institute and Bogor Agricultural University. A Project Report, 2007.

[25] Kreb, C.J. Ecological Methods. Columbia (US), Harper Collins Publisher, 1989.

[26] Kusmana, C. Metode Survey Vegetasi. Bogor (ID), PT. Penerbit Institut Pertanian Bogor, 1997.

[27] Levin, N. and H. Lahav, "Landscape continuity analysis: A new approach to conservation planning in Israel". Landscape and Urban Planning, 79, pp, 53-64, 2007.

[28] Lindborg, R. and O.Eriksson, "Historical landscape connectivity affects present plant Species diversity". Ecology, 85(7), pp, 1840–1845, 2004.

[29] McGarigal K. "Landscape Metrics", 1995. http://www.umass.edu/landeco/pubs/mcgarigal.marks.1995. [7 September 2011]

[30] McGarigal, K. and B.J. Marks, B. J. "FRAGSTATS: spatial pattern analysis program for quantifying landscape structure". Gen. Tech. Rep. PNW-GTR-351, 1995

[31] Nagendra, H., J. Southworth and C. Tucker, "Accessibility as a determinant of landscape transformation in Westeren Honduras". Landscape and Urban Planning, 83, pp, 154 – 167, 2003.

[32] Nikolakaki, P. "A GIS Site Selection process for habitat creation: estimating connectivity of habitat patches". Landscape and Urban Planning, 68, 77-94, 2004.

[33] Perbatakusuma, E.A. and F. Kaprawi, "Kajian spasial lahan kritis berbasis sistim informasi geografis untuk rehabilitasi kawasan koridor satwa liar dan harangan desa di kawasan hutan Batang Toru" Report Project, Konsorsium Ikon Koridor to Sigadis, 2011a

[34] Perbatakusuma, E.A. and A. Damanik, "Pengelolaan sumber daya alam di konsesi usaha perusahaan swasta : penekanan pada kawasan bernilai konservasi tinggi di lansekap hutan Batang Toru – Taman Nasional Batang Gadis". Report Project, Konsorsium Ikon Koridor to Sigadis, 2011b

[35] Saunders, D.A., R.J. Hobb and C.R. Marques, "Biological consequences of ecosystem fragmentation, a review". Conservation Biology, (5), pp, 18-32, 1991.

[36] Saura, S., P. Vogt, J. Velázquezc, A. Hernandoa and R. Tejeraa, "Key structural forest connectors can be identified by combining landscape spatial pattern and network analyses". Forest Ecology and Management, 262, pp, 150–160, 2011.

[37] Schumaker, N. "Using landscape indices to predicts habitat connectivity". Ecology, 77(4), pp, 1210-1225, 1996.

[38] Selman, P. Planning at the landscape scale. New York, NY, Routledge, 2006.

[39] Simone, R.F., T.J. Hawbaker, P.M. Jean. "Effects of roads, topography, and land use on forest cover dynamics in the Brazilian Atlantic". Forest Ecology and Management, 259 (3),pp, 410–417, 2010

[40] Taylor, P.D., L. Fahrig, K. Henein and N.G. Merriam, "Connectivity is a vital element of landscape", 1993.

[41] Tena, A.G., L. Brotons and S. Saura, "Effects of forest landscape change and management on the range expansion of forest bird species in the Mediterranean region". Forest Ecology and Management, 259, pp, 1338–1346, 2010.

[42] Van Lier, H.N. "Sustainable land use planning. An editorial commentary". Landscape Urban Planning, 41, pp, 79-82, 1998.

[43] Verburg, P.H., W. Soephoer, A. Veldkamp, R. Limpiada and V. Espaldon, "Modeling the spatial dynamics of regional landuse the CLUE'S". Model Environmental Management, 30, pp, 391-405, 2002.

[44] Xi, W.M. "Landscape modeling for forest restoration planning and assessment: lessons from the Southern Appalachian Mountains". Journal of Forestry, 106 (4), pp, 191-197, 2008.

# Spatiotemporal distribution of peatland fires in Kapuas District, Central Kalimantan Province, Indonesia

**Achmad Siddik Thoha[1], Bambang Hero Saharjo[2], Rizaldi Boer[3], Muhammad Ardiansyah[4]**

[1]Department of Forestry Science, Faculty of Agriculture, University of North Sumatra, Medan, Indonesia
[2]Department of Silviculture, Faculty of Forestry, Bogor Agricultural University, Bogor, Indonesia
[3]Department of Geophysics and Meteorology, Faculty of Mathematics and Natural Sciences, Bogor Agricultural University, Bogor, Indonesia
[4]Department of Soil Science and Land Resources, Faculty of Agriculture, Bogor Agricultural University, Bogor, Indonesia

## Email address:

siddikthoha@gmail.com (A. S. Thoha), bhsaharjo@gmail.com (B. H. Saharjo), rizaldiboer@gmail.com (R. Boer), ardyaja@gmail.com (M. Ardiansyah)

---

**Abstract:** Peatland fires occur almost every year in Indonesia and their impact is harmful to human life and the environment. The objective of this study was to determine the distribution of peat fires, including hotspot characteristic, fire period and spatial distribution according to biophysical factors and human activities. Characteristics of peatland fires were determined through spatial analysis between hotspots and rainfall and factors affected peatland fires such as land cover, distance from the river, distance from road, spatial plan and peat thickness. The study found that strong indication of peat fires in Kapuas is indicated by the hotspot with a confidence value of more than 50%. August to October is the hotspot increasing time in Kapuas because the rainfall is low. The densest hotspots are generally found in swamp shrubs, close to the river, close to the roads, in the very deep peat, with a spatial planning of peatland conservation area. We concluded that peatland conservation area and unmanaged land in Kapuas District were in vulnerable condition of the fire. We suggest to the government to improve unmanaged land utilization and protect peatland conservation area in Kapuas District.

**Keywords**: Hotspot, Fire Period, Spatial Analysis, Fire Management

---

## 1. Introduction

Peatland fires occur almost every year in Indonesia and their impact is harmful to human life and the environment. A peatland fire in 1997/1998, when the El Nino occurred in Indonesia, contributed 13-40% emissions of global carbon emissions [1,2,3]. In Kapuas district, there are hundred thousand hectares of peatland located in the former of a million hectares of Peatland Development Project, the famous one was Mega Rice Project (MRP), which was proposed since 1995. The MRP opens the peatlands massively including swamp forests with thick peat with the construction of thousand kilometers canals that have an impact on the drying of peat ecosystem. The drying of peat ecosystem makes the area very prone to fire which results in massive carbon emissions [4]. It has been shown by [5] that carbon emissions from fires will be even greater when the peatland is burnt. Fires from tropical peatlands produce carbon dioxide emissions 4-40 times greater than other tropical habitats.

The factors causing peatland fires of various studies are related to climate and human activity. As in the study by [6] found that there is a close relationship between the rainfall data from the satellite to fire hotspot activity in Central Kalimantan. Rainfall anomaly during the dry season in June to October is a critical moment in determining the activity of fire. Human activity factors, for example distance from the village, distance from roads, distance from the canal or river networks contributed the risk of peatland fires while land cover affected the fire in 48% [7,8].

The efforts of forest fire management aimed at reducing the incidence of fires that can affect the lives and livelihoods. Peatland fires characteristic such as the fire period, relationship with the weather factors and the fire distribution indication according to biophysical factors and human activities are needed as suggestions for the development of early warning systems of forest and land fires in certain areas. The objective of this study was to identify the distribution of peatland fires, including hotspot characteristic, periods of fire and the spatial distribution according to biophysical factors and human activities. The results of this study are expected to evaluate the characteristics of peat fires that occur in Kapuas, Central Kalimantan Province by both time and location distribution.

## 2. Material and Methods

### *2.1. Study Area*

The study was conducted in the Kapuas District of Central Kalimantan Province (Fig. 1) from April to December 2012. Data collection and ground check location of peat fires conducted in 15 villages in 6 sub districts in Kapuas. Six sub districts are namely Mantangai, Dadahup, Basarang, Kapuas Timur, Kapuas Murung, and Kapuas Barat. Based on the results of the fire occurrence identification, Kapuas district is one of the areas which is monitored as the place with the frequent forest and land fires, mainly in peatlands. As reported by the report [9] and [10] it was mentioned that the fires in 2002, 2007 and 2011 in the Ex-MRP area interfere human activities and exacerbate the peatland ecosystem damage. The research [7] also mentioned that the Kapuas district is an area that belongs to a risk class or a very high flammability class (extremely risk) in Central Kalimantan because most of the million hectare areas of ex-MRP were in Kapuas.

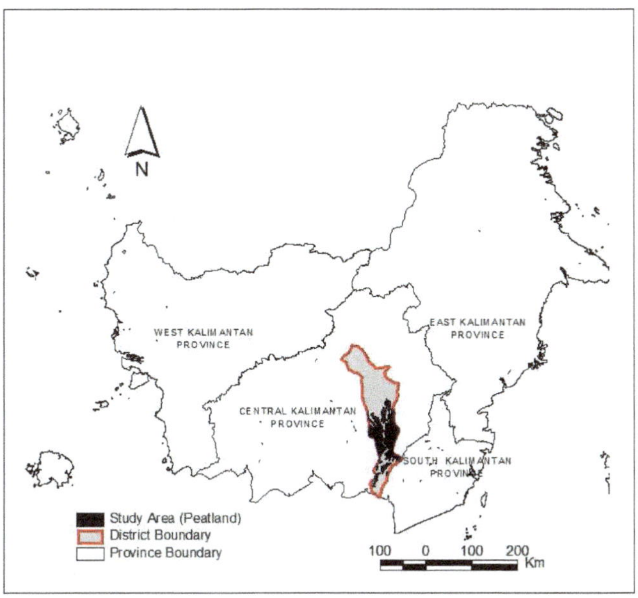


*Figure 1. The map of study area*

### *2.2. Data Collection*

The hotspot data from the Terra/ Aqua satellite by MODIS (Moderate - resolution Imaging Spectroradiometer) sensor in 2001-2012 was obtained from the Fire Information for Resource Management System (FIRMS), which can be accessed free of charge at the link https://earthdata.nasa.gov/data/near-real-time-data/firms/active-fire-data. The map of the district and sub-district administrative boundaries obtained from the Regional Development Planning Agency (Bappeda) of Kapuas. The land cover map, the river network map, the road network map, the spatial plan map of the province and the peat thickness map are collected from the Ministry of Forestry that is used for the analysis of spatial and temporal distribution of fire activity. Hotspot is also used as the basis for determining the location of ground check. Monthly rainfall data of 2001-2011 year sourced from the station of rainfall originating from the Department of Horticultural Agriculture and Food Crops is used to analyze the relationships between hotspots and rainfall.

The main software for doing this study was ArcView 3.3, Microsoft Excel and Minitab. All of analysis uses a personal computer (PC). To obtain the coordinates of the location of the fire in the field, the researcher uses the Global Positioning System (GPS).

### *2.3. Data Analysis*

Determination of hotspots that have a strong indication as the peat fire location is by taking hotspot with confidence values more than 50, more than 70 and more than 90. The hotspots with those confidence values are compared with the forest fire spots and the actual land (2012) from the ground check results. A confidence value is used to separate the weak to the strong indication of fire. The confidence value ranges from 0-100 %. The use of confidence values in separating hotspots and real fire spot will be different in different locations. Comparing hotspots to different confidence values with the results of a ground check of the fire will find the hotspot characteristics which are close to the real fire incident in the field.

The determination of the fire temporal pattern is performed with spatial analysis of monthly hotspots which are in the peatland areas. The hotspot data in peatland areas are separated by months and years of time and through geoprocessing process that is clipping and spatial joint. Monthly hotspots are tabulated and described to see the trend patterns of the fire season. Temporal hotspot analysis is done to see the indication of land and forest fires from time to time and its relationship with rainfall.

The analysis of the indication distribution of the fire location based on the hotspot distribution in the peatland areas based on biophysical and human activities factor which is being overlaid with maps are; land cover, distance from the river, distance from the road, the peat thickness. Each map of factors affecting peatland fires were classified in several classes as shown in Table 1 and Fig.2.

*Table 1. Factors classes used in spatial analysis*

| Factors | Class |
|---|---|
| Land cover | Secondary dry land forest, Shrubs, Dry land agriculture, Water body, Paddy field, Secondary swamp forest, Open land, Swamp shrubs |
| Distance from river | Buffer with interval 1000 m (1km) |
| Distance from road | Buffer with interval 1000 m (1km) |
| Spatial land use plan *) | Limited production forest (HPT), Production Forest (HP), Transmigration (T1), Area for settlement (KPPL), Agriculture land "Handil" (KHR), Water bodies (DS), Area for agricultural and plantation production development (KPP), Area for conservation of hydrology (KH), Area for conservation of flora and fauna (KFF), Deep peat conservation (KGT), Black water conservation (KEAH) |
| Peat depth | Non-peat, Very shallow, Shallow, Medium, Deep, Very deep, Very very deep |

*) These classes are derived from Spatial land-use planning of Kalimantan Tengah Province.

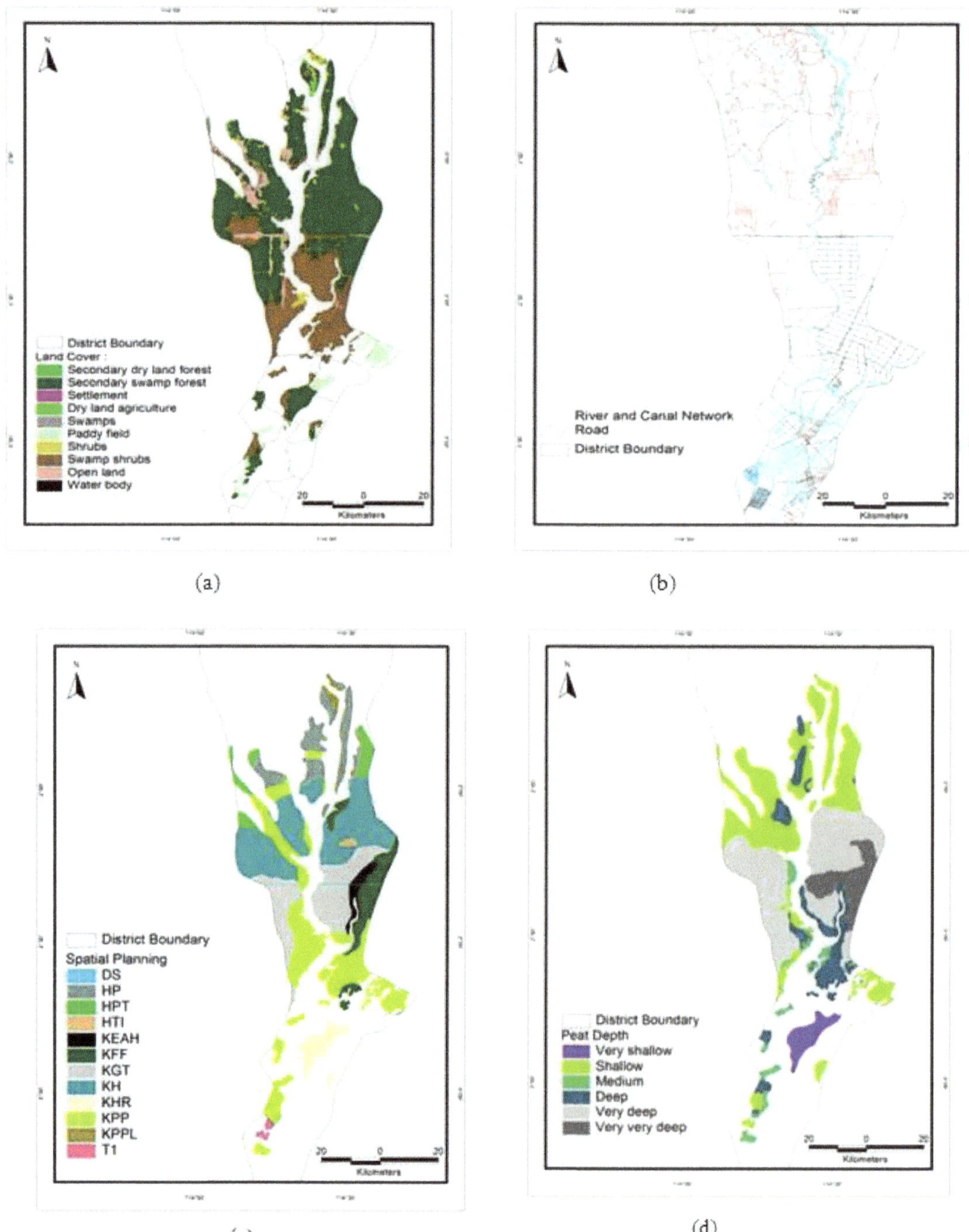


***Figure 2.*** *Map of biophysical and human activities factors; land cover (a), river and road (b), spatial planning (c) and peat depth (d).*

# 3. Results and Discussions

### *3.1. Hotspot Confidence*

The monitoring data and the detection of forest and land fire in Indonesia today still rely on the hotspot data. Hotspot characteristics that have strong relationships with the fires need to be determined because not all the hotspots detected in the peatland are the location of the fire.

***Table 2.*** *The comparison between the fire location based on ground check and hotspot number in peatlands of Kapuas District 2012)*

| No | Sub District | Numbers of fire location in 2012 | Hotspot numbers in 2012 | | |
|---|---|---|---|---|---|
| | | | C 50 | C 70 | C 90 |
| 1 | Basarang | 13 | 25 | 10 | 0 |
| 2 | Dadahup | 6 | 19 | 11 | 7 |
| 3 | Kapuas Barat | 3 | 27 | 11 | 1 |
| 4 | Kapuas Murung | 2 | 12 | 5 | 0 |
| 5 | Kapuas Timur | 1 | 1 | 0 | 0 |
| 6 | Mantangai | 2 | 8 | 4 | 0 |

Description (remarks): CX (hotspots with confidence values more than X %)

Hotspot which has a confidence value of more than 50% (C50) represents all fires result from the ground checks (Table 2). The hotspot with confidence more than 70% and more than 90% have the vacuum of hotspots on the actual fire location in the field. It shows that the hotspot that can be used to indicate a strong relationship with the peat fire incidence in Kapuas district is a hotspot that has a confidence value above 50%. Therefore, a spatial analysis, which is related to peatland fires using MODIS hotspot is a hotspot which data that has a confidence value above 50 %. The confidence level is used to determine the classes of low confidence (> 30 %), nominal-confidence (30%-80%) or high-confidence (> 80 %) in all the fire pixels [11]. The confidence value of a hotspot is highly variable in different places in the world. In Kapuas district, a hotspot with confidence values above 50% quite describes the peat fire in the field.

### *3.2. Peatland Fires Period*

Based on the temporal analysis of 2001-2012 years through a number of hotspots of Terra/ Aqua MODIS, the descriptive relationship of the number of annual hotspots for 12 years in all of Kapuas regions was acquired (Fig. 3).

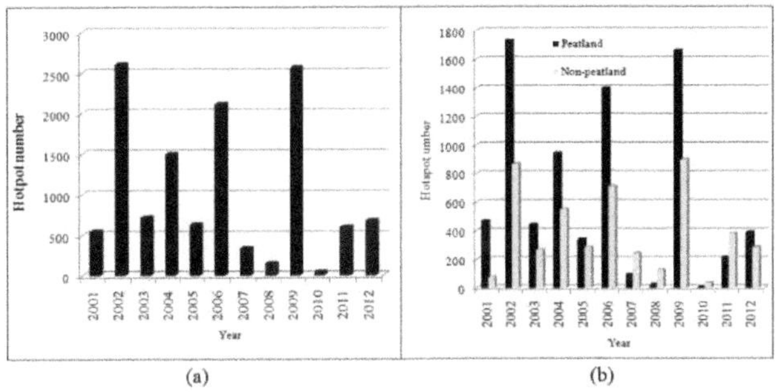


***Figure 3.*** *The Numbers of annual hotspots in Kapuas 2001-2012*

During the 12 -year period as shown in Fig. 3-a, the highest hotspot occurred in 2002 and then repeated in 2006 and 2009. At the height of the fire in 2002, 2006 and 2009 the hotspots were observed in peatland much more than in non-peatland (Fig. 3-b). Peatlands in Kapuas are mostly in the area of ex PLG. Based on the field observations and interviews with the society, the area of ex PLG is almost every year has a case of fire. Peatland become combustible after experiencing repeated fire every year. This phenomenon is reinforced by [12] that repeated fires in 2002, 2006 and 2009 occurred in the area of ex MRP.

The burning peat experiences irreversible drying so that it can no longer absorb water. The burning peat is also difficult to be extinguished and fire can creep under the surface so that it could extend uncontrollably. The dry peat with very low moisture, in the dry season will be a fuel that is very combustible. In addition, the land cover in the peatland is generally a wasteland overgrown by shrubs and grasslands. This is also confirmed by [12] that the dried peat cannot absorb any water when it was watered and it is inflammable. The burning peat generates heat energy greater than the burning wood or charcoal.

The peatland area covers 26.55% of the Kapuas total area. The peatland area is located in the southern part of Kapuas District. A non-peat land is relatively has a lower hotspot density value. The non-peat land covers 73.45% of the Kapuas district, located in the North region of Kapuas with the characteristics of a dry land with the soil minerals, a ramp to steep slope and the land cover is in the form of mixed farms and forests. The peatland condition experiencing drainage has increased the risk of fire in which the fire occurred widely in peatland, especially in the years where the rainfall is below the normal [1].

The increasing hotspot condition in the Kapuas District in 2006 was strengthened by the evidence that in that year the forest and land fire gave impact on the deterioration of air quality in several areas in Central Kalimantan. As shown by [1], in 2006 the air condition is not healthy or harmful, lasting more than 80% of the days during September-November.

The difference of hotspot number observed every year in Fig. 3 related to the rainfall in the area. Based on Fig. 4 calculated from the Terra/ Aqua MODIS data hotspot, it can be seen that the highest peak of hotspot number around the

Kapuas District regions occurred in 2002, 2006 and 2009. When the rainfall condition decreases, hotspots tend to increase. In contrast, the hotspots decreased in the years in which the rainfall increased in 2007, 2008 and 2010.

The rainfall influences much on fuel moisture content, especially in peatlands. As rainfall increases, peatlands store large amounts of water so that the fuel water content increases and it is difficult to burn. When the rainfall decreases, the peat moisture content is to be on the wane. Peat with low moisture content is very susceptible to burn. As in [13,14] explained that rainfall affects the dynamics of moisture content and groundwater levels from peat. Both fluctuations are strongly influenced by the dynamics of rainfall and wet soil. In the seasons with lower rainfall, water level drops to a critical threshold that causes the peat soil is highly flammable [15].

The highest number of hotspots in the peatland occurred in each month from August to October with minimally 100 hotspots (Fig. 4). The monthly hotspots increase when the monthly rainfall decreases. An increasing number of hotspots in August to October is related to people activities that adjust the end of the dry season and the beginning of rainy season. In June-July people who are generally farmers have opened and clean the land for farming preparation. The fuel that has been cut down and cleaned is allowed to dry for one to two months. In the next month from August to October at the end of the dry season, just like their habit, they burn biomass or the waste of the land cleared. Generally, human activities related to land clearing using fire in Central Kalimantan occurred during the dry season from May to September. Farmers clean the land during that time, when the biomass is drier [4].

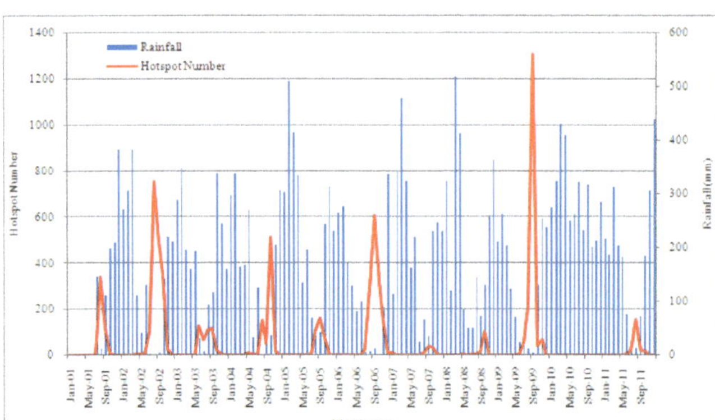


***Figure 4**. The Monthly Hotspot and Rainfall in peatland area in Kapuas 2001-2011*

### *3.3. Peatland Fires Distribution Based on Biophysical and Human Activities Factor*

Spatially, the peat fire activity is indicated by the density of hotspots based on the biophysical aspect and human activity. The more dense the hotspot, supposedly the activity caused the fire is getting bigger. Based on land cover, the highest density of hotspots in peatland area occurs largely in the form of swamp shrubs and open land (Fig. 5). Almost every year the hotspots are found in the swamp shrubs and open land with the large density and number. The areas with the lowest density of hotspots are found in the water body cover. A swamp shrubs is an area that is vulnerable to burning because it provides abundant cargo fuel and flammable in the dry season because it belongs to the fine fuel category. Open lands in the field are found in a form of grasslands, the former of burned land. Based on the interviews with villagers, fires often begin from land cover of open ground because it is not managed or an abandoned land. Open lands will be burned by people to hunting and fishing or being burned because of the negligence to throw cigarette butts.

The areas with swamp shrubs and grassland covers are found in the southern region of Kapuas. This area is generally located in the area of Ex MRP. Currently the area is largely a neglected area that is not managed (unmanaged land). In the displaced areas people burn the land more easily because it becomes an open access area and difficult to be monitored the villagers.

The hotspot density decreases as much as the estrangement distance from the rivers and canals in peatland as shown in Fig. 6. The area that closer to the river more dense hotspots area were detected. Conversely, the farther from the river, the density of hotspot is also lower. Descriptively, the number and density of hotspots will be lower and none at all at a distance of more than 5 km from the river. At a distance of more than 5 km from the river, there are only maximum 8 hotspots were observed during 12 years. This has to do with the people access in using the rivers and canals to carry out various activities like gathering and transporting timber, hunting, fishing and farming. People choose the location near the river and canal to perform many diverse activities. Therefore, the location adjacent to the rivers and canals are vulnerable to fire because it is easily accessed for doing activities that cause fires.

As in shown by [6,16], more than 4000 km canals have been built in the peatland area in Ex MRP which provides access allowing many community activities can be done. Existing of canal causing peatland areas surrounding drier than before and are very prone to burning. In the peat swamp forest in the Sugihan River [8] also found that the closer the rivers and canals the higher a chance the fire will occur.

The density of hotspots is higher by closer to the road (Fig. 7). Roads provide access for people and companies to perform a variety of activities that lead to susceptibility for land to be burned. People open a lot of land by burning it and it is generally close to the roads. Activities cause the land fires are like throwing the burning cigarette butts and open land for land tenure. The analysis by [7] also showed that a significant portion of fires occurred close to the road network, which can also suppose important cause of the fire.

The results of the spatial analysis also found that in the years in which the high rainfall and hotspots are less monitored, the farthest hotspots occur at the location 3 kilometers from the road. As happened in 2001 and 2010, where the number of hotspots is relatively low, the hotspots are observed only at a distance of 3 km from the road.

According to the spatial plan distribution, the highest density of hotspots based on the spatial plan in Kapuas is in KEAH area, KGT and KPP as shown in Fig. 8. Most of high density hotspot areas are located in the southern region of the Kapuas that is the transmigration areas and plots of peat in the area of ex MRP. The lowest density area of hotspots is in transmigration areas (T1). Area of KEAH and KGT are planned in the future as the peat conservation area. These locations have depth peat from very deep to very deep (400 – 1200 cm) as shown Fig.2-c. Transmigration areas based on the results of field checks are residential areas or villages. Land allocation for resettlement and cultivation could be pushed a region prone to fires. As in [17] found that the fires in Jambi Province are caused by factors related to human activities such as the existence of transmigration projects and the allocation of land for specific land uses.

Based on the peat thickness (Fig. 9), the highest density area of hotspots is mostly in a medium peat (100-200 cm) and very very deep (800-1200 cm). The lowest density area of hotspots was found in peat with a very shallow depth. In the year with the highest hotspot density, the depth area of peat was shifted to a very deep. In 2002 the highest density of hotspots is found in the peat medium, but in 2006 and 2009 they shifted to the very deep peatlands. The peatland area is becoming widely used by people for farming. By burning peat farmers get additional natural fertilizer from the rest of peat fires. Generally, the medium peat layer areas are also located in an easy access and close to roads, river and canal networks.

According to various peat thickness classifications greater than 3 m is considered to be deep, while 0.5 m is the minimum thickness used in the Indonesian system of peat classification [18,19,20]. Spatial analysis conducted by [21] found area peat dome are located in Block B, northern Kapuas. This area have very very deep peat thickness that is composed of remnant peat swamp forest 'islands' and large burnt areas.

Hotspot can be used to assess peatland fires indication. Hotspot density is one of the indications that can be used to assess fire activity in a region. Based on the spatial analysis, generally the densest hotspots are located in swamp shrubs, close to the river, close to the roads, in the very deep peat as well as in peat conservation area of spatial plan. It also showed that peatland characteristics above tend to be fire prone areas. In contrast, land with low density hotspot indicates that fire activity was also lower in these areas. In Kapuas District, peatland in the form of a water body, far away from the river, far from the road, the spatial plan of transmigration area and non-peat land are more resistant and lower in its hotspot density. This could indicate that the land with those characteristics is relatively more resistant to fire.

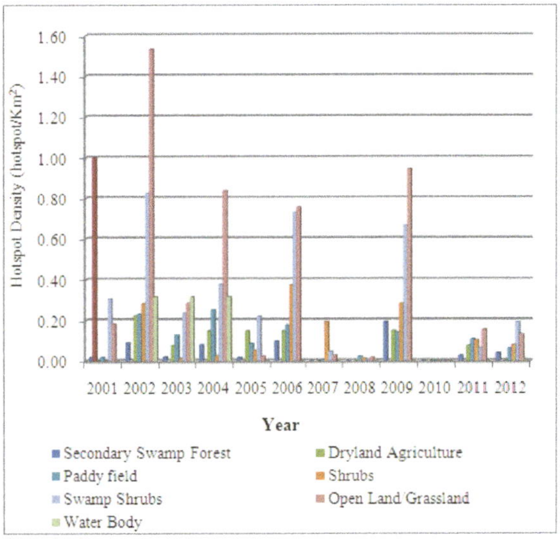


***Figure 5**. The hotspot density in various land covers in Kapuas peatlands 2001-2012*

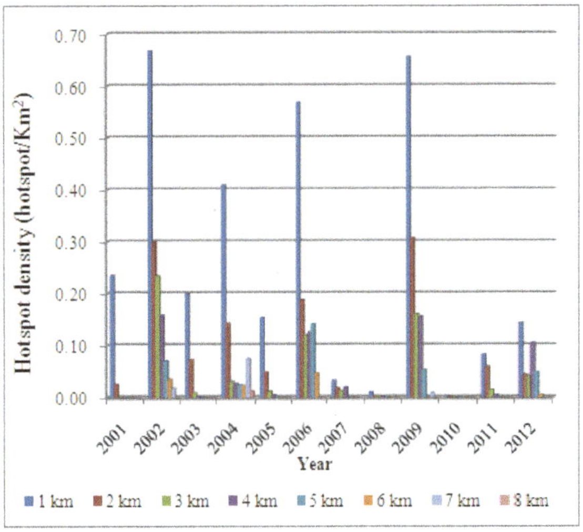


***Figure 6**. Hotspot density based on the distance from river in Kapuas peatlands 2001-2012*

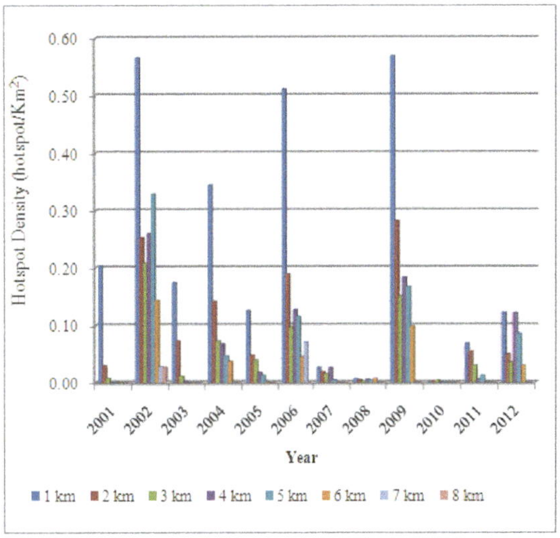


***Figure 7**. Hotspot density based on the distance from the road in Kapuas peatlands 2001-2012*

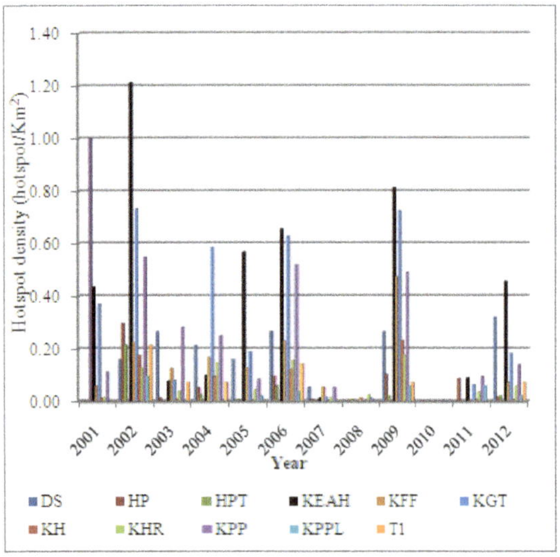


*Figure 8. The hotspot density according to the spatial planning in Kapuas peatlands in the year of 2001-2012*

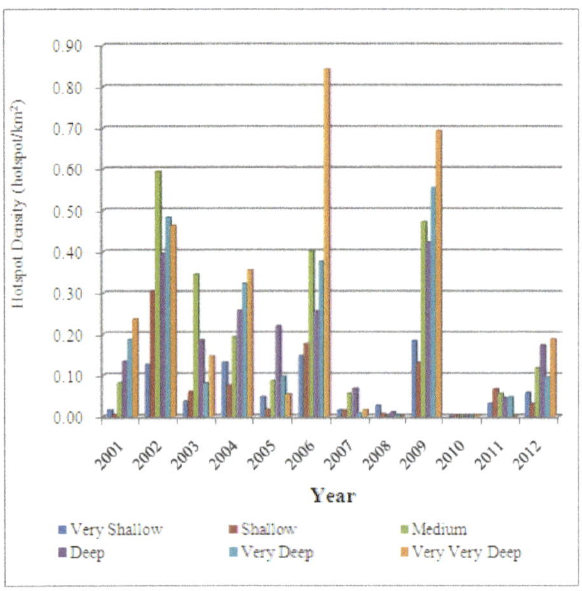


***Fig. 9.*** *The hotspot density at various peat thickness in the peatland of Kapuas in 2001-2012*

## 4. Conclusion

The strong indication of peatland fires in Kapuas District is shown by hotspots derived from Terra/Aqua MODIS satellite with the confidence value more than 50%. The number of hotspot increases in the year where the rainfall is low. Based on the time, August to October is the time of the increasing hotspots in Kapuas. The densest hotspot in peatland of Kapuas District is generally located in the swamp shrubs, close from the river, close from road, in the very deep peat with spatial planning of the peat conservation area. Based on discussion above can be indicated that peatland conservation area and unmanaged land in Kapuas District were in vulnerable condition of the fire. Government needs intervention to improve unmanaged land utilization and protect the peatland conservation area in Kapuas District.

## Acknowledgements

This research was funded by Columbia University and Bogor Agricultural University Partnership to Build Capacity for Adaptation to Climate Risks In Indonesia, USAID Grant No.AID-497-A-11-00011. We also thank to Regional Operation Manggala Agni II of Kapuas Natural Resources Conservation Center (BKSDA) Central Kalimantan for the support and assistance provided to the researcher during the data collection process. Sincerely appreciation is also extended to anonymous reviewer for correction and comments.

## References

[1] Harrison M.E,, S.E. Page, and S.H. Limin, "The global impact of Indonesian forest fires," Biologist, vol. 56(3), pp. 156-163, 2009.

[2] Langner, A and F. Siegert, "Spatiotemporal fire occurrence in Borneo over a period of 10 years". Glob. Change Biol, vol. 15, pp. 48-62, 2009.

[3] Page, S.E., F. Siegert, J.O. Rieley, H.V. Boehm, A.Jayak and S.H. Limin, "The amount of carbon released from peat and forest fires in Indonesia during 1997," Nature, vol. pp. 420, 61-65, 2002.

[4] Hooijer, A., M. Van der Vat, G. Prinsen, , R. Vernimmen, , J. J.Brinkman, and F. Zijl, "Hydrology of the EMRP Area – Water Management Implications for the Peatlands," Technical Report Number 2, Master Plan for the Rehabilitation and Revitalization of the Ex-Mega Rice Project Area in Central Kalimantan. Euroconsult Mott MacDonald and Deltares and Delft Hydraulics. 2008.

[5] Cochrane, M.A., "Fire science for rainforests," Nature, vol. 421, pp. 914-919, 2003

[6] Ceccato, P, I.N,S. Jaya, J.H. Qian, M.K. Tippet, A.W. Robertson and S. Someshwar, "Early Warning and Response to Fires in Kalimantan, Indonesia," IRI Technical Report 10-14, International Research Institute for Climate and Society, Palisades, New York (US), 23 pp, 2010.

[7] Jaya, I.N.S., E.S. Purnama, I. Arianti and J. Boonyanuphap, "Forest fire risk assessment model and post-fire evaluation using remote sensing and GIS: A case study in Riau, West Kalimantan and East Kalimantan Provinces, Indonesia," Paper presented at The Forest Restoration and Rehabilitation Training Course and Workshop in the Viiki Tropical Resources Institute (VITRI) of the University of Helsinki, Finland, 13~19 May 2007, Helsinki (FIN), 2007.

[8] Soewarso, "Development of peatland forest fire prevention using prediction model". [Disertation] Bogor (ID): Graduate School of Bogor Agricultural University (Unpublished), 2003

[9] [MoF] Ministry of Forestry of Indonesia, "Land fire information and data in Operation Area II of Kapuas.," Natural Conservation Agency of Central Kalimantan, Report. Kapuas (ID) (Unpublished), 2011.

[10] [WIIP] Wetland International Indonesia Program, "Community based peatland management in Indonesia," Wetland International Indonesia Programmme,. Report. Bogor (ID), 2007

[11] Giglio, L., J. Descloitres, C.O. Justice, Y.J. Kaufman, "An Enhanced Contextual Fire Detection Algorithm for MODIS." Remote Sensing of Environment, vol. 87, pp. 273-282, 2003.

[12] Agus, F. and I.G.M. Subiksa. "Peatland: Potency for agriculture and its environmental aspects," Soil Research Centre and World Agroforestry Centre (ICRAF), Bogor (ID), 2008.

[13] Syaufina, L., A.A. Nuruddin, J. Basharuddin, L.F. See and Yusof, M.R.M, "The effects of climatic variations on peat swamp forest condition and peat combustibility". Journal of Tropical Forest Management, vol. 10(1), pp.1-14, 2004.

[14] Taufik M, B.I. Setiawan , L.B. Prasetyo , N.H. Pandjaitan, dan Soewarso, "Development of fire danger index at SBA Wood Industries, South Sumatra," Jurnal Penelitian Hutan Tanaman, vol. 8, pp. 215-223, 2011.

[15] Wosten, J.H.M., E. Clymans, S.E. Page, J.O. Riley and S.H. Limin, "Peat–water interrelationships in a tropical peatland ecosystem in Southeast Asia," Catena, vol. 73, pp. 212–224, 2008.

[16] Hecker, J.H, "Promoting Environmental Security and Poverty Alleviation in the Peat Swamps of Central Kalimantan, Indonesia," Commissioned by the Institute for Environmental Security, The Netherlands. Online available in http://www.envirosecurity.org/espa/PDF/IES_ESA_CS_Kalimantan_Case_Study.pdf, 2005, [20 October 2013]

[17] Stolle, F, K.M. Chomitz, E.F. Lambin, and T.P. Tomich. "Land use and vegetation fires in Jambi Province, Sumatra, Indonesia," Forest Ecology and Management, vol. 179, pp. 277-292, 2003.

[18] Radjagukguk, B, "Peat soils of Indonesia: location, classification and problems for sustainability". In: Rieley, J.O., Page, S.E. (Eds.), Biodiversity and Sustainability of Tropical Peatlands. Proceedings of the International Symposium on Biodiversity, Environmental Importance and Sustainability of Tropical Peat and Peatlands, Palangka Raya, Central Kalimantan, Indonesia. Samara Publishing Limited, Cardigan (UK), pp. 45–54, 1997.

[19] Rieley, J.O., A.A. Ahmad-Shah, M.A. Brady, "The extent and nature of tropical peat swamps," In: Maltby, E., Immirzi, C.P., Safford, R.J. (Eds.), Tropical Lowland Peatlands of Southeast Asia. Proceedings of a Workshop on Integrated Planning and Management of Tropical Lowland Peatlands, Cisarua, Indonesia. IUCN, Gland, Switzerland. 294 pp, 1996.

[20] RePPProT, "A National Overview from the Regional Physical Planning Programme for Transmigration." UK Overseas Development Administration and Directorate BINA Programme, Ministry of Transmigration, Jakarta (ID), 1990.

[21] Jaenicke J, J.O. Riley, C. Mott, P. Kimman and F. Siegert, "Determination of the amount of carbon stored in Indonesian peatlands". Geoderma, vol. 147, pp. 151–158, 2008.

# Nutritional evaluation of varying levels of cooked flamboyant seed meal (*Delonix regia*) on the growth performance and body composition of Nile tilapia (*Oreochromis niloticus*) fingerlings

**Gabriel Gana Bake*, Elizabeth Itunu Martins, Suleiman Omeiza Eku Sadiku**

Department of Water Resources, Aquaculture and Fisheries Technology, School of Agric and Agric Technology, Federal University of Technology Minna Minna Niger state, Nigeria

**Email address:**

gabbygana@futminna.edu.ng (G. G. Bake)

**Abstract:** A feeding trial was conducted to evaluate the effect of inclusion level of cooked *Delonix regia* seed meal (CDRM) in the practical diet of *Oreochromis niloticus* fingerlings through their growth performance and nutrient utilization for 56 days. Four iso-nitrogenous and iso-lipidic experimental diets were formulated to contain 35% crude protein and 9.5% lipid, CDRM was included at different varying inclusion levels viz: 0, 10, 15 and 20 and were designated as D1 (0 % inclusion), D2 (10 % inclusion), D3 (15 % inclusion) and D4 (20 % inclusion). Each treatment was allocated to three tanks of twenty *O. niloticus* per tank with initial mean weight of 1.18±0.10g. There was no significant difference (*P*>0.05) in the survival rate of all the treatments, D2 has the highest value in percentage weight gain, specific growth rate, feed intake, protein efficiency and protein retention, while D4 (20% inclusion) had the lowest value in all the growth parameters and feed utilization measured. However, there is no significant difference (P>0.05) between fish fed D3 (15% inclusion) and D1 (control: 0% inclusion). Proximate composition result revealed that increase in the inclusion level of cooked *Delonix regia* lead to an increase in carcass lipid content and a decrease in the moisture content of the fish fed experimental diet. The results indicate 10-15% inclusion of *Delonix regia* seed meal improved growth performance and nutrient utilization of Nile tilapia fingerlings.

**Keywords:** Delonix Regia Seed Meal, Nile Tilapia, Growth Performance

## 1. Introduction

In order to sustain the consistent expansion being witnessed in aquaculture sector worldwide, a matching increase in fish feed production is imperative (Francis et al., 2001). In finfish aquaculture aqua feed production accounts for between 40-65% of total operating costs, depending on the level of aquaculture intensification. (Jauncey and Ross, 1982; Hardy 1989 Lovell, 1989). Protein is the most critical and costly ingredient in aqua feed production feeds and usually protein sources represent about 60% or more in the cost of the feed. Traditionally, fishmeal is preferred dietary protein source for many farmed fish and is appreciated for its amino acid balance, vitamin content, palatability and unidentified growth factors (Tacon, 1993). However, limited supply, increasing demand and high price of fish meal are challenges to the sustainable development of fish culture industry. Wider use of more economical plant or animal protein ingredients is needed for the formulation of cost-effective and eco friendly aquafeeds. Several studies focused on assessing the potential to reduce fish meal level in fish diets in the last two decades Bureau et al. 2000; Kureshy et al. 2000; Webster et al. 2000; Milliamena 2002; Gaylord & Rawles 2005; Wang et al. 2006; Guo et al. 2007). The magnitude of fish meal replacement by economical protein sources varies greatly in literatures because of great variability of fish meal and protein levels used in the basal (control) diets, as well as great variability of chemical composition and nutritive value of the alternate protein ingredients (Ali, et al., 2003; Bureau, et al. 2000;

Bake, et al 2009). Furthermore, considerable emphasis has been focused on the use of conventional plant protein sources, such as soybean (Nyirenda et al., 2000; Koumi et al., 2009), groundnut (Ovie and Ovie 2007), cotton seed (Mbahinzireki et al., 2001; El-Saidy and Gaber, 2003) and rapeseed meal (Burel et al. 2000a,b), however, their scarcity and competition from other sectors for these conventional crops for livestock and human consumption as well as industrial use make their costs too high and put them far beyond the reach of fish farmers or producers of aquafeeds (Fasakin et al., 1999). Therefore, in order to attain a more economically sustainable, environmentally friendly and viable production, research interest has been redirected towards the evaluation and use of unconventional protein sources, particularly from plant products such as seeds, leaves and other agricultural by-products (Siddhuraju and Becker, 2001; Olvera-Novoa et al., 2002; Ali, et al., 2003; Bake et al 2009).

*Delonix regia* a wild plant otherwise called flame of the forest originated from America (Madagascar) but found wild or as ornamental plants in various parts of the world including Nigeria (Purseglove 1994). It is an ornamental, leguminous plant, which produces tones of pods containing seed in the fruiting season (Keay and Stanfield, 1964). These seeds are left unutilized since they are neither consumed by any animal nor utilized for any other medical purposes. *Delonix regia* (Caesaipincidae) is a tree, which is about 12m high with a slightly flat or curving crown. Seeding is abundant and germination is free. The flowers are usually red but sometimes bear a paler orange red flower and flowering is spontaneous, it is one of the most spectacular flowering trees of the tropics. It is widely grown ornamental leguminous plant, which produces 25-40cm long pods containing seed which laboratory analysis has reported to contain crude protein 36.92%, crude lipid 4.17%, crude fiber 11.39% (Grant et al, 1991). Report also showed that the seed extracted exhibited low haemagglutinin activity and contains non-toxic lectins. Flamboyant seed meal compare favorably with mechanically extracted groundnut cake meal in term of crude protein content (NRC, 1993).Heat treatment is employed to dry certain plant product and to improve quality of pelletized feed. Hence, heat treatment can improve the nutritional value of Delonix regia by reducing the hydrocyanic acid and other anti-nutritional factor.

Tilapia is the second most culture fresh water finfish. It is also a common fish in tropical fish farms and its prolific breeding abilities makes it to dominate most ponds and water bodies where the population is not checked. Tilapias are voracious diurnal feeders. Currently, farmed tilapia represents more than 75% of world tilapia production (FAO, 2006). The high exponential growth of tilapia culture in recent years could be attributed to several factors such as their ability to be grown year round with some variability in fish yields depending upon seasonal differences related to the area (Green et al., 1990). It also effectively utilizes alternative feed ingredients such as rice, cocoa, various flours, and soya-nut oil (Jackson et al., 1982 Brown, 1983). Intensification of tilapia culture requires development of suitable feeds for both complete and supplementary feeding in tanks and ponds (Koprucu and Ozdemir 2005). Therefore, this present study was carried out to evaluate the effect and suitability of cooked flamboyant seed meal (Delonix regia) inclusion in the practical diet of Nile tilapia fingerlings through their growth performance and nutrient utilization.

# 2. Materials and Methods Diet Formulation and Preparation

### *2.1. Soybean Meal*

Soybean was obtained from the market and was toasted using a frying pan and allowed to cool before milling with the aid of a grinding machine. Cooking was done to remove the anti-nutritional factor present in the seed.

### *2.2. Cooked Flamboyant Seed Meal (Delonix Regia)*

Mature and dried Flamboyant seedpods were collected from the ornamental garden of the Center for Preliminary and Extra Moral Studies (CPES) of the Federal University of Technology Minna; the seed were collected by manually opening the pods. The seeds were boiled at $100^0$C for 80mins to reduce the effect of hydrocyanic acid (anti-nutritional factor) and were allowed to cool by sun drying and were later grounded to homogeneous powder.

### *2.3. Fishmeal*

The fishmeal used for these experiments was obtained from the market store. The crude protein and lipid content of fishmeal were 61.25% and 11.94%.

All the ingredients were separately milled and mixed with warm water to form consistent dough, which was then pelleted, sun-dried, packed in polyethylene bags and stored. The feed composition table is shown in Table1

***Table 1.*** *Proximate composition of dietary ingredients diets.*

| Ingredients | Fish meal | Soybean meal | Raw *D.regia* meal | Cooked *D.regia* meal | Maize meal | Millet meal |
|---|---|---|---|---|---|---|
| Moisture (%) d.b | 6.21 | 4.57 | 3.74 | 3.24 | 5.09 | 5.63 |
| Crude protein(%) d.b | 61.25 | 48.07 | 23.42 | 25.11 | 9.32 | 8.98 |
| Lipid (%) d.b | 10.70 | 7.04 | 4.60 | 7.30 | 4.20 | 3.56 |
| Ash (%) d.b | 6.69 | 4.35 | 8.79 | 6.48 | 3.84 | 3.60 |

*1 Dry basis

### *2.4. Experimental Diets*

Based on the nutritional requirements of tilapia (NRC 1993), four isonitrogenous and isolipid diets were formulated at 35 % protein and 9.5 % lipids, containing 10-30% toasted flamboyant seed meal of different levels of inclusion Table 2. The experimental diets were designated D1 (0 % inclusion), D2 (10 % inclusion), D3 (15% inclusion) and D4 (20 % inclusion).

***Table 2.*** *Formulation and proximate composition of the experimental diets for Oreochromis niloticus fingerlings.*

| Ingredients | D1 | D2 | D3 | D4 |
|---|---|---|---|---|
| Fishmeal | 384.70 | 345.40 | 325.70 | 302.70 |
| Soybean meal | 220.00 | 220.00 | 220.00 | 220.00 |
| Cooked *delonix regia* meal | 0.00 | 100.00 | 150.00 | 200.00 |
| Maize meal | 47.00 | 47.00 | 47.00 | 47.00 |
| Millet | 47.00 | 47.00 | 47.00 | 47.00 |
| Starch | 107.90 | 55.00 | 55.00 | 55.00 |
| Vitamin premix | 25.50 | 25.50 | 25.50 | 25.50 |
| Mineral | 30.00 | 30.00 | 30.00 | 30.00 |
| Soybean oil | 30.00 | 30.00 | 30.00 | 30.00 |
| Cellulose | 107.90 | 100.10 | 69.80 | 42.80 |
| Total | 1000.00 | 1000.00 | 1000.00 | 1000.00 |
| *Moisture* | 5.25 | 5.00 | 4.92 | 4.81 |
| *Crude protein* | 34.35 | 36.78 | 36.11 | 34.34 |
| *Lipid* | 9.48 | 9.24 | 9.51 | 9.34 |
| *Ash* | 10.83 | 10.12 | 9.92 | 9.88 |

### *2.5. Experimental System and Fish*

The experimental fish, pure- bred *O. niloticus* fingerlings, with a initial mean weight of (1.19±0.03) were purchased from Tagwai Fisheries hatchery, Ministry of Livestock and Fisheries, Minna, Niger state. The fish were transferred in a well-oxygenated water plastic container from hatchery to the Department of Water Resources, Aquaculture and Fisheries Technology laboratory, Federal University of Technology, Minna where the feeding trial was conducted. Upon arrival they were acclimatized in a transitional tank in the laboratory for four days and were fed commercial feed (coppense feed) at 35% crude protein once a day before the experiment commenced.

Fresh filtered dechlorinated tap water was supplied to the flow–through system, which consisted of 15-rounded plastic tank, each with an approximate capacity of 45 liter at the flow rate of 250-300 ml/min. The water quality parameters in the system were monitored weekly, and the ranges were dissolved oxygen 6.8-7.4 mg/l, total ammonia 0.1-0.27 mgN/l, and pH 6.9-7.5. No critical values were detected for nitrite and nitrate.

Three replicates of each treatment using 20 fish per plastic tank were reared on each of the four experimental diets. Feed was manually administered. The fish were fed three times daily at 5% of body weight at 0900, 1200 and 1600 hours. Feeding rate were subsequently adjusted according to their growth rates per plastic tank.

The uneaten and faecal matters were siphoned out of tank every morning before feeding, and 45 minutes after the fish have been fed. The fish were denied feed 24 h prior to sampling. Ten fish were randomly sampled on weekly basis, and weights were measured using a digital electronic weighing balance.

### *2.6. Biochemical Analysis*

About 15 g initial sample and 10 g of final samples from each plastic tank were pooled separately and then homogenized using laboratory mortal and pestle. The major ingredient used for the diet; the formulated diet and the fish body samples were subjected to chemical analysis. Proximate composition analyses were performed according to AOAC procedures (AOAC, 2000). Moisture content was determined by drying samples at 105±2°C until a constant weight was obtained. Dried samples were used for determination of crude fat, protein and Ash contents. Crude fat was measured by solvent extraction method in a soxhlet system where n-hexane was used as solvent. Crude protein content was calculated by using nitrogen content obtained by Kjeldahl method. A conversion factor of 6.25 was used for calculation of protein content (AOAC, 2000).

### *2.7. Evaluation of Growth Parameters*

Growth performance and nutrient utilization were analyzed in terms of weight gain (WG), feed efficiency (FE), specific growth rate (SGR), feed intake (FI), protein efficiency (PER) and protein retention (PR). The following formulas were usedas described by Takeuchi, (1988) as the following equation:

Weight gain (%) = (final weight (g) – initial weight (g) x 100 / initial weight (g)

Feed efficiency (%) = (weight gained (g) / feed fed (g)) x 100

Specific growth rate (%)= (In final weight (g) – In initial weight (g) / feeding period (day) x 100

Feed intake (mg/fish/day) = dry feed (mg) given / number of fish / feeding period (day)

Protein efficiency ratio = wet body gain x 100 / protein intake (g)

Protein retention (%) = protein gain x 100 / protein fed.

### *2.8. Statistical Analyses*

Data were analyzed using one – way analysis of variance (ANOVA) using Statistica 6.0 (Stat-Soft, Inc., USA). Differences between treatments were compared by Tukey's test. Level of significance was tested at $P<0.05$.

## 3. Results

The proximate composition of the major ingredients used in the formulation of the experimental diet is shown in Table 1. Fishmeal had the highest crude protein and lipid contents (61.25 and 10.70%) followed by soybean meal (43.07 and 7.0%) while cooked Delonix regia meal had 25.11 and 8.30% of crude protein and lipid contents, respectively. Though all the experimental diets were similar in terms of their proximate composition among the experimental diets, D2 had the highest crude protein content of 36.93% while diet 4 had the lowest crude protein content of 34.34%. The total lipid was highest in diet 3 (9.51%) while diet 2 had the lowest total lipid content

(9.24%). Ash content was highest in diet 1(10.83%) and lowest in diet 4 (9.88. %) among the test diets (table 2).

There was no feed rejection among the fish fed experimental diets; however the acceptability of the diets differed considerably among the treatments. Table 3 showed that Tilapia fed with 10% inclusion of cooked *Delonix regia* meal (D2) had the highest final body weight gain, specific growth rate, and total feed intake and was significantly higher than the rest of the treatments ($P< 0.05$) while tilapia fed with 20% inclusion of cooked *Delonix regia* meal (D4) had the lowest significant value in all the growth parameters measured. The growth parameter values measured for tilapia fed D1 and D3 were not significantly different from each other but were significantly higher than tilapia fed D4 ($P<0.05$).

Fish fed with D2 had the highest significant value ($P<0.05$) feed efficiency, Protein efficiency ratio and protein retention. Although fish fed D4 had the lowest feed efficiency and protein efficiency ration, it was not significantly different from fish fed D1 and D3 ($P>0.05$) however the protein retention of tilapia fed D1 and D3 was significantly higher ($P<0.05$) than those fed D4.

Variations in the tilapia body composition among fish fed experimental diets at the end of the feeding trial are shown in Table 4.The result of the body composition revealed that except for lipid and moisture fish fed cooked *Delonix regia* meal displayed a marginal increase in the carcass proteinand ash at the end of the experiment and were significantly higher than the initial body composition at the start of the experiment. There was no significant differences in the carcass ash contents among all the treatments fed experimental diets, fish fed D2 having the highest crude protein content value and was significantly higher than the other fish fed other experimental diet ($P<0.05$), however fish fed D1, D3 and D4 do not show any significant difference ($P>0.05$) from each other. Fish fed D4 had the highest significant lipid composition ($P<0.05$) while D1 had the lowest value but was not significantly different from fish fed D2. Fish fed D1 showed the highest moisture among the experimental diets although not significantly different from D2 ($P<0.05$) but were significantly different from D3 and D4.

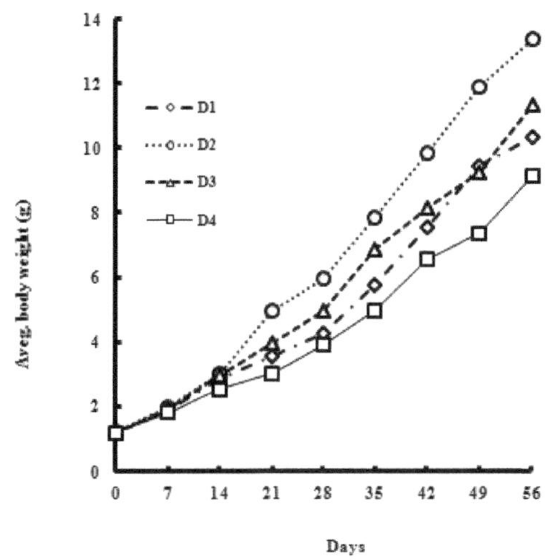


***Fig 1.*** *Growth of O.niloticus fingerlings fed the experimental diets for 56 days.*

***Table 3.*** *Growth performances of O. niloticus fry fed experimental diets for 56 days.*

| Diet code | Body weight (g) Initial | Body weight (g) Final | Weight gain (%) | Survival rate (%) | Specific growth rate (%) | Total feed intake (g) | Feed efficiency | Protein efficiency ratio | Protein retention (%) |
|---|---|---|---|---|---|---|---|---|---|
| D1 | 1.21±0.04 | 10.32±0.01$^{b}$ | 932.00±12.00$^{b}$ | 97.30±0.74$^{a}$ | 3.83±0.01$^{bc}$ | 11.56±0.10$^{b}$ | 0.79±0.01$^{c}$ | 2.28±0.03$^{b}$ | 35.93±0.44$^{b}$ |
| D2 | 1.20±0.07 | 13.40±0.43$^{a}$ | 1240.00±43.59$^{a}$ | 98.22±0.42$^{a}$ | 4.31±0.04$^{a}$ | 12.63±0.10$^{a}$ | 0.97±0.04$^{a}$ | 2.60±0.06$^{a}$ | 41.97±0.91$^{a}$ |
| D3 | 1.17±0.06 | 11.32±0.54$^{b}$ | 1032.67±54.72$^{b}$ | 97.68±0.57$^{a}$ | 4.05±0.06$^{b}$ | 11.65±0.72$^{b}$ | 0.87±0.01$^{bc}$ | 2.39±0.03$^{b}$ | 36.70±0.51$^{b}$ |
| D4 | 1.18±0.02 | 9.13±0.64$^{c}$ | 812.33±64.33$^{c}$ | 97.12±1.05$^{a}$ | 3.41±0.08$^{c}$ | 10.47±0.84$^{c}$ | 0.75±0.01$^{c}$ | 2.14±0.04$^{b}$ | 33.33±0.69$^{c}$ |

*1 Values in the same column with different superscript letters are significantly different ($p<0.05$) from each other.

***Table 4.*** *Proximate composition analyses of whole body O. niloticus (wet basis) fed experimental diets for 56 days*

| Component (%) | Initial | Final*1 D1 | Final*1 D2 | Final*1 D3 | Final*1 D4 |
|---|---|---|---|---|---|
| Moisture | 75.64 | 74.25±1.5$^{a}$ | 73.56±1.3$^{ab}$ | 72.75±0.3$^{c}$ | 72.62±1.4$^{c}$ |
| Protein | 14.94 | 15.39±1.3$^{bc}$ | 16.09±1.7$^{a}$ | 15.67±1.1$^{b}$ | 15.46±1.5$^{b}$ |
| Lipid | 5.19 | 5.08±0.4$^{c}$ | 5.03±0.6$^{c}$ | 6.28±0.5$^{b}$ | 6.68±0.2$^{a}$ |
| Ash | 4.63 | 5.15±0.6$^{a}$ | 5.05±0.4$^{a}$ | 5.22±0.1$^{a}$ | 5.16±0.4$^{a}$ |

*1 Values in the same row with different superscript letters are significantly different ($p<0.05$) from each other (n=3).

## 4. Discussion

This study investigated the possibility of utilizing cooked *Delonix regia* meal in the diet of *O. niloticus* fingerlings. Fish mortality was low and relatively uniform in all the treatments, furthermore all the experimental fish remaining in the tanks were morphologically normal at the end of the feed trial. In our study, the water temperature which ranged between between 25±0.5°C - 29±0.4°C and the dissolved oxygen 6.6-8.3mg/L were within the acceptable range for tilapia culture (Balarin and Halton, 1979).

The proximate composition of cooked *Delonix regia* meal in this present study revealed that the crude protein content was 25.11% (Table 2). This value was higher than the values reported earlier by Balogun *et al.*(2004). The differences might be attributed to differences in environmental conditions such as soil types, harvesting time and the processing conditions. Fagbenro (1999), Francis *et al.* (2001) and Solomon *et al.* (2007) reported that heat treatment substantially reduces and inactivates levels of secondary compounds in the seed meals. Our results tend to agree with their report.

The suitability of inclusion of alternative ingredient in a fish diet in terms of growth performance has been reported to vary highly among fish species and experimental conditions (El-Sayed 1999, Bake *et al.*2009). The percentage survival was excellent throughout the experimental period. This could be attributed to good handling, good water quality management and proper processing and the suitability of cooked *Delonix regia* meal inclusion in the diet of *O. niloticus.* However within the conditions of this experiment, cooked flamboyant seed meal (*Delonix regia*) was acceptable source of protein in the diets of *O .niloticus* fingerlings. The fishes adapted to the diets within 3-4 days of commencement of feeding trial, however the acceptability of the diet varied significantly. Table 3 and Figure 1 showed that acceptability of the diet reduced with the increase in the inclusion level of cooked flamboyant seed meal. The slight lower growth response by tilapia fed 20% cooked *Delonix regia* meal (D4) inclusion can be attributed to reduced palatability of the diet, which could be as a result of low feed intake. The best growth responses were obtained with tilapia fed (D2) 10% inclusion of cooked *Delonix regia.* Although, there was no significant difference between fish fed D1 and D4. Proper processing of ingredients has been reported to have effect on the texture, palatability and taste of experimental diets and are related to the level of plant material incorporated, hence can affect the acceptability of feed and may consequently retard growth of the fish when in high quantity (Fagbenro, 1999; Francis *et al* 2001 and Solomon *et al* 2007). The high FI obtained in this study may also be ascribed to the cooking processing technique employed in processing of the flamboyant seed. This may have reduced the effect of anti-nutritive factors and other secondary compound activities in the seed. Akinmutimi (2006) reported decrease in crude protein content of *Mucuna Pruriens* when cooked for 60 minutes; Balogun (2009) reported that boiled *Delonix regia* seed meal has higher potential of being utilized by *O. niloticus* when boiled for 80 minutes. In our present study cooking of *Delonix regia* seed meal for 120 minutes at 100°C and including it in the diet of Nile tilapia fingerlings at 10-15% significantly improve their growth performance and nutrient utilization.

Olvera-novoa *et al.* 2002 reported that when plant proteins are included in the diet of Nile tilapia at 10-20% they do have better feeding efficiency. Result from our study revealed that the FE, PER and PR decreases with increase in the inclusion level of the cooked *delonix regia* meal. Although there was no much significant difference between D1, D3and D4, D1 and D3were better than D4.This indicates that D4 has high nutritive content and its low intake may be responsible for low in WG and SGR.

From the result of the proximate composition in Table 4, the total lipid of the fish fed experimental diets is related to the body moisture content and also showed that higher-level inclusion of cooked *Delonix regia* seed meal in the diet of Nile tilapia fingerlings as the potential in increasing lipid deposition in its carcass. This result agrees with the findings of Abd El- Hakim *et al.* (2003) that plant protein in the diet of fish has the potential to increase its lipid deposition in the carcass of the fish fed plant protein.

The present study has demonstrated that the cooking processing method of *Delonix regia* seed meal as significant effect on the tilapia diet hence can partially inactivate some of the secondary compound in *Delonix regia* seed meal there by improve its growth performance and feed efficiency. Furthermore, *Delonix regia* seed meal can be suitable as an ingredient in Nile tilapia diet.

## References


[1] Abdelghany, A. E. (2003). Partial and complete replacement of fishmeal withgambusia meal in diets for red tilapia (*Oreochromis niloticus x O.mossambicus*). *Aquaculture Nutrition*9:145–154.

[2] Abd El-Hakim N. F., Hussein M. S. and Abdel-HalimH. A. (2009) Effect ofpartial replacement of soybean meal protein with dehydrated alfalfa meal(*Medicago sativa* L.) on growth performance and feed utilization of maleNile tilapia (*Oreochromis niloticus* L.) fingerlings reared in tanks. *Egypt J. Aquat. Biol. and Fish.*, Vol. 13, No. 2:35 – 52.

[3] Ali, A.; Al-Asgah, N.A.; Al-Ogaily S.M. and Ali, S. (2003). Effect of feeding different levels of alfalfa meal on the growth performance and body composition of Nile tilapia (*Oreochromis niloticus*) fingerlings. Asian. Fish.Soci-16: 1-2.

[4] Bake, G. G., M. Endo, A. Akimoto and T. Takeuchi (2009) Evaluation of recycled food waste as a partial replacement of fishmeal in the diets for first feeding Nile tilapia *Oreochromis niloticus. Fish. Sci.*, 75, 1275- 1283.

[5] Balarin, J.D. and J.P. Halton, 1979. *Tilapia*: A guide to their biology and culture in Africa. University of Stirling, Stirling, Scotland, UK.

[6] Balogun, J.K., Abdullahi, S.A., Auta, J. and Ogunlade (2004) Feed conversion, protein efficiency, digestibility and growth performance of *Oreochromis niloticus* fed *Delonix regia* meal . 19th Annual conference Processing's of the Fisheries Society of Nigeria (FISON) Pp 823-831.

[7] Borgeson, T. L., V. J. Racz, D. C. Wilkie, L. J. White, and M. D. Drew. (2006). Effect of replacement fishmeal and oil with simple or complex mixtures of vegetable ingredients in diets fed to Nile tilapia (*Oreochromis niloticus*).*Aquaculture Nutrition* 12:141–149.

[8] Burel, C., Boujard, T., Tulli, F., Kaushik, S.J., 2000a. Digestibility of extruded peas, extruded lupin, and rapeseed meal in rainbow trout (*Oncorhynchusmykiss*) and turbot (Psetta maxima). *Aquaculture* 188, 285–298.

[9] Burel, C., Boujard, T., Escaffre, A.M., Kaushik, S.J., Boeuf, G., Mol, K., van der Geyten, S., Kuahn, E.R., 2000b. Dietary low glucosinolate rapeseed meal affects thyroid status and nutrient utilization in rainbow trout (*Oncorhynchus mykiss*). *British Journal of Nutrition* 83, 653–664.

[10] El-Saidy, D. M. S. D. and M. M. A. Gaber. (2003). Replacement of fish meal with a mixture of different plant protein sources in juvenile Nile tilapia,*Oreochromis niloticus* (L.) diets. *Aquaculture Research* 34:1119–1127.

[11] El-Saidy, D. M. S. D. and M. M. A. Gaber. (2004). Use of cottonseed mealsupplemented with iron for detoxification of gossypol as a total replacement of fish meal in Nile tilapia, *Oreochromis niloticus* (L.) diets. *Aquaculture Research*35:859–865.

[12] Fagbenro O. A and Davis S. J, (2003). Use of high percentages soybean protein concentrate as fishmeal substitute in practical diets for African catfish, *C. gariepinus* (Burchell 1822): growth, feed utilization and digestibility J. *Applied Aquac.*. 16 (1).

[13] Fapohunda O. O. (2012) Evaluation of Processed Soybean Meal in the Feeding of *Clarias gariepinus* Fingerlings *J Anim Sci Adv*, 2(2): 244-249.

[14] Fasakin, E. A., Balogun, A. M., and Fasuru, B. E. (1999). Use of duckweed, *Spirodela polyrrhiza* L. Schleiden, as a protein feedstuff in practical diets for tilapia, *Oreochromis niloticus* L. *Aquaculture Research* 30:313–318.

[15] Francis, G; Makkar, HPS; Becker, K (2001). Antinutritional factors present inplant-derived alternate fish feed ingredients and their effects in fish.*Aquaculture,* 199, 197-227.

[16] FAO. (2007). The State of World Fisheries and Aquaculture (2006). Fisheriesand Aquaculture Department, FAO, Rome, 162 pp.

[17] Gaber, M. M. (2006). The effects of plant-protein-based diets supplemented with yucca on growth, digestibility, and chemical composition of Nile tilapia (*Oreochromis niloticus*, L) fingerlings. *Journal of the World Aquaculture Society* 37:74–81.

[18] Gatlin III, D.M., Barrows, F.T., Brown, P., Dabrowski, K., Gaylord, G.T., Hardy, R.W.,Herman, E., Hu, G., Krogdahl, Å., Nelson, R., Overturf, K., Rust, M., Sealey, W., Skonberg, D., Souza, J.E., Stone, D., Wilson, R., Wurtele, E., (2007). Expanding the utilization of sustainable plant products inaquafeeds: a review. *Aquaculture Research* 38, 551-579.

[19] Gaylord, T.G. Rawles, S.D., (2005). The modification off poultry by-product meal for use in hybrid striped bass *Morone chrysops x M. saxatilis* diets. *Journalof the World Aquaculture Society* 36, 365–376.

[20] Glencross, BD; Booth, M; Allan, GL (2007). A feed is only as good as its ingredients – A review of ingredient evaluation strategies for aquaculture feeds. *Aquac. Nutr.,* 13, 17-34.

[21] Grant, G., I.J. Moore, N.H. Mckenzie, P.M. Doward, J.C. Stewart, L. Teleck and A.Psuztai, (1991). A Survey of Haemaglutination Properties of Several Tropical Seeds. *Livestock Research for Rural Development*, 3: 1-7.

[22] Green, B.W., D.R. Teichert-Coddington and P.R. Phelps. (1990). Response of tilapia yield and economics to varying rates of organic fertilization andseason in two Central American countries. *Aquaculture* 90: 279-290.

[23] Guo, J., Wang, Y. and Bureau, D.P. (2007). Inclusion of rendered animal ingredients as fishmeal substitutes in practical diets for cuneate drum, Nibea miichthioides (Chu, Lo et Wu). *Aquaculture Nutrition*, 13:81–87.

[24] Hardy, R.W., Sullivan, C.V., (1983). Canola meal in rainbow trout Salmogairdneri production diets. Can. J. Fish. Aquat. Sci. 40, 281 – 286.

[25] Jauncey, K. and B. Ross. (1982). A Guide to Tilapia Feeds and Feeding. Institute ofAquaculture, University of Stirling, United Kingdom.

[26] Jackson, AJ; Capper BS; Matty AJ (1982). Evaluation of some plant protein in complete dietsfor the tilapia *Sarotherodon mossambicus*. Aquaculture 27: 97-109.

[27] Koprucu, K. and Ozdemir, C. (2005). Apparent digestibility of selected feed ingredients for Nile tilapia (Oreochromis niloticus). Aquaculture, 250:308-316.

[28] Koumi, A.R., Atse, B.C. and Kouame, L.P., (2009). Utilization of soya protein as an alternative protein source in Oreochromis niloticus diet: Growth performance, feedutilization, proximate composition and organoleptic characteristics. African Journal of Biotechnology, 8(1):091-097.

[29] Kureshy, N., D. A. Davis, and C. R. Arnold. (2000). Partial replacement of fishmeal with meat and bone meal, fl ash dried poultry by-product meal, and enzyme-digested poultry by-product meal in practical diets forjuvenile red drum. North American Journal of Aquaculture 62:266–272.

[30] Mbahinzireki, GB; Dabrowski, K; Lee, KJ; El-Saidy, D; Wisner, ER (2001). Growth, feed utilization and body composition of tilapia (*Oreochromis sp.*) fed with cottonseed meal-based diets in a recirculating system. *Aquac. Nutr.*7, 189-200.

[31] National Research Council, (NRC) (1993). Nutrient Requirements of Fish.NationalAcademy Press, Washington, DC, 128 p.

[32] Nyirenda, J., Mwabumba, M., Kaunda, E. and Sales, J.,( 2000). Effect of Substituting Animal Protein Sources with Soybean Meal in Diets of Oreochromis karongae (Trewavas, 1941). Naga, The ICLARM Quarterly, 23(4):13-15.

[33] Ovie S. O., and Ovie S. I. (2007). The effect of replacing Fishmeal with 10% of Groundnut cake in the diets of *H. longifilis* on its Growth, Food conversion and Survival. J. Appl. Sci. Environ. Manage.Vol. 11(3) 87 – 90.

[34] Purseglove, J.W. (1994) Tropical crops: Dicotyledons. Longman, London, U.K. p 122.

[35] Richter, N., P. Siddhuraju, and K. Becker. (2003). Evaluation of nutritional quality of moringa (*Moringa oleifera* Lam.) leaves as an alternative protein source for Nile tilapia (*Oreochromis niloticus* L.). Aquaculture 217:599–611.

[36] Siddhuraju, P. and K. Becker (2003) Comparative nutritional evaluation of differentially processed mucuna seeds (*Mucuna pruriens* (L.) DC. var. *utilis* (Wall ex Wight) Baker ex Burck), on growth performance, feed utilization and body composition in Nile tilapia (*Oreochromis niloticus* L.). Aquacult. Res.,34, 487-500.

[37] Tacon, A. G. J. (1993). Feed ingredients for warm water fish. Fishmeal and other processed feedstuffs. FAO Fish. Circ. No. 856, FAO, Rome, Italy, 64 pp.

[38] Takeuchi, T. (1988) Laboratory work- Chemical evaluation of dietary nutrients. In:*Fish nutrition and Mariculture*. JICA Textbook, the General Aquaculture Course. (ed.by T. Watanabe). Kanagawa International Fisheries Training Centre, JapanInternational Cooperation Agency, Tokyo, pp. 179-233.

[39] Tiamiyu, L. O., Solomon, S. G., and Oketa, E. J. (2006). Effects of different boilingtime of soyabean (Glycine Max (L) Merril) on growth performance of tilapia(*Oreochromis niloticus*) fingerlings. *Journal of Aquatic Sciences* 21(1): 15-18.

[40] Wang, Y., Guo, J., Bureau, D.P., Cui, Z., (2006). Effects of dietary protein and energy levels on growth, feed utilization and body composition of Cuneate drum, Nibea miichthioides. *Aquaculture* 252, 421–428.

[41] Webster, C. D., K. R. Thompson, A. M. Morgan, E. J. Grisby, and A. L. Gannam.(2000). Use of hempseed meal, poultry by-product meal, and canola meal in practical diets without fishmeal for sunshine bass (*Morone chrysops x M. saxatilis). Aquaculture*188: 299–30.

# Analysis of rules in use the formation of committee boundary (PTB) state forest area in Indonesia

**Pernando Sinabutar[1], Bramasto Nugroho[2], Hariadi Kartodihardjo[2], Dudung Darusman[2]**

[1]Forest Center for Area Stabilization Region XII, Tanjungpinang, Indonesia
[2]Forest Management Department, Bogor Agricultural University, Bogor, Indonesia

**Email address:**
damesinab@yahoo.co.id (P. Sinabutar), bramasto2001@yahoo.co.id (B. Nugroho), hkartodihardjo2000@yahoo.com (H. Kartodihardjo), ddarusman@yahoo.com (D. Darusman)

**Abstract:** Forest area gazettment in Indonesia is done through the appointment process, boundary demarcation, mapping and determination of forest area. The process was organized by the Committee Boundary (PTB) with the involvement of several agencies in the area are coordinated by Forest Center for area stabilization (BPKH). This study analyzed the interaction of PTB in decision making and the effectiveness of the rules in-use the formation of PTB. Data were collected through in depth interviews, participant observation and document review, and then analyzed by using content analysis based IAD (Institutional Analysis and Development) Framework. The results show there has been a disparity in the interaction. Types of rules in-use the formation of PTB among others position rules, boundary rules, authority rules, agregation rules, information rules, scope rules, and pay-off rules have not been able to direct of interaction. The formation of PTB has not noticed the involvement of agencies setting, the setting position and authority, setting rights and obligations, and accountability arrangements. For improvements, the configuration of the seven types of rules that can be used because the results of the analysis showed that the seven types of rules that have found substantial weaknesses in the formation of PTB.

**Keywords:** Forest Area Gazettment, PTB, Rules In-Use, Interaction

## 1. Introduction

Characteristics of state forests as common pool resources (CPRs) requires effective arrangements that must be respected and enforced [37, 28]. It was intended to ensure ownership and control. It is a form of guarantee of legitimacy and is one of the important factors in the management of forests [10]. Legitimacy will be served to increase the successful management of natural resources including forests [37, 15], encouraging management to be effective CPRS [23] and encourage agricultural investment and boost economic growth in rural areas [38].

In Indonesia, the guarantees the ownership and control of state forest land is obtained through a confirmation process that starts from the forest area designation, boundary demarcation, mapping and forest area stipulation [32 article 12, 33 article 15, 22] the implementation is delegated to the Committee Boundary (PTB). One of the duties and authorities [21] is to determine the steps to resolve the rights of third parties stretch along the borders and in the forest. Duties and authority are confirmed also in the [22] which not only determines the steps to resolve, but also finishing third party rights (article 23). This means that the duty and authority of the PTB is to realize the legitimacy. However, the Government has not been able to provide the assurance of the control and management [14, 10], even the policy has been raising potential conflict since the very beginning [4]. This proves that the Government had a difficulty in achieving the legitimacy [1, 6, 25, 29].

Some studies suggested some activities which are able to guarantee the legality and legitimacy, for example [24] with claims verification activities, [4] to propose a formal legal approach through the scheme rural forest, community forest and social forestry, and [2] to build partnerships with the government to improve the ownership regime and [3] to strengthen the capacity of the community. The practice in Indonesia, some of the less implementable proposals for the implementation of the boundary represented by the PTB does not yet have the capacity and capability. The interaction of the actors in decision-making is still a "passive" and be done unilaterally for the appointment of involved members elements are not based on skill. The rules in-use in forming the PTB have not noticed yet the elements such as norms, sanctions, and responsibilities, this case brings it in to difficult enforcement. In fact, sanctions and enforcement are needed in the management of natural resources [9]. Enforcement of rules, without sanctions can be useless institution [7]. Therefore, knowledge of the rules of formation of PTB interesting to study. This study aims to find substantial weaknesses of the formation of the PTB, then formulate the rules overhaul to improve the effectiveness of PTB in decision making. This study will use the IAD (Institutional Analysis and Development) framework from Ostrom, because in the exogenous variable group a rules in-use are components that affect the structure of the action and the behavior situation of the actors in interaction.

## 2. Material and Methods

### *2.1. Study Area*

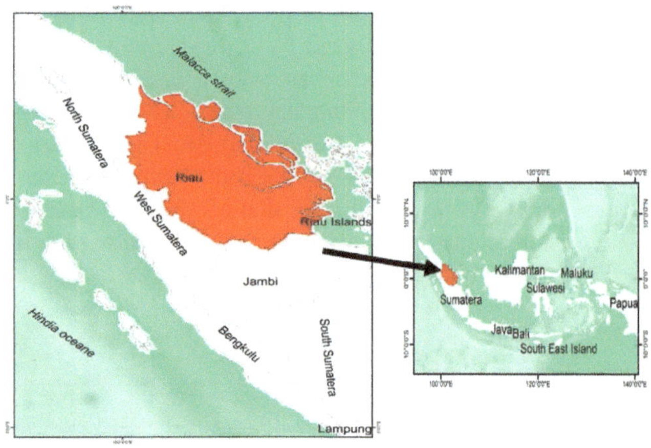


***Figure 1.*** *Study area*

This research was conducted in the Province of Riau, Indonesia (Figure 1) from June 2013 to December 2013, with the following considerations: (1) Riau Province has never done synchronization between the Provincial Spatial Plan (RTRWP) with Forest Land Use by Consensus (TGHK); (2) Riau Province has issued regulation are not approved by the Ministry of Forestry (MoF) in 1994, but guided to develop Spatial Planning District/City (RTRWK); and (3) Riau Province has the performance of forest area gazettment is still low.

### *2.2. Material*

This study uses multiple rules as material analysis, among others regulation of the Minister for Forestry Number: P.47/Menhut-II/2010 concerning the formation of PTB; regulation of the Director General for Forestry Planning Number: P.6/VII-KUH/2011 concerning technical instructions forest area gazettment and Riau Governor's decision Number: Kpts.662/v/2011 concerning the establishment of forest PTB district/city in the province of Riau including regulation of the Minister for Forestry Number: P.44/Menhut-II/2012 jo. P.62/Menhut-II/2013 concerning forest area gazettment. The analysis will be restricted to the articles related to the formation of the PTB, pay close attention at the implications for the behavior of the participants were formed, as well as the performance or impact.

### *2.3. Data Analysis*

This study uses a qualitative approach [11, 36] with the case study method [34]. Data were collected through in-depth interviews (in depth interviews), participant observation (participant observation) and a review of documents [12]. Informant determined by purposive, namely PTB and partly by snowball sampling technique [36]. Validation or testing the validity of the data is done by triangulation of sources and techniques [36]. Then, the data were analyzed using content analysis (content analysis) based on the IAD framework [17]. [16] divides the rules into seven types of rules, namely: position rules, boundary rules, authority rules, agregation rules, information rules, scope rules, and pay-off rules. This classification is directly related to the elements of the structure of the action situation, namely the participants, positions, actions of authority, control, information, results or impact, and cost-benefit. The linkage between the elements of rules in-use with action elements of the situation shown in Figure 2

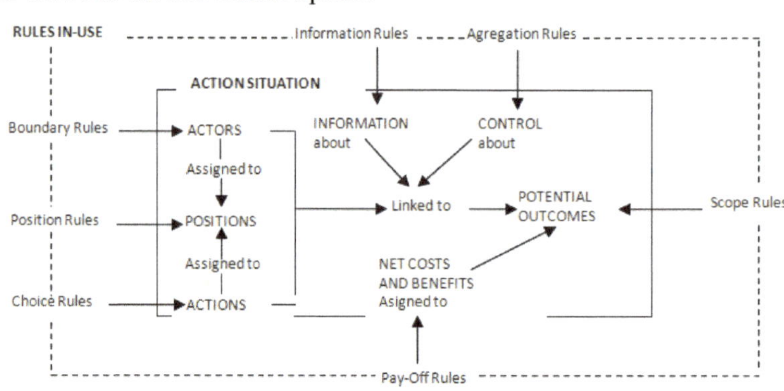


***Figure 2.*** *Linkage rules in- use to action situation [16]*

# 3. Results and Discussions

### *3.1. Interaction and the Role PTB of Forest Area Gazettment*

Structure of PTB put Regents (Bupati) as chairman, Forest Service District/City (Dinas Kehutanan Kabupaten) as secretary, BPKH, Regional Development Planning Board (Bappeda) District/City, Provincial Forestry Office (Dinas Kehutanan Provinsi), National Land Agency (BPN) District/City, subdistrict (Camat) and village chief/headman (Kepala Desa/Lurah) as an element members. The results of mapping [30], is factually put Regents and BPKH as key players and other members of the element as a marginal actor (context setters). The placement of elements that negate the role of other members in their interaction. Regents power among other legitimate power and coercive power is used to force (power to coerce) as ± 75% element PTB members are officers determined by the Regents. Although, structurally BPKH and BPN are not chosen by the Regents, but in practice these institutions are subject to the decision of the Regents. While BPKH have the power to dominate the activity boundary as in a power that is legitimate power, expert power and access to resources (knowledge, information, finance and technology) as consequence of their duties and functions. Regents of power and domination BPKH result of policy decisions into a single (one size fits for all policy) without control and the involvement of other members. As a result, the interaction only "agreed" to sign the Minutes Boundary (BATB) forest area and maps the results of the boundary. There are at least five descriptors. First, from the beginning, the boundary has been positioned to complete physical and accountable budget allocation received, so there is not opened-space coordination and negotiation.

Second, communities attributes such as knowledge, values and norms, the level of homogeneity and policy assessment gazettment unequal (unbalanced), and limited. It is providing an opportunities to make unilateral boundary. Third, they are forcing power and domination that weaken the interaction. When the power is performed, then the dominance of determining the shape of legitimacy, whereas the ideal is the legitimacy determining the power [18]. The power has reduced the involvement of elements of other members. Fourth, the boundaries do not open the claim transparently, so the results are not recognized and will be prone to be overlapped the control, licensing, or use of, even lead to resistance and conflict. Fifth, the rules in-use has not imposed sanctions in enforcement, so it does not force responsibility. Sanctions and enforcement are needed in the management of natural resources [9]. Enforcement of rules, without sanctions can be enforced is useless institution [6, 7]. Based on these findings, the role of PTB in the forest area gazettment has not effectively produce the legality and legitimacy, because the rules in-use is not efficient due to conflict with the norms. The rules will not be efficient if it is contrary to the norm, it can even lead to high transaction costs in enforcement [8].

### *3.2. Analisis of Rules In-Use*

*Rules in-use* will affect each variable in the action situation [17, 26, 35, 16]. In the IAD framework [17], the rules used will affect the structure and behavior of the actors in the interaction. The situation is built on seven action rules are used to define the actions that occur [16, 26, 27]. If there is no common sense when rules are formulated, there will be confusion about what actions are required, permitted, or prohibited [16]. Therefore, understanding the rules in-use in an institution is very important. In other words, the performance of an institutional rules can be determined through the analysis of the seven rules in-use configuration.

The performance of state forest area gazettment is determined by the interaction of PTB on the action situation. This interaction will ensure that the forest area that has been delineated to comply with the decision and is also recognized by the parties. The effectiveness of these interactions will be affected by the rules in-use, among others, the rules that define the elements involved members, position and authority, rights and obligations, financing, and accountability rules. The ability of the Government to establish by the PTB is held by considering the satisfied elements of any of the rules which will affect the results of the forest area gazettment. In this case, policy analysis can examine how far the composition rules have been consistent, whether there is a gap between the goal to reality. In the forest area gazettment in Riau Province, there has been a gap between the establishment of forest area with boundaries. For that, it is needed to find substantial weaknesses in the rules used in the formation of PTB. The results described below.

#### *3.2.1. Position Rules*

Position rules create positions for participants and determine how many can hold a position [5]. At the national level, [21 (article 5, article 7 and article 9)] it is set the position of either PTB outer boundary, the boundary function and regional boundaries and water conservation. At the level of Riau Province, the Governor determines the position of the PTB in all districts/cities including duties and authority as approved [31]. There is no defined mechanism in determining the structure of PTB. PTB positioning is done automatically, by placing the Regent as chairman and ranking officials in each work unit as a component member. All elements members are part of the government, so there is no balance in decision-making. The Chairman position politically gives him power to influence other members to sign BATB and boundary map results. Coordination is facilitated BPKH ineffectively and banged by protocol and echelon levels. As a result, coordination is not effective and decision taken unilaterally by BPKH. Society as the recipient of the impact of the boundary is only represented by the village chief/headman. The facts found in the position rules, namely:

(1) PTB is still following the rules above it normatively. There should be an explanation for the position and contribution too, for example, in terms of what the contribution of the chairman and secretary, and so are other members , not the cumulative contribution;
(2) duties and authorities are still being done normatively as the rules above them It should explain the duties and authority of the members of the appropriate elements as their duties and functions. For example, BPN is responsible for providing and giving information/evidence documents ever issued rights. Sub-district and village chief/headman in charge of ensuring forest boundaries in his village and the position of people who have been using the forest area to be delineated. Bappeda assess the forest with the spatial plans. Likewise, the role/duties of other members;
(3) no clear job description and setting mechanism. The position is attached to the structural position. Similarly, tasks and authority, there is no specification of the role which they are responsible for. For example, who responsible for the accuracy of the identification and inventory of third party rights is and who responsible for researching written documents related to evidence of third party rights is, and the others;
(4) examination of the work monitoring mechanisms have not been effective boundary. The announcements and field observation are held to fulfill administrative responsibilities only.

### *3.2.2. Boundary Rules*

Boundary rules specify the set of criteria (e.g., qualifications or eligibility of a member or a director) that participants are required to fulfill to hold positions, and they directly affect the "participants" in the action situation [5]. It sets the determining process of the qualified persons to be the members of PTB in any position including the criteria and mechanisms. [21, 13, 22]. In fact they have not set that condition. Although article (5), (7) and (9) [21] set membership, but the requirements that must be met somebody a member yet determined because the member is assigned automatically officials attached to structural or other official authorized to be appointed as members. Each ranking officials in organizational units automatically become members of the PTB. Supposedly, the rules below (Governor's or Regents decisions) are able to establish criteria and requirements of somebody become a member because of regulations above it give space to determine the members. According to the rules above them PTB members consists of the elements that can be interpreted to choose a competent member of the appropriate expertise. As a result, PTB do not have right to demand the responsibility. Substitution members due to mutations and ended his tenure is not associated with membership in the PTB. Substitution takes place automatically without handing over duties and responsibilities.

### *3.2.3. Authority Rules*

Authority rules regulate the types of actions / formal authority for each position. In other words, these rules regulate the rights and obligations of the PTB as the organizer of the boundary and make the determination of the Minister of Forestry. Rights and obligations are still normative. Rules were more focused set obligations, while the rights of each member is not regulated. This means that the rules are less demanding responsibilities. The main obligation of the rule is to determine the steps to resolve the problems related to land rights / land along the route of the line and in the forest area. Throughout its implementation, completion mechanism has not been determined. PTB does the boundary based on the instruction of the head BPKH as the member, so that there is the demarcation done unilaterally. Determination is the authority of the Minister of Forestry. Unfortunately, the authority has not been applied, it has been seen since 2009 and 2013 there has been no determination, even since colonial times there are 7648.35 Km (68.74%; 98 forest groups) which have not been determined, but has been already delineated.

### *3.2.4. Agregation Rules*

This rule governs the mechanism of decision-making, whether the individual/group or organization may decide an action individually or collectively, it is necessary to do consensus. This rule will regulate the things that lead to results boundaries accepted or rejected for the set. Setting process will be done if it meets the provisions required by Directorate of Gazettment, Stewardship and Tenure of Forest Areas (DPPTKH) among other technical provisions, the provisions of juridical and completeness of documentation. At the PTB, the setting is set in the decision making [21 article 12], namely: (1) meeting led by chairman, (2) in case of the chairman is unable to attend the meeting may delegate to officers designated by a power of attorney and was given full authority to make decisions and sign documents boundary, (3) in case of the members unable to attend the meeting, the member may delegate to an authority appointed by a power of attorney and was given full authority to make decisions and sign documents boundary, and (4) meeting shall be valid if attended by at least two-thirds of the members. Poorly the procedure of making the decision was not implemented, and it is worse in some cases, BATB and the boundaries map were signed by the chairman or other member in their own office room. It showed it was not the collective decision-making, but on their own without giving information each other.

Furthermore, article (13) states: (1) decision of the meeting shall be valid if approved and signed by the chairman, secretary, chief BPKH and the other committee members; (2) the members who do not approve of the meeting's decision, the concerned member shall submit the reasons set forth in the document boundaries; (3) in terms of meeting the decision was not approved and signed by the chairman, secretary, or head BPKH, then head BPKH report to the Governor through the head of the provincial forest service (fourteen working days); (4) based on the provincial forestry department head reports, the governor can approve and reject the results of the meeting (twenty-one working days); (5) in the governor rejected, then head BPKH (twenty-one working days) to report to Directorate General of Forestry Planning (DitjenPlan); and (6) maximum of twenty-one working days, DitjenPlan on behalf of the Minister has to make a decision. Unfortunately, the decision setting has not translated yet into operational rules under it, such as [13, 31].

The results of the boundaries have not been set because it has not complied with DPPTKH. Elements of administration including evidence are claimed disable to meet BPKH. Claims dominated by the written evidence is not right, so BPKH it difficult to meet the evidence. This rule does not apply sanctions, including the results of the boundary can not be determined. The terms have not fulfilled the determination yet are not fixed as necessary and it goes on and on. As a result there is 7648.35 km (68.74%; 98 forest groups) which have not been determined, but already delineated.

### *3.2.5. Information Rules*

Information rules specify information regarding what outcome is expected to be produced when actions are taken by the participants in positions. This rule directly affects the information for PTB on the situation action. [21] it has not been set, although [22] and operational rules that [13] has been set up on the information that. The information is intended to announce the results of temporary boundary pole to the public. Then to observe of the field situation to find out whether the rights of third parties which have not been resolved when the boundary pole announced. Installation of notice boards at every distance of ± 1 km by writing the function of the forest and group's name. Lastly, BATB forest area and boundaries results map which have been signed should be submitted to the manager and the manager is responsible for carrying out the maintenance and securing boundaries forest.

In fact, the announcement was just given to the village chief/headman head with the minutes of evidence. While the field observation was never carried out, although there is minute of evidence to prove the events administratively, but it is only for fulfilling administrative responsibilities. PTB has not formulated a mechanism announcement and field observation, as a result this activity has not been done in earnest. The boundaries result are not submitted to the manager immediately, there is no socialization when they submit it. This rule has not been set up grievance mechanisms/procedures for community complaints (claims). There is only a general arrangement of PTB obligation to respond and resolve disputes caused by third-party claim. The rules also do not set penalties, either delay or truth of boundary demarcation, the responsibility is at a minimum.

### *3.2.6. Scope Rules*

Scope rules set conditions/circumstances as a must (mandatory), not allowed (forbidden), and may be (allowed). This rule determines the desired outcome of PTB action in situations action. Authorizes the formation of PTB is the Minister delegated to the Governor and administrative preparations made by BPKH through the head of the provincial forestry service [21 article 2, paragraph (1), (2), and (3)]. The article explained that the implementation of the boundary in each district/city is the PTB is responsible to the Minister through the Governor (21 article 2 paragraph (4) and article 3). PTB is given space to determine the steps to resolve the land rights along the route of the line and in the forest. However, this is almost never been done. The identification and inventory of the rights of third parties not involved BPN, sub-district and village heads who really understand the real field situation. As a result, the activity is only reported the manners normatively. In addition, there is no job description in the implementation of the boundary. There is no setting provision and maximum percentage of area that can be removed ("out") of the forest area and the minimum that can be incorporated ("in") into a forest area. Subjective determination made by the officer/interpreter measure. The result of the interaction of PTB is qualified forest area, which has the legality and legitimacy. The fact, the interaction has produced as much as 85.37% boundaries, but which has only 16.63% of legality and not legitimacy.

### *3.2.7. Pay-Off Rules*

Pay-off rules set up costs and benefits, including incentives and sanctions for the community. PTB financing charged to the State Budget (APBN) or other funding valid sources and not binding [21 article 14 paragraph (1)]. While the license boundaries of financing borne by the permit holder or applicant [21 article 14 paragraph (2)]. Some issues that have not been regulated in the pay-off rules, namely: (1) there are no rules to prioritize the allocation of funding for the acceleration boundary, for example through the Local Budget (APBD) or other funding that does not bind; (2) there is no incentive setting for regional who does good quality boundaries; and (3) no arrangements of sanctions to members who neglect and cover up the truth of the information.

Some of the rules used to conclude that the rules in-use had not been a whole set position, membership and mechanisms to resolve the rights of third parties as a boundaries major problem. Based on the interpretations of researchers supported by the facts, rules are used (rules in use) would open space to BPKH to perform boundary element unilaterally because the members consider that the task is a task BPKH boundaries. Besides that, the PTB are not given space by the chairman or BPKH to negotiate based on the information they hold. In other words, these rules have not been able and effective to direct PTB to interact. Table 1 shows the results of content analysis to the three rules that are used in the formation of PTB.

***Table 1.*** *Abstraction three rules of content analysis results are used in the formation of PTB*

| Component analysis of rules in-use | Rules in-use | | |
|---|---|---|---|
| | **Regulation of the Minister for Forestry Number: P.47/Menhut-II/2010** | **Regulation of the Director General for Forestry Planning Number: P.6/VII-KUH/2011** | **Riau Governor's decision Number: Kpts.662/v/2011** |
| Position rules | adjust the position and duties of PTB | not set the position and duties of PTB | set position, and the PTB task normatively |
| Boundary rules | PTB set membership, but does not specify the criteria and mechanisms for determining the member | no set membership PTB | PTB set membership, but does not specify the criteria and mechanisms for determining the member |
| Authority rules | set authority PTB | authority to regulate the establishment of forest area | set authority PTB |
| Agregation rules | set of decision-making procedures | not govern the procedures of decision-making | set of decision-making procedures |
| Information rules | not organize information resulting from the implementation of the boundary | there are arrangements in organizing information through the announcement of the boundary delineation and field trips | not organize information resulting from the implementation of the boundary |
| Scope rules | regulate the authority to determine steps to resolve the issues of land rights, regulate the formation of PTB delegation of authority, and organize boundaries | not regulate the authority steps to resolve land rights. | regulate the authority to determine steps to resolve issues over land rights |
| Pay-off rules | arranging financing the implementation of the boundary | not arrange financing the implementation of the boundary | arranging financing the implementation of the boundary |

### *3.3. To Improve Rules In-Use of PTB*

Renewal policies are basically fixing norms, rights, and limits set. The renewal does not mean to quit sociological processes whereas the agreement achieve collectively. This is where the challenges will emerge, especially for those who are usually too rigid in bureaucracy and only see the rule as black - white side. [19]. Enforcement the country's forests as the CPRS through the confirmation assigning of PTB as the boundaries organizers is predicted will not be effective. Assignment policies will prioritize the technical and administrative aspects, while the problem lies on the institution that organizes the boundary, then the social and political issues. The other thing is the budget allocation is not based on social issues, but based on the length of the boundary and its use is restricted for one year. Determination of element members are attached to the structural officials, and the results of the boundary is not opened to the public, including a claim disclosure is not held transparently, it may reduce the gazettment performance. The success of the forest area gazettment will be determined by the organizers of the boundary. While the implementation of the boundary will be determined by the rules they use. Rules in-use in the formation of the PTB have not noticed values, norms and sanctions. As a result, the rules were not effectively direct PTB to interact in decision making.

***Table 2.*** *To improve rules in-use of PTB*

| Component to improve rules in-use | Criteria used to improve rules in-use | Form of to improve rules in-use |
|---|---|---|
| Position rules | establish a position based on the duties and functions, rather than on the structure of the post. | preparing job descriptions and position within the organization boundaries |
| Boundary rules | avoid direct appointment by the members of the institutions involved | develop criteria and mechanism for setting member |
| Authority rules | strengthening the authority of PTB in solving social problems field. | formulate mechanisms for resolving claims and rights of third parties |
| Agregation rules | establish decision-making mechanism is not tiered as long as it is implemented. | develop Standard Operating Procedures (SOP) decision-making |
| Information rules | specifies the type of information needed to facilitate decision-making, for instance announcement and review of field | enforcement of the rules by setting penalties, if the decision is not directed to achieve legality and legitimacy. |
| Scope rules | clarify the results to be achieved of interaction | develop a plan of action of any interaction (criteria and indicators of success forest boundary) |
| Pay-off rules | open space is not only a top-down financing, but it involves the area of financing | proposed financing through state and local budgets (APBN and APBD) |

Based on these findings, it can be concluded that the rules used have led to boundary done unilaterally, do not transparently reveal claims, and let the forest inaugural mechanism done out of the rules. The forest area gazettment reach the physical and legality targets more than the legitimacy, although the legality of the target itself is still low. The results show, from a total of 11126.35 km forest area in Riau Province, there is a determination throughout 1850.67 km (16.63%; 22 forest groups), but has been delineated 9499.02 km (85.37%; 120 forest group). Factual results of the boundary can not be used to minimize the conflict (claims), in fact the claims continue increasing over time. The performance improvement requires change in the rules at the level of the organizers. Rules in-use the formation of PTB has not been effective, so that the interaction that occurs in the decision-making tends to be "passive". The repairing rules used in the formation of PTB shown in Table 2.

## 4. Conclusion

State forest's characteristics as CPRs requires an effective setting up, respected and able to enforce the parties who are entitled and not entitled (non-excludable) to fulfill the realization of legality and legitimacy. To achieve this goal, the Government commissioned PTB to organize the demarcation as part of the confirmation process of forest area. As a result, only 16.63% have been legalistic, although it has been laid out along 85.37% limit. The results show that the rules in-use to form the PTB but this situation open a certain an opportunity BPKH to do demarcation unilaterally, they do not reveal the factual claims transparently and dominate the execution boundaries. The component of rules in-use are positition rules, are boundary rules, authority rules, agregation rules, information rules, scope rules and pay-off rules have not been able to direct the PTB to interact in decision making. In the implementation of the rule in-use PTB applied them normatively. As a result, PTB would rather pursuit physical complexion than open a space as well as doing coordination and negotiation of claims reveal transparently to solve social problems field. Therefore, the repair is required in every component of the rules in-use.

## Acknowledgement

The authors extend deeply acknowledgment to the Forestry Education and Training Center Republics Indonesia for scholarship and support to budgeting and accomplish this research. Sincerely appreciation is also extended to anonymous reviewer for correction and comments.

## References

[1] Afiff ,S. Tinjauan atas konsep "tenure security", dengan beberapa rujukan pada kasus-kasus di Indonesia. J Ilmu Sosial Transformatif, 20, pp, 227-249, 2005.

[2] A.K. Dev Roy, K. Alam, J. Gow. A Review of the role of property rights and forest policies in the management of the Sundarbans Mangrove Forest in Bangladesh. J. Forest Policy and Economics, 15, pp, 46–53, 2011. doi : 10.1016/j.forpol.2011.08.009.

[3] A.N, Ansoms, I. Wagemakers, M. Walker, M., and J. Murison. Land contestation at the micro scale: struggles for spacein the African Marshes. J. World Development, 54, pp, 243-252. doi : 10.1016/j.worlddev.2013.08.010, 2014

[4] A. Nugraha. Mengakhiri rezim konflik kehutanan: Momentum pasca keputusan MK. Dalam: Darurat hutan Indonesia. Mewujudkan arsitektur kehutanan Indonesia. Banten: Wana Aksara, 2013.

[5] A. Sarker, T. Itoh, R. Kada, T. Abe, M. Nakashima, G. Herath. User self-governance in a complex policy design for managing water commons in Japan. Journal of Hydrology, 510, pp, 246-258, 2014. doi:10.1016/j.jhydrol.2013.12.034.

[6] B. Nugroho. Land Rights of Community Forest Plantation Policy: Analysis from an Institutional Perspective. J Management Hutan Tropika, 17(3), pp, 111-118, 2011.

[7] B. Nugroho. Reformasi institusi dan tata kepemerintahan: faktor pemungkin menuju tata kelola kehutanan yang baik. Dalam: Kartodihardjo H, editor. Kembali ke Jalan Lurus: Kritik Penggunaan Ilmu dan Praktek Kehutanan Indonesia. Yogyakarta (ID): Nailil Printika, 2013.

[8] B. Voltan, P. Sebastian, F. Markus. Analysis Co-managing common-pool resources: Do formal rules have to be adapted to traditional ecological norms?. J. Ecological Economics, 95, pp, 51-62, 2013. doi.org/10.1016/j.ecolecon.2013.08.010

[9] C.C. Gibson, J.T. Williams, E. Ostrom. Local Enforcement and Better Forests. J. World Development, 33(2), pp, 273-284, 2005. doi:10.1016/j.worlddev.2004.07.013.

[10] Contreras-Hermosilla, A. and C. Fay. Strengthening forest management in Indonesia through land tenure reform: issues and framework for action. Bogor: Bogor World Agroforestry Center dan Forest Trends, 2006

[11] Creswell, J. W. Research Design Qualitative & Quantitative Approaches. Sage Publications, Inc, 1994.

[12] Denzin, N.K. Intervretive Biography: Qualitatif Research Method. Series 17. Sage Publication, 1989.

[13] Direktorat Jenderal Planologi Kehutanan. Peraturan Direktur Jenderal Planologi Kehutanan Nomor: P.6/VII-KUH/2011 tentang petunjuk teknis pengukuhan kawasan hutan, 2011.

[14] E. Effendi. Rationalizing the facts: Forest zone rationalization in the context of local spatial planning and development programs. Indonesia: Paper for the world bank, 2002.

[15] E. Ostrom, and E. Schlager The Formation of Property Right. Di dalam : Hanna S, Folke C, dan Maler KG. 1996, editor: Rights to Nature. Ecological, Economic, Cultural, and Political Principles of Institutions for the Environment. Washington DC: Island Press, 1996.

[16] E. Ostrom. Background on the Institutional Analysis and Development Framework. J Policy Studies, 39(1), pp, 7-27, 2011.

[17] E. Ostrom. Understanding Institutional Diversity. Princeton and Oxford: Princenton University Press, 2005.

[18] Haryatmoko. Etika Politik dan Kekuasaan. Cetakan ketiga Jakarta: Penerbit Kompas, 2014.

[19] H. Kartodihardjo. Kesatuan Pengelolaan Hutan (KPH) dalam politik pembaruan kebijakan. Jakarta: Bahan diskusi koalisi pembaruan kebijakan kehutanan (KPKK), 2007.

[20] J.W, Creswell. *Research Design Qualitative & Quantitative Approaches*. Sage Publications, Inc, 1994.

[21] Kementerian Kehutanan. Peraturan Menteri Kehutanan Nomor: P. 47/Menhut-II/2010 tentang pembentukan Panitia Tata Batas (PTB), 2010.

[22] Kementerian Kehutanan. Peraturan Menteri Kehutanan Nomor: P. 44/Menhut-II/2012 jo. P.62/Menhut-II/2013 tentang pengukuhan kawasan hutan, 2012.

[23] K. Kitamura, and R.A. Clapp. Common property protected areas: community control in forest conservation. J. Land Use Policy, 34, pp, 204-212, 2013. doi : 10.1016/j.landusepol.2013.03.008.

[24] Komisi Pemberantasan Korupsi (KPK). Pendekatan klaim verifikasi untuk percepatan penetapan kawasan hutan secara definitif. Jakarta: Unit Kerja Presiden Pengawasan dan Pengendalian Pembangunan (UKP4), 2013.

[25] M.A. Safitri. Keadilan agraria di kawasan hutan: menafsirkan tanggung jawab negara terhadap reforma agraria. Dalam: Hakim I dan Wibowo L R, editor. Jalan terjal reforma agraria di sektor kehutanan. Bogor: Puspijak Kementerian Kehutanan, 2013.

[26] M.D. McGinnis. An introduction to IAD and the language of the Ostrom workshop: a simple guide to a complex framework. J Policy Studies, 39(1), pp, 169-183, 2011a.

[27] M.D. McGinnis. Networks of adjacent action situations in polycentric governance. J Policy Studies, 39(1), pp, 51-78. doi: 10.1111/j.1541-0072.2010.00396.x, 2011b

[28] M.J. Mutenje, G.F. Ortmann, S.R.D. Ferrer. Management of non-timber forestry products extraction: Local institutions, ecological knowledge and market structure in South-Eastern Zimbabwe. J. Ecological Economics, 70(3), pp, 454-461, 2011. doi: 10.1016/j.ecolecon.2010.09.036.

[29] M. Sirat, L. Situmorang, G. Galudra, C. Fay, G. Pasya. Kebijakan pengukuhan kawasan hutan dan realisasinya. Bogor, Indonesia: World Agroferestry Centre, 2014.

[30] M.S. Reed, A. Graves, N. Dandy, H. Posthumus, K. Hubacek, J. Morris, C. Prell, C.H. Quinn, L.C. Stringer. Who's In And Why? A Typology Of Stakeholder Analysis Methods For Natural Resource Management. Journal of Environmental Management, 90, pp, 1933–1949, 2009.

[31] Pemerintah Provinsi Riau. Keputusan Gubernur Provinsi Riau Nomor: Kpts.662/v/2011 tentang pembentukan PTB kawasan hutan kabupaten/kota di wilayah Provinsi Riau, 2011.

[32] Republik Indonesia. Undang-undang Nomor 41 tahun 1999 tentang kehutanan, 1999.

[33] Republik Indonesia. Peraturan Pemerintah Nomor 44 tahun 2004 tentang perencanaan kehutanan, 2004.

[34] R.K. Yin. Studi kasus, desain & metode. Terjemahan dari: Case Study Research: Desain and Methods. Jakarta, Indonesia: PT. Raja Grafindo Persada, 2014.

[35] R.J. Oekarson, and R.B. Parks. The Study of Local Public Economies: Multi-organizational, multi-level institutional analysis and development. J Policy Studies, 39(1), pp, 147-167, 2011.

[36] Sugiyono. Memahami Penelitian Kualitatif. Bandung (ID): Alfabeta, 2010.

[37] U.J. Frey, and H. Rusch. Modeling Ecological Success of Common Pool Resource Systems Using Large Datasets. J. World Development, 59, pp, 93–103, 2014. doi.org/10.1016/j.worlddev.2014.01.034.

[38] V. Sivalai., Khantachavana, C.G. Turvey, R. Kong, X. Xia. On the Transaction Values of Land Use Rights in Rural China. Journal of Comparative Economics, 41, pp, 863–878, 2012. doi : 10.1016/j.jce.2012.11.003.

# Methane emission on intensive rice farming with water frequency and fertilizer management in North Sumatera

**Khadijah EL Ramija[1, 2], Zulkifly Nasution[1, 3], M. Zarlis[1, 4], Retno Widiastuti[1, 5]**

[1]Study Program of Natural Resource and Environmental Management, Graduate School, University of Sumatera Utara, Jln. Maas, Medan, Indonesia

[2]Assessment Institute for Agricultural Technology (AIAT), North Sumatera, and the Indonesian Agency For Agricultural Research And Development (IAARD), Ministry Of Agriculture

[3]Departement of Agroecotechnolgy, Faculty of Agriculture University of Sumatera Utara, jln Prof A. Sofyan No 3, Medan Indonesia, 20155

[4]Department of Mathematics, Faculty of Mathemathics and Natural Sciences, University of Sumatera Utara, Jln. Bioteknologi No. 1, Medan, Indonesia

[5]Department of Biology, Faculty of Mathemathics and Natural Sciences, University of Sumatera Utara, Jln. Bioteknologi No. 1, Medan, Indonesia

## Email address:

Khadijahramija@yahoo.co.id (K. E. Ramija), zulkifli100000@lycos.com (Z. Nasution), m.zarlis@yahoo.com (M. Zarlis), retnows2002@yahoo.com (R. Widiastuti)

---

**Abstract:** Rice cultivation during four planting seasons by using cultivation system improvements, especially in the management provision of water and fertilizer, other than increasing rice production can also reduce methane emission which is often rumored as a cause of global warming. Efforts to reduce methane emission from paddy fields must be done because of the impact of ecological damage caused by climate change due to global warming. It is estimated that by the year 2100 the average surface temperature of the earth will increase up 2 to 3° C. The experiment was conducted in the village of Purbaganda, Pematang Bandar District, Simalungun over four planting seasons from July 2011 until June 2012. The research design used in the study was split plot design which was organized into groups based on the difficulty of obtaining an ideal environmental uniformity in the field. Watering System treatment as main plot factor (A) and fertilization as subplot factor (B), with three replications. The treatments in the main plot were intermittent and continuous irrigation system, conducted to determine the amount of methane emissions in each planting season. For subplot, the fertilization treatments were based on laboratory analysis of soil, and Fertilization Recommendation of the Minister of Agriculture regulation No. 40 OT.140/2007. These were then combined with probiotic fertilization. The results showed that the pattern of methane emission varies in each treatment. The average methane emission was highest in treatment A1B1 with 338.50 kg $ha^{-1}$ per season, and lowest in A2B6 treatment with 63.25 kg $ha^{-1}$ per season. A2B6 treatment that used fertilization according to laboratory analysis with probiotic fertilization experienced intermittent irrigation process. The higher dosage of fertilizer N in treatment A1B1 led to higher methane emission than treatment A2B6. N fertilizer in rice fields can increase methane emissions due to increased rice growth, which was the source of methane biomass that increased the emission lines. Interaction between flooding and fertilization treatments that gave the highest emission during the four planting seasons was A1B1 and the lowest was A2B6. Comparison between the interaction of A1B1 and A2B6 on methane emission results was significantly different (DMRT test, P = 0.05).

**Keywords:** Methane Emission, Intensive Rice Cultivation, Intermittent and Fertilization

---

## 1. Introduction

According to Intergovernmental Panel on Climate Change [10], sources of greenhouse gases (GHGs) from agricultural activities are grouped as follows: 1) enteric fermentation; 2) livestock waste management; 3) agricultural burning; 4) burning pastures; 5) agricultural lime usage; 6) urea usage; 7) direct and indirect emissions of $N_2O$ from the soil; 8) indirect $N_2O$ emissions from livestock waste; and 9) irrigated land. Enteric fermentation process emitted methane while the management of livestock waste produced emissions of methane and $N_2O$.

The sources of $N_2O$ emissions from agricultural land were from N fertilizer, crop residue management, organic materials, and land conversion that led to mineralization of soil organic matter [27]. $CO_2$ emissions came from liming and urea fertilizing, while non-$CO_2$ emissions came from burnt crop residues such as straw (rice, corn, sugarcane) and from the burning process done at the time of conversion. Irrigation emitted methane due to anaerobic process that occurred in decomposition of organic matter in waterlogged rice soils and released into the atmosphere by plants [5] [30]. Methane gas volume from wetlands were affected by the planting season, type of irrigation, organic and non-organic fertilizers, soil type, temperature, and varieties [12].

Climate change is the most serious challenge facing the world today. Global warming caused by greenhouse gases (GHG) such as carbon dioxide ($CO_2$), methane, and nitrous oxide ($N_2O$) was often associated with agriculture. Wetlands are a significant source of methane due to the waterlogged soil condition that facilitated the formation of methane. The considerable size of agricultural areas, especially in developing countries, was identified as one of the major sources and contributors to the increase in atmospheric methane concentrations [24]; [11]. On a national scale, contribution of GHG emissions to the total paddy soil are still quite high. Therefore, efforts to reduce methane emissions from paddy soil must still be done. Mitigation measures chosen should not sacrifice aspects of production and is local-specific. In addition, mitigation efforts priority need to be directed to wetlands ecosystem that have a high potential for methane emissions, which is on irrigated land. Wetland rice fields are an important source of methane, a potent greenhouse gas [24] [11]. The first field measurements were done in California [4]; [28], followed by extensive studies in Spain [17] and Italy [7] ; [14].

Methane emission from agricultural land was estimated at 100 Tg year$^{-1}$ [25]; [18]. Indonesia that has 6.8% of the world's agricultural area allegedly contributed as much methane emissions as 3.4 to 4.5 Tg year$^{-1}$ Tg (1 Tg = $10^{12}$ g). Based on this data, wetlands are not a main contributor to the increase in global methane emissions.

Increased cropping index (ICI) is a more practical and realistic step in increasing acreage and harvest towards a sustainable rice production because it does not require a relatively large cost and can increase farmers' income, regardless of various problems related to changes in environmental quality such as the quality of soil, water, GHG especially methane and the development of pests and diseases in rice. Research on patterns of change in environmental quality to an increase in central rice production was due to the change to 400 Rice Farming Index (Rice Cropping Index -400 ) cropping intensity to support sustainable rice production.

Increased intensity of cropping to increase rice productivity conducted by Rice Integrated Crop Management approach (ICM) is an alternative to intensive management of rice on irrigated land. ICM components include integrated pest management, integrated nutrient, water, and weed treatment which have been proven to increase yield rice to 1 t ha$^{-1}$. In Tamil Nadu, India, ICM implemented in 2002-2004 planting season increased yields by 1.5 t ha$^{-1}$ [2]. Results of upland rice cultivation with ICM approach reached 4.3 t ha$^{-}$[22]. In Pinrang, South Sulawesi, agricultural intensification with ICM increased farmers' income by IDR 1,066,504 ha$^{-1}$, or 20.7% higher than that without ICM [1]. If ICM could suppress GHG emissions, this system would be an ideal rice crop management because in addition to saving agricultural inputs and increasing grain yield and farmers' income, it also reduces GHG emissions so the system becomes a sustainable and environmentally-friendly approach to rice cultivation. Wetland area, which has been rated as one emitter of greenhouse gases, will raise some arguments on its feasibility as sustainable wetland management.

This too will have an impact on the decline of paddy system benefits. Therefore, it is necessary for an in-depth study of GHG emissions from this wetland system, especially methane emissions during four planting seasons on main plots that used different irrigation treatments (continuous and intermittent irrigation). The subplot treatment used fertilization with varied levels. This study aims to analyze the level of methane emission and determine how much GHG emission of methane on technical irrigated wetland during the four seasons.

## 2. Materials and Methods

The research was conducted in July 2011 to July 2012 in the village of Purbaganda, District Bandar Simalungun Causeway which lies at 03$^0$05'03" and 99$^0$13'45.9" with climate type D1 according to Oldeman and ZOM 7 classification, with average rainfall at 1785 mm/year. The experimental design used in the study was split plot design which was organized into groups. This grouping is based on the difficulty of obtaining ideal environmental uniformity in the field. Watering System treatment as main plot factor (A) and fertilization as subplot factor (B with three replications. Treatment of irrigation systems was intermittent and continuous irrigation system (*continuous flooding*) in the main plots, conducted to determine the amount of methane emission in each planting season. Subplot treatment (fertilization treatment) was based on laboratory analysis of soil, and Fertilization Recommendation of the Minister of Agriculture regulation No. 40 OT.140/2007. These were then combined with probiotic fertilization. The treatments carried out were as follows:

1. Main Plot (A) with two Irrigation Systems:
    1.1. A1 = Irrigation Continuous Flow (*continuous flooding*),

1.2. A2 = Intermittent (*control flooding*),

2. Sub Plot (B) with eight fertilization systems:
   2.1. B1 = Recommended Fertilization, Minister of Agriculture regulation No. 40 OT.140/2007,
   2.2. B2 = Recommendation Fertilization with Laboratory analysis,
   2.3. B3 = Regulation of 40 (100% dose) + probiotics,
   2.4. B4 = Regulation of 40 (70% dose) + probiotics,
   2.5. B5 = Agriculture minister rules 40 ( 40% dose) + probiotics,
   2.6. B6 = Laboratory analysis (100% dose) + probiotics,
   2.7. B7 = Laboratories Analysis (70% dose) + probiotics and
   2.8. B8 = Laboratory analysis (40% dose) + probiotics.

As many as 16 treatments were repeated with 3 replications in order to obtain 48 experimental plots. Linear equations of the experimental design was in accord with Gomez and Gomez (1994). Observation data was collected from plant height and productive tillers, plant biomass, soil pH and methane emission in the field which was measured by using a 50 cm x 50 cm x 103 cm flexiglass box. Gas sampling was conducted at 06.00 am using a 5 ml syringe from *polypropilen* with extraction interval of 6 minutes. To obtain the linearity of methane concentration increase in one unit of time, four extractions were carried out in each gas sampling at minute 6, 12, 18, and 24. In one growing season, gas samples were extracted 3 times per phase.

Phase 1 was done at 1 month days after planting (dap), phase 2 at 2 months, and phase 3 at 3 months dap. This sampling was continued throughout the four planting seasons. Methane concentration in each unit of time was measured using gas chromatography equipped with flame ionization detector with a column using N porapak. Injector temperature was 110$^{o}$C and column temperature was 92$^{o}$C. Flux (F) of methane emitted from an area of rice soil was calculated using the equation adopted from [8] as follows.

Formula:

$$F = \frac{dc}{dt} \times \frac{Vch}{Ach} \times \frac{mW}{mV} \times \frac{273{,}2}{(273{,}2 + T)}$$

Where:

F: flux of methane (mg/m$^2$/day)

dc/dt: differences in the concentration of methane per time (ppm/minute)

Vch: volume box (m$^3$)

Ach: area of box (m$^2$)

mW: wide box methane (g)

mV: constant volume of methane (22,411)

T: average temperature during sampling ($^{o}$C)

Value of 273.2: Kelvin temperature constant.

Data of methane emissions and plant parameters were analyzed using ANOVA (analysis of variance), and Duncan's Multiple Range Test (DMRT) was used to see the significant differences between treatments. The software used for statistical test was Statistical Analysis System (SAS) version 9,0. Correlation analysis was performed to examine the relationship between plant parameters with methane emissions and was done with Microsoft Excel 2007.

# 3. Results and Discussion

### *3.1. Methane Emissions*

Methane emission patterns vary in each treatment as shown in Figure 1. The highest average value of methane emissions occurred in A1B1 treatment in the amount of 349 kg ha-[1] per planting season. The lowest occurred in A2B6 with 58.33 kg ha-[1] per planting season. A1B1 treatment was in accord with Fertilizer Recommendation by Minister of Agriculture regulation No. 40 of 2007 (112.5 N + 27 P2O5 + 30 K2O) with continuous flooding process that led to the creation of anaerobic conditions that are very suitable for methanogenic bacteria which produced methane.

In A2B6 treatment that used fertilization according to laboratory analysis (100% dose = 90 N + 40.8 + 22.3 P2O5 K2O) with probiotic fertilization, an intermittent irrigation process occurred. A1B1 used a higher dose of N than A2B6 that led to a higher methane emission than A2B6. N fertilizer in wetlands can increase methane emission due to increased rice growth as the source of methane in the form of biomass that increased the emission lines [3].

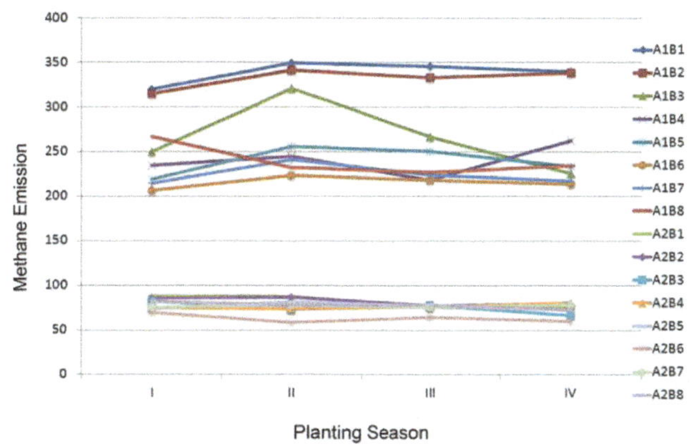


***Figure 1.*** *Methane emissions for four planting seasons*

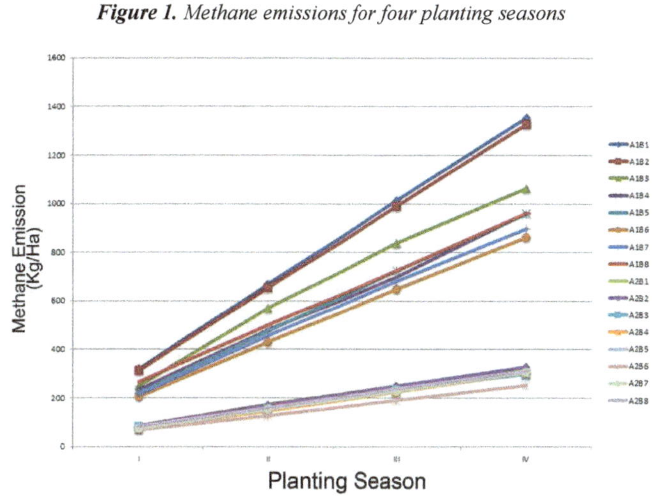


***Figure 2.*** *Cumulative methane for four planting seasons*

Methane emission patterns during four planting seasons showed the tendency of increased methane emission due to increased intensity of cultivation with continuous flooding. However, in intermittent irrigation, emissions tended to be constant. Figure 2 shows the cumulative increase in methane emissions during the four seasons. The highest cumulative increase in methane emission in each planting season occurred in treatment A1B1 with average emissions of 338.5 kg ha-$^{1}$ per season. The lowest was in treatment A2B6 (intermittent and fertilization by laboratory analysis as well as the addition of probiotic fertilizer).

Total methane emission during four planting seasons in one year is presented in Figure 3. Cumulatively, over the four planting seasons, the highest total emission was 1354 kg ha-$^{1}$ found in treatment A1B1, while the lowest was 253 kg ha-$^{1}$ in treatment A2B6. This result is consistent with the results by [29] which suggested that the soil condition with continuous flooding emits relatively higher emission than muddy and intermittent irrigation.

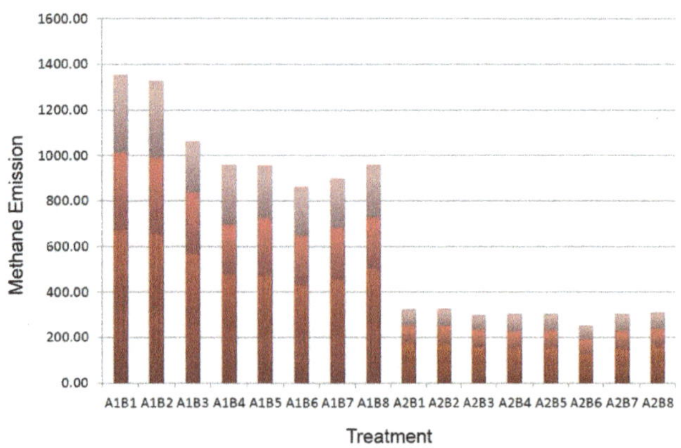


***Figure 3.*** *Total methane emissions in each treatment during four planting seasons*

In A2B6 treatment that used fertilization according to laboratory analysis with probiotic fertilization, an intermittent irrigation process occurred. This is consistent with studies that showed intermittent treatment produced lower emission than continuous flooding. Drying created aerobic condition in the soil and methanotroph bacteria that oxidized methane into $CO_2$, therefore, lots of methane were oxidized before being released into the atmosphere [29]. Furthermore, [21] proposed that of all the methane produced in the soil only 16.6% were emitted and the rest was oxidized.

The low emission from intermittent irrigation was caused by the increase in soil oxidation-reduction value. Therefore, indirect reductive decomposition did not occur. Intermittent treatment was intended to regulate land condition to become dry and waterlogged consecutively. In addition to saving water, intermittent irrigation can give roots a chance to receive air, allowing it to grow deeper. Intermittent irrigation provides benefits to agricultural land such as preventing iron, preventing accumulation of $H_2S$ which can impede root development, activate beneficial microbes, reduces collapse rate, reduces the number of unproductive tillers, harmonize ripening of grain, accelerates harvest time and eases the immersion of fertilizer in the soil. According to [20], the total emissions of methane with various rice crop management are very different, namely: total methane emissions in inundated non-ICM was 289.93 kg/ha/season; non-ICM with intermittent irrigation was 57.87 kg/ha/ season; ICM with intermittent irrigation was 78.33 kg/ha/season; and ICM with continuous flooding was 347.03 kg / ha / season.

***Table 1.*** *The value of F towards methane emissions over four planting seasons*

| Treatment | Planting season I | | Planting season II | | Planting season III | | Planting season IV | |
|---|---|---|---|---|---|---|---|---|
| A | 2374.64 | ** | 418.29 | ** | 2320.66 | ** | 1651.66 | ** |
| B | 8.00 | ** | 13.61 | ** | 13.20 | ** | 25.00 | ** |
| AxB | 4.64 | ** | 8.67 | ** | 11.95 | ** | 19.15 | ** |

** = significantly different at 1% level;
A = flooding system; B = fertilization; AxB = interaction between the fertilization and flooding factors

Based on the F value in Table 1, it can be seen that treatment flooding system (A), fertilization (B) as well as the interaction between irrigation and fertilization treatments (AxB) on methane emission results were significantly different at 1% level. This suggests that the different flooding treatments (A1 and A2) had a significant effect on the results of methane emissions. Similarly, the different fertilizer treatments (B1, B2, B3, B4, B5, B6, B7 and B8) and the interaction of flooding and fertilization treatments (A1B1, A1B2, A1B3, A1B4, A1B5, A1B6, A1B7, A1B8, A2B1, a2b2, A2B3, A2B4, A2B5, A2B6, A2B7, and A1B8) also affected the results of methane emissions significantly.

From the interaction between flooding and fertilization treatments (AxB), it can be seen that the highest emission during the four planting seasons was in A1B1 and the lowest was in A2B6. Comparison between A1B1 and A2B6 interaction on methane emission results was significantly different. To see the influence of fertilization and irrigation on methane emissions during four planting seasons can be seen in Table 2.

***Table 2.** Effect of Fertilization and Irrigation on methane emissions during four planting seasons*

| Treatment | PSI | | PS II | | PS III | | PS IV | |
|---|---|---|---|---|---|---|---|---|
| A1B1 | 320.00 | a | 349.00 | a | 345.67 | a | 339.33 | a |
| A1B2 | 314.67 | a | 341.33 | a | 333.00 | a | 338.00 | a |
| A1B3 | 250.00 | bc | 320.67 | a | 266.67 | b | 225.33 | c |
| A1B4 | 234.67 | bcd | 244.67 | b | 218.33 | c | 262.33 | b |
| A1B5 | 219.00 | cd | 256.00 | b | 250.33 | c | 233.67 | c |
| A1B6 | 206.33 | d | 223.33 | b | 218.00 | c | 214.00 | c |
| A1B7 | 214.67 | cd | 241.00 | b | 224.00 | c | 217.67 | c |
| A1B8 | 266.67 | b | 232.33 | b | 226.67 | c | 234.33 | c |
| A2B1 | 87.67 | e | 87.33 | c | 75.33 | d | 75.00 | d |
| A2B2 | 85.67 | e | 86.33 | c | 77.67 | d | 79.00 | d |
| A2B3 | 82.33 | e | 73.33 | c | 78.00 | d | 66.33 | d |
| A2B4 | 74.67 | e | 74.33 | c | 76.00 | d | 80.67 | d |
| A2B5 | 74.00 | e | 81.33 | c | 77.00 | d | 72.33 | d |
| A2B6 | 69.67 | e | 58.33 | c | 64.67 | d | 60.33 | d |
| A2B7 | 75.33 | e | 77.33 | c | 75.67 | d | 78.33 | d |
| A2B8 | 82.00 | e | 77.33 | c | 78.67 | d | 72.33 | d |

Equal numbers followed by the same letter are not significantly different at P = 0.05 DMRT

According to Table 2 it can be seen that the effect of fertilization on continuous flooding treatment (A1) were significantly different for methane emissions, while intermittent irrigation treatment (A2) for four planting seasons did not significantly affect methane emissions.

In the first planting season, fertilizer treatment B1 gave the highest methane emissions (320.00 kg ha-1) followed by fertilization B2 (314.67 kg ha-1). Statistically, the total methane emission from B1 and B2 was not significantly different (DMRT; P = 0.05), but significantly different to other fertilizers, i.e. B3, B4, B5, B6, B7 and B8.

In the second planting season, fertilizer treatment B1 also gave the highest methane emissions (349.00 kg ha-1) followed by B2 and B3. The total emission from all three was not significantly different, but significantly was to other fertilizers, i.e. B4, B5, B6, B7 and B8.

In the third planting season, fertilizer treatment B1 again gave the highest methane emissions (345.00 kg ha-1) followed by B2 (333.00 kg ha-1). While both treatments showed nearly equal methane emissions, they were significantly different compared to B3, B4, B5, B6, B7 and B8.

The fourth planting season was the same as the previous one, with fertilizer treatment B1 emitting the highest methane (339.33 kg ha-1) followed by fertilization B2 (338.00 kg ha-1). The total was almost equal, but significantly different with B3, B4, B5, B6, B7 and B8.

In general, during the four cropping seasons, i.e. A1B1 treatment with continuous flooding that followed the Fertilization Recommendation by Minister of Agriculture regulation No. 40 of 2007 showed the highest methane emissions, while the lowest was in intermittent water treatment A2B6 with fertilization laboratory analysis (100% dose) and probiotic fertilizer. The effect of fertilization on methane emissions during the four planting seasons can be seen in Table 3 below.

***Table 3.** Effect of fertilization on methane emissions during the four planting seasons*

| Fertilizing | PS I | | PS II | | PGS III | | GS IV | |
|---|---|---|---|---|---|---|---|---|
| B1 | 203.83 | a | 218.17 | A | 210.50 | a | 207.17 | a |
| B2 | 200.17 | a | 213.83 | A | 205.33 | a | 208.50 | a |
| B3 | 166.17 | bc | 197.00 | A | 172.33 | b | 145.83 | c |
| B4 | 154.67 | bc | 159.50 | Bc | 147.17 | c | 171.50 | b |
| B5 | 146.50 | c | 168.67 | b | 163.67 | bc | 153.00 | c |
| B6 | 138.00 | c | 140.83 | C | 141.33 | c | 137.17 | c |
| B7 | 145.00 | c | 159.17 | Bc | 149.83 | bc | 148.00 | c |
| B8 | 174.33 | b | 154.83 | Bc | 152.67 | bc | 153.33 | c |

Equal numbers followed by the same letter are not significantly different at P = 0.05 DMRT

The effect of fertilizer treatment on methane emission during the four planting seasons showed that B1 fertilization that followed the Fertilization Recommendation by Minister of Agriculture (112.5 N + 27 P2O5 + 30 K2O) generally produced the highest methane emissions. The lowest emission produced was in B6 that utilized laboratory analysis (100% dose N = 90 + 22.3 + 40.8 P2O5 K2O) fertilizer with probiotics. Based on the DMRT for the four planting seasons, the methane emission from B1 was significantly higher than B6. This indicated that fertilization affected the activity of methanogenic/methanotroph bacteria in forming methane emissions.

### 3.2. Correlation of Maximum Number of Tillers with Methane Emission

Correlation of the results of the methane component can be seen in Table 4 below.

***Table 4.** Correlation of the Component of Methane Emissions*

| Yield Component | Methane Emissions | |
|---|---|---|
| | A1 | A2 |
| Productive tillers | 0.4236 | -0.2522 |
| Maximum number of tillers | 0.2966 | -0.3863 |

Correlation with the total number of productive tillers with total methane emission showed a positive correlation (r = 0.4236) in the flooding, while intermittent treatment showed a negative correlation (r = -0.2522). The same applies in the relationship that showed the maximum number of tillers. In flooding, the correlation was positive (r = 0.2966), while in intermittent treatment it was negative (r = -0.3863). These mean that an increase in the number of productive tillers or maximum number of tillers would increase the total methane emission by flooding. This is in line with observations by [26] which showed that the number of tillers can increase the density and number of aerenchyma vessels, increasing the transport capacity of methane. According to [23], the more root exudates formed, the higher the methane emission will be.

In contrast, in intermittent treatment, an increase in the number of productive tillers or maximum number of tillers will not increase the total methane emissions. According to [21], intensification by using flooded system (anaerobic) will hamper the biological properties of the soil in addition to increasing greenhouse gas emissions and inhibit the development of rice plants root system. Biological diversity is very limited under anaerobic conditions. Aerobic soil biota can not develop and it is estimated that only 25% of rice plant roots were well-developed. Cropping system with aerobic (moist) root system production was at least 3-4 times greater compared to flooded system. Optimal development of the root system, supported by biodiversity in the soil can increase rice yield potential.

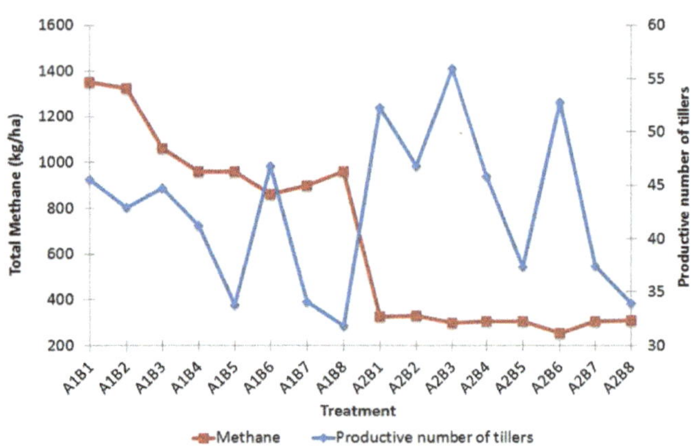


***Figure 4.*** *Maximum Number of tillers in correlation with Total Methane Emissions*

Figure 4 shows the relationship between the maximum number of tillers with the total amount of methane. The increase in the number of tillers did not increase the total methane emission, depending on the treatment. This can be seen in treatment A1B6, A2B1, A2B3 and A2B6. Treatment A1B6 and A2B6 are B6 fertilization treatments at the same dosage (90 N + 22.3 + 40.8 P2O5 K2O) with probiotic fertilizer. Therefore, it can be concluded that B6 fertilization produced the lowest total methane emissions with continuous or intermittent flooding. The maximum number of tillers during four planting seasons in one year was the highest in treatment A2B3 in the amount of 105 tillers, followed by A2B6 with 104 tillers. However, for total emission, A2B6 produced the lowest methane emissions at 253 kg ha-$^{1}$, while A2B3 produced 300 kg ha-$^{1}$ methane emissions.

Based on the previous explanation that the lowest methane emissions was from treatment A2B6, it is necessary to see its correlation with the maximum number of tillers with the total methane produced in treatment A2B6 (Figure 5).

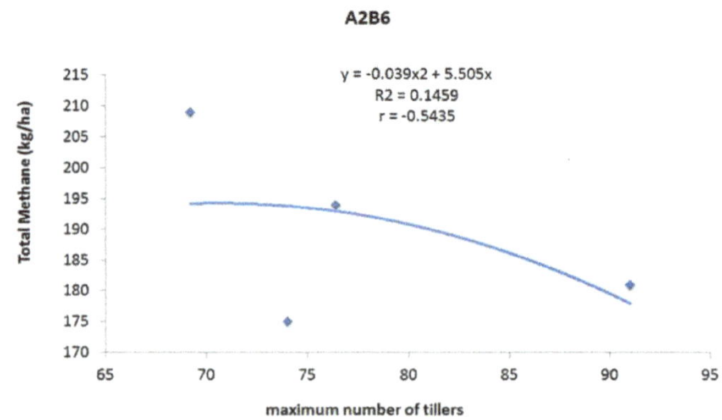


***Figure 5.*** *Maximum Number of Tillers in correlation with Total Methane treatment A2B6*

From Figure 5, it can be seen that the correlation between the maximum number of tillers with the total emission was a strong negative correlation (r = -0.5435). The functional relationship between the maximum number of tillers with the total methane was R2 = 0.1459. This indicates that the maximum number of tillers affects the outcome of methane emissions. In A2B6 treatment, if the number of productive tillers increased, the emission actually declined.

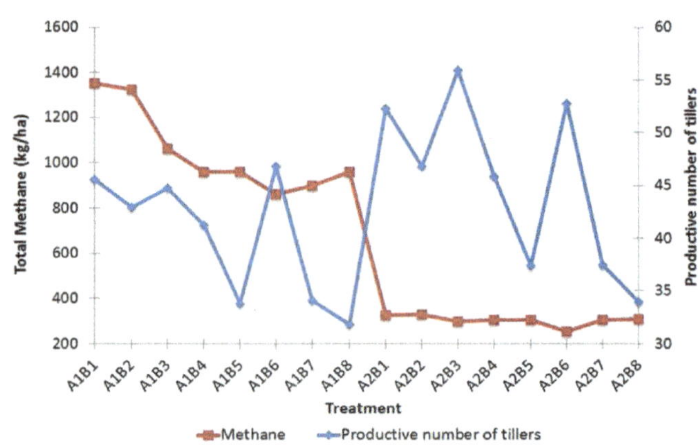


***Figure 6.*** *The number of productive tillers with Total Methane Emissions*

From Figure 6 can be seen the correlation between the maximum number of tillers with the total emission in each treatment. The graph shows that the increase in the total or number of productive tillers did not increase the total methane emission, depending on the treatment. The highest number of productive tillers during four planting seasons in a year was in treatment A2B3 with 56 tillers followed by A2B6 with 53 tillers.

### *3.3. Correlation of Productive Number of Tillers with Methane Emission*

Based on the previous explanation that the lowest methane emissions was from treatment A2B6, it is necessary to see its correlation with the number of productive tillers with the total methane produced in treatment A2B6 (Figure 7).

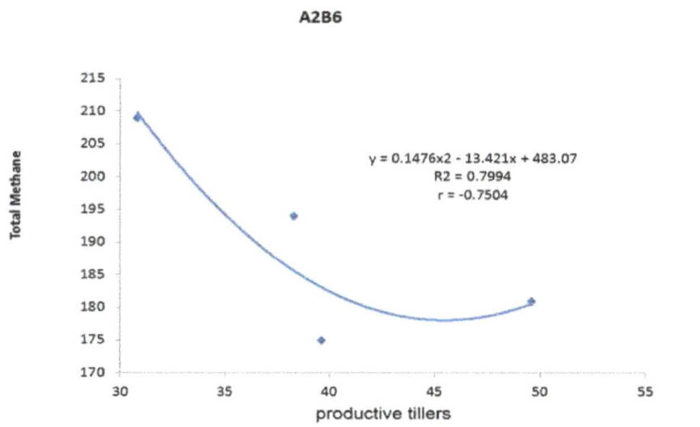


***Figure** 7. Total Productive Tillers relationship with Total Methane treatment A2B6*

From Figure 7, it can be seen that the correlation between the total number of productive tillers with the total e mission was a strong negative correlation (r = -0.7504). The functional relationship was R2 = 0.7994, greater than the maximum number of tillers with R2 = 0.1459. This indicates that the maximum number of tillers affects the outcome of methane emissions.

## 4. Conclusions and Recommendations

Intensive rice cultivation by increasing cropping intensity does not degrade the quality of soil and water quality and methane emission can be reduced by up to 66.05% with ICM approach. Methane emissions during planting season with four treatments showed that A1B1 with continuous flooding that followed the Minister of Agriculture regulation on Fertilization Recommendation No. 40 of 2007 produced the highest methane emissions. The lowest emission was in A2B6 with intermittent flooding that followed fertilization laboratory analysis (100% dose) with probiotic fertilizer. The number of productive tillers was more influential on the outcome of methane emissions than the maximum number of tillers. In A2B6 treatment, if the number of productive tillers increased, the emission actually declined.

## References

[1] Arafah. 2005. Assessment Intensification rice field base on Integrated Crop Management in Pinrang Regency. South Sulawesi Assessment Institute for Agricultural Technology. http://jabar. litbang.deptan. go.id/html.tp.html [05 Jan 2007].

[2] Balasubramanian, V. 2006. Integrated Crop Management (ICM): Field evaluation and lesssons learnt *In* Sumarno, Suparyono, A.M. Fagi, M.O. Adnyana (*Eds*.). Rice Industry, Culture and Environment. Book 1. Indonesian Center for Rice Research. p. 33-42.

[3] Cheng-Fang, L., Dan-Na, Z., Zhi-Kui, K., Zhi-Sheng. Z., Jin-Ping, W., Ming-Li. C., and Cao Cou-Gui, C. 2011. Effects of Tillage and Nitrogen Fertilizers on methane and $CO_2$ Emissions and Soil Organic Carbon in Paddy Fields of Central China. National Tecnology Project for High Food Yield of China. On line : http://www.ncbi.nlm.nih.gov/pmc/articles/PMC3344821/

[4] Cicerone, R. J., And J. D. Shetter. 1981. Sources of atmospheric methane: measurements in rice paddies and a discussion. *Journal of Geophysics Research* 86C: 7203–7209.

[5] Conrad, r., and f. Rothfuss. 1991. Methane oxidation in the soil surface layer of a flooded rice field and the effect of ammonium. *Biology and Fertility of Soils* 12: 28–32. (Cicerone & Shetter, 1981; Cicerone *et al.*, 1983)

[6] Gomez, A.K. And A.A. Gomez. 1994. Statistical Procedures For Agricultural Research. Translation. Sjamsuddin, E and S. B. Yustika (1995). Prosedure Statistics to Research. Puslisher University of Indonesia Press. Jakarta. 689 p.

[7] Holzapfel-P Schorn A, Conrad & Seiler W, 1986. Production of Methane in rice Padies. Fems icrobiol. Ecol 31: 343-351

[8] (IAEA). 1993. Manual on measurement of methane and nitrous oxide emission from agricultural. Vienna.

[9] IPCC-Intergovernmental Panel on Climate Change. (1997a). *Guidelines for National Greenhouse Gas Inventories*: Workbook, Chapter 4.3, Agriculture: Rice Cultivation. OECD, Paris, France.

[10] IPCC-Intergovernmental Panel on Climate Change. (2006). Guidelines for National Greenhouse Gas Inventories: Reference Manual, Chapter 4.3, *Agriculture: Methane emission from rice cultivation*. OECD, Paris, France.

[11] IPCC (International Pannel For Climate Change) . 1994 . Climate change 1994. Cambridge University Press, Cambridge, UK.

[12] Minami, K. (1995) The effect of nitrogen fertilizer use and other practices on methane emission from flooded rice. *Fertilizer Res*. 40:71-84.

[13] Makarim, A. K., A. Bey, R. Boer and P. Setyanto (1995). Methane *gas from lowland rice fields in Indonesia:* Development and research directions (in Indonesian). Central Research Institute for Food Crops, Bogor, Indonesia.

[14] Schütz H., A. Holzapfel-Pschorn, R. Conrad, H. Rennenberg and W. Seiler (1989). A three-year continuous record on the influence of daytime season and fertilizer treatment on methane emission rates from an Italian rice paddy field. *Journal of Geophysical Research 94: 16405-16416.*

[15] Schütz H., Seiler W & Lonrad R (1994) Methane Production And Emmision From Four Reclaimed Pristine Wetlands of South Eastern United States Soil sciense Am. J. 58: 1270-1275

[16] Setyanto P, A. Karim Makarim and Mudzakkir Fagi, 1997. Methane Emission From Rainfed Rice Fields at Jakenan, Central Java as affected by Organic Matter and Water Condition.

[17] Seiler, W., A. Holzapfel-Pschorn, R. Conrad And D. Scharfe (1984). Methane Emission From Rice Paddies. *Journal Of Atmospheric Chemistry* 1: 241-268.

[18] Seiler, W., Conrad, R., Scharffe, D., 1984. Field studies of methane emission from termite nests into the atmosphere and measurements of methane uptake by tropical soils. Journal of Atmospheric Chemistry 1, 171-186.

[19] Setyanto, P. 2004. Methane emission and its mitigation in rice field under different management practices in Central Java. Disertasi Fakultas Pascasarjana University Putra Malaysia.

[20] Setyanto P and Rina K, 2008. Management systems rice field low metan emissions. A Journal. A Journal Agricultural Reseach Food Crop. Volume 27.

[21] Setyanto, P. 2004. Mitigation methane of rice field. In F. Agus *et al.* (*Eds*.). The low land and management technology. The central research and development of Soil and agroklimat. Bogor. P. 287-294.

[22] Toha, H.M., Permadi K., Prayitno, and I. Yuliardi. 2005. The increase production of upland rice with integrated crop management and integrated resources. Indonesian Center for Food Crops Reseach and Development. Seminar. Bogor. July, 2005

[23] Wihardjaka, 2001. Methane emission in low land irigation with aplication of several organic matter. Agrivitas 23 (1):

[24] Wang M, Shangguan X& Ding A (1996). Methane in riceagriculture in from Cicerone *et al.*, 1983 Atmosphericcirculation to Global Change. The Institute of Atmospheric Physics, Chinese Academic of Sciences, p 647-659. China Meteorological Press Beijing.

[25] Yagi, K. and K. Minami. 1990. Effects of organic matter applications on methane emisión from Japanese paddy fields. *In* A.F. Bouwan (*Eds*.). Soil and the greenhouse effect. John Wiley and Sons. New York. p.467-473.

[26] Aulakh, M.S., R. Wassamann, H. Rennenberg, and S. Fink. 2000. Pattern and amount of Aerenchyma realate to variable methane transport capacity of different rice cultivars. Plant Biology 2: 182-194.

[27] Bouwman, A. F., Boumans L.J.M., Batjes N.H., 2002. Emissions of $N_2O$ and from fertilized fields: summary of available measurement data. Global Biogeochem Cycles, 16: art.no. 1058

[28] Cicerone, R. J., And J. D. Shetter and C.C. Delwiche, 1983. Seasonal variation of methane flux from California rice paddy. J. Geophys. Res., 88:11,022-11,024

[29] Setyanto P., Suharsih, A,K Makarim, and J. Sasa. 1999. Inventory of emissions and mitigation of methane on rice field in Jakenan. Symposium on Food Crops IV, Bogor, 22-24 November 1999.. Indonesian Center for Food Crops Reseach and Development (ICFORD). Bogor

[30] Nouchi I, Mariko S, Aoki K (1990) Mechanism of methane

[31] transport from the rhizosphere to the atmosphere through rice plants.Plant Physiology, 94, 59–66.

# Wetland management in Bangladesh: A study on Beel Bakar

**Shoumo Khondoker[1, 3], Md. Lokman Hossain[2, *], Khondoker Azizul Hoque Moni[3]**

[1]Department of Fisheries and Marine Bioscience, Jessore University of Science and Technology, Jessore 7400, Bangladesh
[2]Department of Global Change Ecology, Faculty of Biology, Chemistry and Geosciences, University of Bayreuth, D-95440 Bayreuth, Germany
[3]Ulashi Sreejony Sangha, Jessore 7400, Bangladesh

**Email address:**
shoumo100429@gmail.com (S. Khondoker), lokmanbbd@gmail.com (M. L. Hossain), uss.jessore@gmail.com (K. A. H. Moni)

**Abstract:** Being a riverine country, Bangladesh maintains a large volume of water by its different channels like river, canal, beel, haor, baor, lake and so on. Beel is one of the ancient types of water body in the context of Bangladesh. Among all the types of water bodies existing in Bangladesh, beel is obviously a fascinating one due to its biodiversity and ecological importance in local and national perspective as it contains and passes significant amount of water connecting with river channel which ultimately goes to the sea. The Beel Bakar is a small beel in the southern part of Jessore District of Bangladesh. This study initiated to reveal the existing situation of the aquatic biodiversity of the Beel Bakar along with identifying the past and present water management system, the socio economic aspect of the beel dependent fishermen and possible future steps. The study was carried out through mix group discussion, focus group discussion, key informants interview, direct observation and household survey. The study revealed that once this beel was rich in biodiversity and important for local fisheries. Though a considerable part of the beel and its floodplain is under gher culture, still there is scope to ensure protection and to enhance livelihoods of fisher community. In spite of the degradation of natural habitats of fishes, still 19 species of fishes were observed in the beel as well as some non fish organisms like snail, crab, mussel, frog, migratory and native birds. The study also found that excessive gher culture practice and possession of non fishermen over the water body are the main constrains of the beel. Most of the fishermen live in semi concrete house and vast majority of them have debt to bank or cooperatives. Experts and local fishermen both suggested ensuring the access of real fishermen in Beel Bakar and the Mukteswari River to improve their circumstances.

**Keywords:** Wetland, Beel, Ecosystem, Fisheries, Floodplain, Water Management

## 1. Introduction

Wetlands play a crucial role in maintaining the ecological balance of ecosystems. However, exponential growth of population, unplanned land use, intensive agriculture, overfishing, siltation, pollution, rapid urbanization, lack of intra-sectoral coordination, lack of awareness, etc. play a tremendous impacts on both biotic and abiotic components of wetlands [1]. Nishat [2] pointed out that the degradation of wetlands in Bangladesh were mainly due to: increase of population and expansion of human habitats; expansion of agriculture and subsequent conversion of wetlands through drainage into rice fields; flood control and irrigation project for enhancement of agricultural productivity; national, local and rural infrastructures like ill-planed roads; narrow culvert etc.; over-felling of wetland trees; over-grazing by livestock; over-fishing and associated disturbances; siltation due to degradation of watershed areas which are often trans-boundary in nature; indiscriminate control/ regulation / use of water flows of main river systems in upper riparian; and pollution of water due to industrial, urban, agrichemical and other types of pollutants including pollution from trans-boundary sources. Wetland provides habitat for a variety of resident and migratory waterfowl, a significant number of endangered species, and a large number of commercially important species. Almost 50% of the country's people are directly dependent on wetlands resources [3]. Bangladesh is estimated to possess seven to eight million ha of wetlands in the form of permanent rivers and streams (0.48 Mha), estuarine and mangrove swamps (0.61

Mha), shallow lakes and marshes (0.12-0.29 Mha), large reservoirs (90,000 ha), small ponds and tanks (0.15-0.18 Mha), shrimp ponds (0.09-0.115 Mha) and seasonally flooded flood plains (5.77 Mha) [2].

Beel is referred to surface water body that accumulates surface runoff water through internal drainage channel [4]. Beels are important wetlands and regarded as valuable fish and wildlife habitat [4,5]. In the active floodplains of the Surma-Meghna, the Brahmaputra-Jamuna, and the Ganges-Padma river systems, there are several large and small beels. In Bangladesh, there are thousands of beels of different sizes. Most of these large beels have shrunk quite considerably in recent decades [4,5]. Akonda, [6] and Khan *et al.,* [7] recorded that total area of beels and haors in Bangladesh is 1142 ha.

Beels and wetlands have not yet received proper attention in our country. In the 2011-12 fiscal year beel fisheries provided 85,208 metric tons natural and culture fish production where production was 746 kg/ha [8]. This plays a significant role in fulfilling the demand of Small Indigenous Species (SIS) and other viable species. These fisheries aspect connects directly with the livelihood of a huge number of fishermen and other people's dependant on the beel. Livelihood of real fishermen is immersed with debts, sorrows and many social and economic restraints. Though there are few initiatives but they are not adequate and in many cases not practical due to influence of vested people and as those approaches are not participatory. Usually beel dependent people are two types: farmers and real fishermen. Former one is based on individual property ownership pattern whereas, later one is based on common property management. Value wise paddy is produced more in the beel compared to production of natural fisheries. Price wise value of fish produced from the beel exceeds the value of rice production. Again cost for agricultural activities is huge thus net benefit is very marginal compared to its investment whereas, the cost of management and harvest in natural variety is very low compared to its nominal investment. Many parts of the floodplain of the beel have been converted to gher (a small confined area mainly used for aquaculture but sometimes agricultural product during dry season). There are benefits and detriments of gher fish culture over the natural fisheries and agriculture. Direct income and scope for employment are available there, thus some people manage ghers and living better life but their number is very few. This system has led to negative impact on the environment too. Livelihoods of the real fishermen of this beel dependent are very measurable. Some of them tried to adjust with other work but could not cope very nicely. Only few fishermen were able to take lease of land and doing fish culture. Most of the fishermen still prefer that they get an ensured access of the Mukteswari River which is their last resort. The study of Ahmed, *et.al.,* [9] ; Nishat, [2]; Gnauck and Islam, [10]; Islam, [11]; Islam and Gnauck [12]; Islam and Gnauck, [13]; Islam and Gnauck, [14]; Islam, [15]; Islam and Gnauck, [16] ; Akonda, [6]; Khan *et al*., [7]; and Barbier *et al*., [17] depicts the wetlands resources in Bangladesh, their present situation, causes of degradation and management aspects. No comprehensive study yet undertaken to assess the past and present situation of Beel Bakar. Therefore, it is necessary to assess the present situation, existing livelihood, fisheries resources as well as the current management practice of beel Bakar so as to summarize the problem and recommend prospects. The study attempts to assess the present and past status of Beel Bakar, Monirampur, Jessore, Bangladesh. The study also reveals the state of the livelihoods of real fishermen in the context of converting the floodplain to the fish culture gher.

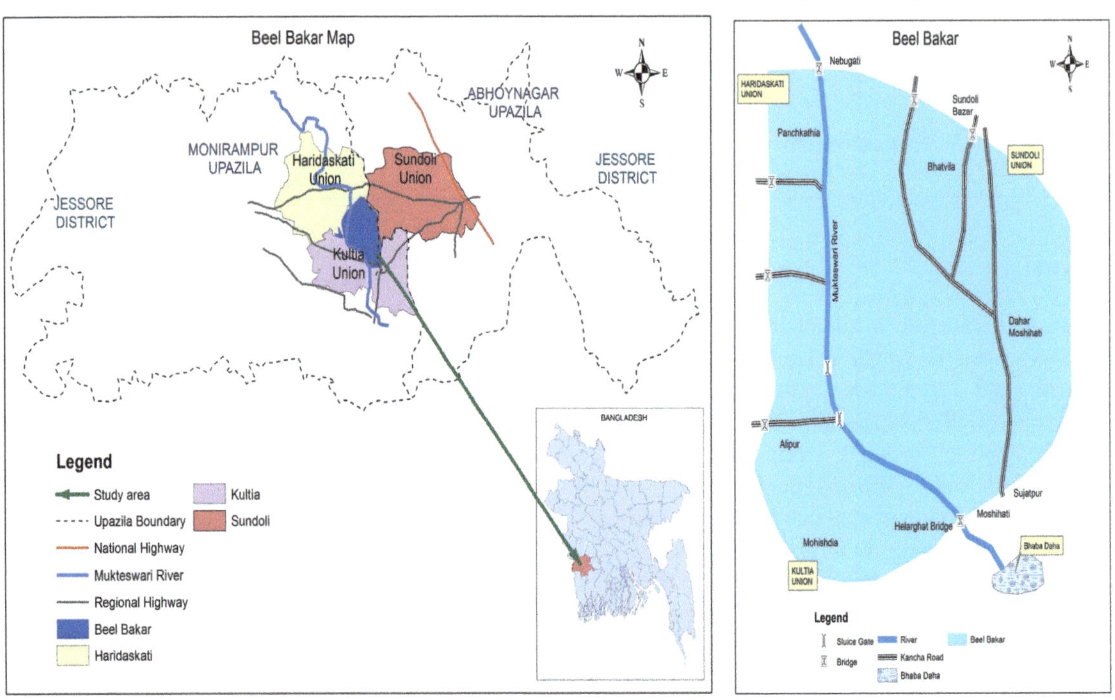


***Figure 1.*** *Study area (Beel Bakar covering the surrounding area of Mukteswari River)*

### *1.1. Study Area*

The Beel Bakar is one of the important beel under the Dakatia Beel Cluster and one of the first beels under the Mukteswari River system. The beel is situated mostly north of the Moshihati Bazar and south of the Monirampur-Abhoynagar road. The Mukteswari River runs through it from north to south and western side of the river belongs to Monirampur Upazilla (sub-district) and eastern side is under Abhoynagar Upazila. Haridashkati and Kultia are two unions (group of nearby villages) in which the beel is situated. To choose this beel as study site is relevant in this way that, it represents the maximum beel of Jessore-Khulna region and having similar ecosystem pattern, but it needs several interventions which should be come out from this initiative. Moreover, geographically it is a very important beel as it has connection with people of both Monirampur and Abhaynagar Upazilla of Jessore district, Bangladesh. With the present changing situation, it is difficult to identify the beel area though major part is reasonably deep and goes under water during wet season. In dry season, water remains in the river and canals and also in low depressions. The floodplain and catchment area is quite big. It is estimated that the Beel Bakar is having an area of about 1750 ha including its floodplain.

## 2. Methodology

The study was designed based on Focus Group Discussion (FGD), Key Informant Interview (KII), direct observation and household survey. Three FGDs were conducted with local fishermen and two with farmers. One mix group discussion with diverse professions was also accomplished to get some discrete data. Six KIIs were performed with two social workers, a teacher, an old gher operator and two gher owners. Forty fishermen's households were surveyed to know their socio economic condition in present days. To get proper data from this survey, question about their daily income pattern, family size, age group, alternative income source, ownership of land, household type, diet, loan, sanitation, drinking water, mobile phone use, medical service and some property related information were gathered which eventually resulted in a portrait of fishing community's life in study area.

## 3. Result and Discussion

### *3.1. FGD and KII Findings*

#### *3.1.1. Past and Present Water Management*

To support Green Revolution water management interventions were achieved through number of activities in this area also. Construction of 21 vent flap gate type sluice gate known as Bhabadaha Sluice gate was made on the Mukteswari River. Apart from it, numbers of sluice gate for water control were constructed upstream and laterally to adjoining areas. Such structures in the downstream were considered to prevent entry of saline water. With a longer period impact of stopping tidal water gradually invited deposition of silts and eventually introduced water logging to particular *beel* and later extended over to most of the *beel* and river network of the region. After having the bitter experience of longer period water logging, with the introduction of Tidal River Management (TRM) concept, remedy was observed. Now like many other *beel*, *beel* Bakar becomes waterless in dry season. However, conversion of many areas of the *beel* into gher system limited the return to previous system with a result of inequity of the resources in the form of advantages went to non target people. Many local farmers suggested that, if water release properly through the Bhabadaha Dam (sluice gate), the situation would be nice. They also opined for an additional option of water to divert to Bhairab River (River near Noapara of Jessore district) through Fultala area. They also suggested for re-excavation of Amdangar Khal and Boyrar Khal. It should be noted that unless excavated soil is properly managed, excavation or re-excavation would bring no positive outcome.

#### *3.1.2. Situation of Gher and Aquaculture*

Initially Gher was done in water logged condition by few people in the *beel* periphery. Later some people introduced it to trap wild fishes. Now there are three types of ghers:

- Seasonal ghers, mostly done fish culture in wet season and paddy in the dry season usually without any trench.
- Seasonal ghers but, mostly with trench so that, it is possible to fish is in considerable part. This type has two different patterns namely; i) Relatively medium and smaller sized; ii) Very big in size, mostly combining several plots
- Perennial ghers where scope for paddy cultivation hardly available or operators are mostly interested in fish culture.

Apart from gher fish culture, there are number of perennial and seasonal ponds where fish culture is going on. Water retention in the ponds of north-western areas was found poor and suitable for seasonal fish culture.

#### *3.1.3. Status of Fishermen Livelihoods in the Context of Changed Situation*

*Earning and Debt:* Most of the fishermen earn 100 to 200 BDT per day only. Thus, monthly average income stands 3000 to 6000 BDT (40-80 US$) which is very poor. They cannot manage to maintain their family not even to meet basic requirements. Loan becomes a part and parcel of their life due to low income. Threshold capacity to take loan and loan recovery (returning) are also in very measurable position. Before returning a loan another loan is required.

### *3.2. Household Survey Findings*

Fishermen villages are conspicuous and isolated. Two villages are in the western side, one in the north and others are in the eastern side. It was observed that all the respondent fishermen are Hindu and lead a low profile life style, although they have a satisfactory water and sanitary system.

#### *3.2.1. Religion, Family Size, Age Group and Daily Income*

As per household survey, it was found that traditional fishermen are mostly Hindu in religion. Maximum (53%) respondents (fishermen) are within 31-50 years old, followed by 40% above 50 years and 7% range 20-30 years (Fig. 2). The study also revealed that majority (65%) of the respondents are very poor whose, average daily income ranged 100-200 BDT. Only 17% respondents earn above 300 BDT per day (Fig. 3). Very big family size (more than 6 family members) is 15% (lowest) whereas, small household size (1-4 members) is 25%. However, medium household size (5-6 members) is maximum (60%).

### *3.2.2. Ownership of Land, House Type, Alternative Income Source and Sources of Loan*

Majority (68%) of the respondents live in semi concrete house followed by 17% in concrete and 15% in earthen house respectively (Fig. 4). The study revealed that 100% respondents have own land to construct their house. Most of respondents depend on loan from NGOs and bank. Highest (40%) respondents receive loan from NGOs, followed by 15% from bank and 7% from both bank & NGOs. However, significant (38%) proportions of the respondents are not depending on loan from any sources (Fig. 5).

### *3.2.3. Alternative Income Source and Status of Fish and Meat in Consumption*

In rainy season major part of the farmers and local people work in fish ghers and do fishing in little open beel portion. In addition to that many of them work as day laborers, rickshaw/van puller etc. Usually they earn 100-150 BDT per day. About 93% respondents take fish nearly every day. Only 2% household takes fish a day or two per week (Fig. 6). Reversely, not a single household take meat every day (Fig. 9). The study also revealed that fish from beel in the diet is still not diminished. Twenty percent fish in their consumption comes from beel, but most of them (47%) come from river source (Fig. 10).

### *3.2.4. Status of Drinking Water, Use of Mobile Phone, Sanitary Latrine and Medical Service*

The study revealed that all the households use underground water for drinking, no one use water from river or tanks. No one of the respondents was found who does not have mobile phone set (Fig. 11). Most (72%) of the households have ring slab toilet, whereas, 25% use concrete and hygienic sanitary latrine. Only 5% household uses service of health complex whereas, 40% gets medical treatment from local practitioner (usually pharmacy person) (Fig. 12).

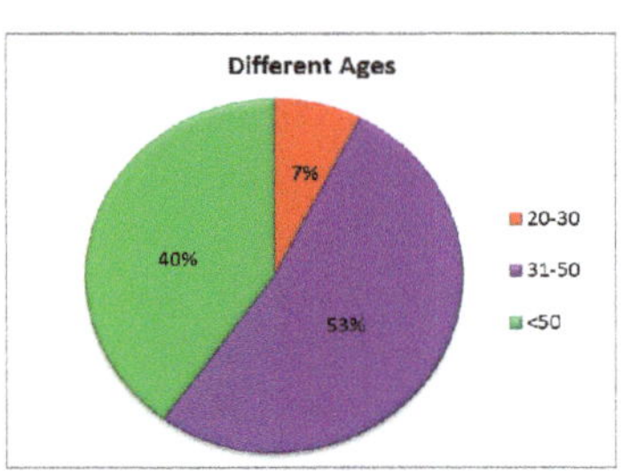


***Figure 2.*** *Age group of the respondents in Beel Bakar*

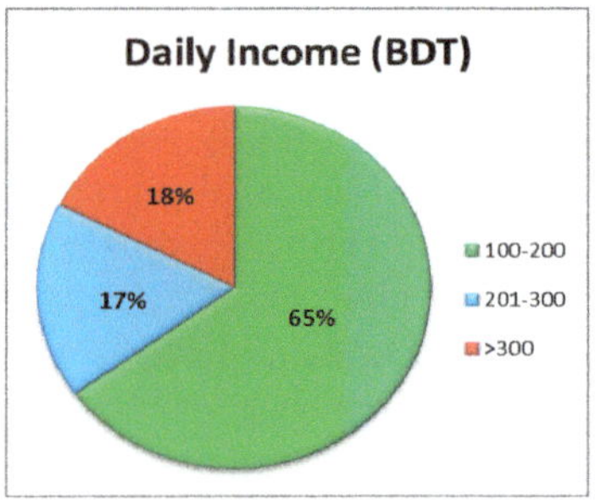


***Figure 3.*** *Daily income of the respondents in Beel Bakar*

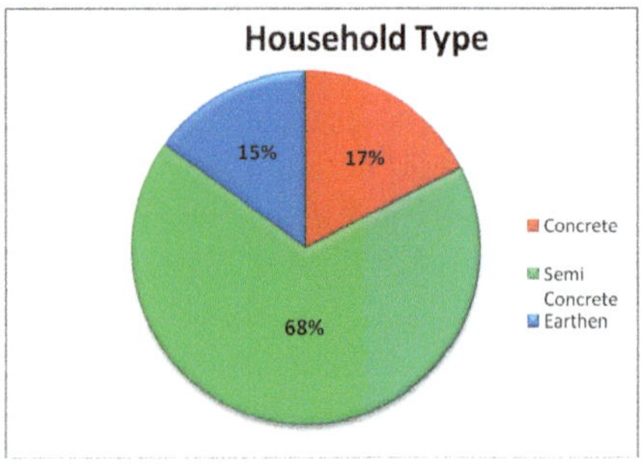


***Figure 4.*** *House type of the respondents in Beel Bakar*

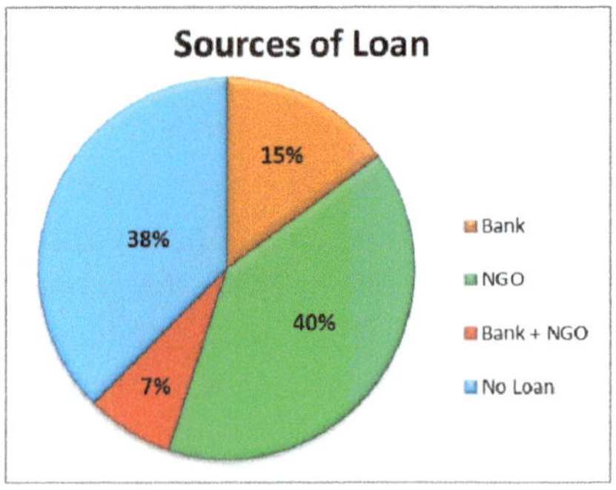


***Figure 5.*** *Sources of loan receive by the respondents in Beel Bakar*

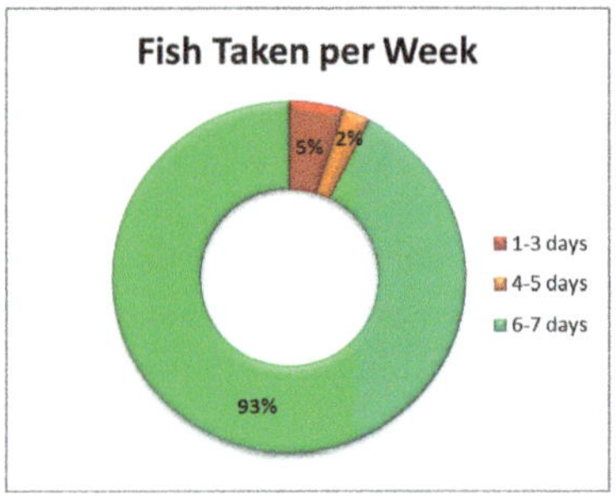


***Figure 6.*** *Fish consumption pattern of the respondents in Beel Bakar*

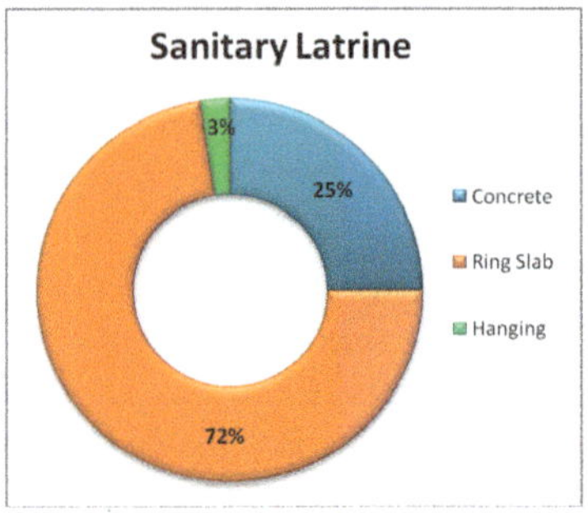


***Figure 7.*** *Household's sanitary latrines use in Beel Bakar*

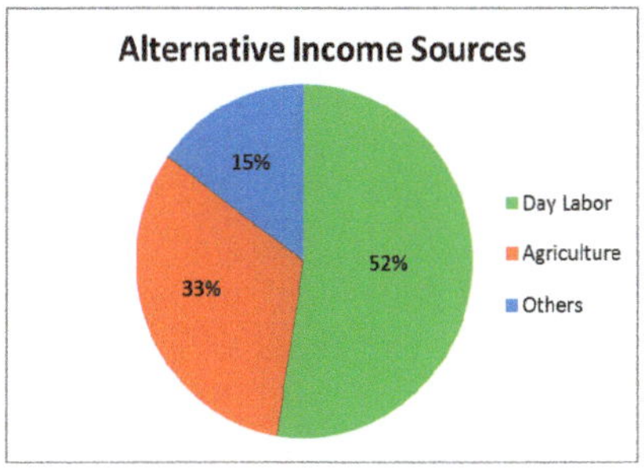


***Figure 8.*** *Household's alternative income sources in Beel Bakar*

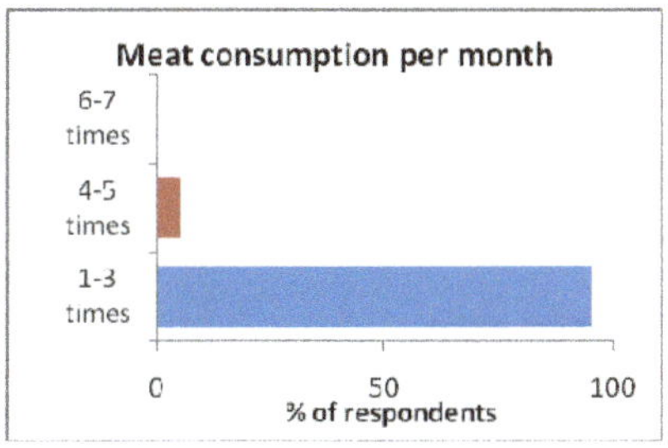


***Figure 9.*** *Household's meat consumption pattern in Beel Bakar*

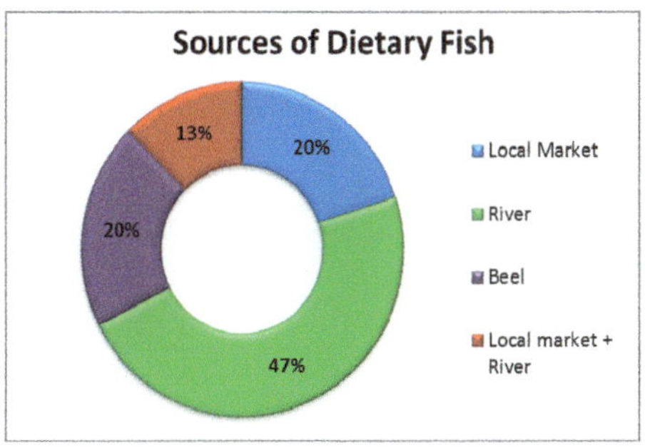


***Figure 10.*** *Household's sources of dietary fish in Beel Bakar*

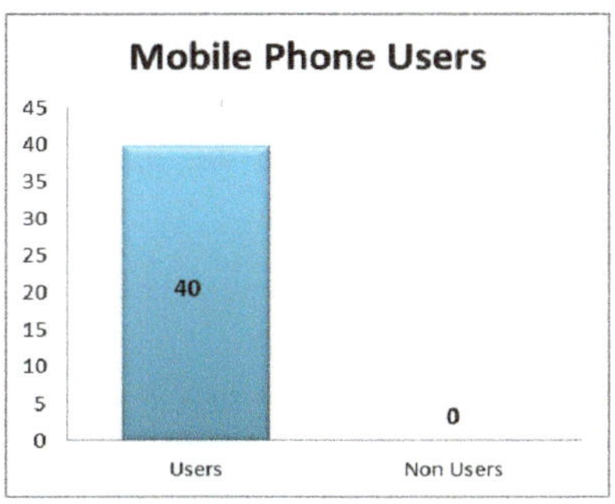


***Figure 11.*** *Mobile phone users in Beel Bakar*

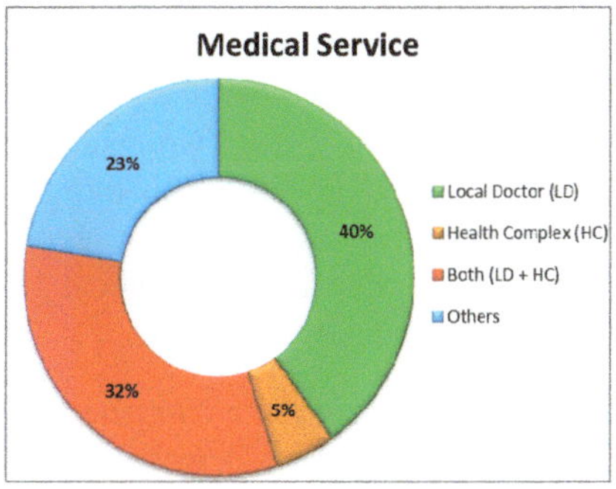


***Figure 12.*** *Sources of medical service in Beel Bakar*

### *3.3. Limiting Factors for Propagation of Beel Fisheries in Bangladesh*

Gher system is the main barrier for the propagation of *Beel* fisheries and maintaining fish biodiversity. For this, proper management cannot put in place. Introduction of different interventional tools and activities cannot be established like fish sanctuary, habitat restoration and swamp forestry development. Each gher has its own boundary and it stops regular water flow and also restricts flow of nutrition. So, fish cannot move in a natural manner. Sometimes it cannot reach its breeding ground. The over population of some aquatic weeds also help to fall the water quality and eventually the fish propagation. This is associated with making the water stagnant and transparent which is a resultant of excessive man-made intervention.

#### *3.3.1. State of Conflict and Possible Resolution: Fishers and Farmers*

Usually conflict remains within two professional groups namely; fishers and farmers. Surprisingly, in Beel Bakar there is no conflict between them. This is because traditional fishers are cornered so much that society does not count their sayings. On the other hand, farmers are also cornered by fish culturist.

#### *3.3.2. State of Conflict and Possible Resolution: Fishers and Aquaculturists*

Fishermen are silent but do have long term conflict with gher fish culturist. This is because they are grabbing most of the floodplain where fish grow. Encroachment to the river is also alarming. Many side canals are interrupted by ghers in such a way that fish movement, breeding, grazing are hardly possible. Fishermen are not in a position to raise their issues in front of the fast growing fish farmers and outside investors as it appears that there is no issue. It is apparent that, more serious conflict or contradiction is going to take place between environment and the present trend of development.

#### *3.3.3. Beel Fisheries and Fishermen's Livelihoods*

Every *beel* needs to maintain its open water fisheries for ensuring livelihoods of traditional fishermen. Control of the resources should not go to the person who has the control over resources other than *beel* also. Khas boundary needs to be demarcated and encroachment needs to be stopped.

#### *3.3.4. Fish Fry Release and Future Access Status*

Fishermen claimed that the fish fry releasing in open water by the line agency is being diverted by the influential to their fish ghers instead of open *beel* and rivers. They demanded that program should be implemented in the river. A matter of concern also envisaged that situation has come to such a position that in case fish fry is released in the river, the influential may capture the position or restrict real fishers in managing resources in the river. Thus, there may be a risk of losing the ruminant accessibility of the water body.

### *3.3.5. Reasons of Degradation of Fish and Fisheries Resources Including Habitat Quality*

- Role of insecticide/pesticide in disappearing fish
- Affect of pesticides and fertilizers
- Limiting fish breeding and propagation
- Leaving no fish to breed next year
- Agricultural practices without consideration of fishermen issues

***Figure 13.*** *Gher operation in Beel Bakar*

***Figure 14.*** *Present Scenario of Beel Bakar*

### *3.3.6. Problems of the Farmers*

- Now-a-days paddy production is good than earlier, but the production cost is comparatively high.
- Gher practice makes a large portion of the beel busy in fish farming. This hampers the agriculture production of the beel area.
- Gher operators are rich people, thus expansion of gher farming is replacing small scale agriculture.

### *3.3.7. Gher Culture Status, Impacts and Solution*

The rapid expansion of the gher construction should be controlled considering its carrying capacity in the *beel* by maintaining ecosystem functionality. As many people are interested in expansion of gher but significant number of people is now aware of the impacts of gher on beel ecosystem. Solution is needed balancing demand and ecosystem based approach.

Local level planning should come with an outcome of planned zoning of the *beel* with recommended extent and pattern of ghers if needed. Following points may be considered as solution:

- Ghers should not be continuous or clusters. There should be adequate intervals.
- Ghers should be constructed with appropriate set back distance from the bank of the river or canal.

*Local level problem with gher culture*

- Gher owner hires the labors and guards from outside of beel area. People from South (Keshabpur and Monirampur town side) mainly regulate the gher business.
- Local people are deprived of the chance of getting employment from gher.
- The people, who own the ghers, are not real fishermen. This is why; they don't think of the natural habitat of the fish and kill the unwanted species (natural species) for culture purpose.
- Due to gher culture, agricultural production has declined also.

### *3.3.8. Fish Migration and Movement Status*

Fish migration and movement takes place up and downstream of the river and also laterally. Due to intervention and with number of fixed ghers and net/ban migration and movement have seriously hampered. Lateral movement and migration also significantly affected due to construction of ghers without any spacing. As grazing area also encroached, number of fish in the river drastically dropped. Few sluice gates are constructed within the *Beel* Bakar. These are not for paddy cultivation rather, in virtue for the benefit of the gher operators. These sluice gates and ghers constructed in front of bridge/culvert restrict not only fish movement but also contribute in changing water management.

### *3.4. Factors need to be addressed for Propagation of Beel Fisheries in Beel Bakar*

### *3.4.1. Need for Beel Based Water Management Planning*

Water management should focus natural fisheries and to some extent for agriculture and least for aquaculture/gher farming. Department of Environment (DoE) along with Upazila or District authority should do this planning in participatory manner involving relevant experts. The local level planning should identify and execute how much and which areas of the beel should be identified and take necessary measures to implement it.

### *3.4.2. Need for Beel Based Environmental Planning*

Unless *beel* based environmental planning is done in accordance of the fisheries planning, sustainability of the resource management can be worsen especially in the context of climate change. We do have coastal land use planning and even for greater Dhaka city. District Water Resources Assessment made by LGED by engaging number of consulting firm was a good initiative but could not achieve much due to lack of commitment. DoE, IUCN or relevant organizations should get the power of exercise to do swamp forestry development on land which does not belong to *khas land* (the land that is owned by the government). Thus, redefining *beel* boundary based on ecosystem and supported by or in accordance to ecological boundary is required to revert encroachment trends.

### *3.4.3. Beel Based Conservation of Fishermen's Rights*

In the river dominated areas, fishermen's rights are river based like the marine fishers go to sea and coast for catching fish. Similarly, in the upland areas where scope of fishing in the river is limited but there are number of *beels*, rights of fishermen should be established *beel* based. Thus fishermen should get access to habitat area, resources especially natural fisheries.

### *3.4.4. National Context of the Demand*

The problems, issues, challenges are little different in this *beel* compared to many other *beels.* But this is very important to know the status and process of changes of this beel as in coming days many *beels* may become like this. For this reason the study findings would be useful in the national context. On the other hand, demand to promote livelihoods of the fishermen is not the demand of this *beel* only but it should be a national demand.

## *3.5. Base Situation*

### *3.5.1. Hydrology and Morphological Set Up*

The Mukteswari River flows through this *beel*. There are number of *khals* (small pass way of water) and small tributary rivers also. Rain water comes from the upstream through the river and other inlets and rainfall to the catchment are main source of water. Tidal water used to reach the *beel* earlier but due to different interventions it is not reaching. The slope of the *beel* is from the north-west and north-east to the south. Thus southern part of the *beel* and floodplain are deeper and northern, north-west part are shallow. Few parts of the *beel* are with perennial water and rest is seasonally flooded. Once upon a time (about 15 years ago even) it had normal *beel* characteristics, later experienced with longer period water logging and recently water logging is over due to TRM process in the adjoining *beel* and downstream part. The *beel* and its floodplain are having later connectivity too apart from northern side. In the downstream it is connected with the same river but a different name called Hari River. Water has to pass through the 21 vent BWDB Bhabadaha sluice gate. Ultimately the water flows to the Sundarbans and to the coast.

### *3.5.2. Biological Set Up*

Once, the *beel* was full of water based resources along with paddy, wildlife and birds. Bird trapping by poachers was a regular scene which indicates richness of the *beel* in terms of biodiversity. Fisheries were one of the important sectors from which people used to get fishes from the natural sources. Presently, due to change of water management pattern morphology, hydrology and culture pattern have significantly changed. Influential people made ghers for shrimp and white fish culture. Part of the *beel,* specially its floodplain earlier and now supports grazing of livestock. In the natural system, it prevailed much better earlier invited number of migratory and resident birds. This site was also famous for amphibians, reptiles and mammals. Around 20 fish species are still common [18]. In recent past, number of fish species was about 70. Species which are not available now a days are Boal, Pabda, Sarpunti, Magur. snail, crab, kuchia, frog, snakes are still observed in the *beel.*

***Table 1.*** *Commonly available fishes in Beel Bakar*

| Sl No | Local Name | Scientific Name | Order |
|---|---|---|---|
| 01 | Puti | Puntius spp. | Cypriniformes |
| 02 | Rui | Labeo rohita | Cypriniformes |
| 03 | Catla | Catla catla | Cypriniformes |
| 04 | Mrigel | Cirrhinus cirrhosus | Cypriniformes |
| 05 | Kalbaus | Labeo calbasu | Cypriniformes |
| 06 | Bata | Labeo bata | Cypriniformes |
| 07 | Shol | Channa striatus | Channiformes |
| 08 | Koi | Anabas testudineus | Perciformes |
| 09 | Shing | Heteropneustes fossilis | Siluriformes |
| 10 | Tilapia | Oreochromis mossambicus | Perciformes |
| 11 | Taki | Channa punctatus | Channiformes |
| 12 | Tengra | Mystus spp. | Cypriniformes |
| 13 | Baim | Mastacembelus spp. | Mastacembeliformes |
| 14 | Bele | Glossogobius guiris | Perciformes |
| 15 | Mola | Amblypharyngodon mola | Cypriniformes |
| 16 | Dhela | Osteobrama cotio | Cypriniformes |
| 17 | Kholisha | Colisa fasciatus | Perciformes |
| 18 | Chang | Channa gachua | Channiformes |
| 19 | Small prawn | -- | -- |

Fishing ghers that were common in this *beel* are Current jal (net), Vasal jal (lift net), Goga jal, Khewla jal, Arende jal, Koach, Kachari, Khuni, Charo, Nesa, Borsi, Gaze, Pata, Chari etc (Table 2). Aquatic vegetation including water lily and other species (kolmi, helencha, hogla pata) are still present but once used to maintain a healthy food chain. This was mainly dependent on non fish organisms as important steps of the food chain and food web. Both terrestrial and water dependent plants and reeds are still found. Pitali which is one of the swamp plants growing widely. Common birds of this *beel* are Pankouri, Stork, Finge, Shalik, Doyel, Dongur, Vilabile, Bakchu, Ghu ghu, King Fisher etc. Once the agriculture was with harmony with natural fisheries and wildlife but now it has been transforming to Boro paddy (a native variant of paddy) within the ghers rapidly. Both agriculture and culture fisheries are with rapid acceleration threatening biodiversity and livelihoods of poor people especially the traditional fishermen. In addition of main cereal crops (rice, wheat) non-cereal crops and vegetables like Pumpkin, Cucumber, Cabbage, Broccoli, Cauliflower, Chili, Kohlrabi, Bitter gourd, Petsai etc. are also seen to be grown by the *beel* areas. Farmers use chemical fertilizers for crop production. Fertilizers used in the study area are Urea, TSP, Zinc and Potash. Inorganic fertilizers are used in the *beel* are for IRRI crop. Main crop in the beel is IRRI. But in early days the *beel* was flourished with different vegetables and native paddy variants like Aush, Amon, Boro etc. But due to the introduction of hybrid crops and low production of native crops, farmers became less interested in the production of these crops. In *Beel* Bakar, no crops were cultivated during 1986 to 2009 due to water logging. From 2010-11 the cultivation was again started. But 75% of the total beel remains under water all the year round. BRI 28 gives production of 20-25 *mound* (a local unit which represents 40 kg) per bigha (42 decimal) land. About 30-35 mounds paddy per bigha is produced with hybrid variety like

SL 8, Hira, Sathi, Rajkumar. Fishes cultured in the ghers of *Beel* Bakar are golda, rui, silver carp, grass carp, black carp, common carp, big head, tilapia, mrigel and some other Indian major carps and Chinese carps.

*Table 2. Fishing ghers used in Beel Bakar*

| Gher Name (Local) | Major species caught |
|---|---|
| Vasal Jal (Lift net) | All kind of fishes (small to large) |
| Khewla Jal | Punti, Taki, Royna etc. |
| Goga Jal | Mainly big fishes |
| Kachari Jal | Rui, Shol etc. (Mainly medium size species) |
| Khuni | Chingri (Shrimp), Puti etc. |
| Charo (Fish trap) | Koi, Taki, Gutia etc small species. |

### 3.5.3. Social Set Up

The *Beel* Bakar is surrounded by many villages namely Mohishdia, Alipur, Poradanga, Padmanathpur, Parualia, Dharmatala, Panchkathia in the western side and Hazirhat, Nebugati, Kucholia, Sundali in the north. Villages of the east are Dhopadi and Hargacha and in the south Moshihati. There are some villages situated within the *beel* but mostly in its eastern floodplain are Bhatbhila, Danga Mashihati, Dohor Moshihati, Bariadaha, Baje Kultia. Common people are dependent on agriculture, fisheries, aquaculture and small business mainly. Gher fish culture of the beel is mainly controlled by rich and influential people. However, along the catchment boundary especially in the northwest part, medium and small farmers also doing gher fish culture within their small plots. Fishermen belong to several villages. In Mohishdia village, there are about 65 fisher families in the Rajbongshi para (a cluster of inhabitants) alone. Their livelihoods are under extreme threat and vulnerable due to drastic shrinking of open water to aquaculture. There are many non-traditional fishermen. In wet season mostly all person gets involved in fishing for livelihoods and subsistence in nature.

### 3.6. Proposed Interventions

### 3.6.1. Fish Sanctuary by the Fishermen Community

Usually local people in most part of Bangladesh are afraid of conservation as the idea they first get is "Restriction" giving them flavor denying access. Concept of conservation practically means getting resources at a certain level for longer time period instead of higher quantity but for short period. The study was also confronted with this situation in *Beel* Bakar area. Some of the respondents opined that if conservation effort is taken then where they will catch fish. It seems that all development they have been acquainted is planned and made by officials and imposed to them and the process is not participatory. It was discussed with them in detail and they opined that if the planning process of conservation is made by taking their opinion or process, itself allow them to take decision where to conserve, when to and what to. Then they would be happy and that effort certainly benefits them. It appeared that with full participation of fishermen dependent on the Muteswari River and the *bee*l can establish Community Based Fish Sanctuary. They are clear about the concept that if they can harvest the outcome of the conservation then no problem in allowing conservation.

### 3.6.2. Fish Habitat Restoration of Beel and River

Restoration of fish habitat mainly consists of re-excavation with proper slope and ensuring management of excavated soil. Apart from this, encouraging regeneration of swamp forest to support fish is important aspect of the habitat restoration. Sometime introduction of recently disappeared fish species considered as part of the process. Though total environment of the *beel* has been marching towards materialistic development, still there is scope for restoration of the river and adjoining floodplain or *beel* part also. The process should be designed and monitored by fish habitat restoration expert and executed by fishermen, local organization and related field agencies. A decision on planning process for maintaining a setback distance for gher making from the river and *khal* even if that is part of private property should come up. So this habitat restoration should be friendly to the ecosystem and environment to benefit fishermen and local people ultimately. Fishermen should not be treated as competitor to the resources of the *beel* rather it is an advantage of getting service of fishermen at nominal cost in managing resource generation and recommended harvesting process.

## 4. Conclusion and Recommendations

### 4.1. Conclusion

In Bangladesh, about 11% populations are involved with fisheries sector and this sector contributes 4.39% to the national GDP. This field deserves more attention by both government and private investors to conserve biological resources and ensure food security. The study reveals that that *Beel* Bakar desires a better management approach and for its long term effectiveness it should be participatory, mainly with the involvement of local community. Although it lost its diversity in both flora and fauna, the existing species can be conserved by the conservation of *beel*, its floodplain and Mukteswari River. As around 8450 families are directly and indirectly dependent on the *beel,* it is also important to restore the wetland for ensuring livelihood of surrounding communities.

### 4.2. Recommendations

- Need to establish rights of the fishermen. Main contradiction with this issue is to ensure the access of real fishermen into Beel resource. The non fishermen/ fish businessmen need to be stopped who degrade the natural part of the beel.
- Recommended river stretch for conservational effort (also suggested by fishermen) is Lokhaidanga, Kultia Bridge to Helar Ghat Bridge, Helar ghat to Bandhar Matha and Lohadanga.

Recommendations suggested by the fishermen are:

- Fisher folk believe that, they have lost possession over the beel. Now as their only hope they want the river (Mukteswari) free from all side effects and have the access year round.
- They also want to ensure the total fish fries in the river allotted for this area by the Government so that they can harvest/capture them when they are mature.
- Government released fish should be captured by the real fishermen only.
- Access of the non fishermen community need to be restricted in the beel and river.
- It has strongly recommended to keep a setback distance from the river bank to construct gher and also to keep required gap from a cluster of gher to another (not to make continuous) to reduce impact on natural fisheries and aquatic environment.

## Acknowledgements

The authors extend their profound thanks to Ulashi Sreejony Sangha (USS) for giving a scope to conduct this study under CSRL approach. The authors are grateful to OXFAM in Bangladesh for providing financial support to USS in beel movement and the protection of beel dependent people's rights. In addition, special thanks go to Dr. Munir Ahmed, Executive Director, TARA, Bangladesh for his sincere effort during the whole study. Moreover, the authors are thankful to Dipshikha Sangastha (DSS) and its research team (especially Prokash Chandra Dhar, Executive Director, and Mrs. Jyotsna Mollick, Coordinator) for providing necessary information, logistic support during each field trips, household survey and focus group discussions.

## References

[1] Byomkesh, T., Nakagoshi, N., Shahedur, R.Md., 2008. State and management of wetlands in Bangladesh. *Landscape and Ecological Engineering,* 5(1):81-90. DOI: 10.1007/s11355-008-0052-5.

[2] Nishat A., 1993. Fresh water wetlands in Bangladesh: Status and issues. In: Nishat A, Hossain B, Roy M K, Karim A, eds. Freshwater Wetlands in Bangladesh - Issues and Approaches for Management. Dhaka: IUCN, 9–22.

[3] Islam, S.N., 2010. Threatened wetlands and ecologically sensitive ecosystems management in Bangladesh. *Frontiers of Earth Science in China.* 4(4):438-448. DOI:10.1007/s11707-010-0127-0.

[4] Banglapedia, 2012. National Encyclopedia of Bangladesh, Asiatic Society of Bangladesh, Dhaka, Bangladesh. Available from URL: http://www.banglapedia.org/HT/B_0429.htm.

[5] Bangladesh Water Development Board (BWDB), 2014. http://www.bwdb.gov.bd

[6] Akonda, A.W., 1989. 'Bangladesh' in DA Scott ed, A Directory of Asian Wetlands, IUCN, Switzerland, 1989.

[7] Khan S M, Haq E, Huq S, Rahman A A, Rashid S M A, Ahmed H (1994). Wetlands of Bangladesh. Dhaka: Holiday Printers Limited, 1–88.

[8] Department of fisheries (DoF), 2013. DoF Magazine, Department of Fisheries, Government of People's Republic of Bangladesh, Dhaka, Bangladesh.

[9] Ahmed, I., Deaton, B.J., Sarker, R., Virani, T., 2008. Wetland ownership and management in a common property resource setting: A case study of Hakaluki Haor in Bangladesh. *Ecological Economics,* 68(1-2):429-436. DOI: 10.1016/j.ecolecon.2008.04.016.

[10] Gnauck A, Islam S. N., 2004. The Threatened Mangrove wetlands ecosystems in Bangladesh: A case study on the Sundarbans. In: Proceedings of 7th INTECOL International Conference on Wetlands 2004, Utrecht. The Netherlands: Universiteit Utrecht Press, 109

[11] Islam S N, 2007. Water salinity and thresholds for the protection of mangrove wetland ecosystems in the Sundarbans Region. In: Ganuck A, ed. Modellierung und Simulation von Ökosystemen (*Modelling and Simulation of Ecosystems Journal*). Aachen: Shaker Verlag, pp.187–201

[12] Islam S N, Gnauck A., 2007a. Effects of salinity intrusion in mangrove wetlands ecosystems in the Sundarbans: an alternative approach for sustainable management. In: Okruszko T, Jerecka M, Kosinski K, eds. Wetlands: *Monitoring Modelling and Management.* Leiden: Taylor & Francis/Balkema, pp.315–322.

[13] Islam S N, Gnauck A., 2007b. Salinity intrusion due to freshwater scarcity in the Ganges catchment: a challenege for urban drinking water and mangrove wetland ecosystems in the Sundarbans region, Bangladesh. In: Proceedings of 6th World Wide Workshop for Young Environmental Scientists 2007, Paris. Paris: CEREVE Enpc, 20–30.

[14] Islam S N, Gnauck A., 2008. Mangrove wetland ecosystems in Ganges-Brahmaputra delta in Bangladesh. *Front Earth Sci China,* 2(4): 439–448.

[15] Islam S N., 2008. Cultural landscapes changing due to anthropogenic influences on surface water and threats to mangrove wetlands ecosystems: a case study on the Sundarbans, Bangladesh. Dissertation for the Doctoral Degree. Cottbus: Brandenburg University of Technology, 1–179

[16] Islam S N, Gnauck A., 2009a. Threats to the Sundarbans mangrove wetland ecosystems from transboundary water allocation in the Ganges basin: a preliminary problem analysis. *International Journal of Ecological Economics & Statistics,* 13(W09): 64–78.

[17] Barbier, E., Acreman, M., and Knowler, D., 1997. Economic valuation of wetlands: a guide for policy makers and planners. Ramsar Convention Bureau, Gland.

[18] Rahman, A.K.A., 2005. Freshwater fishes of Bangladesh. Zoological Society of Bangladesh, Dhaka 366pp.

# Influence of small cereal intercropping and additive series of seed proportion on the yield and yield component of lupine *(Lupinus Spp.)* in north western Ethiopia

**Yayeh Bitew**

Agronomist, Adet Agricultural Research Centre, Institute of Amhara Agricultural Research, Bahir Dare, Ethiopia

**Email address:**
yayehbitew@gmail.com

**Abstract:** Small cereals as a main crop and lupine as a minor crop are food crops often traditionally grown in association in North Western Ethiopia. The experiment was conducted on intercropping of lupine *(Lupinus albus L.)* with wheat *(Triticum aestivum)*, barley *(Hordeum vulgar)* and finger millet *(Eleusine coracana)* in 2009 at Adet Agricultural research station. The treatments were sole wheat at a seed rate of 175 kg/ ha, sole barley at a seed rate of 125 kg/ ha, sole finger millet at a seed rate of 30 kg/ ha, sole lupine at a seed rate of 90 kg/ ha, and an additive series of 25, 50 and 75 % of the sole lupine seed rate combined with the full cereal seed rate to determine the effect of small cereal intercropping in additive series on the yield and yield component of lupine. The experimental design was a completely randomized block with nine intercropping and four sole cropping systems in three replications. Lupine was planted in rows after establishment of main crops. SAS software's were used to compute the analysis of variance. Maximum lupine seed proportion was superior to the lowest when intercropped with wheat and finger millet. The lowest population density resulted in reduced agronomic attributes of lupine. Intercropping higher proportion of lupine with wheat and finger millet did help much in increasing grain yield and biomass yield of lupine without affecting cereal crop yield. However, nearly complete dominance of barley over lupine at all seeding ratios leads to absolute reduction in grain yield. The lupine-finger millet mixture at the 50:100 and 75:100 seeding ratio and lupine-wheat mixture at the 75:100 seeding ratio had a higher yield advantage of intercropping for exploiting the resources of the environment compared with the other intercropping systems.

**Keywords:** Intercropping, Wheat, Barely, Finger Millet, Lupine, Seeding Ratio

## 1. Introduction

Intercropping is the cropping system involving the growing of two or more crops in the same piece of land at the same time or relayed which could compute for growth resources for certain growth period. It provides valuable ecosystem services such as improved pest control (Mitchell et al., 2002), increased resource use efficiency (Hauggaard-Nielsen et al., 2001) in crop livestock mixed farming system. However, in many parts of Ethiopia, farmers traditionally harvest mainly once in a year on sole crop basis even in high rain fall areas. Moreover, in the past much research efforts have been directed towards improving technology for sole cropping. Such traditional farming did not insure the production of adequate food for a family (Nigusie, 1994). Different crops are grown traditionally in mixtures by small farmers to satisfy dietary needs, spread the period of peak demand for labor and minimize the risk associated with climate conditions (Tilahun eta al, 2012).

Intercropping cereal with a legume, however, is relatively the most common in most parts of the country. Indeed, the traditional objective has been to produce a full yield of cereal (as much as with a sole crop) while the associated legume yield is considered as additional yield (Tilahun eta al, 2012)

Lupine (Lupinus Spp.) is one of the major highland food legumes grown in Ethiopia. It is traditionally gown as intercrop with cereals and oil crops by low input farmers and is restricted to low-income classes, to times of drought (Jansen, 2006). Finger millet, wheat, and barley are the third, fourth and fifth stable cereal crops grown in west Gojam, respectively (CSA, 2007). Farmers grow it as traditional undefined additive system of intercropping in which lupine used as minor crop and cereals as major crop. They grow it for the strategies to overcome the shortage of arable land and attribute several crops for diversification of crop products and for maintenance and improvement of soil fertility (Allege and Steven, 1987).

Moreover, lupine normally has greater growth duration than cereals, so that when grown together at the same time, cereals utilize resources earlier than lupine. Another possible explanation for such intercrop yield advantage is that the tape root system of lupine could exploit water and nutrients from deeper soil layers than cereals (Jansen, 2006). Therefore, this cropping system may help improve productivity of low external input farming or scarce in natural resources, unreliable rain fall and poor soil fertility.

However, management of cereal intercropped with lupine to maximize their complementarity and to minimize competition between them follows simple natural principles, and its practice is limited only by the imagination of farmers. Therefore, the objective of this paper was to determine effect of small cereal intercropping in additive series on the yield and yield component of lupine (Lupinus Spp.)

# 2. Materials and Methods

## *2.1. Description of the Study Area*

The study was conducted in the 2009 rain fed cropping season at Adet Agricultural Research station (AARC), North Western Ethiopia. It is located between 11°17' N latitude and 37°43′ E longitude with an altitude of 2240 m.a.s.l (AARC, 2002). According to Gonder soil testing laboratory center (2009), the soil characteristics of experiment site were clay as shown Table 1.The study area receives a uni-modal rainfall which extends early June to late September with regard to its monthly distribution June, July and August are the three important months with high rain fall and more or less uniform spatial distribution (Aleligne and Steven, 1987). According to Adet Metrological station (2009), the total annual rainfall during the experimental growing season was 975.3 mm which is less than the 23 year average total annual rainfall (1253.4 mm) (Figure 1). The mean monthly minimum and maximum temperatures during the growing season were 11 $^0$c and 27.2 $^0$c which is greater than the 23 year average mean monthly minimum (9.1 $^0$c) and maximum (25.7 $^0$c) temperatures (Figure 2).

***Table 1.*** *Physico-chemical properties of the soil.*

| Chemical soil properties | | Mechanical properties | |
|---|---|---|---|
| PH | 6.06 | Sand (%) | 28.00 |
| OC (%) | 2.47 | Clay (%) | 46.72 |
| Total N (%) | 0.18 | Silt (%) | 25.28 |
| Av.P (ppm) | 1.98 | Class | clay |
| CEC | 37.97 | | |

Notes: CEC: Cation exchange capacity measured in cmol (+)/kg soil (NHAc), Av.P: Available phosphors in ppm and OC: organic carbon

## *2.2. Field Experimental Design*

Plots were laid out in randomized complete block design (RCBD) with three replications. Spacing between plots and replications were 0.5 and 1 m, respectively. There were nine intercropping and four sole cropping systems in additive series in the experiment (Table 2). The plot size was 12m$^2$ (2m*6m). Sole lupine was common to all lupine-cereal combinations for comparison purpose.

## *2.3. Sowing Method and Management Practices*

The experiment was conducted in rain fed season. Pure stands of lupine, wheat, barley and finger millet as well as nine lupine-cereal mixtures in three seeding ratios in additive series (25 %, 50 % and 75 % of recommended lupine seed rate with full cereal seed rates) were planted. Sole cropping of lupine, wheat, barley and finger millet were planted at a recommended seeding rate of 90, 175, 125 and 30 kg/ha, respectively. In sole cropping, lupine was planted in an inter-row space of 30cm; and wheat, barley and finger millet were broadcasted. In the intercropping system, first lupine row was established in the inter-row spacing of 120, 66 and 35 cm for the 25 %, 50 % and 75 % seed proportion, respectively, and full cereal components were broadcasted. Lupine was planted after establishments of cereal crops. For all intercropping systems space between lupine plants were 5 cm.

All plots were received a basal application of Diammonium phosphate (DAP) at the rate of 100 kg/ha at planting. For cereal components, 100 kg/ha Urea was applied except the sole lupine treatment assuming the lupine was benefit from self-fixed nitrogen. One third basal and two third top-dressed application of UREA were applied during planting time and at tillering stage of sole and intercropped cereals, respectively.

## *2.4. Data Collected*

Agronomic attributes of lupine includes plant height (cm), number of branches, and number of pods per plant, number of seeds per pod, grain and biomass yield (kg/ha), harvest index (%) and thousand Seed weight (gram). Moreover, to determine the competitive ability of the component crops partial land equivalent ratio was conducted using the formula developed by de Wit (1960). To find more realistic comparison of the yield advantage of intercropping over monocropping in terms of time taken by component crops in the intercropping systems Area time equivalent ratio (ATER) was calculated by formula developed by Heibsch (1980).

## *2.5. Data Analysis*

Data were statistically subjected to analysis of variance (ANOVA) using JMP-5 (SAS, 2002). Combined analysis of variance was performed over the three lupine-cereal combinations to determine yield and yield component of lupine, partial land and Area time equivalent ratio (ATER). In all the comparisons, the level of significance was set at α

= 0.05. Mean comparison for the treatments were computed using each pair Tukey-HSD test for parameters found to be significantly different at a given level of significant

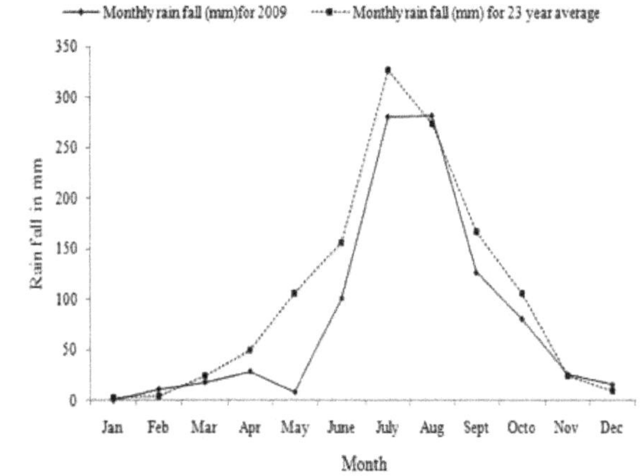


*Figure 1. Mean monthly rainfall (mm) of the study area for 23 year average and 2009 cropping season.*

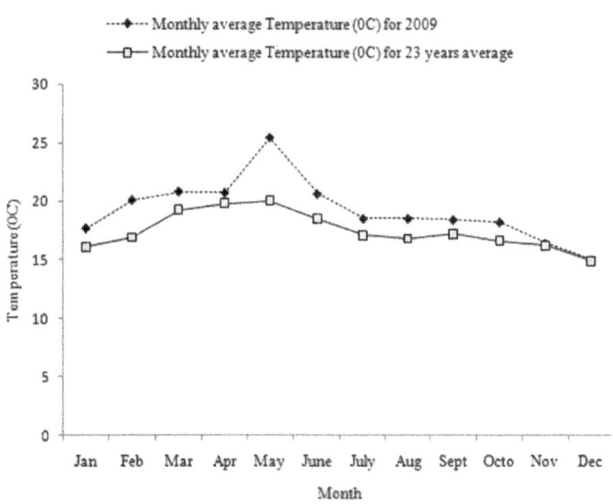


*Figure 2. Mean maximum and minimum air temperature (Tc⁰) of the study area for 23 year average and 2009 cropping season.*

*Table 2. Additive series of component crop production (Treatments)*

| Number | Crop species | Cereal Proportion (%) | Lupine (%) | Total population (%) |
|---|---|---|---|---|
| 1 | wheat | 100 | 0 | 100 |
| 2 | wheat | 100 | 25 | 125 |
| 3 | wheat | 100 | 50 | 150 |
| 4 | wheat | 100 | 75 | 175 |
| 5 | Barley | 100 | 0 | 100 |
| 6 | Barley | 100 | 25 | 125 |
| 7 | Barley | 100 | 50 | 150 |
| 8 | Barley | 100 | 75 | 175 |
| 9 | Barley | 100 | 0 | 100 |
| 10 | Finger millet | 100 | 25 | 125 |
| 11 | Finger millet | 100 | 50 | 150 |
| 12 | Finger millet | 100 | 75 | 175 |
| 13 | Lupine | 0 | 100 | 100 |

# 3. Results and Discussion

## *3.1. Lupine Growth and Yield Components*

### *3.1.1. Plant Height (cm)*

Plant height appeared to be significantly (P<0.01) affected due to differences in seed ratio, associated crop types and cropping system. Though, sole lupine showed the highest plant height than the other seeding ratios, plant height increases from 25:100 (101.17 cm) to 75:100(106.89 cm) seeding ratios in lupine-cereal combinations. The study revealed that the overall highest mean plant height (cm) was recorded in sole lupine (133.88 cm) but statistically on par with lupine-finger millet association at 75:100 (133.63 cm) seeding ratios probably because of the presence of intra-specific competition particularly struggle for light in the former cropping system and presence of both intra-specific competition in the highest minor crop proportion and inter-specific competition between the component species in the latter cropping system (Table 3). The overall shortest plant height was recorded in lupine-barley intercropping systems at 75:100 seeding ratio (60.43 cm) but, it was not significantly differ from the other seeding ratios of the same combination. The reason might be due to the fact that barley causes nearly complete dominance of growth resources over lupine in all seeding ratios in the early stage of the minor crop. The other possible reason could be shading and lodging effect of barley over lupine. In lupine-finger millet (133.63 cm) and lupine-wheat (126.6 cm) combinations the highest plant height was recorded at the highest seeding ratios (Table 3). However, lupine-finger millet combination (128.97 cm) showed higher plant height than lupine-wheat (122.49) and lupine-barley (60.49 cm) combinations. Moreover, except lupine-barley combinations at all seeding ratios, as the companion crop proportion increases in each combination plant height had favoured to grow (Table 3). This agrees with the results obtained by Muoneke and Asiegbu (1997) in maize-okra mixture that high population densities induced higher plant height.

### *3.1.2. Branch Per plant*

Results of the present study also showed that crop types, seed ratios and cropping systems were significant effect (P<0.01) on branch per plant, pod/plant and seed/pod (Table 3). Highest branch per plant was recorded in sole cropped (5.48) than intercropped seeding ratios (3-4.31). This result was in agreement with Rahman et al. (2008), who reported that branch per plant were maximum and minimum in sole and intercropped plants, respectively in mustard–lentil intercropping system. However, the experiment revealed that as seeding ratios increases branch per plant also increases (Table 3). With regarding lupine-cereal combination, lupine-finger millet (5.37) combination showed higher branch per plant as compared to other combinations (Table 3). Thus, the overall highest mean branch per plant was recorded in lupine-finger millet intercropping at 75:100 seeding ratio (5.67), but it was statistically on par with lupine pure stand (5.48), while the overall lowest mean branch per plant was recorded in lupine-barley intercropping at 75:100 (2.61) seeding ratio but statistically on par with the other seeding ratios of the same combination (Table 3). In lupine-wheat combination the highest branch per plant was recorded at the highest seeding ratio (75:100) (4.64). Except lupine-barley combinations, there was a general increase in branch per plant as the companion crop proportion increases in the case of lupine-wheat and lupine-finger millet combinations. This might be due to the fact that comparatively less tillering ability, slow and long growth period of wheat in general and finger millet in particular causes slow uptake of growth resources as compared to barley. However, this positive effect was more pronounced in lupine-finger millet combination (Table 3).

### *3.1.3. Pod per Plant*

Among the cropping systems, the overall highest lupine pod/plant were remarkably observed in sole lupine cropping system (40.25) as compared to intercropped with cereals in all seeding ratios(28-31) (Table 3). But, it was statistically on par from lupine-finger millet at 75:100 (39.61) seeding ratio. Reduction in number of pods due to intercropping was also reported by Galal et al. (1979), who intercropped soybean in maize. The combined lowest pod/plant (16-16.9) and seed/pod (2.24-2.36) were obtained from all seeding ratios of lupine-barley intercropping system which were statistically on par with each other (Table 3). All seeding ratios in lupine-wheat combinations gave statistically different pod/plant in which the highest pod/plant was recorded at the highest seeding ratios (37.69). The same is true for lupine-finger millet combination (Table 3). However, each lupine-finger millet, lupine-wheat and lupine-barley combinations showed significantly different pod/plant in which the former combination (36.5) showed the highest pod/plant as compared to the others. As the companion seeding ratios increases on each combination, its pod/plant also increases (Table 3). However, this result was inconsistent with Mcgibbon and Waltkin (1980), who reported that high plant population (lupine) caused a reduction in lupine pod set and consequently in seed yield.

The overall highest pod/plant (40.25) and seed/pod (6) in sole lupine might be due to the absence of other species and presence of intra-specific competition between lupine crops which favoured efficient utilization of growth resources. This corroborate with the results of Ghosh (2004), who stated pod yield of groundnut were lower in groundnut-cereal (maize, sorghum, and pearl millet) intercropped than in monoculture. In general, the lowest agronomic attributes of lupine in lupine-barley combinations might be due to higher competitive advantage, shading and lodging effect of barley over lupine. This was consistent with Venkateswarlu (1984), who reported that a wrong choice of cereals and grain legumes causes a great risk of loss of yield and reduction in grain yield and quality. This illustrated by combining an early ripening crop like barley or pea, with a late ripening crop like wheat or faba bean was increase the risk considerably as the early crop can lose its grains while waiting for the late crop to ripen (Venkateswarlu, 1984).

### *3.1.4. Seed per Pod*

The study indicated that the highest and the lowest seed/pod were recorded at sole lupine (6.08) and lupine-cereal combination at 25:100 seeding ratios (4.26) respectively. Intercropping lupine with finger mille (3.84) gave the highest seed per pod than with wheat and barley. The overall highest seed/pod was recorded in sole cropped (6.09) (Table 3). But, it was statistically on par from lupine-finger millet at all seeding ratios and lupine-wheat combinations at 75:100 seeding ratios (5.74). The overall lowest seed/pod was recorded in all lupine-barely combinations at all seeding ratios (Table 3). The latter cropping system was disagreement with Mandal et al (2010), who reported that number of seeds per pod for mungbean was more with rice-mungbean combination as compared to sole mungbean, though pod/plant increases in the reverse situation.

Generally, results of the present study showed that intercropping lupine with finger millet had less effect on growth and yield component of lupine as compared to wheat and barley. However, intercropping barley with lupine had adverse effect on those agronomic parameters as compared to wheat. This is mainly due to wheat in general barley in particular having fast growth and more extensive root system, particularly a larger mass of fine roots and nutrient utilize characteristics (Carr et al., 2004) (high competitive advantage) as compared to legumes. According to the results of the present study, except lupine-barley intercropping systems, as the companion crop seeding ratios increases on the constant full cereal seed rate, lupine growth and yield components also increases (Table 3). This situation probably due to intra-specific competition between lupine crops at higher proportion causes efficient utilization of growth resources and hence, hastens lupine growth and yield component parameters.

### *3.1.5. Thousand Seed Weight (Gram)*

Thousand seed weight was significantly affected by the cropping ratios ($P<0.01$) and associated crop types ($P<0.05$) (Table 3). Sole lupine (336.67 gram) showed lower TSW than lupine-cereal combination at 25:100 seeding ratio (398.36 gram). In contrast to the other yield components, the highest TSW was recorded in lupine-barley combinations at 25:100 (407 gram) seeding ratio but statistically on par from the other seeding ratios in the same combination and lupine-wheat combination at all seeding ratios (Table 3). Lupine in lupine-barley followed by lupine-wheat combinations was slow and less vegetative growth showed higher mean seed weight. The highest TSW of lupine in lupine-barely could be due to large seed size in a very low plant population which could be attributed to large accumulation of assimilate due to long maturity period. Similarly, the highest TSW of lupine in lupine-wheat and lupine-barley combinations as compared to lupine-finger millet combinations might be attributed to early maturity of wheat and barley leaves the resources for long maturity of lupine to continue growth (Table 3).

The lowest TSW was recorded in sole cropped (336.56 gram) followed by lupine-finger millet combination at 75:100 (365.33 gram) and 50:100 (379.67 gram) seeding ratios. This could be due to higher intra-specific competition causes reduction of assimilates for the former cropping system and the late maturity of finger millet that caused little difference in plant growth result in higher demand of resources at the same periods causes reduction of assimilates and might have resulted to poor seed filling of lupine. This result was corroborated with the findings of Bismillah and Khaliq (2004), who reported that sole cropped soybean showed lower thousand seed weight as compared to soybean-cotton, based intercropping systems. However, it was disagreement with the report obtained by Mandal et al. (2010), who noted that thousand seed weight for mungbean was higher in pure stand crops as compared to intercropped with rice. Further increase in seeding ratios in each lupine-cereal combinations TSW of local lupine decreases, though, all seeding ratios had statistically on par with each other in lupine-wheat and lupine-barley combinations (Table 3).

***Table 3.*** *Mean Effect of crop proportion and crop types on growth and yield components of lupine in lupine-cereal intercropping system at AARC, North Western Ethiopia in 2009.*

| Treatment and statistics | PH (cm) | B/P | PO/P | SE/PO | TSW(gram) |
|---|---|---|---|---|---|
| Seeding ratios (Lupine: Cereal) | | | | | |
| 25:100 | 101.17 c | 3.97c | 28.45 c | 4.26c | 398.36a |
| 50:100 | 104.46bc | 4.35 b | 30.17 b | 4.51bc | 395.33a |
| 75:100 | 106.89 b | 4.31 b | 31.18 b | 4.68b | 387.11a |
| Sole lupine | 133.88 a | 5.48a | 40.25a | 6.08a | 336.67b |
| Lupine-Cereal combination | | | | | |
| Lupine-barely | 78.84 c | 3.46c | 22.36 c | 3.26 c | 389.17a |
| Lupine-finger millet | 130.20 a | 5.37a | 38.67 a | 5.84 a | 368.25ab |
| Lupine-wheat | 125.76 b | 4.75b | 36.49 b | 5.55 b | 380.67b |
| Cropping system | | | | | |
| Sole lupine | 133.88a | 5.48a | 40.25a | 6.08a | 336.67b |
| 25 % Lupine+100 % wheat | 119.76bc | 4.42c | 32.25c | 4.84b | 401.67a |
| 50 % Lupine+100 % wheat | 121.13c | 4.63bc | 35.79bc | 5.50ab | 400.00a |
| 75 % Lupine+100 % wheat | 126.60abc | 4.64bc | 37.69ab | 5.74a | 391.33a |
| 25 % Lupine+100 % barely | 60.46d | 2.83b | 16.00d | 2.25c | 407.00a |
| 50 % Lupine+100 % barely | 60.59d | 2.92b | 16.97d | 2.37c | 406.33a |
| 75 % Lupine+100 % barely | 60.43d | 2.61b | 16.22d | 2.35c | 406.67a |
| 25 % Lupine+100 % f/millet | 121.63c | 4.83b | 37.08ab | 5.68a | 393.33a |
| 50 % Lupine+100 % f/millet | 131.66ab | 5.50a | 37.75ab | 5.65a | 379.67ab |
| 75 % Lupine+100 % f/millet | 133.63a | 5.67a | 39.61a | 5.93a | 363.33ab |
| LSD (0.05) | 5.24 | 0.29 | 1.84 | 0.38 | 28.81 |
| CV (%) | 2.82 | 3.44 | 16.60 | 4.58 | 4.90 |

Notes: Values (means) connected by different superscript letters are significantly ($P<0.05$) different within columns according to Tukey- HSD tests. PH (cm): Plant height in cm; B/P: Branch per plant; PO/PL: Pod per plant; SE/PO: Seed per pod and TSW: Thousand seed weight in gram, respectively

## *3.2. Lupine Biomass, Grain Yield and Harvest Index*

### *3.2.1. Grain Yield*

The overall yield reduction of lupine yield below the optimum in all cropping systems was observed probably due to poor competitor of quick growing cereal crops, moisture deficiency during the reproductive period (Figure 1) and low phosphors availability (Table 1). Jansen (2006) reported that lupine is drought-tolerant due to their deep roots, but is sensitive to moisture during the reproductive period and P deficiency.

Nevertheless, the present study indicated that seeding ratios, crop types and cropping systems significantly affect grain yield of lupine ($P<0.01$). Though, different seeding ratios of lupine were planted, a substantial reduction in grain yield of associated lupine crop was observed as compared to lupine alone perhaps due to poor competitor of quick growing cereal crops. The reason also confirmed by Jansen, (2006), who stated lupine prefers disturbed sites and poor soils, where there is less competition from other species. It was performed better when grown as a sole crop (2852 kg/ha) than when grown in a mixture (Table 4). Yield reduction could be attributed to inter-specific competition for space, growth resources between the components species and genetical difference between cereals which, in turn made the reduction of plant population, number of branches/plant, pods/plant, seeds/pod, and other yield attributes that sufficiently reduced yields in all crop combinations. This corroborates the findings of Sheri et al. (2008), who reported a yield depression of intercropped lupine and faba bean when intercropped with higher seed rate of wheat. Similarly, Ofori and Stern (1987) reported a yield depression of cowpea as a result of intercropping with maize.

Although, there was a general reduction in the yield of lupine as a result of intercropping, highest grain yield among the combination was recorded in lupine-finger millet intercropping (1749 kg/ha) followed by lupine-wheat (1407 kg/ha) intercropping systems as compared to lupine-barley (757 kg/ha) combination (Table 4). Thus, combined mean highest grain yield was obtained from lupine-finger millet intercropping at 75:100 seeding ratio (1812 kg/ha), whilst the lowest grain yield was recorded in lupine-barley combinations at 25:100 seeding ratio (50 kg/ha). However, lupine-barley combinations at all seeding ratios were statistically on par with each other.

However, absolute lupine yield reduction was observed in lupine-barley intercropping at all seeding ratios (<100 kg/ha). In addition to the other reasons, it might be due to shading and lodging effect of barley. This result was in agreement with Jones and McCown (1983), who reported that Caribbean stylo (Stylosanthes hamata cv. Verano), produced little seed in an intercrop with maize due to its failure to flower in the shade of a full maize canopy. The former reason has also supported by Challa and Bakker (1998), who reported that growth and yield of crops are related to the amount of solar radiation received during the growing period. Higher loss in lupine yield due to barely intercropped were also in agreement with Knudsen et al. (2001), who reported that lupine-barely intercrops did not show intercropping advantage. To sum up, except lupine-barley intercropping, higher lupine yield was obtained when high proportion of lupine was intercropped with full finger millet (1812 kg/ha) and wheat (1558 kg/ha) seed rates as compared to low proportion of lupine was intercropped with full finger millet (1459 kg/ha) and wheat (785 kg/ha) (Table 4). These findings are also parallel to those of ATER (Table 5). The main characteristics defining the differences in yield of lupine were mainly effect of cereals and seed proportion. In descending order, barley, wheat and finger millet were taller and grew faster during the early growing period. Lupine was suppressed during initial stage but later except in lupine-barley combinations it grew taller and took over wheat and finger millet utilizing the remaining growth resources during the final part of the growing season. This difference in plant growth might have resulted in demand of resources at different periods allowing sharing of resources.

### 3.2.2. Biomass Yield

Lupine biomass yield was significantly affected (P<0.01) by seeding ratios, crop types and cropping systems (Table 4). Highest lupine seeding ratio intercropped with full cereal seed rate gave the highest biomass yield (Table 4). The combined mean highest biomass yield was recorded in sole cropping (12250 kg/ha) as compared to all lupine-cereal cropping systems. The combined lowest biomass yield was recorded in lupine-barley intercropping at 25:10 (333 kg/ha) but statistically on par from the other seeding ratio of the same combination (Table 4). This result was corroborated with Bismillah and Khaliq (2004), who reported that sole cropped soybean showed higher biological and seed yield as compared to soybean-cotton based intercropping systems. The highest biomass yield in lupine-wheat (6667 kg/ha), lupine-barley (1250 kg/ha) and lupine-finger millet (10667 kg/ha) was recorded at 75:100 seeding ratio as compared to the other respective seeding ratios. This means, the biomass yield increased in line with cropping ratios in each combinations (Table 4).

Comparatively, biomass production was very low in lupine-barley at all seeding ratios than lupine-finger millet and lupine-wheat combinations. Similarly, biomass production of lupine was lower in lupine-wheat than lupine-finger millet intercropping (Table 4). Biomass production of lupine was influenced by the duration of the growing season of the component crops, the rate of growth, cropping ratios and crop types of the component crops. Barley in particular wheat in general showed fast growth characteristics than finger millet, which reduces more the biomass of local lupine. The results revealed that in lupine-barley combinations the new intervention cropping ratios (50:100 and 75:100 seeding ratios) were not significantly different from the farmers' cropping ratios of ≤ 25:100 seeding ratios based on optimum seed rate. However, in lupine-wheat and lupine-finger millet combinations, the farmers' cropping ratios was significantly different from the new intervention cropping ratios (Table 4)

Thus, the highest seed rate of lupine used in the study in lupine-barley combinations did not help much to increase the grain and biomass yield of lupine but increase the grain and biomass yield of the same crop in lupine-wheat and lupine-finger millet combinations. The latter two intercropping systems disagree with Prasad and Brook (2005), who reported that biomass and grain yield of soybean were least at greater density and greatest at the lowest density, whilst biomass and grain yield of maize increased. In general, the above findings on grain yield and biomass yield suggest that difference in seeding ratios and crop types of the component crops could affect yield and yield attributes of lupine.

### 3.2.3. Harvest Index (HI)

The harvest index of lupine was significantly influenced (P<0.01) by seeding ratios, crop types and cropping system (Table 4). Sole cropping showed the highest harvest index (23.73 %) as compared to lupine-cereal combination at 50:100 seeding ratios. However, the latter seeding ratio did not significantly differ from the remaining seeding ratios (Table 4). Lupine-wheat combination gave the highest harvest index (20.18 %) were as lupine-barley combination gave the lowest harvest index (14.28 %).

The combined mean highest harvest index was recorded in sole cropped lupine (23.73 %) which was not statistically differed from lupine-wheat combinations at 75:100 seeding ratios (23.46 %) as compared to the other crop combinations in different seeding ratios (Table 4). This is mainly due to high grain yield and low biomass yield of lupine in the respective cropping system, since; HI is a derived parameter of economic yield (grain yield) and biomass yield. This result was corroborated with Bismillah and Khaliq (2004), who reported that sole cropped soybean showed higher HI as compared to soybean-cotton, based intercropping system.

The combined mean lowest harvest index was recorded in lupine-barley combination at 75:100 seeding ratios (5.82 %) followed by lupine-finger millet combination at 25:100 (11.76%) and lupine-barley combination at 50:100 (11.73%) seeding ratios. In intercropping, low harvest index could also be attributed to low plant population. The highest harvest index in lupine-wheat, lupine-barley and lupine-finger millet was recorded at 75:100 (23.46 %), 25:100 (15.8 %) and 50:100 (18.56 %) seeding ratios, respectively (Table 4). Except lupine-barley combinations, as the companion seeding ratio increased in each lupine-cereal intercropping, HI also increased.

In general, highest harvest index of lupine was obtained from the highest grain yield and the lowest biomass yield in sole cropped, lupine-wheat combination at the highest cropping ratios and in lupine-finger millet combinations at 50:100 seeding ratio, whilst lowest harvest index value was obtained from the lowest grain and lowest biomass yield in the highest cropping ratios in lupine-barley combinations (Table 4).

### *3.3. Partial Land Equivalent Ratio and Area Time Equivalent Ratio (ATER)*

#### *3.3.1. Partial Land Equivalent Ratio*

The present experiment showed that except partial LER of wheat in lupine-wheat combinations, partial LER of barley and finger millet in lupine-barley and lupine-finger millet combinations and partial LER of lupine in all lupine-cereal combinations at all seeding ratios (<1.00) were lower than sole cropped of each crop species (1.00) (Table 5). Within the intercropping, the combined highest partial LER of lupine was recorded in lupine-finger millet combinations at 75:100 seeding ratio (0.63528), while the lowest was recorded in lupine-barley combination at 25:100 seeding ratios (<0.5) but statistically on par with the other seeding ratios of the same combinations (Table 5). A comparison of the data presented in Table 5 also show that partial LER of lupine increased as its proportion increased in all lupine-cereal combinations probably due to efficient utilization of nutrients created by intra-specific competition.

This more explained in which the partial LER of lupine was higher than 0.5 only in lupine-finger millet intercropping at 75:100 (0.635) and 50:100 (0.512) seeding ratios and in lupine-wheat intercropping at 75:100 (0.546) seeding ratios while, in most cases, partial LER of cereals at all seeding ratio was higher than 0.5 which indicates that there was an advantage for cereals and a disadvantage for lupine in these intercropping systems. Thus, the results ascertain that cereals were the major contributor to the mixture yield which also confirms farmers' justification of growing the lupine as an intercrop. This was in agreement with Chen et al. (2004), who reported that partial LER for cowpea was lower than 0.5 which indicated advantage for cotton in intercropping. While, partial LER for sorghum was more than 0.5, it indicated disadvantage for cotton in intercropping.

***Table 4.*** *Mean effect of crop proportion and crop types on grain and biomass yield, and harvest index of lupine in lupine-cereal intercropping systems at AARC, North Western Ethiopia in 2009.*

| Treatment and statistics | Grain yield (kg/ha) | Biomass yield (kg/ha) | Harvest -index (%) |
|---|---|---|---|
| Seeding ratios (Lupine: Cereal) | | | |
| 25:100 | 459d | 3432c | 14.92b |
| 50:100 | 768c | 4444c | 15.53b |
| 75:100 | 1146b | 6194b | 15.41b |
| Sole lupine | 2852a | 12250a | 23.76a |
| Lupine-Cereal combination | | | |
| Lupine-barley | 757c | 3597c | 14.28b |
| Lupine-finger millet | 1749a | 9562a | 17.76ab |
| Lupine-wheat | 1413b | 6581b | 20.18a |
| Cropping system | | | |
| Sole lupine | 2852a | 12250a | 23.76a |
| 25 % Lupine+100 % wheat | 456e | 3667ef | 13.90abc |
| 50 % Lupine+100 % wheat | 785de | 4861de | 16.31abc |
| 75 % Lupine+100 % wheat | 1558ef | 6667cd | 23.46a |
| 25 % Lupine+100 % barley | 50f | 333f | 15.80abc |
| 50 % Lupine+100 % barley | 59f | 556f | 11.73bc |
| 75 % Lupine+100 % barley | 69f | 1250f | 5.82c |
| 25 % Lupine+100 % f/millet | 872d | 7417cd | 11.76bc |
| 50 % Lupine+100 % f/millet | 1459c | 7916bc | 18.56ab |
| 75 % Lupine+100 % f/millet | 1812b | 10667ab | 16.96ab |
| LSD (0.05) | 198 | 1723 | 6.29 |
| CV (%) | 10.82 | 16.21 | 26.00 |

Notes: Values (means) connected by different superscript letters are significantly (P<0.05) different within columns according to Tukey- HSD tests. F/millet: finger millet

***Table 1.*** *Partial land equivalent ratios and Area time equivalent ratios of grain yield in lupine-cereal intercrops at three seeding ratios (additive series) at AARC, North Western Ethiopia in 2009.*

| Treatments and statistics | Partial LER | | ATER |
|---|---|---|---|
| | Lupine | Cereal | |
| Sole lupine | 1.000a | - | 1.000ab |
| Sole wheat | - | 0.990ab | 0.990b |
| Sole barley | - | 1.000ab | 1.000ab |
| Sole finger millet | - | 1.000ab | 1.000ab |
| 25 % Lupine+100 % wheat | 0.160e | 1.174a | 1.049ab |
| 50 % Lupine+100 % wheat | 0.275e | 1.038ab | 1.075ab |
| 75 % Lupine+100 % wheat | 0.546c | 0.944ab | 1.313a |
| 25 % Lupine+100 % barley | 0.017f | 0.748b | 0.391c |
| 50 % Lupine+100 % barley | 0.021f | 0.765ab | 0.402c |
| 75 % Lupine+100 % barley | 0.024f | 0.868ab | 0.455c |
| 25 % Lupine+100 % f/millet | 0.306d | 0.791ab | 1.111ab |
| 50 % Lupine+100 % f/millet | 0.512c | 0.813ab | 1.378a |
| 75 % Lupine+100 % f/millet | 0.635b | 0.659b | 1.372a |
| LSD (0.05) | 0.05 | 0.24 | 0.23 |
| CV (%) | 6.16 | 15.75 | 12.79 |

Notes: Values (means) connected by different superscript letters are significantly (P<0.05) different within columns according to Tukey- HSD tests. F/millet: finger millet

### *3.3.2. Area Time Equivalent Ratio (ATER)*

ATER values showed an advantage of 37.87 % and 37.19 % in lupine-finger millet combination at 50:100 and 75:100 seeding ratios, respectively and 31.34 % in lupine-wheat at 75:100 seeding ratio followed by 11% in lupine-finger millet combination at 25:100 seeding ratios than the other mixtures and all sole cropped (Table 5). This could be due to the reason that intercropping systems can actually give more efficient total resource exploitation and greater overall production than sole crops (compatible intercrops) (Reddy et al., 1994), optimum mixture ratios and staggered planting arrangements particularly planting dates. Similarly, compared with corresponding sole crops, yield advantages have been recorded in many non-legume-legume intercropping systems, including groundnut-cereal fodders (Ghosh, 2004), barley-pea (Chen et al., 2004), and faba bean-barley (Trydemanknudsen et al., 2004), bean-wheat (Hauggaard-Nielsen et al., 2001).

Whereas, all lupine-barley intercropping showed values less than 1.00 and pure stands thus indicated the disadvantage. Lupine-barley intercropping resulted in 55-61 % disadvantage with maximum disadvantage at 25:100 seeding ratio (61 %) with similar response to the other seeding ratios. This was in agreement with Banik et al. (2000), who reported that intercropping reduced the yield of mustard-pea, mustard-lentil, and mustard-gram mixtures over sole-cropping. In other words the present experiment showed that the higher seeding ratio gave the highest yield advantage which was differ from what farmers currently used (≤25 % lupine seed rate with full cereal seed rate).

## 4. Conclusion

The present study demonstrated that lupine agronomic attributes were less affected when intercropped with finger millet than wheat and barley. Intercropping higher proportion of lupine with wheat and finger millet did help much in increasing grain yield and biomass yield of lupine without affecting main crop yield. However, nearly complete dominance of barley over lupine at all seeding ratios leads to absolute reduction in grain yield and ATER as compared to sole cropped. The lupine-finger millet mixture at the 50:100 and 75:100 seeding ratio and lupine-wheat mixture at the 75:100 seeding ratio had a higher yield advantage of intercropping for exploiting the resources of the environment compared with the other intercropping systems. This also indicated that the area planted to monocultures would need 37.78 %, 37.19 %, and 31.3 % more lands, respectively than the area planted to intercropping for the two crops to produce the same combined grain yield.

## Acknowledgment

The authors duly acknowledge the financial support provided by the rural capacity building project implemented in Ethiopia.

## References

[1] AARC (Adet Agricultural Research Center) (2002). Summary of research achievements (1987-2001), Adet, Ethiopia

[2] Adet Meteorological Station (2009). Climatic data. Amhara Regional State of Ethiopia; Bahir Dar

[3] Aleligne, K. and Steven, F. (1987). Initial results of informal survey Adet farming system, Gojam region,

[4] Banik, P., Samsal, T., Ghosal, P.K. and Bagchi, D.K. (2000). Evaluation of mustard and legume intercropping under 1:1 and 1:2 row replacement series system. Journal of Agronomy and Crop Science 185: 9- 14.

[5] Bismillah, M.K. and Khaliq, A. (2004). Production of soybean (Glycine max L.) as cotton based intercropped. Journal of Research Science 15(1): 79-84.

[6] Carr, P.M., Horsley, R.D and Poland, W.W. (2004). Barley, oat and cereal-pea mixtures as dry land forages in the Northern Great Plains. Journal of Agronomy 96: 677–684.

[7] Central Statistical Agency (2007). Agricultural sample survey. 2006/2007 Volume I Report on area and production of crops. Statistical bulletin no. 388 .July 2007 Addis Ababa.

[8] Challa, H. and Bakker, J. (1998). Potential production within the greenhouse environment. In: Enoch, Z. and Stan hill, G. (Eds.) Ecosystems of the world. The greenhouse ecosystem. Amsterdam: Elsevier. pp. 333-348.

[9] Chen, C., Westcott, M., Neill, K., Wichman, D. and Knox, M. (2004). Row configuration and nitrogen application for barley-pea intercropping in Montana. Journal of Agronomy 96:1730-1738.

[10] de Wit, C.T. (1960). On competition.Verslag Land-bouwkundige Onderzoek No.66 (8):1-82.

[11] Ghosh, P.K. (2004). Growth, yield, competition and economics of groundnut/cereal fodder intercropping systems in the semi-arid tropics of India. Field crop Research 88:227-237

[12] Gonder soil testing laboratory center (2009), Gonder, Ethiopia

[13] Hauggaard-Nielsen, H., Ambus, P. and Jensen, E.S. (2001). Interspecific competition, N use and interference with weeds in pea-barley intercropping. Field Crop Research 70: 101-109.

[14] Hiebsch, C.K. (1980). Principles of intercropping. Effect of N fertilization and crop duration on equivalency ratios in intercrops versus monoculture comparisons, PhD thesis, North Carolina State University, Raleigh, N. C., USA.

[15] Jansen, P. C.M. (2006). Lupines albus L.In. Brink, M. & Belay, G. (Editors) PROTA (Plant Resources of Tropical Africa) Wageningen, Netherlands.

[16] Jones, R.K. and McCown, R.L. (1983). Research on no-till tropical legume farming strategy. In: Proceedings of Eastern Africa -ACIAR Consultation on Agricultural Research. Nairobi, Kenya: 18-23.

[17] Knudsen, M.T., Havggaard-Nielsenh, J.B. and Jensen, E.S. (2001).Comparison of inter-specific competition and N-use in pea- barely, faba bean-barely and lupine-barely intercrops. Abstract.

[18] Mandal, B.K., Dhara, M.C., Mandal, B.B., Das, S.K. and Nandy, R. (2010). Effect of intercropping on the Yield Components of Rice, Mungbean, Soybean, Peanut and Blackgram. Journal of Agronomy and Crop Science 162: 30-34.

[19] McGibbon, R. and Watkin, W. (1980). Effects of Plant and Canopy Density on Seed Yield and Oil Content in White Lupin ( Lupinus albus). Experimental Agriculture 16: 409-414

[20] Mitchell C.E Tilman D. and Groth J.V. (2002). Effects of grassland plant species diversity, abundance, and composition on foliar fungal disease. Ecology 83: 1713-1726.

[21] Nigusei, T. (1994). Performance of maize/bean intercropping systems under low and medium rainfall situations, unpublished M.Sc. thesis, Alemaya University of Agriculture, Ethiopia.

[22] Ofori, F. and Stern, W.R. (1987). Cereal and legume intercropping systems. Advanced Agronomy 41:41

[23] Prasad, R. B. and Brook, R.M. (2005). Effect of varying maize densities on intercropped maize and soybean in Nepal. Experimental Agriculture 41: 365-382. Abstract. Accessed in, March, 2010 at: mailto:afs082@bangor.ac.uk

[24] Rahman, M.M., Awal, M.A., Amin, A. and Parvej, M.R. (2009). Compatibility, growth and production potentials of mustard/lentil intercrops. International Journal of Botany 5: 100-106. Accessed in, March, 2010 at:.http://scialert.net/fulltext/?doi=ijb.2009.100.106&org=11

[25] Reddy, K.C., Visser, P.L., Klaij, M.C.and Renard, C. (1994). The effects of sole and traditional intercropping of millet and cowpea on soil and crop productivity. Experimental Agriculture 30: 83–88.

[26] SAS Institute Inc. (2002). JMP-5 Statistical Software, Version 5.Cary, NC, USA.

[27] Sheri, M.S., Jane, R.k., Ken, J.L. and Neli, H.K. (2008). Forage potetial of intercropping barely with faba bean,lupine and field pea.Published in Segoe,Canada. Accessed in, December, 2009 at: htt://agron.scijournals.org

[28] Tilahun T., Minale L. and Alemayehu A. (2012). Role of maize (Zea mays L.)- Fababean (Vicia faba L.) intercropping planting pattern on productivity and nitrogen use efficiency of maize in northwestern Ethiopia highlands. International Research Journal of Agricultural Science and Soil Science 2(3):102-112

[29] Trydemanknudsen, M., Hauggaard-Nielsen, H., Jornsgard, B. and Steen jensen, E. (2004). Comparison of inter-specific competition and N use in pea–barley, faba bean–barley, and lupine–barley intercrops grown at two temperate locations. Journal of Agricultural science 142:617–627

[30] Venkateswarlu, J. (1984). Nutrient management in semi-arid red soils. In: Nutrient management in dry lands with special reference to cropping systems and semi-arid red soils. All India Coordinated Research Programmers for Dry land Agriculture, Hyderabad, India. Part 2:1-56.

# 27

# Forest offence types, trends and lapses: Case study of the Nkawie Forest District of Ghana from 2005-2010

**Mercy A. A. Derkyi[1], Ton Dietz[2]**

[1]Department of Forest Science, School of Natural Resources, University of Energy and Natural Resources, P.O. Box 214 Sunyani, Brong Ahafo Region of Ghana
[2]African Studies Centre, PO Box 9555, 2300 RB Leiden, The Netherlands

**Email address:**
mercy.derkyi@uenr.edu.gh (M. A. A. Derkyi), dietzaj@ascleiden.nl (T. Dietz)

**Abstract:** The documented analysis of the forest offences revealed eight forest offence types with prevalent cases being chainsaw milling, illegal logging and illegal farming. Between 2005 and 2010, the district recorded 121 offences with more offences occurring in the on-reserve areas than in the off-reserve areas. The timber species, which was most exploited illegally is *Triplochiton scleroxylon* (wawa), a scarlet star-rated species, which is under imminent economic threat. Analyses of the official records revealed lapses in record keeping since there are no references to the outcomes of most offences, especially those relating to cases under investigation by the Police and Regional Forest Services Division. A call for capacity development of the frontline staff of the Forest Services Division at the District levels is paramount especially in detecting, collecting, entering and analyzing forest offences cases and outcomes as well as training in forest investigation and incident management. Furthermore, equipping the district offices with the requisite technological equipment like computers and accessories, software and real-time Global Positioning System (GPS) trackers will enhance effective monitoring and data management of forest resources especially when Ghana is committed in ensuring timber legality under the Voluntary Partnership Agreement.

**Keywords:** Forest Offences, Nkawie Forest District, Chainsaw Milling, Illegal Logging, Law Enforcement

## 1. Introduction

The fundamental components of forest governance arrangements are attributed to policy approaches and instruments, which regulate forest management to protect environmental and other forest values especially in an era where environmental and natural resource crime has significant adverse effects on ecosystems, livelihoods, and the global economy [1, 2, 3].

In Ghana, the forest sector has an extensive range of laws that govern the management of forest resources. In practice, however, it appears that the enforcement of such laws is compounded by political and institutional challenges, which result in the prevalence of illicit forest activities that eventually lead to environmental damage, revenue loss and the destruction of livelihood sources of forest fringe communities [4, 5, 6]. Three weaknesses that undermine the regulatory and legislative instruments of Ghana forest sector are (i) non-compliance with the regulations by the sector itself, (ii) too many regulations governing the forestry sector, and (iii) low penalties for default [6]. Furthermore, weak forest law enforcement is attributed to (i) the remoteness of forested areas, (ii) the absence of a road network, or a deficient road network, which impedes movement, and (iii) a closed canopy that hinders direct observations [7]. Nevertheless, Tasmania's forest practice systems in their Forest Practices Act has recorded higher compliance and enforcement through well-designed and implemented co-regulatory approaches with emphasis also on training and education and with enforcement powers like penalties as last resort [8]. In 2008, when Ghana committed itself to the Non-Legally Binding Instrument on All Types of Forests (NLBI), one of the four key policy objectives stakeholders recommended for urgent policy consideration in order to start realizing sustainable forest management was strengthening law enforcement. This was in line with the United Nations Forestry Forum (UNFF) NLBI policy measure resolution 6(n), which states, 'Review, and as needed, improve forest-related legislation, strengthen forest law enforcement and promote good governance at all levels in order to support sustainable forest management, to create an enabling environment for forest investment and to combat and eradicate illegal practices according to national legislation, in the forest and other related sectors' [9]. In an effort to achieve this policy objective, one of the strategies proposed is to document forest offences in the NLBI monitoring and evaluation framework as a baseline for monitoring illegal forest activities and as a means of checking whether existing measures have led to a minimization of these offences [10]. It is also worth mentioning that Ghana's voluntary partnership agreement (VPA) with the European Union is in line with this objective through enforcement of law via legality of assurance and compliance systems [11].

Against this background, this paper aims to contribute insights that will support the national objective to strengthen law enforcement by analyzing forest offences types, trends and identify institutional challenges in the Nkawie forest district for policy consideration.

The specific questions addressed in this paper are:

1. What politics govern the Nkawie Forest District regarding institutional and legislation frameworks?
2. What types of forest offences prevail, how are the trends from 2005-2010 and the lapses thereof?

The paper is based on secondary analysis of forest politics in Ghana and the analyses of official forest offence records at the Nkawie Forest District from 2005-2010.

# 2. Methodology

## 2.1. Study Area

The Nkawie Forest District is one of seven forest districts in the Ashanti region under the jurisdiction of the Forest Services Division of the Forestry Commission. The Forest District headquarters is at Nkawie, about a forty-five minute drive from Kumasi, the capital of the Ashanti Region. The forest district covers six political districts, namely Atwima Nwabiagya, Atwima Mponua, Atwima Kwawoma, Bosomtwi, Ahafo-Ano South and Ahafo-Ano North. The Nkawie Forest District shares common boundaries with eight other forest districts, namely Bechem and Offinso Forest Districts in the north, Juaso Forest District in the east, Kumawu Forest District in the northeast, Goaso, Bibiani and Sefwi Wioso Forest Districts in the west, and Bekwai Forest District in the southwest.

Six forest reserves are managed by the Nkawie Forest District covering a total area of 1,019.72 $km^2$. These are Asenayo, Desiri, Jimira, Offin Shelter, Tano-Offin and Tinte Bepo forest reserves. All the reserves are vested in the Golden Stool (i.e. the Asantehene) under the care of some stools who act as caretakers, as shown in Table 1.

***Table 1.*** *The forest reserves, total area and perimeter in the Nkawie Forest District*

| Forest Reserve | Total area ($km^2$) | Total perimeter (km) |
|---|---|---|
| Asenanyo | 227.92 | 100.26 |
| Desiri | 150.95 | 63.46 |
| Jimira | 62.85 | 49.78 |
| Offin Shelterbelt | 60.23 | 40.26 |
| Tano-Offin | 402.23 | 117.90 |
| TinteBepo | 115.54 | 64.62 |
| Total | 1019.72 | 436.28 |

Source: Nkawie Forest Services Division info-sheet (2010).

## 2.2. Data Collection and Analysis

The study gathered data using secondary and official data sources. The secondary data sources were gathered from official documents of the District FSD and the Forestry Commission laws to understand the politics governing the forest resources in the Nkawie forest district and the laws governing forest offences in Ghana. The study further employed unobtrusive method in collecting data from the official records of forest offences from 2005 to 2010. This kind of data collection offers an opportunity for longitudinal research to analyze the trends and connection over time [12]. The variables of interest were: i) type of offences recorded during the period and its distributions; ii) the species in relation to timber related offences; iii) size of cleared forest lands in terms of land use offences such as forest farming. Other variables included the occurrences of the offences with regards to tree species, species names, offences outcomes and the ecological importance of tree species harvested in respect to the star rating system. Ghana's star rating system indicates conservation priorities for timber species as follows: green = no threat of extinction; pink = significantly exploited but no threats yet to their economic future; red = current rates of exploitation present a significant danger of economic extinction; scarlet = under imminent threat of economic extinction [13].

The secondary data were content analysed while the official data sources were analysed using thematic coding based on the variables using SPSS. Data were descriptively analyzed and inferences made such as offence types and number of offence cases from 2004-2010 as shown in Table 2.

***Table 2.*** *Occurrence of forest of offence types in the district from 2005-2010*

| Offence type | Number of offence cases from 2005-2010 | | | | | | Total |
|---|---|---|---|---|---|---|---|
| | 2005 | 2006 | 2007 | 2008 | 2009 | 2010 | |
| Sawn lumber (chainsaw milling) | 24 | 9 | 9 | 8 | 1 | 2 | 53 |
| Illegal logging | 12 | 9 | 6 | 9 | 7 | 7 | 50 |
| Illegal farming | 4 | 0 | 1 | 1 | 1 | 2 | 9 |
| Others* | 2 | 0 | 1 | 2 | 1 | 3 | 8 |

*Including charcoal burning, canoe carving, illegal mining (galamsey), human-induced wildfire and deliberate destruction of plantations.

# 3. Results

### *3.1. Politics Governing the Forest Resources*

The Nkawie Forest District, which is headed by a manager and two assistant managers, is responsible for the management of forest reserves at the micro level. The range supervisors and forest guards are the frontline officials in direct and frequent contact with the local communities and the timber operators. As far as forest management is concerned, the District FSD monitors and regulates the harvesting of forest resources, especially timber, in reserve and off-reserve areas. With regard to protection, the District FSD protects the reserves by maintaining forest reserve boundaries, preventing wildfire and taking steps to keep out illegal loggers, chainsaw millers and farmers especially in hill sanctuaries, convalescence areas, globally significant biodiversity areas (GSBAs) and along river banks. It is the mandate of the forest guards to keep reserve boundaries clear and patrol the reserve in order to prevent illegal activities. In some of these activities, the District FSD collaborates with forest-fringe communities. The range supervisors are in charge of regulating timber exploitation in a designated forest area. Using the tree information form (TIF), they conduct stock surveys and measure felled trees to estimate tree volume. Additional responsibilities are issuing log conveyance certificates to contractors and conducting post-harvesting timber inspections.

An important core activity of the District FSD is forest plantation development. This activity is carried out in collaboration with District Assemblies when plantations are established in the administrative districts, with local communities under the modified taungya system and with plantation developers under the commercial plantations scheme [14]. Another activity is forestry education, which encompasses public awareness creation. This is done in collaboration with other entities such as Radio Stations, District Assemblies, the Ghana National Fire Service and some NGOs. Topics addressed in educational and sensitization programmes include, but are not restricted to, wildfire prevention, afforestation and reforestation. The District FSD also provides technical advice to individuals or institutions who wish to engage in forestry activities such as tree nursery and plantation establishment. The Nkawie Forest District was one of the districts in which the Voluntary Partnership Agreement (VPA) with the EU to combat illegal logging was piloted as well as the site for the implementation of the Wildfire Management Act [14].

Three sets of legislation of relevance to forest offences are: (i) the Timber Resources Management Act, 1997 (Act 547) and its amended Act 617; (ii) the Timber Resources Management Regulation, 1998, (L.I. 1649), and (iii) the Forest Protection Decree 1974 NRCD 243 as amended by the Protection Amendment Act 2002, Act 624; these are presented in Box 1.

***Box 1.*** *Laws governing forest offences and their sanctions*

Box 1a. The Timber Resources Management Act, 1997(Act 547), Section 17 stipulates that:

'(1) Any person charged with the management or protection of a forest resource by virtue of his employment in any institution of government who(a) By any act or omission in the performance of his duties facilitates the breach of any provision of this Act; or(b) condones or connives with any other person in breach of a provision of this Act, commits an offence and is liable on summary conviction to a term of imprisonment of not less than 6 months and not exceeding two years without the option of a fine.

(2) Any person who(a) harvests timber to which this Act applies without a valid Timber Utilization Contract; or(b) operates a vehicle to carry, haul, evacuate or transport timber harvested in contravention of this Act*, or (c) offers for sale, sells or buys timber harvested in contravention of this Act; or(d) stocks timber harvested in contravention of this Act; or(e) carries, hauls or evacuates by non-mechanical means any timber harvested in contravention of this Act, or (f) is the owner of a vehicle or not being the owner causes to be operated a vehicle to carry, haul, evacuate or transport timber harvested in contravention of this Act**, commits an offence and is liable on summary conviction to imprisonment for a term of not less than 6 months and not exceeding 2 years.

(3) Where a person is convicted under subsection (2) the court shall order the confiscation to the State of any tool, equipment and machinery involved in the commission of the offence; and the court shall order to be confiscated and sold any timber harvested in the commission of the offence.

(4) Notwithstanding the right of the court to sentence a person convicted under subsection (2) of this section to imprisonment, the court may in lieu of sentence of imprisonment impose in respect of the offences specified in-(a) subsection (2) (a) or (b), a penalty in the sum of 1000% of the market value of the timber involved in the commission of the offence; or(b) subsection (2) (c) or (d), a penalty in the sum of 500% of the market value of the timber involved in the commission of the offence; or(c) subsection (2) (e), a penalty in the sum of 100% of the market value of the timber involved in the commission of the offence.'

* As amended by the Timber Resources Management (Amendment) Act, 2002 (Act 671), s. 4(a).

** As inserted by the Timber Resources Management (Amendment) Act, 2002 (Act 617), s. 4(b).

Source: [15]

Box 1b. Timber Resources Management Regulation, 1998, (L.I. 1649) Regulation (41) deals with the offences and declares that:

'(1) Any person who (a) obstructs an authorized person of the Forest Products Inspection Bureau or the Forestry Department in the exercise of his duties contrary to sub-regulation (2) of regulation 15; or (b) is found in possession of any timber product that is not properly marked or numbered contrary to sub-regulation (3) of regulation 20; or (c) is found moving or in possession of any timber without a conveyance certificate in respect of the product contrary to sub-regulation (1) of regulation 24; or (d) being the owner of a chainsaw fails to register the chainsaw with the relevant District Assembly contrary to regulation 28(1); or (e) uses a chainsaw not registered with the District Forest Officer to fell any tree or saw any timber contrary to regulation 29(1); or (f) fails to mark the stump of a tree he fells with the timber registration number contrary to regulation 33; or (g) sells or offers for sale lumber cut with a chainsaw contrary to regulation 32(2); or (h) being a landowner permits an unregistered chainsaw to be used on his land for felling trees or sawing timber contrary to regulation 34; or (i) fells or is found in possession of a restricted species of timber contrary to sub-regulation (1) of regulation 40, commits an offence and is liable on summary conviction to a fine not exceeding ¢5 million* (equivalent to GH¢ 500 or US$ 333)[1] or to imprisonment for a term not exceeding twelve months or to both, and where the offence continues to a fine not exceeding ¢100,000.00 (GH¢ 10 or US$ 7) for every day or part of a day during which the offence continues.

(2) Where a person fells any tree or saws any timber with an unregistered chainsaw the court shall, in addition to the punishment imposed under sub-regulation (1) of this regulation order that the chainsaw or the tree or timber products in respect of which the offence was committed be forfeited and disposed of as the court may direct. (3) Any officer who- (a) declares a false measurement of timber contrary to sub-regulation (1) of regulation 23; or (b) falsely calculates stumpage fees; or (c) assists a contractor to under declare tree volumes used in the calculation of stumpage fees; commits an offence and is liable on summary conviction to a fine not exceeding ¢5 million or imprisonment for a term not exceeding 12 months and shall be liable to be dismissed from the public services.'

* The new Ghana cedi and US dollar equivalents of the old Ghana cedi were added by the authors.

Source: [15]

Box 1c. Forest Protection Decree 1974 NRCD 243 as amended by the Protection Amendment Act 2002, Act 624

1. "Any person who in a Forest Reserve without the written consent of the competent forest authority

(a) fells, uproots, lops, girdles, taps damages by fire or otherwise damages any tree or timber;

(b) makes or cultivates any farm or erects any building;

(c) causes any damage by negligence in felling any tree or cutting or removing any timber;

(d) sets fire to any grass or herbage, or kindles a fire without taking due precaution to prevent its spread;

(e) makes or lights a fire contrary to any order of the Forestry Commission;

(f) in any way obstructs the channel of any river, stream, canal or creek;

(g) hunts, shoots, fishes, poison water or sets traps or snares;

(h) subjects any forest produce to any manufacturing process or collects, conveys or removes any forest produce; or

(i) pastures cattle or permits any cattle to trespass;

commits an offence against the state and is liable to summary conviction to a fine not exceeding 500 penalty units or to imprisonment not exceeding 2 years or to both, except that for a second or subsequent offence under this section the offender shall be liable on summary conviction to a fine of not less than 250 penalty units or to imprisonment not exceeding 3 years or to both.

2. Any person who

(a) knowingly counterfeits or fraudulently uses upon timber or standing a mark or indicates that the timber or tree is the property of any person;

(b) without the written consent of a Forest Officer alters, defaces or obliterates a mark placed on any timber or tree; or

(c) alters, moves, destroys or defaces any boundary mark of any Forest Reserve;

commits an offence and is on summary conviction liable to a fine not exceeding 500 penalty units or to imprisonment not exceeding 2 years or to both, except that for a second or subsequent offence under this section the offender shall be liable on summary conviction to a fine of not less than 250 penalty units or to imprisonment not exceeding 3 years or to both.

A 'competent forest authority' means a forest officer with the rank not below Assistant District Manager who is able to take decisions on behalf of the Executive Director of the Forest Services Division or the Chief Executive of the Forestry Commission."

Source: [15]

[1]Exchange rate was then $1equivalent to GH 1.5

### *3.2. Forest Offence Types in the District from 2005-2010*

A review of the forest offences records in the Nkawie Forest District indicated a total of 121 offence cases in the six forest reserves and the off-reserve areas from 2005-2010. Of these cases, twenty (17%) occurred in the off-reserve area, whereas 101 (83%) occurred in the six forest reserves.The most frequently recorded offence type was chainsaw milling (44%) followed by illegal logging (41%)[2] and illegal farming (7%). The other cases (8%) were charcoal burning (1%), canoe carving (2%), illegal mining (2%), human induced wildfire (2%) and intentional destruction of established timber plantations (1%).

### *3.3. Trend of the Forest Offences from 2005-2010*

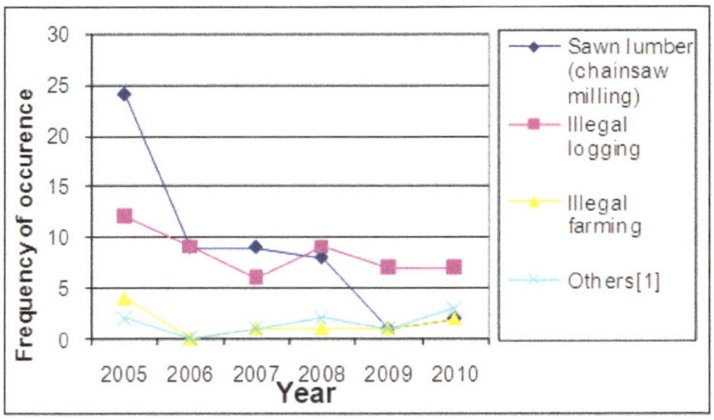


***Figure 1.*** *Trend in the number of recorded forest offences in Nkawie Forest District*

Figure 1 shows the various offences in the specific years. In 2005 the Nkawie Forest District recorded 24 chainsaw milling cases. In the subsequent years, fewer offences were recorded, a trend that is also evident for the other offence types. In 2005, four cases of illegal farming were recorded, with no such incidences in 2006 and incidental incidences in subsequent years.

#### *3.3.1. Offence Related to Chainsaw milling (Number of Cases =53)*

Of the 53 recorded chainsaw milling offences cases, 79% occurred in the six forest reserves, with the Tano-Offin forest reserve recording the highest number of cases (n=17), and 21% were reported from the off-reserve area. Generally, those who were arrested were fined by the FSD, while the lumber, chainsaws and vehicles carrying the illegal lumber were confiscated. Confiscated lumber is generally auctioned. In some cases, offenders absconded while others were arrested by the police or the FSD military team. The majority of cases (n=35) were dealt with through police investigation and administrative means available to the FSD. In sixteen of the recorded cases, the culprits absconded, although the FSD managed to confiscate the lumber and some of the chainsaw machines. Only two cases were sent to the district court for judgment, where the offenders were sentenced to a fine of GH¢ 250 ($167) and GH¢ 400 ($267) respectively in 2005. The records showed that offence cases related to chainsaw milling are charged under Act 547 sec 17 (2) (a) and (b) and its amended Act 624 Sec 1(a) (h) (Box 1a).The three most common trees felled by these offenders were *Triplochiton scleroxylon* (Wawa), *Piptadeniastrum africanum* (Dahoma) and Khaya spp. (African mahogany), which fall under the scarlet and red star rating respectively (Table 3). A total of 102 tree species were reported to be sawn during these periods. In thirteen cases, both on and off-reserve, records did not indicate the tree species types felled by the culprits.

***Table 3.*** *Types of tree species sawn into lumber (2005-2010) (n=40 cases)**

| Scientific name | Trade/local name | No. of trees/logs felled | Star rating** |
|---|---|---|---|
| *Triplochiton scleroxylon* | Wawa | 40 | Scarlet |
| *Piptadeniastrum africanum* | Dahoma | 17 | Red |
| *Khaya spp.* | African mahogany | 9 | Scarlet |
| *Mansonia altissima* | Oprono | 8 | Red |
| *Aningeria spp.* | Asenfena | 6 | Scarlet |
| *Milicia excels* | Odum | 5 | Scarlet |
| *Nesogordonia papaverifera* | Danta | 5 | Pink |
| *Chrysophyllum albidum* | Akasaa | 3 | Red |
| *Terminalia superba* | Ofram | 3 | Red |
| *Entandrophragma angolense* | Edinam | 2 | Scarlet |
| *Turraeanthus africanus* | Avodire | 1 | |
| *Ceiba pentandra* | Onyina | 1 | Red |
| *Terminalia ivorensis* | Emire | 1 | Scarlet |
| *Antiaris toxicaria* | Kyenkyen | 1 | Red |
| *Cola gigantean* | Waterpuo | 1 | Pink |
| *Entandrophragma cylindricum* | Sapele | 1 | Scarlet |

#### *3.3.2. Offence Related to Illegal Logging (Number of Cases =50)*

From 2005-2010, fifty cases of illegal logging were reported. The majority of the cases (82%) were recorded in the on-reserve areas with Tinte Bepo reserve recording most cases and Desiri forest reserve having no reported cases in this offence type.Similar to illegal chainsaw milling, Act 547 sec 17 (2) (a) and (b) is the main law that applies to the arresting, prosecuting and fining of perpetrators of this illicit activity. In three of the cases, the offenders were arrested and prosecuted in the law courts. The records indicate that two culprits attracted fines of between GH¢ 200 and GH¢ 300 (US$ 133 and US$ 200). The number of logs confiscated

[2]*Illegal logging refers to the removal of trees from the forest without a Timber Utilization Contract or Timber Utilization Permit; chainsaw milling refers to the on-site conversion of logs into sawn wood for commercial purposes using chainsaws [16].*

was three logs of kyenkyen (*Antiaris toxicaria*) in the Tano-Offin reserve and five logs (i.e.bako (*Tieghemella heckelii*), papao (*Afzelia Africana*) and African mahogany (*Khaya spp.*)) in the Offin Shelterbelt in 2005. The third case occurred in the Tano-Offin reserve, which involved the exploitation of four logs of wawa (*Triplochiton scleroxylon*) (Table 4). The case was sent to court. The offence record mentions a second hearing in 2005, but no outcome of that hearing was stated in the report. According to the records, thirty-one cases were under police investigation, of which 15 referred to culprits who had absconded. The FSD used administrative means to deal with the other 16 cases. In other words the offenders were arrested, fined and some logs auctioned. Some of the cases involved legal timber operators who were fined by the FSD mainly because their vehicles carried illicit logs. In these cases, the FSD deflated the tyres of the vehicles in order to prevent movement. As in the case of chainsaw milling, wawa was the most felled tree species. In this offence type wawa was followed by odum (*Milicia excels*) and onyina (*Ceiba pentandra*) as shown in Table 4. Over these years, the total number of trees recorded in the district in relation to illegal logging was 413, comprising twenty-seven timber species. The star ratings of these species were scarlet (eleven species), red (eight species), pink (five species) and green (two species) (Table 4).

***Table 4.*** *Tree species exploited under illegal logging (2005-2010*

| Scientific name | local name | Trade name/ No. of trees/logs felled | Star rating |
|---|---|---|---|
| *Triplochiton scleroxylon* | Wawa | 168 | Scarlet |
| *Milicia excels* | Odum | 37 | Scarlet |
| *Ceiba pentandra* | Onyina | 28 | Red |
| *Piptadeniastrum africanum* | Dahoma | 22 | Red |
| *Antiaris toxicaria* | Kyenkyen | 21 | Red |
| *Pterygopta macrocarpa* | Koto | 19 | Scarlet |
| *Nesogordonia papaverifera* | Danta | 15 | Pink |
| *Terminalia superba* | Ofram | 13 | Red |
| *Khaya spp.* | African mahogany | 13 | Scarlet |
| *Entandrophragma utile* | Utile | 11 | Scarlet |
| *Tieghemella heckelii* | Bako | 10 | Scarlet |
| *Celtis mildbreadi* | Esa | 9 | Pink |
| *Aningeria spp.* | Asenfena | 8 | Scarlet |
| *Entandrophragma cylindricum* | Sapele/ Penkwa | 9 | Scarlet |
| *Mansonia altissima* | Oprono | 5 | Red |
| *Chrysophyllum albidum* | Akasaa | 5 | Red |
| *Albizia ferruginea* | Awiemfosamna | 4 | Scarlet |
| *Daniella ogea* | Hyedua | 4 | Scarlet |
| *Petersianthus macrocarpus* | Esia | 3 | Pink |
| *Cylicodiscus gabunensis* | Denya | 2 | Pink |
| *Lovoa trichiliodes* | Dubinibiri | 2 | Red |
| *Afzelia africana* | Papao | 2 | Red |
| *Klainedoxa gabonensis* | Kroma | 1 | Green |
| *Albizia zygia* | Okoro | 1 | Green |
| *Distemonanthus bentamianus* | Bonsamdua | 1 | Pink |

Under the chainsaw milling and logging offence types, eight and eleven scarlet species respectively had been exploited between 2005 and 2010. This contributes to a loss of timber revenues to the state since these trees were taken illegally and not all of the culprits were arrested and fined for the state to generate revenue. Equally exploited species are those of red and pink star rating, which face potential economic extinction. Only two species (i.e. *Klainedoxa gabonensis* and *Albizia zygia*) of the trees felled under illegal logging are less valuable species with a green star rating. Illegal logging and chainsaw milling thus create a potential of losing high economic value timber species if access to these resources is not well regulated by the mandated institution.

### *3.3.3. Offence Related to Illegal Farming in the Forest Reserves (Number of Cases =9)*

Nine cases of illegal farming were recorded in four of the reserves namely Tano-Offin, Desiri, Asenanyo and Tinte Bepo. The total recorded area cleared for farming was 166 ha over the period of 2005-2010. There was one case in the Tano-Offin forest reserve in which the area of the cleared forest was not stated but covered large tracts in six compartments. From these nine cases, one offender was sent to the Nyinahin community tribunal and was fined GH¢ 300 (US$ 200). Four culprits were arrested and handed over to the police with confiscated chainsaws. In some cases, the culprits were arrested and handed over to the regional FSD office for investigation, while their farms were destroyed. The legislation for the prosecution of offenders were the NRCD 243 (amended Act 642) sec 1 (1b) (see Section 7.3.2) and Act 547 sect 17. Sub. Sec. 2 (a) (Box 1c &1a). There were also four cases in which the offenders could not be apprehended but the farms were destroyed. In such instances, surveillance was put in place to arrest any culprit. In some reserves where national plantations are being established, the cleared areas are being replanted to replace lost stocks. The outcomes of the nine offences are presented in Table 5.

*Table 5. Illegal farming offence outcomes from 2005-2010*

| Offence outcome | Number of cases* |
|---|---|
| Culprit sent to court with fine | 1 |
| Cases under police investigation | 2 |
| Cases resolved by administrative means (destruction of farms, converting it into a national forest plantation) | 6 |

* The reporting system does not indicate names of the culprits.

### *3.3.4. Other Offence Types (Number of Cases=9)*

The other offence types are made up of four categories that occurred sporadically. These were offences related to canoe carving (n=2), charcoal burning (n=2), human-induced wildfire incidence (n=2), deliberate destruction of plantations (n=1) and illegal mining (n=2). With regard to canoe carving cases occurred in both the Tano-Offin Globally Significant Biodiversity Area (GSBA) and Asenayo forest reserve in 2010 with the carving of 10 canoes whose owners absconded. The Tano-Offin GSBA offence case was charged under the NDPC 243 sec (1) (a) (amended Act 642), whereas that of Asenayo was under Act 547 Sec. 17 sub-sec 2(a). Since the culprits absconded, the matter was reported and preparations made for the auctioning of the canoes. Two cases of charcoal production were reported in the offence records of 2005 in the Asenayo and Desiri forest reserves. The culprit in one of the cases was handed over to the police for prosecution. The outcome of the investigation since 2005 is not reflected in the offence records.The last offence type in the next category is human-induced wildfire. The legislation applicable to this offence is the PNDC law 229. The two cases reported were all human-induced, but the culprits were never arrested. The wildfires were brought under control with the help of people from the forest-fringe communities. Another offence type was related to the intentional destruction of national forest plantations based on a personal grudge by a farmer. This led to the arrest of the offender and prosecution under Act 624 Sec 1 Sub-sec. (1) (a) in the district court. As in most cases, the outcome of the court case was not indicated in the reporting system. The last offence type in this category was illegal mining, which took place in the Offin-Shelterbelt and destroyed about 4 ha of forestland. In one of the cases, the offenders had to appear in court and were fined GH¢ 300 ($200) whereas in the other case, which resulted in the clearing of about 0.8 ha of forest land, the offenders were sent to court. There was, however, no mention of the court outcomes.

## 4. Discussion

The Nkawie Forest District is endowed with six forest reserves categorised into different management regimes. The accessing, processing and conveying of forest resources without a permit constitutes a criminal offence as stipulated in the Act 547 (amended 617), LI.1649 and NRCD (amended 642). Thus, every illicit activity in forest reserves and off-reserve areas especially related to timber resources is treated as a criminal offence. Such a situation makes it difficult to assess what is criminal and not criminal when it comes to people's livelihood needs. There is an indication that compliance with law enforcement is difficult especially when offences are entangled with issues of tenure [7]. However, is it a crime to access forest resources to meet one's livelihood needs? It has been argued that such questions make it difficult for local people to acknowledge that what the law says is a crime and they perceive that as injustice [17]. According to the same author, they respond to the perceived injustice by 'a culture of conspiracy' and by violating forest laws, 'as either an act of defiance or a desperate attempt at achieving subsistence' [17:16]. As indicated in the literature, policy implementation is always challenged because it is not just a mechanical translation of stated goals into activities, but rather influenced and mediated by multi-stakeholder perceptions and attitudes [18].Lessons could be drawn from Tasmania's Forest Practices Act, which uses a co-regulatory approach complemented by training and education, self-monitoring and reporting with enforcement powers being the last resortto improve compliance and enforcement in the sector [8]. The documented forest offences revealed eight forest offence types with prevalent cases being chainsaw milling, illegal logging and illegal farming. These offences confirm the conflict types reported by the local communities around Tano-Offin forest reserve in relation to forest and tree livelihood conflicts [19].

Between 2005 and 2010, the district recorded 121 offences with more offences occurring in the on-reserve areas than in the off-reserve areas. These contrast with the findings in a similar study in three forest districts in the Brong Ahafo region where more offences tend to occur in off-reserve areas than in on-reserve areas [20]. The difference between the two studies could be due to the different location of the study areas with peculiar needs of inhabitants and ease of access to on-reserve and off-reserve forests. Nonetheless, both studies have highlighted the fact that the timber species, which is most exploited illegally is *Triplochiton scleroxylon* (wawa), a scarlet star-rated species, which is under imminent economic threat.

There is empirical evidence that large numbers of cases are resolved by the FSD through the imposition of fines and the subsequent release of vehicles associated with the particular forest offence. For example, the FSD uses several approaches to resolve illegal farming offences within forest reserves. The illegal forest farms are often destroyed and the subsequent measure involves three strategies: the culprit is sent to the police station and the matter ends there, or the offence is dealt with by the FSD officials at the regional or district level, or the offender is arrested and prosecuted in court and fined. The use of administrative sanctions in civil law to deal with natural resource offences without judicial involvement is acceptable in countries like the United States [4], but is harder to implement in developing countries. An alternative procedure is what is called 'compounding', in which case an executive body (like the FC) is empowered to levy fines without involvement of the court [7:149]. In order for the FC to legally exercise its administrative powers, the FC needs to be empowered by an administrative Act to guide its actions and enable it to have a standard prosecuting or fining mechanism across its jurisdiction in the country. As this study made clear, most of the cases reported in court had either no evidence or inadequate evidence since the offenders absconded, making it difficult to present such cases in the law court under criminal penalty. Evidence of the violation of law is an essential ingredient before one can be sentenced to a fine or imprisonment. Article 19 (2) (c) of the Constitution of Ghana clearly states that '*A person charged with a criminal offence shall be presumed to be innocent until he is proven or has pleaded guilty'*. Assigning prosecuting powers to the FC could help solve the impunity of forest offences, provided measures are taken to prevent non-transparent charges and corruption.

Analyses of the official records revealed lapses in record keeping since the outcomes of offences, especially regarding cases under police and regional FSD investigations are not indicated. This makes it difficult to assess whether actions taken against these offenders actually have a positive impact on the level of offences in the district. Even though there has been a decline in recorded forest offences from 2005 to 2010, this decline may be due to the (in) efficiency of the FSD monitoring system or to the FC's incapacity to detect the offenders and record the offences. A call for capacity development of FC officials, both staff trained at university level and field officers may be essential as has also been recommended by forest governors and experts [21]. In South Carolina in the United States, for instance, the Forestry Commission officers are trained and certified in criminal justice in addition to training in forest law, forest investigation and incident management. Law enforcement therefore becomes part of the training of the forest manager and somehow this is missing in the current training of forest managers in Ghana's forest sector. The legal department of the FC could extend its mandate from representing the FC in legal matters to advocacy for the FC to establish a prosecuting system[3] as well as building capacities of FC officials in legal issues and conflict management. There is also a need for officials of the legal department to have knowledge of natural resource management in addition to knowledge of law, so that they have a better and more in-depth understanding of the negative environmental effects of forest offences and how best to quantify them during presentation of cases in the law courts. Furthermore, equipping the districts offices with the requisite technological equipment like computers and accessories, software and real-time Global Positioning System (GPS) trackers will enhance effective monitoring and data management of forest resources especially when Ghana is committed in ensuring timber legality under the Voluntary Partnership Agreement.

[3]The Ghana Forestry Commission is now having a prosecution training programme for its staff, which is a positive step in minimizing forest offences (GNA, March, 19th, 2013).

## 5. Conclusion

This study revealed that there is a broad range of legislation, which governs the forest sector Ghana in ensuring effective governance processes. Nevertheless, the implementation of these laws is challenged because of the prevalence of forest offences and low fines attached to these offences, coupled with institutional challenges of properly recording these offences. There is a clear indication of sizeable number of scarlet, red and pink timber species illegally exploited between 2005 and 2010, which contributes to environmental damage in addition to loss of timber revenues to the state. Illegal logging, chainsaw milling and illicit farming thus contribute to losing high economic value timber species if access to these resources is not effectively regulated. However, this will only be achievable if the mandated institution pays much attention to the capacity development of its frontline staff in terms of training and logistics to facilitate monitoring of the resources and data management. Furthermore, tertiary institutions in charge of training natural resource practitioners could contribute to minimizing forest offences through integrating law enforcement courses such as forest investigation, prosecution procedures and incident management in the curriculae in addition to engaging law practitioners as guest lecturers to make law enforcement an integral component of the training of the natural resource manager.

## Acknowledgements

The authors are grateful to Tropenbos International (TBI) who funded this research, which was carried out under the TBI Ghana programme. Thanks to Dr. Mirjam Ros-Tonen and Dr. Kyereh Boateng for their guidance to make the first author's PhD a reality. Appreciation also goes to the Nkawie Forest District for providing the data on forest offences. The authors also acknowledge the valuable comments provided by two anonymous reviewers on the earlier draft of this paper.

## References

[1] C.L. McDermott, B. Cashore and P. Kanowski, (2010). Global Environmental Forest Policies: An International Comparison. Earthscan, London, 2010, p.372.

[2] B. Cashore, Key components of good forest governance, Exlibris. ASEAN-German ReFOP Project, the Analysis and Making of Regional Public Policy. Discussion Paper No.6, July 2009 (2009).

[3] K. Sander, J. Lee, V. Hickey, V. B. Mosoti, J. Virdin and W.B. Magrath, Conceptualizing maritime environmental and natural resources law enforcement – The case of illegal fishing. *Environmental Development (In press)*. Available online 10 August, 2013- DOI: 10.1016/j.envdev.2013.08.002

[4] E. Larbi, J. Yankson, E.S. Annag, B.K. Acheampong and F.A. Koomson, Judicial attitude towards forest related offences in the Nkawie Forest District of Ghana. Unpublished LLB thesis, Kwame Nkrumah University of Science and Technology (KNUST), Kumasi, Ghana. 2011.

[5] E. Nutakor, E. Marfo and P.O. Tutu, Socio-political constraints to the enforcement of forest laws: A case study of chainsaw operations in Ghana. *Ghana J. Forestry* 27: 24-36, 2011.

[6] K.A. Oduro, V.K. Agyeman andK. Gyan, Implementing timber legality assurance regime in Ghana: A review of stakeholders concerns and current institutional constraints *Ghana J. Forestry* 27(2):1-10, 2011.

[7] L.C. Christy, C.E. DiLeva, J.M. Lindsay and P.T. Takoukam, Forest law and sustainable development: Addressing contemporary challenges through legal reform. Law, Justice and Development series. Washington, D.C.: World Bank, 1997, p.206.

[8] G.R.Wilkinson, M. Schofield and P. Kanowski, Regulating forestry — Experience with compliance and enforcement over the 25 years of Tasmania's forest practices system. *Forest Policy and Economics* 40, 1-11, 2014.

[9] UNFF-United Nations General Assembly, Resolution adopted by the general assembly on Non-legally binding instrument on all types of forest. Sixty–second session agenda item 54, 2008.

[10] Forestry Research Institute of Ghana (FORIG) (2010), Development of a monitoring and evaluation system for non-legally binding instrument (NLBI) on all types of forests in Ghana, 2010, p.141.

[11] Ghana-EU VPA, Annual Report: Implementing the Ghana-EU Voluntary Partnership Agreement, 2011, p.19.

[12] A. Bryman, Social Research Methods. Oxford University Press. 3rd edition, 2008, p.

[13] W. D. Hawthorne and M. Abu-Juam, Forest protection in Ghana with particular reference to vegetation and species. Gland, Switzerland, and Cambridge, UK: IUCN, 1995, p.220.

[14] NFSD Info-sheet on the Nkawie Forest Services Division, 2010 (unpublished).

[15] Ghana Forestry Commission (undated) http://76.12.220.51/library.php?id=15 (accessed 6 October 2011)

[16] E. Marfo, Chainsaw milling in Ghana: context, drivers and impacts. Wageningen, the Netherlands: Tropenbos International, 2010, p.80.

[17] K.S. Amanor, Equity in forest benefit sharing and poverty reduction. In: K.S. Nketiah, J.A.S. Ameyaw and B. Owusu, eds, Equity in forest benefit sharing: Stakeholder Views, pp. 15-23, 2005. Tropenbos International–Ghana (2005). Workshop proceedings 3, October 29, 2004.Wageningen, the Netherlands: Tropenbos International.

[18] S. Vasan, Ethnography of the forest guard contrasting discourses, conflicting roles and policy implementations. *Economic and political weekly, special articles* 5, 4125-4133, 2002.

[19] M. Derkyi, M.A.F.Ros-Tonen, B. Kyereh, B. and T. Dietz, Emerging Forest Regimes and Livelihoods in the Tano Offin Forest Reserve, Ghana: Implications for Social Safeguards. *Forest Policy and Economics* 32: 49-56, 2013.

[20] Abugre, S. and F.P. Kazaare, Trends and status of illegal timber logging in three forest districts of BrongAhafo region of Ghana. *Ghana Journal of Forestry* 26: 89-100, 2010.

[21] M.A.A Derkyi, Fighting over forest: interactive governance of conflicts over forest and tree resources in Ghana's high forest zone. African Studies Collection; issue: 41, Leiden, 2012, p.319.

# 28

# Sustainable harvesting of *Alstonia boonei* De Wild. (Apocynaceae) in the community forest of Mbeth II, East Region of Cameroon

**Jean Lagarde Betti[1], Joseph Ambara[2]**

[1]Department of Botany, Faculty of Sciences, University of Douala, BP 24 157 Cameroon
[2]Ministry of Environment and Nature Protection, Cameroon

**Email address:**
lagardeprunus@gmail.com (J. L. Betti), ambara.joseph@yahoo.fr (J. Ambara)

**Abstract:** This paper aims to analyze the traditional usages and, assess the abundance of stems and the mass of the barks of *Alstonia boonei* for its use as a medicinal plant in the community forest of Mbeth II, East region of Cameroon. Ethnobotanical surveys were conducted among the community of Mbeth II and in city markets in Bertoua and Yaoundé between January and April 2012. The management inventory was conducted in July 2012 to assess the abundance of *A. boonei* stems in the forest. During the inventory, samples of barks were collected from mature trees and the mass of the stem barks of *A. boonei* was estimated using an indirect method. *A. boonei* was cited by 26 informants in the village and markets in the treatment of six ailments. Malaria, typhoid fever and intestinal worms were largely cited. The stem bark is confirmed for its activity against malaria/fever. The density (number of stems/ha) of *A. boonei* obtained in the Community forest of Mbeth II was 0.89 stems/ha. The density of exploitable stems was therefore 0.744 stems/ha. The average volume of fresh bark for an exploitable tree is 0.0356 $m^3$, while the average fresh mass of the bark of an exploitable tree was 34.4 kg. Considering that for all trees above 30 cm diameter at breast high (considered as a minimum exploitable diameter for the harvesting of barks), only two quarters of the bark are taken from the main stem up to the first branch, the average sustainable mass of *Alstonia boonei* in Mbeth II is 17.2 kg. Results discussed in this paper can be summarized, in a specific document called "the simple management plan of *A. boonei* as a medicinal plant in the community forest of Mbeth II". The glaring development challenge at the back ground of what precedes is the urgent need to achieve the recommendations proposed in such a document.

**Keywords:** Sustainable Harvesting, *Alstonia Boonei*, Community Forests, Simple Management Plan, Fresh Mass, Management Inventories, Exploitable Tree, Cameroon

## 1. Introduction

The tropical rainforests of the Congo Basin cover an area of approximately 170 million hectares, and constitutes the world's second largest tract of rainforests after that of the Amazon River Basin. Cameroon belongs to the Congo Basin and takes the third position in terms of biodiversity richness behind the Democratic Republic of Congo and Gabon. The total surface area of the forests in Cameroon is 19 916 000 ha, representing about 42% of the national territory [1].

More than 25 years ago, Cameroon Government decided with the help of the international Community, to tackle the general problematic of sustainable forest development. The Government therefore first focused its efforts on the knowledge of the timber resource of the forest zone of the country. A national forest inventory has therefore been conducted with the assistance of the Canadian cooperation (ACDI), comprising 7 phases. Four out of these phases have yet been finalized by the years 80, for a total forest bloc of 14 000 000 hectares, with the north limit situated at about 4th parallel. In fact, the basis work undertook within the national inventory conducted during the years 1980 [2; 3] led to the elaboration of principal norms and technical tools for the management of the forest domain. These tools include: (1) the zoning plan of the forest area (phases 1-4 of national inventory) which led to the division of the forest zone in two main domain types, namely the permanent domain and the non-permanent domain, and (2) all norms related to the interventions in the forest milieu (production forests to be précised).

The permanent domain [4] is composed of two areas including (1) the state-owned forests, which belong to the State, (2) and communal forests which belong to the private domain of the council. The state-owned forests are themselves divided into production forests, protected areas, and forest reserves. Production forests are the most important in terms of surface area, 64% of the permanent domain, and 40% of the meridional zone. Production forests are mainly composed of big forest concessions. Each forest concession is composed of one or many forests of more than 5,000 ha, called the forest management unit (FMU).

The non-permanent domain comprises the community forests, sales of standing volume (small forest concessions of not more than 2,500 ha), and mining zones. It is also composed of lands affected for agricultural and other agroforestry activities [4]. Rigorous management schemes are not applied to the non-permanent domain.

Cameroon is considered as one of the most advanced in terms of forest sector policy in the Central Africa region [5; 6]. This means that, Cameroon is one of the first countries to have produced and started implementing a good and coherent forest code in the sub-region, after the summit of the world on sustainable development held in Rio de Janeiro (Brazil) in 1992 [4]. Compared with past policies (the last one promulgated in 1981), the 1994 forest policy makes the involvement of rural populations lawfull, partners and stakeholders in its implementation, notably by the ownership of community forests and hunting zones. This innovation is intended to encourage people to better protect their wildlife and vegetation cover. In forest management, one of the concerns of the state has always been the involvement of the surrounding population to enable them have a fair and equitable share from forest exploitation. Although the populations enjoy usufruct rights, they do not have a substantial share from the commercial exploitation of forest resources. A better implication of local communities in forest activities may not only lead to the sustainable use of forest resources, but also help those communities through the revenues gained to enhance the struggle for poverty alleviation.

The 1994 forest law and its application tools allowed the beginning of many experiences of community forestry, conceding to communities the right of managing the resources of their own territories. A manual of procedures of attribution together with standards of the management of community forests was produced in 1998, and revised in 2009. This manual constitutes a fair juridical basis of the application of the concept of community forestry. It also represents an important tool showing clearly the modalities and the practical steps of the acquisition and management of community forests. A community forest is a block of 2 500 ha to 5 000 ha of forest, that the community requests from the government. Community forests are assigned to the sustainable production of the wood and other resources (for e.g. non timber forest resources) in respect to the conditions that allow the preservation of ecological, environmental, and sociological functions of the forests. To own a forest, the community must: (1) organize itself in a Common Initiative Group (GIC), (2) produce a map of the forest requested or desired, (3) organize a meeting of collaboration in presence of the forest administration's officers and the civil authorities, (4) describe activities already done in the forest, (5) designate a resource person who will be responsible for the management of the forest, (6) address a letter of application to the forest administration, giving details on the specific objectives that the group intend to rich through the exploitation of the forest. When the community forest is attributed based on the elements listed above, the community is now requested to produce the Simple Management Plan (SMP) of the forest before its exploitation. This Simple Management Plan is the only document that leads to the signature of the common agreement of twenty five years between the forest administration and the community.

The first community forest was attributed by the Government in 1997, about 27 years ago. By December 2010, some 378 forests were reserved as community forests, covering about 142 470 ha. Community forests are largely used for the harvesting of timber. Little community forests are used for the harvesting of food, medicine, material culture and items for many other material and spiritual benefits, that are generally called Non-Timber Forest Products (NTFP).

This paper aims to analyze the traditional usages, assess the abundance of stems and the mass of the barks of *Alstonia boonei* for its use and trade as a medicinal plant in the community forest of Mbeth II in the East region of Cameroon.

## 2. Material and Method

### *2.1. Study Site*

Mbeth II is a village located in the Diang subdivision, the Lom and Djerem Division, in the East region of Cameroon. Mbeth II is situated at 25 km to Bertoua, the capital of the east region. The village is situated between 04°35' - 04°36' latitude nord, and 31°44' – ... longitude east. The community forest of Mbeth II is vast of 5 000 ha.

The climate is an equatorial and Guinean type, characterized by four in-equal seasons: a big dried season from December to mi-March, a small rainy season from mi-March to May, a small dried season from June to August, and a big rainy season from September to November. The average temperature is 24.3°C, with the maximum at 25.8°C in March. The average annual rainfall varies between 1 000 and 1 500 m. Soils are iron soils type.

The community forest of Mbeth II is located in the Guinean Congolese floristic region, in the low and medium altitude, in the domain of dense and rain semi-deciduous forest of Sterculiaceae (*Sterculia spp* and *Cola spp*) and Ulmaceae. Situated closed to the forest-savannah contact zone, Mbeth II forest presents phytogeographic features belonging to the two small domains or sectors including the Guineo-Soudanian sector with its different savannah stands and the typical semi-deciduous forest sector [7]. This forest has already been subjected to forest logging by the year 1970, which explains the general feature of old secondary forests observed in the field. Principal timber species include: *Triplochiton scleroxylon, Ceiba pentandra, Uapaca guineensis, Sterculia rhinopetala, Ricinodendron heudelotii, Alstonia boonei, Afzelia africana, Entandrophragma cylindricum, Hymenocardia lyrata.* Typical secondary forests of *Musanga cecropioides* and *Albizia spp* also exist in former farms and fallows.

Mbeth II hosts about 964 inhabitants, with a density of 8 inhabitants/km2. Agriculture, poaching, pit sawing and gathering NTFPs are the main activities conducted in that village. Cash tree crops include cocoa and coffee. The slash and burn type of cultivation system is practized, with maize, groundnuts, cassava, bananas and cocoyam being the main crops. Both timber and NTFP are found in Mbeth II. Some years ago, pit sawing practized mainly by young people was the unique way of exploiting timber resource, due to the lack of community forests and other forms of logging permits. Tradesmen used to buy the wood at a very low price (6 US$/tree or 2US$/m3) from each villager. The same wood was sold at Bertoua market at 50 – 60 US$/ $m^3$, which was not profitable to the villagers [8]. In 2011, the community of Mbeth II decided to better valorize their surrounding forest. They organized themselves within a Community Initiative Group (CIG) called "BWOK" and applied to obtain their community forest. The certificate of the measurements of the forest and the preliminary two years convention were signed by the Minister of Forestry and Wildlife in 2011 and 2012 respectively, allowing the community to own their 5 000 ha of forest. By doing this, the Government intended to halt illegal logging in this village.

The genus *Alstonia* belongs to the Apocynaceae family. It comprises about 40 species and has a pantropical distribution. It is a large deciduous tree, up to 45 m tall and 1.2 m in diameter; bole often deeply fluted to 7 m, small buttresses present; bark greyish-green or grey, rough; slash rough-ranular, ochre-yellow, exuding a copious milky latex; branches in whorls. *Alstonia boonei* is found in dry, peripheral, semi-evergreen Guineo-Congolian forest and transitional rainforest. In Cameroon, the plant occurs elsewhere in similar habitats and in swamp and riverine forests. *A. boonei* requires large amounts of light and colonizes gaps in the forest. It has plenty of natural regeneration in young secondary forest. Growth is rapid, and it is not uncommon for an annual increment of 1.8 m to occur in the sapling stage. It grows in a succession of crowns and should not be pruned but left to develop secondary crowns, which will later kill off the lower ones. Mature trees are often damaged by wind and decay but are fast growing and coppice readily from the base. The tree snaps easily in strong wind and therefore should not be planted near buildings [9; 10]. The sapwood, which is not differentiated from the heartwood, is very wide, up to 200 mm, soft, and light in weight when dried. The wood weighs about 400 kg/cubic m. nearly yellowish-white when freshly cut, the timber darkens on exposure. It has a low luster and no characteristic odour or taste. The wood can be glued, stained and polished satisfactory. Export prospects are doubtful, although it has a local potential for stools, carvings, domestic utensils, toys, masks, canoes, horns, light carpentry, boxes [11]

### 2.2. Methods

The method used in this study consists of ethnobotanical surveys, assessment of the abundance and stock, and biomass determination.

#### 2.2.1. Ethnobotanical Survey

Ethnobotanical surveys were conducted among the community of Mbeth II and in city markets in Bertoua and Yaoundé between January and April 2012.

The method used among the community is known as the "method for the popular pharmacopoeia", consisting of gathering data on the popular use of medicinal plants in a given area [12]. Following this method, the data for this study were obtained from direct interviews with the Mbeth II people. The survey aimed at identifying plants used in the popular pharmacopoeia among local people. The household was considered as the sample unit. In each household who accepted to answer to our questions, data were mostly recorded from adult women (mothers), because they usually knew the plants better than men and younger people. They provided useful and first-hand information on the popular use of medicinal plants. During the survey, we made enquiry "as to what ailments were treated by which plant species" rather than asking "which plants were used to treat which ailments". For each health problem cited, the name of the plants and the plant parts used were carefully recorded. The vernacular names of the plants were recorded as much as possible, and we tried to collect the plants mentioned by the informants. The final identification of plants was made at National Herbarium of Cameroon (YA) with the help of Dr. Onana and Mr. Paul Mezili. Voucher herbal specimens, collected in three samples are kept at the YA. The therapeutic statements were made of a specific disease, a symptom or a physiological effect. Information on the diagnosis of ailments was provided through a semi-structured interview of nurses or local health officials. In this paper, anti-malarial plants refer to the plants used for treating malaria or fever on a broad scale. To classify the data according to the world system, the ailments were grouped according to the classification proposed by the World Health Organization and adapted by the African Unity Organization (AUO) for the Cameroonian pharmacopoeia [13].

The second ethnobotanical survey was conducted from March to April 2012 in the Bertoua and Yaoundé markets. The survey aimed in identifying plant parts which were sold at the market for their medicinal purpose. The method used is the same as used in 2002 in Yaoundé markets [14]. Bertoua is suitable since it is the big market situated closed to Mbeth II village. Youndé is far from Mbeth II, at about 350 km in the South west. But Yaoundé is the political capital of Cameroon and was identified as one of the main point which receives forest products harvested in the Centre, East, and South regions of Cameroon [14]. The data collected among sellers in markets can easily be compared with those recorded among people living in Mbeth II village. For any plant species sold at the market, the vernacular names of the plant and the name and ethnic group of the seller were recorded. Then we asked the seller questions concerning plant parts used and their therapeutic indication. These indications (therapeutic statements or ailments or judgments) were made on a specific disease, a symptom or a physiological effect. The plant parts were then weighed and the prices were recorded. We bought one sample of the plant parts for information set and from each seller. The plants identified were kept in the National Herbarium of Cameroon (YA).

### *2.2.2. Assessment of the Abundance and Stock of A. Boonei in the Community Forest of Mbeth II*

#### *2.2.2.1. Sampling Design*

The assessment of the abundance of *Alstonia boonei* was conducted in July 2012. It consisted of gathering qualitative and quantitative data on the stems of *A. boonei* found in the Community forest of Mbeth II. This is a mixture of primary and secondary forests, extended on 5 000 ha. We used a Global Positioning System (GPS) and vegetation maps to settle our sampling plots. The sampling was systematic and stratified to 1 degree when the statistical unit is the plot. The samples or plots of 0.5 ha are distributed systematically throughout the entire population. The systematic disposal of plots allows to assume that the intensity of sampling for each stratum is proportional to its area in the forest.

In practice, sampling was carried out along straight and continuous axes or transects. These transects are oriented along a predetermined cardinal direction but are systematically arranged in such a way that they are mostly parallel, equidistant and perpendicular to the general direction of the drainage. Rectangular plots arranged along a transect are contiguous and measure 250 m in the direction of the transect (length) and 20 m in the direction perpendicular to the transect (width). This gives a surface area of 0.5 ha for each plot. The mapping activity allowed us to distinguish four main forest types inside the perimeter of the Mbeth II community forest including: the primary forest (891 ha), the secondary forest (3 961 ha), the temporary drained forest (100 ha), and the permanent drained forest (48 ha). Primary forests or less perturbed forest are forests where human activities are less observed. Secondary forests are forests which have already been subjected to forest logging activities more than 30 years ago. Temporary drained forests are forests which are drained during the rainy seasons, and dried during the dried season and suitable for logging activities. Primary, secondary and temporary drained forests totalize 4 952 ha and constitute the production forests. These are the forests where the community of Mbeth II will harvest forest products. Permanent drained forests are forests which are drained all the year and cannot be suitable for timber or non-timber harvesting activities. These areas are often considered as vulnerable ecosystems, un-productive forests which are devoted to the conservation/protection targets.

Table 1 presents the sampling design which was used. Each line opened is presented with its number of plots and the corresponding sampling area. A total of 274 plots of 0.5 ha each were inventoried in 20 lines. A total of 137 ha out of the 5 000 ha was totally covered by forest inventories giving an average sampling intensity of 2.74%, which is good and in conformity with the national standards (for surface areas ≤ 5 000 ha, the sampling rate should be at least = 2%).

Figure 1 illustrates the sampling design described above. The distance between two consecutive lines is 900 m.

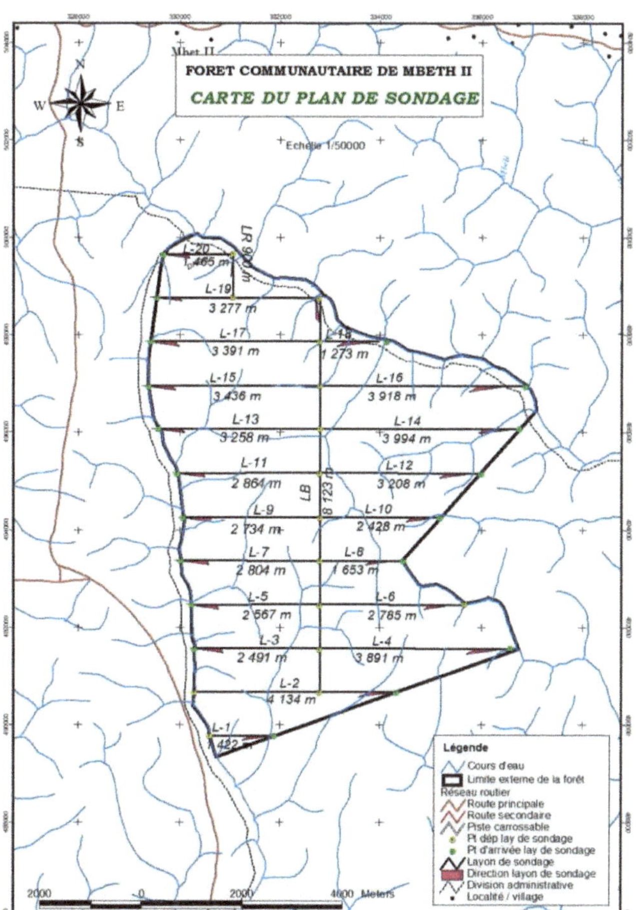


***Figure 1.*** *Sampling map of the inventory of Alstonia boonei used for the community forest of Mbeth II.*

*Table 1. Number of lines, plots and the sampling rate used to assess the abundance of Alstonia boonei in the community forests of Mbeth II.*

| Line number | Number of plots of 0.5 ha | Sampling area per line (ha) | Sampling rate (%) |
|---|---|---|---|
| L1 | 7 | 3.5 | 0.07 |
| L2 | 20 | 10 | 0.2 |
| L3 | 12 | 6 | 0.12 |
| L4 | 19 | 9.5 | 0.19 |
| L5 | 12 | 6 | 0.12 |
| L6 | 12 | 6 | 0.12 |
| L7 | 14 | 7 | 0.14 |
| L8 | 8 | 4 | 0.08 |
| L9 | 13 | 6.5 | 0.13 |
| L10 | 12 | 6 | 0.12 |
| L11 | 14 | 7 | 0.14 |
| L12 | 16 | 8 | 0.16 |
| L13 | 16 | 8 | 0.16 |
| L14 | 19 | 9.5 | 0.19 |
| L15 | 16 | 8 | 0.16 |
| L16 | 19 | 9.5 | 0.19 |
| L17 | 16 | 8 | 0.16 |
| L18 | 6 | 3 | 0.06 |
| L19 | 16 | 8 | 0.16 |
| L20 | 7 | 3.5 | 0.07 |
| Total | 274 | 137 | 2.74 |

#### 2.2.2.2. Implementing the Sampling Design

The inventory consists of two steps: line opening/transect cutting and counting. This is done by a production team of about 20 people composed of two sub-teams of ten people each including the lining sub-team and the counting sub-team.

#### 2.2.2.3. Line Opening/Transect Cutting

This step consists of opening or cutting according to a defined magnetic direction, corridors or alleys of 1.5 m width. These corridors are clearly cleaned by cutting shrubs, vines and branches that obstruct the passage. They are identified by marks. "Layons" constitute the reference system which will be used for counting by the subsequent team. It is during the line opening that details on topography, habitat types, rivers and the corrected horizontal distance of the transect (after reading the slopes) are given. It is also during this stage that the sample plots are identified and numbered. The data collected are recorded on specific sheets.

#### 2.2.2.4. Counting

*Alstonia boonei* is a high tree species which can reach 45 m height and 1.2 m in diameter. In Cameroon, the minimum exploitable diameter (MED) for *A. boonei* has been fixed by the forest administration at 50 cm for timber logging purpose. In the field, villagers prefer harvesting the barks on trees that are easily accessible, with diameter less than 40 cm. In this paper, we adopted the minimum diameter of 30 cm for stem barks, as fixed by the forest administration for *Prunus africana*, another important medicinal plant in Cameroon. This is, only trees with diameter equal or high than that value (30 cm) are authorized to be harvested by villagers.

The counting step includes all operations relating to dendrological and dendrometric records. During the counting, several operations are made including: identification of stems of trees, the measurement of stems with diameter at breast height (dbh = 1.50 m) >= 20 cm.

### 2.2.3. Estimation of the Average Mass of the Bark of an Exploitable Tree

#### 2.2.3.1. Selection of Trees

This part of the study started two weeks later after the team in charge of assessing the abundance of *Alstonia boonei* trees has finalized their work. The sheets of the sub-team in charge of counting were used to select the best trees that can allow to better appreciate the mass of the stem bark of *Alstonia boonei* trees. Practically, trees were selected according to their diameter, their accessibility, their healthy and their conformity (only straight stems were selected). Once a given stem was identified in a counting sheet, we noted its geographical data (latitude and longitude) and we went with the global position system (GPS) in the field to identify it and take appropriated measures of the stem. A total of 37 *Alstonia boonei* trees with diameter at breast high $\geq 30$ cm were sampled.

To estimate the mass of the stem barks of *Alstonia boonei*, we used an indirect method. In fact, a direct method would require diameter tape and bark thickness measurements at the critical heights (breast high, first branch, etc), with some degree of verification through destructive sampling. This can yield précised data compared to indirect method. We preferred indirect method for two main reasons: (1) it is non-destructive, and (2) it can easily be applied in the field by foresters during forest management inventories.

We proceeded in three steps to estimate the mass of the stem barks: (1) establishment of the relation linking the diameter at breast high (DBH) of each tree with the thickness of its stem bark, (2) establishment of the relation linking the volume of the stem bark and its mass (weigh), and (3) determination of the mass of the fresh bark for exploitable (mature) trees.

#### 2.2.3.2. Relation Diameter – Thickness of Volume of the Stem Bark: Volume Based Tariff (VBT)

The volume based tariff (VBT) is a mathematical formula which gives the unit volume of a given tree according to different variables. These variables can be the diameter, the circumference, or the height,. The tariff is more valid for the area where the samples were collected [15]. The volume of the tree can be estimated with or without the stem bark.

To estimate the volume of the stem bark of *Alstonia boonei*, we used the VBT developed for *Alstonia boonei* in the same area during the fourth phase (phase IV) of the national forest inventory. This phase was conducted from 1989 to 1990 on a total area of 2 355 313 ha. The VBT for *Alstonia boonei* under (without) the stem bark was established as follow [2; 3]:

$V = aDb$

V = Volume of the tree with the stem bark;

D = diameter of the tree with the stem bark.

a = 0.00461 ; b = 2.166608, a and b are parameters of the formula.

The diameter of the tree with the stem bark (D) is: D = Dns + 2t

With Dns = diameter of the tree without the stem bark;

t = thickness of the stem bark.

In the same order, the volume of the tree with the stem bark is V = Vns + Vs;

Vns = volume of the tree without the stem bark;

Vs = volume of the stem bark.

Vs = V – Vns

Vs = aDb – a(D – 2t)b

Vs = a[Db – (D – 2t)b]

#### *2.2.3.3. Relation Linking the Volume of the Stem Bark and its Mass = Cubic Mass*

The cubic mass of an entity is the ratio of the mass of that entity and the volume occupied by that mass. Samples of the fresh bark were collected in all trees of the sample used to establish the volume based tariff (see above). For each sample we noted the length (cm), the width (cm), and the thickness (mm). The thickness of the bark was measured using the calliper rule. Collecting the samples was not easy since *Alstonia*'s bark is more stick to the cambium. The three measures including the length, the width, and the thickness allowed us to obtain the volume of the bark sample. We weighted the sample and found the equivalent mass. Equivalencies were made between the average volume of the samples and their corresponding fresh mass/weight. From those equivalencies, we deduced the cubic/volumic mass of *Alstonia boonei* barks in the community forest of Mbeth II. The cubic mass is Cm = m/V with m = mass in kilogram (kg) and V = volume of the stem bark in cubic meter (m3).

#### *2.2.3.4. Mass of the Fresh Bark for Exploitable Trees*

Having the volume of the bark for each tree, we estimated the mass using the formula of the cubic mass (m = V x Cm). The average mass of all the exploitable trees was considered as the mass of an exploitable tree of *Alstonia boonei* in Mbeth II.

### *2.2.4. Data Analysis*

Data were analyzed using the Microsoft Excell 2010 software. For analysis, we determined the number of citations of *Alstonia boonei* in inquiries, the price per gram, the number of stems in different diameter classes and the density (number of stems/surface area). The total number of stems (stock) of the forest is the product of the density by the surface area.

## 3. Results

### *3.1. Ethnobotanical Survey*

***Table 2.*** *Informants interviewed in Mbeth II and markets of bertoua and Yaoundé.*

| Order number | village | Code informant | age | market |
|---|---|---|---|---|
| 1 | Mb | Mmb1 | 51 | |
| 2 | Mb | Fmb1 | 55 | |
| 3 | Mb | Mmb2 | 57 | |
| 4 | Mb | Fmb2 | 80 | |
| 5 | Mb | Fmb3 | 81 | |
| 6 | Mb | Mmb3 | 58 | |
| 7 | Mb | Fmb4 | 52 | |
| 8 | Mb | Mmb4 | 61 | |
| 9 | Mb | Mmb5 | 62 | |
| 10 | Mb | Fmb5 | 54 | |
| 11 | Mb | Fmb6 | 55 | |
| 12 | Mb | Fmb7 | 63 | |
| 13 | Mb | Fmb8 | 60 | |
| 14 | Mb | Fmb9 | 49 | |
| 15 | Mb | Fmb10 | 57 | |
| 16 | Mb | Fmb11 | 58 | |
| 17 | Br | Mbr1 | 43 | Mandjou |
| 18 | Br | Fbr1 | 39 | Sous-prefecture |
| 19 | Br | Mbr2 | 36 | Marché central |
| 20 | Br | Mbr3 | 38 | Marché central |
| 21 | Ya | Fya1 | 32 | Garre |
| 22 | Ya | Fya2 | 33 | Garre |
| 23 | Ya | Mya1 | 60 | mvogmbi |
| 24 | Ya | Mya2 | | Mokolo |
| 25 | Ya | Mya3 | 64 | Mokolo |
| 26 | Ya | Mya4 | 63 | poste centrale |

City/village: Br = Bertoua; Mb = Mbeth II; Ya = Yaoundé,

Code informant: the first letter indicates the gender (M = male, F = female), the second and third letters indicate the city/village, and the number indicates the order number of the informant in the gender and the city,

A total of 26 persons provided information on the use of *Alstonia boonei* as a medicinal plant including 16 informants in Mbeth II, 6 in Yaoundé and 4 in Bertoua (table 2). In Yaoundé, the informants are distributed in four markets, while in Bertoua, they are distributed in three markets. The local name used varies from one place to another. In Mbeth II, it is called "Kougeu", in Bertoua, sellers use its trade name which is "Emien", while in Yaoundé markets, and sellers call it "Ekouk".

***Table 3.*** *Citations of medicinal plants in Mbeth II and markets of Bertoua and Yaoundé*

| Order n of citations | Code_ Informant | Ailment | Latin name | Associated (Ass) plant species | Plant part | Mass (gram) | Price of selling (FCFA) | Ratio Price/mass |
|---|---|---|---|---|---|---|---|---|
| 1 | Mmb1 | Ma/Fe | *A. boonei* | *C. frutescens* | Bark | | | |
| 2 | Mmb1 | Ma/Fe | *C. frutescens* | *Ass_with Alstonia* | fruit | | | |
| 3 | Mmb1 | Ma/Fe | *A. boonei* | *C. frutescens* | Bark | | | |
| 4 | Mmb1 | Ma/Fe | *C. frutescens* | *Ass_with Alstonia* | fruit | | | |
| 5 | Fmb1 | Ma/Fe | *A. boonei* | | Bark | | | |
| 6 | Mmb2 | Ma/Fe | *A. boonei* | *A. chlorantha* + *S. magnificum* + *P. nitida* | Bark | | | |
| 7 | Mmb2 | Ma/Fe | *A. chlorantha* | *Ass_with Alstonia* | Bark | | | |
| 8 | Mmb2 | Ma/Fe | *S. magnificum* | *Ass_with Alstonia* | Bark | | | |
| 9 | Mmb2 | Ma/Fe | *P. nitida* | *Ass_with Alstonia* | Bark | | | |
| 10 | Mmb2 | Ma/Fe | *A. boonei* | *A. chlorantha* + *S. magnificum* + *P. nitida* | Bark | | | |
| 11 | Mmb2 | Ma/Fe | *A. chlorantha* | *Ass_with Alstonia* | Bark | | | |
| 12 | Mmb2 | Ma/Fe | *S. magnificum* | *Ass_with Alstonia* | Bark | | | |
| 13 | Mmb2 | Ma/Fe | *P. nitida* | *Ass_with Alstonia* | Bark | | | |
| 14 | Fmb2 | Worms | *A. boonei* | | Bark | | | |
| 15 | Fmb3 | Ma/Fe | *A. boonei* | | Bark | | | |
| 16 | Fmb3 | Ma/Fe | *B.* | *Ass_with Alstonia* | leaf | | | |
| 17 | Mmb3 | Worms | *A. boonei* | | Bark | | | |
| 18 | Mmb3 | Ma/Fe | *A. boonei* | *B. pilosa* | Bark | | | |
| 19 | Mmb3 | Ma/Fe | *B.* | *Ass_with Alstonia* | leaf | | | |
| 20 | Fmb4 | Ma/Fe | *A. boonei* | | Bark | | | |
| 21 | Mmb4 | Ma/Fe | *A. boonei* | | Bark | | | |
| 22 | Mmb5 | Typh | *A. boonei* | | Bark | | | |
| 23 | Fmb5 | Ma/Fe | *A. boonei* | | Bark | | | |
| 24 | Fmb5 | Ma/Fe | *B.* | *Ass_with Alstonia* | leaf | | | |
| 25 | Fmb6 | Typh | *A. boonei* | | Bark | | | |
| 26 | Fmb7 | Ma/Fe | *A. boonei* | | Bark | | | |
| 27 | Fmb8 | Ma/Fe | *A. boonei* | *B. pilosa* | Bark | | | |
| 28 | Fmb8 | Ma/Fe | *B.* | *Ass_with Alstonia* | leaf | | | |
| 29 | Fmb9 | Ma/Fe | *A. boonei* | | Bark | | | |
| 30 | Fmb10 | Jaund | *A. boonei* | | Bark | | | |
| 31 | Fmb11 | Worms | *A. boonei* | | Bark | | | |
| 32 | Mbr1 | Typh | *A. boonei* | | Bark | | | |
| 33 | Mbr1 | Ma/Fe | *A. boonei* | | Bark | | | |
| 34 | Mbr1 | Ma/Fe | *A. boonei* | | Bark | | | |
| 35 | Fbr1 | Ma/Fe | *A. boonei* | | Bark | 300 | 100 | 3 |
| 36 | Fbr1 | Typh | *A. boonei* | | Bark | 300 | 100 | 3 |
| 37 | Fbr1 | Ma/Fe | *A. boonei* | | Bark | 300 | 100 | 3 |
| 38 | Mbr2 | Typh | *A. boonei* | | Bark | | | |
| 39 | Mbr3 | Ma/Fe | *A. boonei* | | Bark | 300 | 300 | 1 |
| 40 | Mbr3 | Worms | *A. boonei* | | Bark | 300 | 300 | 1 |
| 41 | Mbr3 | Typh | *A. boonei* | *P. nitida* | Bark | 300 | 300 | 1 |
| 42 | Mbr3 | Typh | *P. nitida* | *Ass_with Alstonia* | fruit | | | |

| Order n of citations | Code_ Informant | Ailment | Latin name | Associated (Ass) plant species | Plant part | Mass (gram) | Price of selling (FCFA) | Ratio Price/mass |
|---|---|---|---|---|---|---|---|---|
| 43 | Mbr3 | Typh | *A. boonei* | | Bark | | | |
| 44 | Mbr3 | Diabetis | *A. boonei* | | Bark | | | |
| 45 | Fya1 | Ma/Fe | *A. boonei* | | Bark | 1000 | 1000 | 1 |
| 46 | Fya2 | Worms | *A. boonei* | | Bark | 1000 | 1000 | 1 |
| 47 | Mya1 | Jaund | *A. boonei* | | Bark | 1000 | 1000 | 1 |
| 48 | Mya1 | Jaund | *A. boonei* | | Bark | 1000 | 1000 | 1 |
| 49 | Mya1 | Ma/Fe | *A. boonei* | | Bark | 1000 | 1000 | 1 |
| 50 | Mya1 | Ma/Fe | *A. boonei* | | Bark | 1000 | 1000 | 1 |
| 51 | Mya1 | Typh | *A. boonei* | | Bark | 1000 | 1000 | 1 |
| 52 | Mya1 | Typh | *A. boonei* | | Bark | | | |
| 53 | Mya2 | Ma/Fe | *A. boonei* | | Bark | 500 | 500 | 1 |
| 54 | Mya2 | Snake | *A. boonei* | | Sap | 500 | 500 | 1 |
| 55 | Mya3 | Worms | *A. boonei* | | Bark | | | |
| 56 | Mya3 | Worms | *A. boonei* | | Bark | | | |
| 57 | Mya3 | Ma/Fe | *A. boonei* | | Bark | 1000 | 1300 | 0.77 |
| 58 | Mya3 | Ma/Fe | *A. boonei* | | Bark | 1000 | 1300 | 0.77 |
| 59 | Mya3 | Typh | *A. boonei* | | Bark | | | |
| 60 | Mya3 | Typh | *A. boonei* | | Bark | | | |
| 61 | Mya4 | Ma/Fe | *A. boonei* | | Bark | 1000 | 1000 | 1 |
| 62 | Mya4 | Ma/Fe | *A. boonei* | | Bark | 1000 | 1000 | 1 |

Ailment: Ma/Fe = malaria/fever; worms = intestinal worms ; Typh = typhoid fever ; Jaund = jaundice; Snake = snake bite,
Price of selling: 1 USD = 494.618
C. frutescens : Capsicum frutescens ; A. boonei ; A. chlorantha : Annickia chlorantha;
S. magnificum : Schumanniophyton magnificum; P. nitida: Picralima nitida; B. pilosa: Bidens pilosa

A total of 62 citations were prescribed with *A. boonei* in the treatment of six ailments (table 3). The distribution of citations in different ailments is as follow: malaria/fever (38 citations; 61.3%), typhoid fever (12; 19.4%), intestinal worms (7; 11.3%), jaundice (3; 4.8%), and diabetes and snake bite (1; 1.6%) each. Malaria, typhoid fever and intestinal worms are cited in all the three localities.

*Alstonia boonei* is cited in association with five other plant species in recipes: *Anickia chlorantha, Bidens pilosa, Capsicum frutescens, Picralima nitida,* and *Schumanniophyton magnificum*. The plant part used is the stems bark (98% of citations).

The average price per gram varies from one city to another. In Yaoundé, one gram of the bark of *Alstonia boonei* is sold at 0.964 FCFA. In Bertoua, the same quantity of the bark is sold at 2 FCFA. The average price per gram for the two cities is 1.29 FCFA

### *3.2. Abundance of Alstonia Boonei in the Community Forest of Mbeth II*

Figure 2 illustrates the specific curve of *A. boonei* stems found in the community forest of Mbeth II. It is interesting to note that all diameter classes are represented in this figure which shows a good regeneration of this tree species in the forest.

A total of 122 stems of *Alstonia boonei* were counted in the 137 ha of the forest serving as the sample in the whole forest of Mbeth II. This gives a density of 0.89 stems/ha. One hundred and two (102) out of those stems have reach the exploitable diameter (30 cm) adopted for the use of bark. The density of exploitable stems is therefore 0.744 stems/ha.

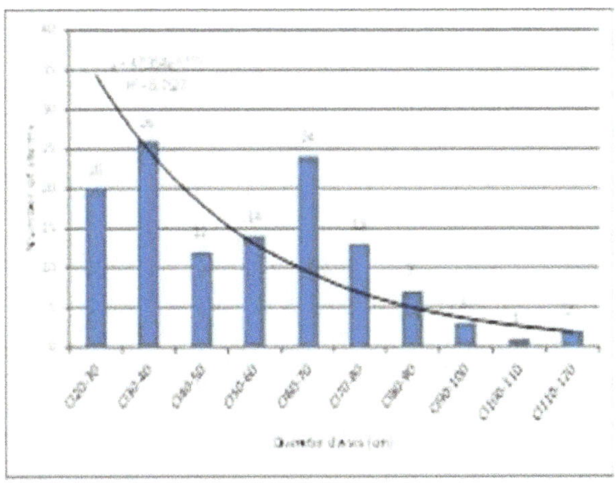

***Figure 2.*** *Distribution of number of stems of Alstonia boonei in different diameter classes (class 20 -30 cm; cl30-40; cl40-50; ...cl110-120 cm) in the community forest of Mbeth II: Y = number of stems, X = diameter classes*

### *3.3. Exploitable Stock of A. Boonei*

In Mbeth II, primary, secondary and temporary drained forests totalize 4 952 ha and constitute the production forests. Thus, the total number of *Alstonia boonei* stems estimated for the community forest of Mbeth II is 4 408 stems. The number of exploitable trees is 3 664 and constitutes the exploitable stock of *Alstonia boonei* in the community forest of Mbeth II.

### 3.4. Relation Diameter – Thickness of the Volume of the Stem Bark: Volume Based Tariff

The 37 sample trees measured are shown in table 4 together with the thickness of the bark and the corresponding estimated volume of the bark for the whole tree. The average volume of a stem bark of an exploitable *A. boonei* tree is 0.0356 ± 0.01 $m^3$.

### 3.5. Relation Volume – Mass of the Stem Bark: Cubic Mass

Table 5 presents for each sample, its volume and the corresponding fresh mass. The average cubic mass of a given *Alstonia* tree is 966.555 ± 0.015 Kg/m3. This means that one cubic meter of the fresh bark corresponds to 966.555 kg.

***Table 4.*** *Relation thickness – volume of the bark*

| N° | Diameter (D) in cm | Thickness of the bark (t) in mm | a | b | Vs = a[Db - (D – 2t)b] (m3) |
|---|---|---|---|---|---|
| 1 | 43.79 | 14.67 | 0.000461 | 2.166608 | 0.0240 |
| 2 | 70 | 11 | 0.000461 | 2.166608 | 0.0312 |
| 3 | 90 | 12 | 0.000461 | 2.166608 | 0.0456 |
| 4 | 94.33 | 12.5 | 0.000461 | 2.166608 | 0.0502 |
| 5 | 45 | 10 | 0.000461 | 2.166608 | 0.0169 |
| 6 | 67.5 | 12.5 | 0.000461 | 2.166608 | 0.0339 |
| 7 | 88.6 | 8 | 0.000461 | 2.166608 | 0.0299 |
| 8 | 117.4 | 12.66 | 0.000461 | 2.166608 | 0.0656 |
| 9 | 86 | 16.83 | 0.000461 | 2.166608 | 0.0606 |
| 10 | 84 | 16 | 0.000461 | 2.166608 | 0.0560 |
| 11 | 65 | 12.17 | 0.000461 | 2.166608 | 0.0316 |
| 12 | 73 | 12.33 | 0.000461 | 2.166608 | 0.0367 |
| 13 | 78 | 12 | 0.000461 | 2.166608 | 0.0386 |
| 14 | 50.7 | 8.83 | 0.000461 | 2.166608 | 0.0172 |
| 15 | 56.7 | 7.83 | 0.000461 | 2.166608 | 0.0174 |
| 16 | 112.6 | 11.5 | 0.000461 | 2.166608 | 0.0568 |
| 17 | 96.4 | 8.33 | 0.000461 | 2.166608 | 0.0343 |
| 18 | 92.4 | 13 | 0.000461 | 2.166608 | 0.0509 |
| 19 | 84.6 | 13 | 0.000461 | 2.166608 | 0.0459 |
| 20 | 66.4 | 12.3 | 0.000461 | 2.166608 | 0.0328 |
| 21 | 105.4 | 11.35 | 0.000461 | 2.166608 | 0.0519 |
| 22 | 47.4 | 11.8 | 0.000461 | 2.166608 | 0.0212 |
| 23 | 76.25 | 11.85 | 0.000461 | 2.166608 | 0.0371 |
| 24 | 34.2 | 10.6 | 0.000461 | 2.166608 | 0.0130 |
| 25 | 108.3 | 12.76 | 0.000461 | 2.166608 | 0.0602 |
| 26 | 48.3 | 12.7 | 0.000461 | 2.166608 | 0.0233 |
| 27 | 38.7 | 9 | 0.000461 | 2.166608 | 0.0128 |
| 28 | 61.3 | 11 | 0.000461 | 2.166608 | 0.0267 |
| 29 | 53.4 | 11 | 0.000461 | 2.166608 | 0.0227 |
| 30 | 64.25 | 13 | 0.000461 | 2.166608 | 0.0333 |
| 31 | 56.45 | 11.25 | 0.000461 | 2.166608 | 0.0248 |
| 32 | 84.35 | 12 | 0.000461 | 2.166608 | 0.0423 |
| 33 | 37.4 | 9.4 | 0.000461 | 2.166608 | 0.0128 |
| 34 | 95 | 8 | 0.000461 | 2.166608 | 0.0324 |
| 35 | 68.47 | 10.67 | 0.000461 | 2.166608 | 0.0295 |
| 36 | 71.46 | 12.65 | 0.000461 | 2.166608 | 0.0367 |
| 37 | 103.2 | 13.3 | 0.000461 | 2.166608 | 0.0593 |
| Max | 117.4 | 16.83 | 0.000461 | 2.166608 | 0.0656 |
| Min | 34.2 | 7.83 | 0.000461 | 2.166608 | 0.0128 |
| Mean | 73.4121622 | 11.62 | 0.000461 | 2.166608 | 0.0356 |
| Standard deviation | 2.03 | 0.00 | 0.00 | 0.01 | |

a = 0.00461 ; b = 2.166608; Vs = volume of the stem bark

***Table 5.** Ratio Mass/volume for the bark of A. boonei in the community forest of Mbeth II*

| N° | Diameter (cm) | Thickness (mm) | Width (cm) | Length (cm) | Volume = V (m3) | Mass = M (Kg) | Ratio (M/V) |
|---|---|---|---|---|---|---|---|
| 1 | 43.79 | 14.67 | 16.67 | 23.67 | 0.0006 | 0.53 | 916.133 |
| 2 | 70 | 11 | 17.25 | 25 | 0.0005 | 0.44 | 927.536 |
| 3 | 90 | 12 | 21 | 32 | 0.0008 | 0.76 | 942.460 |
| 4 | 94.33 | 12.5 | 22.25 | 25.5 | 0.0007 | 0.68 | 958.802 |
| 5 | 45 | 10 | 14.3 | 27 | 0.0004 | 0.4 | 1036.001 |
| 6 | 67.5 | 12.5 | 26 | 32.75 | 0.0011 | 1.04 | 977.099 |
| 7 | 88.6 | 8 | 19 | 24.67 | 0.0004 | 0.36 | 960.171 |
| 8 | 117.4 | 12.66 | 20.4 | 25.5 | 0.0007 | 0.64 | 971.798 |
| 9 | 86 | 16.83 | 17.33 | 27.25 | 0.0008 | 0.78 | 981.016 |
| 10 | 84 | 16 | 24 | 36.75 | 0.0014 | 1.28 | 907.030 |
| 11 | 65 | 12.17 | 20.25 | 26.83 | 0.0007 | 0.62 | 937.822 |
| 12 | 73 | 12.33 | 18.63 | 30.83 | 0.0007 | 0.64 | 903.210 |
| 13 | 78 | 12 | 17.83 | 29.5 | 0.0006 | 0.58 | 918.739 |
| 14 | 50.7 | 8.83 | 12.83 | 25.33 | 0.0003 | 0.3 | 1044.636 |
| 15 | 56.7 | 7.83 | 12.33 | 29.5 | 0.0003 | 0.26 | 912.272 |
| 16 | 112.6 | 11.5 | 14.25 | 16.45 | 0.0003 | 0.27 | 1001.579 |
| 17 | 96.4 | 8.33 | 16.33 | 29.83 | 0.0004 | 0.375 | 923.867 |
| 18 | 92.4 | 13 | 23.63 | 36.25 | 0.0011 | 1.26 | 1131.742 |
| 19 | 84.6 | 13 | 17.8 | 32.6 | 0.0008 | 0.72 | 954.446 |
| 20 | 66.4 | 12.3 | 17.3 | 31.5 | 0.0007 | 0.58 | 865.299 |
| 21 | 105.4 | 11.35 | 16.5 | 28.56 | 0.0005 | 0.52 | 972.358 |
| 22 | 47.4 | 11.8 | 21.3 | 34.6 | 0.0009 | 0.84 | 965.921 |
| 23 | 76.25 | 11.85 | 16.2 | 20.25 | 0.0004 | 0.36 | 926.071 |
| 24 | 34.2 | 10.6 | 9.4 | 26 | 0.0003 | 0.26 | 1003.613 |
| 25 | 108.3 | 12.76 | 23.4 | 31.7 | 0.0009 | 0.94 | 993.121 |
| 26 | 48.3 | 12.7 | 16.45 | 28.4 | 0.0006 | 0.58 | 977.552 |
| 27 | 38.7 | 9 | 11.4 | 24.52 | 0.0003 | 0.248 | 985.668 |
| 28 | 61.3 | 11 | 14.59 | 26.33 | 0.0004 | 0.42 | 993.918 |
| 29 | 53.4 | 11 | 19.64 | 33.8 | 0.0007 | 0.72 | 986.011 |
| 30 | 64.25 | 13 | 22.13 | 28.36 | 0.0008 | 0.78 | 956.229 |
| 31 | 56.45 | 11.25 | 13.47 | 24.2 | 0.0004 | 0.35 | 954.405 |
| 32 | 84.35 | 12 | 14.5 | 21.6 | 0.0004 | 0.38 | 1011.069 |
| 33 | 37.4 | 9.4 | 12.3 | 16.5 | 0.0002 | 0.18 | 943.530 |
| 34 | 95 | 8 | 25.17 | 33.75 | 0.0007 | 0.68 | 1000.736 |
| 35 | 68.47 | 10.67 | 10.5 | 21.63 | 0.0002 | 0.22 | 908.340 |
| 36 | 71.46 | 12.65 | 20.47 | 24.2 | 0.0006 | 0.62 | 989.391 |
| 37 | 103.2 | 13.3 | 15.6 | 24.5 | 0.0005 | 0.52 | 1022.966 |
| Max | 117.4 | 16.83 | 26 | 36.75 | 0.0014 | 1.28 | 1131.742 |
| Min | 34.2 | 7.83 | 9.4 | 16.45 | 0.0002 | 0.18 | 865.299 |
| Mean | 73.41 | 11.62 | 17.63 | 27.50 | 0.001 | 0.571 | 966.555 |
| Standard deviation | 2.03 | 0.00 | 3.52 | 3.503 | 0.015 | | |

## *3.6. Mass and Total Stock of Fresh Barks of Exploitable (mature) Trees*

The average volume of fresh bark for an exploitable tree is 0.0356 m3, while the cubic mass is 966.555 m3/kg. This gives the average fresh mass of the bark of an exploitable tree of 34.4 kg.

The rotation (time between two consecutive harvesting steps) adopted by the Cameroon forest administration for logging in community forests is 25 years. This is, for each harvesting plot delimitated, to cut trees above 50 cm (Minimum exploitable diameter fixed for timber logging of *Alstonia*), and return 25 years later in the same plot to move the reconstituted stems. The surface area of an annual plot is 5 000 /25 = 200 ha and the five years bloc or sector is about 5 000/5 = 1 000 ha. The community forest of Mbeth II has been divided in five sectors as shown in figure 3.

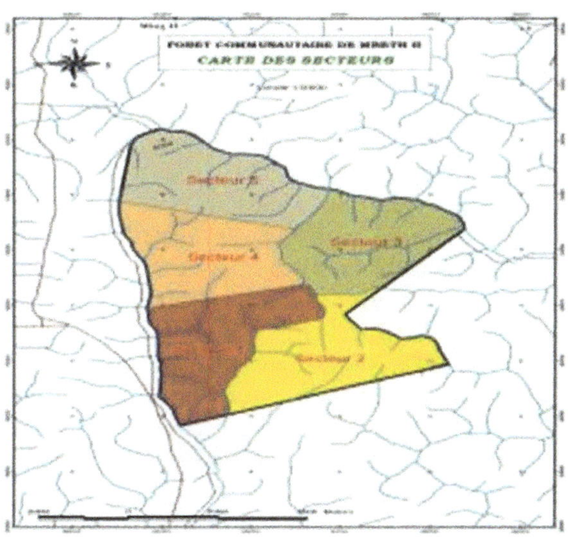

***Figure 3.** The five sectors (five years blocs) of the community forest of Mbeth II.*

In this paper, we adopt the rotation of 10 years and the mid-rotation of 5 years for the harvesting of the stem barks. We consider them for sustainable purposes, and it is recommended to harvest trees of at least 30 cm of diameter at breast high, move the two quarters (1/4) opposite sides of stem bark, and return 5 years later to move the remaining two quarters opposite sides on the same tree, or return 10 years latter to move the same two quarters opposite sides on the same tree. This means that, in each mid-rotation, the harvesting of the bark will be done in the whole five years plot (bloc = 1 000 ha). The harvesting of the two quarters opposite sides gives an average fresh mass of a mature tree of 34.4/2 = 17.2 kg. The total number of mature trees estimated in the community forest of Mbeth II is 3 664 trees. This gives a total stock of the fresh bark of *Alstonia boonei* of 63 020.8 kg. The stock of the fresh bark of *Alstonia boonei* to be harvested in each sector in the community forest of Mbeth II is 63 020.8/5 = 12 604.16 kg. Knowing that the price per gram of the fresh bark of *Alstonia boonei* in markets is 1.29 FCFA, the total bark harvested in each sector (1000 ha) will be sold at about 16 259 366.4 FCFA which is about 32 872.5732 USD.

# 4. Discussion

### *4.1. Traditional Use and Pharmacological Activity of Alstonia Boonei*

*Alstonia boonei* is cited in the treatment of six ailments including: malaria/fever, typhoid fever, intestinal worms, jaundice, diabetes and snake bite. Malaria, typhoid fever and intestinal worms are cited in all the three localities. In Makokou, Gabon, *Alstonia boonei* was cited for treating 14 ailments [12]. Malaria/fever and intestinal worms are commonly cited in the two countries. The usages against malaria/fever are mentioned in 9 countries: Cameroon, Congo-Brazzaville, Gabon, Ivory Coast, Equatorial Guinea, Nigeria, Senegal, Togo, and Democratic Republic of Congo [12]. The usages against intestinal worms are mentioned in four countries including Cameroon, Congo-Brazzaville, the Central African Republic and the Democratic Republic of Congo [12]. The Mbuti and Efe hunter-gatherers based in the Ituri forest, Democratic Republic of Congo apply the white sap on snakebites and wounds, while the bark-decoction is taken for stomach disorder and malaria [16].

*Alstonia boonei* is cited in the markets of Bertoua and Yaoundé for treating jaundice, diabetes, malaria, typhoid fever, and intestinal worms. The treatment against jaundice, malaria, and typhoid fever has also been cited by sellers in the Yaoundé markets in 2002 [14].

The stem bark of *A. boonei* contains antipyretic (against fever) properties [17; 18; 19] and has revealed a good activity (IC50 < 4 µg/ml) against *Plasmodium falciparum* FcB1/Colombia [20]. The prescriptions of *A. boonei* against snake bites can be explained by its analgesic [17]. or anti-inflammatory [17; 19; 21; 22; 23] properties. *Alstonia boonei* is largely cited against intestinal worms; but the anthelminthic properties of this plant were not established [24].

*Alstonia boonei* is cited in association with three other plant species for treating malaria/fever: *Anickia chlorantha, Picralima nitida,* and *Schumanniophyton magnificum. Anickia chlorantha* and *Picralima nitida* are confirmed in the literature for their activity in the treatment of malaria [25].

The average price per gram of the stem bark of *A. boonei* in the two markets is 1.29 FCFA (n = 19 sellers). That of the Yaoundé markets only is 0.96 FCFA (n = 13). Surveys conducted in 2002 revealed that the average price per gram for *A. boonei* in the Yaoundé markets was 1.64 FCFA [14].

### *4.2. Abundance, Stock, and Biomass of Alstonia Boonei*

Figure 3 illustrates the distribution of *A. boonei* trees in different diameter classes or the specific curve. Almost all diameter classes are represented, which shows no specific problem of regeneration for that tree specie in the Mbeth II community forest. The density obtained is 0.89 stems/ha, which is twice compared to the 0.4 stems/ha obtained in the Makokou forest in Gabon [12]. *A. boonei* is a light demanding tree species [9]. As most of light demanding tree species such as Ayous (*Triplochiton scleroxylon*) or Afrormosia (*Pericopsis elata*), the number of small stems is often low in primary or non-perturbed forest, due to the lack of light. A high number of stems of small diameter found in Mbeth II community forest may be caused by the fact that, those forests have already been subjected to forest logging in past years.

The average thickness of stem bark of an exploitable tree of *Alstonia boonei* in Mbeth II is 11.62 mm. this value is high than those obtained for *Prunus africana* in mount Cameroon (8.49 mm), but less than those obtained with *Prunus* in Adamaoua Cameroon (13.01 mm), 12.1 mm in the Kibira national park in Burundi, and Bioko island in Equatorial Guinea (13.01 mm) [26; 27; 28; 29].

The average ratio mass/volume of *Alstonia boonei* in Mbeth II is 966.55 kg/m3. This value is less than the 1014.80 kg/m$^3$ and 1013.16 kg/m3 obtained for *Prunus africana* in Adamaoua and mount Cameroon [26; 27] respectively. Considering that for all trees above 30 cm DBH, only two opposite quarters of the bark are taken from the main stem up to the first branch, the average sustainable mass of *Alstonia boonei* in Mbeth II is 17.2 kg. This value is very less than those obtained for *Prunus africana* in different sites including 50 kg in mount Cameroon, and 69.3 kg in Adamaoua Cameroon [26; 27]. The difference can be linked to the ratio mass/volume of the bark which is high for *Prunus africana.* In fact, the bark of *Alstonia boonei* is granular [10]. If we consider the mid- rotation of 5 years and the price of the stem barks practized in the markets of Bertoua and Yaoundé, the community forest of Mbeth II can produce a total of 12 604.16 kg fresh mass of stem bark every five years in each five years bloc of forest (1 000 ha) and which can be sold at about 16 259 366.4 FCFA or 32 872.5732 USD. These funds can be used to enhance local development if well managed. The advantage of harvesting only two quarters of the bark from the main stem up to the first branch is that it leaves the tree alive, compared to its exploitation as timber.

Cameroon's current forest policy introduced statutory involvement of rural populations, partners and stakeholders in its implementation, notably by the ownership of community forests and hunting zones [4]. This is intended to encourage people to become better stewards of wildlife and forests. Community forests and hunting zones are used by local communities, who can sign specific agreements with the forest administration and private sector partners to harvest their forest resources. Excepted for the north west region where people use community forest for the exploitation of *Prunus africana* barks, most of the community forests found in the meridional zone of Cameroon are assigned for the exploitation of timber resource. Mbeth II is one of the scarce community in the meridional zone, which is enhancing and promoting the use of community forests for non-timber forest products.

## 5. Conclusion

Ethnobotanical surveys and inventories conducted on *Alstonia boonei* in Mbeth II and in the markets of Bertoua and Yaoundé revealed the medicinal and economic importance of the stem bark of that tree species. The stem bark is confirmed for its activity against malaria/fever. *Alstonia* can no longer only be harvested and sold by villagers as a timber, but also for its medicinal value. The usage for its medicinal value can be more sustainable since it avoids cutting the trees, it keeps the trees alive. Results discussed in this paper can be summarized according to the manual of procedures of attribution of community forests, in a specific document called "the simple management plan of *A. boonei* as a medicinal plant in the community forest of Mbeth II". The glaring development challenge at the back ground of what precedes is the urgent need to implement the simple management plan including: (1) conducting the systematic inventory of exploitable stems of *A. boonei* (stems with diameter high than 30 cm) in the first five years bloc, (2) for each mature tree identified, harvesting only the two opposite quarters of the bark from the main stem up to the first big branch and begin cutting at a height of approximately 1 m above the ground, (3) conducting further studies to analyze the activity of the stem bark against typhoid fever and diabetes, (4) and also to study the total market chain of *A. boonei's* usage as a commercial medicinal plant. The bark should be removed with machetes, since they seem to be the tools the workers handle best, giving them better control so as not to damage the cambium. Use of any tool that damages to the cambium should not be permitted.

## Acknowledgment

We thank all the villagers who collaborated with us in this study. This work has been conducted with the financial assistance of the Cameroon Government and the International Tropical Timber Organization (ITTO) under the TFL-PD 014/09 « Reduction of illegal logging through better governance of community forests in the Lom et Djerem Division, East region, Cameroon" in one hand and the financial assistance of the Cameroon Government and the Japanese Cooperation (JICA) under the "FOSAS Forest-Savanna Sustainability Project, Cameroon

## References

[1] FAO « Rapport sur les forêts 2011 : La situation des forêts dans le bassin amazonien, le bassin du Congo et l'Asie du Sud-Est », Brazzaville, 2011J.

[2] CENADEFOR-CTFT, « Inventaire des ressources forestières du Sud Cameroun (1ère tranche) ». Centre National de Développement des Forêts (CENADEFOR) et Centre Technique Forestier Tropical (CTFT). CTFT, 45 bis, avenue de la belle Gabrielle, 94 130 Nogent – Sur – Marne, France, 1983.

[3] CENADEFOR-CTFT, « Inventaire des ressources forestières du Sud Cameroun (2ème tranche). » Centre National de Développement des Forêts (CENADEFOR) et Centre Technique Forestier Tropical (CTFT). CTFT, Département Forestier du CIRAD, 45 bis, avenue de la belle Gabrielle, 94 736 Nogent – Sur – Marne, CEDEX, France, 1985.

[4] Ministry of Environment and Forestry, "A Compendium of official instruments on forest and wildlife management in Cameroon". Imprimerie nationale, Yaoundé, 1996, p 185.

[5] J-C. Carret, « La réforme de la fiscalité forestière au Cameroun: débat politique et analyse économique », Bois et Forêts des Tropiques, 2000, n 264 (2).

[6] C. J. Tieguhong, J. L. Betti, "Forest and Protected Areas in Cameroon: Progress has been significant but challenges remain". ITTO Tropical Forest Update, 2008, 18/1, pp: 6 – 9.

[7] R. Letouzey, « Notice de la carte phytogéographique du Cameroun au 1/50 000 ». Institut de la carte internationale de la végétation. Toulouse-France et IRA de Yaoundé-Cameroun, 1985, 24 p.

[8] J. L. Betti, K. S. Bobo, "Illegal sawnwood in the East Province of Cameroon: an analysis of the economic behind illegal sawnwood trade". ITTO Tropical Forest Update, 2007, 17/3, pp: 3 – 5.

[9] J. P. Kwaku Adotey, G. Etormam Adukpo, Opoku Boahen Yaw, F. Ato Armah. "A Review of the Ethnobotany and Pharmacological Importance of Alstonia boonei De Wild (Apocynaceae)". ISRN Pharmacol. 2012; 2012: 587160.

[10] J. Vivien and J. J Faure, Arbres des forêts denses d'Afrique centrale. Les presses de l'Imprimerie Ediprint 53940 – Saint Berthevin, France. 2011, 945 p.

[11] http://www.worldagroforestry.org/treedb2/AFTPDFS/Alstonia_boonei.pdf.

[12] J. L. Betti, O. D. Yongo, D. Obiang Mbomo, D. Midoko Iponga and A. Ngoy. "Ethnobotanical and floristical study of Alstonia boonei De Wild. (Apocynaceae) in the Makokou region, Ogoué-Ivindo province, Gabon". International Journal of Agricultural Sciences, 2013, Vol. 3 (3).

[13] E. Adjanohoun, N. Aboubakar, K. Dramane, M. E. Ebot, J. A. Ekpere, E.G. Enow-Orock, D. Focho, Z. O. Gbilé, A. Kamanyi, J. Kamsu Kom, A. Keita, T. Mbenkum, C. N. Mbi, A. L Mbiele, I. L. Mbome, N. K. Mubiru, W. L. Nancy, B. Nkongmeneck, B. Satabié, A. Sofowora, V. Tamze and C. K. Wirmum, "Contribution to Ethnobotanical and Floristic Studies in Cameroon", 1996, CSTR/OUA.

[14] J. L. Betti, "Medicinal Plants Sold in Yaoundé Markets, Cameroon", African Study Monographs, 2002, 23 (2): 47-64.

[15] College of Forest Engineers of Quebec. "Forestry Handbook". The Laval University Press, 1996, p. 1428

[16] H. Terashima and M. Ichikawa, "A comparative ethnobotany of the Mbuti and Efe hunter-gatherers in the Ituri Forest, Democratic republic of Congo". African Study Monographs, 2003, 24 (1, 2), 1-168.

[17] O. A Olajide, S. O Awe, J. M. Makinde, A. I. Ekhelar, A. Olusola, O. Morebise and D.T. Okpako, "Studies on the anti-inflammatory, antipyretic and analgesic properties of Alstonia boonei stem bark. Journal of Ethnopharmacology", 2000, 71 (1-2): 179-186.

[18] D. Oliver-Bever, "Medicinal plants in tropical West Africa". Cambridge Univ. Press, 1986, p. 375.

[19] P. O Osadebe, "Anti-inflammatory properties of the root bark of Alstonia boonei." Nigerian Journal of Natural Products And Medicines, 2002, (6): 39-41.

[20] N. Zirihi Guede, « Recensement des plantes médicinales antipaludiques du Département d'Issia (Côte d'Ivoire) et étude descriptive de celles ayant une très bonne activité antiplasmodiale ». Revue de Médecines et Pharmacopées Africaines, 2004, (18): 99-103.

[21] G. Kweifio-Okai, « Antiinflammatory activity of a Ghanaian antiarthritic herbal preparation : II ». Journal of Ethnopharmacology, 1991, 33 (1-2): 129-133.

[22] G. Kweifio-Okai, D. Bird, B. Field, R. Ambrose, A. R. Carrol, P. Smith and R. Valdes "Antiinflammatory activity of a Ghanaian antiarthritic herbal preparation : III". Journal of Ethnopharmacology, 1995, (46): 7-15.

[23] A. Rajic, G. Kweifio-Okai, T. Macrides, R. M. Sandeman, D. S. Chandler and G. M. Polya, "Inhibition of serine proteases by anti-inflamatory triterpenoids". Planta Med.2000, 66 (3), 206-210.

[24] B. B. Fakae, A. M. Campbell, J. Barett, I. M. Scott, P. H. Teesdale-Spittle, E. Liebau and P.M. Brophy, "Inhibition of glutathione S-transferases (GSTs) from parasitic nemathodes by extracts from traditional Nigerian medicinal plants". Phytother. Res., 2000, 14 (8), 630-634.

[25] J. L Betti, "Ethnobotanical study of Picralima nitida (Apocynaceae) in the Dja biosphere reserve, Cameroon". In: X. van der Burgt, J. van der Maesen & J.M. Onana (eds), Systematics and Conservation of African Plants, Royal Botanic Gardens, Kew, 2010, pp. 321 – 332.

[26] J. L. Betti and J. Ambara, "Mass of Prunus africana stem barks on Tchabal mbabo and Tchabal Gang Daba Mountain Forests, Cameroon". African Journal of Environmental Science and Technology. 2013. (7): 204 - 221

[27] J. L. Betti and J. Ambara, "Mass of Prunus africana stem barks on the mount Cameroon forest". International Journal of Biodiversity and Conservation, 2011, 3 (7): 267 – 279.

[28] J. L. Betti, M. Feruzi, J. Rushemeza, A. Rurantije, L. Nziguiyimpa, J. Ahishakiye Mamarivo, E. Buvyiruke, J. C. Ndayishimiye, C. Hakizimana and E. Bantegeyahaga "Exploitable stock of Prunus africana Stems in the Teza Forest, Kibira National Park, Burundi", International journal of agriculture innovations and research, 2013, 3 (2); 317-326.

[29] T. Sunderland and T. C. Tako. The exploitation of Prunus africana on the island of Bioko, Equatorial Guinea Report for the People and Plants Initiative, WWF – Germany and the IUCN/SSC Medicinal Plants Specialist Group, June 1999.

# PERMISSIONS

All chapters in this book were first published in AFF, by Science Publishing Group; hereby published with permission under the Creative Commons Attribution License or equivalent. Every chapter published in this book has been scrutinized by our experts. Their significance has been extensively debated. The topics covered herein carry significant findings which will fuel the growth of the discipline. They may even be implemented as practical applications or may be referred to as a beginning point for another development.

The contributors of this book come from diverse backgrounds, making this book a truly international effort. This book will bring forth new frontiers with its revolutionizing research information and detailed analysis of the nascent developments around the world.

We would like to thank all the contributing authors for lending their expertise to make the book truly unique. They have played a crucial role in the development of this book. Without their invaluable contributions this book wouldn't have been possible. They have made vital efforts to compile up to date information on the varied aspects of this subject to make this book a valuable addition to the collection of many professionals and students.

This book was conceptualized with the vision of imparting up-to-date information and advanced data in this field. To ensure the same, a matchless editorial board was set up. Every individual on the board went through rigorous rounds of assessment to prove their worth. After which they invested a large part of their time researching and compiling the most relevant data for our readers.

The editorial board has been involved in producing this book since its inception. They have spent rigorous hours researching and exploring the diverse topics which have resulted in the successful publishing of this book. They have passed on their knowledge of decades through this book. To expedite this challenging task, the publisher supported the team at every step. A small team of assistant editors was also appointed to further simplify the editing procedure and attain best results for the readers.

Apart from the editorial board, the designing team has also invested a significant amount of their time in understanding the subject and creating the most relevant covers. They scrutinized every image to scout for the most suitable representation of the subject and create an appropriate cover for the book.

The publishing team has been an ardent support to the editorial, designing and production team. Their endless efforts to recruit the best for this project, has resulted in the accomplishment of this book. They are a veteran in the field of academics and their pool of knowledge is as vast as their experience in printing. Their expertise and guidance has proved useful at every step. Their uncompromising quality standards have made this book an exceptional effort. Their encouragement from time to time has been an inspiration for everyone.

The publisher and the editorial board hope that this book will prove to be a valuable piece of knowledge for researchers, students, practitioners and scholars across the globe.

# LIST OF CONTRIBUTORS

**Beugré Grah Avit Maxwell and Gnakri Dago**
Laboratoire de Nutrition et de Sécurité Alimentaire, U.F.R. des Sciences et Technologie des Aliments, Université Nangui Abrogoua, 02 BP 801 Abidjan 02, Côte d'Ivoire

**Kimsé Moussa**
Laboratoire de Biologie et de Cytologie Animale, UFR des Sciences de la Nature, Université Nangui Abrogoua, 02 BP 801 Abidjan 02, Côte d'Ivoire

**Gbogouri Grodji Albarin**
Laboratoire de Biochimie et de Microbiologie, Université Jean Lorougnon Guedé, BP 150 Daloa, Côte d'Ivoire

**Mekonnen Asrat, Heluf Gebrekidan and Bobe Bedadi**
School of Natural Resources Management and Environmental Sciences, Haramaya University, Haramaya, Ethiopia

**Markku Yli-Halla**
Environmental Soil Science, University of Helsinki, Helsinki, Finland

**Wakene Negassa**
Department of Plant Soil and Microbial Sciences, Michigan State University, East Lansing, USA

**Manifas Zubayr**
Forestry Ministry of Indonesia, Jakarta, Indonesia

**Dudung Darusman, Bramasto Nugroho and Dodik Ridho Nurrochmat**
Department of Forest Management, Bogor Agricultural University, Bogor, Indonesia

**Nenibarini Zabbey**
Department of Fisheries, Faculty of Agriculture, University of Port Harcourt, PMB 5323, East-West Road, Choba, Rivers State, Nigeria

**Ijeoma Favour Vincent-Akpu**
Department of Animal and Environmental Biology, Faculty of Biological Sciences, College of Natural and Applied Sciences, University of Port Harcourt, PMB 5323, East-West Road, Choba, Rivers State, Nigeria

**Ibisime Etela**
Department of Animal Science, Faculty of Agriculture, University of Port Harcourt, PMB 5323, East-West Road, Choba, Rivers State, Nigeria

**Teppei Hirata and Shiro Tamaki**
Department of Information Engineering, Faculty of Engineering, University of the Ryukyus, Nishihara town, Japan

**Takeshi Miyagi and Yasunori Nagata**
Department of Electrical and Electronic Engineering, Faculty of Engineering, University of the Ryukyus, Nishihara town, Japan

**Tsutomu Omatsu and Tetsuya Mizutani**
Research and Education Center for Prevention of Global Infectious Disease of Animals, Faculty of Agriculture, Tokyo University of Agriculture and Technology, Fuchu-shi, Tokyo, Japan

**Elias Dadebo, Solomon Sorsa and Kassaye Balkew**
Department of Biology, Hawassa University, P. O. Box 5, Hawassa, Ethiopia

**Negesse Kebtineh**
Department of Wild Life, Wetland and Fishery Management, Wondo Genet College of Forestry and Natural Resources, Hawassa University, Shashemene, Ethiopia

**Md. Redwanur Rahman and Sheikh Helena Bulbul**
Institute of Environmental Science, University of Rajshahi, Rajshahi, Bangladesh

**Girolimetto Daniela and Venturini Virginia**
Centro de Estudios Hidro Ambientales, Facultad de Ingenieria y Ciencias Hidricas, Universidad Nacional del Litoral, C. C. 217, Santa Fe, (3000) Argentina

**Leigh A. Winowiecki**
Soils Research Area, International Center for Tropical Agriculture (CIAT), Nairobi, Kenya

**Matthew P. Whelan, Marilyn Villalobos and Eduardo Somarriba**
Tropical Agricultural Research and Higher Education Centre (CATIE), Turriabla, Costa Rica

**Paul A. McDaniel**
Plant, Soil and Entomological Sciences Department, University of Idaho, Moscow, Idaho, USA

**Bereket Haileselassie and Mehretab Haileselassie**
Mekelle Soil Research Center, Tigray Agricultural Research Institute, Mekelle, Ethiopia

**Dawit Habte**
Mekelle Soil Research Center, Tigray Agricultural Research Institute, Mekelle, Ethiopia

**Gebremedhin Gebremeskel**
Mekelle Agricultural Research Center, Tigray Agricultural Research Institute, Mekelle, Ethiopia

**Emmanuel Asiedu - Darko**
Council for Scientific and Industrial Research-Plant Genetic Resources Research Institute, Bunso Eastern Region, Ghana

**Hanna Artuti Ekamawanti**
Tropical Silviculture Major, Post-graduate School of Bogor Agricultural University, Bogor, Indonesia
Faculty of Forestry, Tanjungpura University, Pontianak, Indonesia

**Yadi Setiadi**
Laboratory of Forest Biotechnology, Research Center for Biological Resources and Biotechnology, Bogor Agricultural University, Bogor, Indonesia

**Didy Sopandie**
Department of Agronomy and Horticulture, Bogor Agricultural University, Bogor, Indonesia

**Dwi Andreas Santosa**
Magister Study Program on Soil and Environmental Biotechnology, Department of Soil and Land Resources, Bogor Agricultural University, Bogor, Indonesia

**Siti Latifah**
Forestry Program Study, Sumatera Utara University, Medan, Indonesia

**Teodoro Reyes Villanueva, Myrna Gregorio Carandang and Nathaniel Cena Bantayan**
Forest Resources Management, University of the Philippines Los Banos, Los Banos, Philippines

**Leonardo M. Florece**
Environmental Science, University of the Philippines Los Banos, Los Banos, Philippines

**Fisseha Hadgu, Heluf Gebrekidan and Kibebew Kibret**
School of Natural Resource Management and Environmental Science, Haramaya University, Haramaya, Ethiopia

**BirruYitaferu**
Amhara Regional Agricultural Research Institute, Amhara National Regional State, Ethiopia

**Bokossa Hervé Kouessivi Janvier and Fiogbé Emile Didier**
Laboratory of Research on Wetlands (LRW), Department of Zoology, Faculty of Science and Technique, University of Abomey-Calavi, Benin

**Saïdou Aliou**
Integrated Soil and Crop Management Unit (ISCM), Laboratory of Soil Sciences, Department of Crop Science, Faculty of Agronomic Sciences, University of Abomey-Calavi, Benin

**Kossou Dansou**
Laboratory of Plant Biology, Department of Crop Science, Faculty of Agronomic Sciences, University of Abomey-Calavi, Benin

**Mohammad Feizal Daud, Fauziah I., Mohd Rasdi Z., Fairuz K., Ismail R., Mohd Hanysyam M. N. and Norazliza R.**
Faculty of Plantation and Agrotechnology, Universiti Teknologi MARA (UiTM) Melaka, Jasin Campus, Melaka, Malaysia

**Abu Zarim U.**
Malaysian Agriculture Research and Development Institute (MARDI) Cameron Highlands Research Station, Pahang, Malaysia

**Syed Abdul Rahman S. A. R.**
Taman Agro Al Mashoor, Cameron Highlands, Pahang, Malaysia

**Badal Ahmed Hassan and Olavi Luukkanen**
Viikki Tropical Resources Institute, Department of Forest Sciences, University of Helsinki, Helsinki, Finland

**Edinam K. Glover**
Faculty of Law, University of Helsinki, Helsinki, Finland

**Ramni Jamnadass**
World Agroforestry Centre (ICRAF), United Nations Avenue, Gigiri, Nairobi, Kenya

**Ben Chikamai**
Kenya Forestry Research Institute, Nairobi, Kenya

**Sukadaryati**
Ph D Student of Forestry Faculty of Bogor Agricultural University, Jalan Raya Dramaga, Kampus IPB Dramaga, Bogor, Indonesia

**Samsuri**
Forestry Department, Sumatra Utara University, Medan, Indonesia

**I. Nengah Surati Jaya**
Forest Management Department, Bogor Agricultural University, Bogor, Indonesia

**Cecep Kusmana**
Silviculture Department, Bogor Agricultural University, Bogor, Indonesia

**Kukuh Murtilaksono**
Soil and Land Resources Department, Bogor Agricultural University, Bogor, Indonesia

**Achmad Siddik Thoha**
Department of Forestry Science, Faculty of Agriculture, University of North Sumatra, Medan, Indonesia

**Bambang Hero Saharjo**
Department of Silviculture, Faculty of Forestry, Bogor Agricultural University, Bogor, Indonesia

**Rizaldi Boer**
Department of Geophysics and Meteorology, Faculty of Mathematics and Natural Sciences, Bogor Agricultural University, Bogor, Indonesia

**Muhammad Ardiansyah**
Department of Soil Science and Land Resources, Faculty of Agriculture, Bogor Agricultural University, Bogor, Indonesia

**Gabriel Gana Bake, Elizabeth Itunu Martins and Suleiman Omeiza Eku Sadiku**
Department of Water Resources, Aquaculture and Fisheries Technology, School of Agric and Agric Technology, Federal University of Technology Minna Minna Niger state, Nigeria

**Pernando Sinabutar**
Forest Center for Area Stabilization Region XII, Tanjungpinang, Indonesia

**Bramasto Nugroho, Hariadi Kartodihardjo and Dudung Darusman**
Forest Management Department, Bogor Agricultural University, Bogor, Indonesia

**Khadijah EL Ramija**
Study Program of Natural Resource and Environmental Management, Graduate School, University of Sumatera Utara, Jln. Maas, Medan, Indonesia
Assessment Institute for Agricultural Technology (AIAT), North Sumatera, and the Indonesian Agency For Agricultural Research And Development (IAARD), Ministry Of Agriculture

**Zulkifly Nasution**
Study Program of Natural Resource and Environmental Management, Graduate School, University of Sumatera Utara, Jln. Maas, Medan, Indonesia
Departement of Agroecotechnolgy, Faculty of Agriculture University of Sumatera Utara, jln Prof A. Sofyan No 3, Medan Indonesia, 20155

**M. Zarlis**
Study Program of Natural Resource and Environmental Management, Graduate School, University of Sumatera Utara, Jln. Maas, Medan, Indonesia
Department of Mathematics, Faculty of Mathemathics and Natural Sciences, University of Sumatera Utara, Jln. Bioteknologi No. 1, Medan, Indonesia

**Retno Widiastuti**
Study Program of Natural Resource and Environmental Management, Graduate School, University of Sumatera Utara, Jln. Maas, Medan, Indonesia
Department of Biology, Faculty of Mathemathics and Natural Sciences, University of Sumatera Utara, Jln. Bioteknologi No. 1, Medan, Indonesia

**Shoumo Khondoker**
Department of Fisheries and Marine Bioscience, Jessore University of Science and Technology, Jessore 7400, Bangladesh
Ulashi Sreejony Sangha, Jessore 7400, Bangladesh

**Md. Lokman Hossain**
Department of Global Change Ecology, Faculty of Biology, Chemistry and Geosciences, University of Bayreuth, D-95440 Bayreuth, Germany

**Khondoker Azizul Hoque Moni3**
Ulashi Sreejony Sangha, Jessore 7400, Bangladesh

**Yayeh Bitew**
Agronomist, Adet Agricultural Research Centre, Institute of Amhara Agricultural Research, Bahir Dare, Ethiopia

**Mercy A. A. Derkyi**
Department of Forest Science, School of Natural Resources, University of Energy and Natural Resources, P.O. Box 214 Sunyani, Brong Ahafo Region of Ghana

**Ton Dietz**
African Studies Centre, PO Box 9555, 2300 RB Leiden, The Netherlands

**Jean Lagarde Betti**
Department of Botany, Faculty of Sciences, University of Douala, BP 24 157 Cameroon

**Joseph Ambara**
Ministry of Environment and Nature Protection, Cameroon

# Index


**A**

Agency Theory, 18-19, 21-22, 24-25
Agent-agent Problems, 23
Agregation Rules, 183-184, 186, 189
Agricultural Technologies, 84
Agro-ecology, 26, 29, 105, 108
Alstonia Boonei De Wild, 226, 237
Alternate Wetting and Drying
Irrigation, 48
Aquatic Agricultural System, 26
Arbuscular Mycorrhizal Fungi, 88, 92, 94-95
Asymmetry Effect, 124

**B**

Batang Toru Watershed, 160-162, 164-165
Beel Bakar, 199-206
Beel Fisheries, 200, 203-204
Bonny River Estuary, 30-31
Boro Rice, 48-49
Bread Wheat, 8, 11, 14-16, 74, 76, 81-82

**C**

Cameroon, 226-230, 235-238
Cassava Flour, 1-2, 4
Central Somalia, 138
Chainsaw Milling, 217-218, 221-223, 225
Chicken Diet, 1
Coastal Erosion, 28
Cooked Flamboyant Seed Meal, 176-177
Costa Rica, 65-67, 72-73
Crop Allocation, 65-67
Crop Parameter Response, 11
Crop Productivity, 8-9, 70, 72, 216
Crop Suitability, 67, 69-70
Crop Varieties, 84-86

**D**

Decomposition Process, 117-121
Deforestation, 18, 25, 28, 96, 139, 147-149, 164, 166
Delonix Regia Seed Meal, 176, 180
Desertification, 28, 138-140, 147-150
Diet Composition, 41, 43, 116-118, 120-121
Dredging Activities, 28
Dystri-haplicnitisols, 10

**E**

Eco-friendly Stimulant, 151, 153-158
Ecological Indicators, 160
Ecological Services, 26
Ecosystem Restoration, 29, 150
Ecosystem Services, 26, 29-32, 148, 208
Environmental Changes, 27, 138, 140, 145, 147
Environmental Degradation, 18, 26-29, 32, 141, 144-145, 148-149
Ethnobotanical Survey, 228-229, 231
Eucalyptus, 9, 96-103, 131-137
Eucalyptus Plantations, 96-97, 132, 135-136
Evapotranspiration, 49, 55-56, 62-64, 138-139
Exploitable Tree, 226, 230-231, 235-236

**F**

Farmer Support System, 34
Fertile Acidic Soils, 9, 15
Fertilizer Management, 75, 191
Food Production, 1, 28-29, 34
Forest Area, 18-19, 21, 23-24, 183-189, 219, 226
Forest Area Gazettment, 183-185, 188
Forest Management, 18, 21-23, 25, 97, 102-103, 131, 137, 149-150, 153, 175, 217-219, 227
Forest Offences, 217-219, 221, 223-224
Forest Plantation, 96, 131, 137, 189, 223
Forest Resources, 18, 25, 97, 137, 141, 146, 217-219, 223-224, 227, 237
Freundlich Model, 104, 108-109, 111-112

**G**

Gas Flaring, 27-28, 32
Goat Utilizing, 34
Green Economy, 26-27, 29, 32-33
Growth of Chickens, 1

**H**

Hotspot Confidence, 171
Hydrogen Cyanide, 1-4

**I**

Infectious Diseases of Livestock, 34
Intensive Rice Farming, 191
Intercropping, 124-125, 128-130, 208-216
Irrigation Treatments, 48-49, 192

Ivory Coast, 1-2, 236

L
Lake Ziway, 41-47
Land-use Characteristics, 141
Landscape-scale Agricultural Production, 65
Langmuir Model, 104, 108-109, 112
Local Soil Knowledge, 65-66, 70, 73
Lupine Biomass, 212-213
Lupine Growth, 210-211

M
Man-made Environmental Problems, 27
Management Inventories, 226, 230
Meat Industry, 34
Merchantable Volume, 96, 98-102, 131-132
Mercury Accumulation, 88, 90, 93-94
Mercury Stress Resistances, 88
Methane Emission, 191-198
Methionine, 1-2, 4-5
Mineral P Fertilizer, 8-9, 11-12, 14-15
Mineralization Rate, 116
Mining Activity, 18

N
Nauclea Orientalis Seedlings, 88
Niger Delta Communities, 26
Nitrogen Fertilizers, 197
Nutrient P Uptake, 14
Nutrient Release, 116-117, 120
Nutrient Use Efficiency, 74, 76, 79

O
Oil Exploration, 27-28
Oil Spills, 27, 31-32
Ontogenetic Diet Shifts, 45

P
Peatland Fires Period, 171
Phosphorus Adsorption, 104-106, 111, 113-114
Phosphorus Fertilizers, 74
Phosphorus Fractionation, 108
Pig Manures, 116, 120
Pipeline Crossing, 28
Plant Materials, 46, 125
Plutella Xylostella, 124, 126, 129-130
Polyculture System, 124
Poultry Diets, 2
Principal-agent Problems, 22
Principal-agent Relationship, 18, 20, 22, 24
Principal-principal Problems, 23

R
Red-belly Tilapia, 41
Remote Sensing, 55, 62-63, 166, 174-175

S
Seeding Ratio, 208, 210-211, 213-215
Small Cereal Intercropping, 208-209
Soil Sampling, 68, 105
State Forest Area, 183, 185
Sustainable Harvesting, 226
Sustainable Rural Livelihoods, 29

T
Tritcum Aestivum, 74
Tropical Forest Landscape, 160

U
Use of Forest Area, 18

V
Venturini Et Al.'s Method, 55-56, 62

W
Waste Discharges, 28
Water Management, 48, 53-55, 62, 130, 174, 199, 201, 204-205
Water Stress Estimation, 55, 62
Water use Efficiency, 48-49, 53
Web System, 34-35, 37, 39
Wetland Management, 192, 199
Wheat Bran, 1-4, 116-117, 120